Ecological Aspects of
Social Evolution

Ecological Aspects of
Social Evolution

---•---

Birds and Mammals

EDITED BY
Daniel I. Rubenstein
AND
Richard W. Wrangham

PRINCETON UNIVERSITY PRESS

PRINCETON, NEW JERSEY

Library of Congress Cataloging in Publication Data will
be found on the last printed page of this book

ISBN 0-691-08439-4 (cloth)   0-691-08440-8 (pbk.)

This book has been composed in Linotron Times Roman

Clothbound editions of Princeton University Press books
are printed on acid-free paper, and binding materials are chosen
for strength and durability. Paperbacks, although
satisfactory for personal collections, are not
usually suitable for library rebinding

Printed in the United States of America by Princeton
University Press, Princeton, New Jersey

7   6   5   4   3   2

Contents

•

v

Contents

Preface

•

THIS BOOK had its origins in Cambridge, England in 1980. After participating in a conference on sociobiology and behavioral ecology (King's College Sociobiology Group, 1982), we felt that so much of the exciting current work on the evolution of social behavior focuses on specific aspects that there is a risk of the big picture's being lost—what accounts for species differences in the overall pattern of social organization? Exploring the larger picture has its dangers, however, because questions tend to become less tractable as they grow. Nevertheless, we felt it would be useful to put together a book in which the ecological principles generating different kinds of animal societies are analyzed in similar ways, and we judged that useful comparisons could be made across taxonomic groups. To test the water we organized a one-day symposium on "Ecological Aspects of Social Evolution" at the American Society of Zoologists' Annual Meeting in Philadelphia in December 1983. The response to our invitation to participate was excellent and encouraged us to publish a volume that would also include several authors who were unable to attend the symposium.

In Chapter 1 we acknowledge many individuals that have molded our thinking. Richard Alexander has been a particularly important influence because for many years he has insisted on looking for broad general rules for understanding social evolution. As a professor and colleague to each of us he has been consistently provocative and insightful. It is a pleasure to acknowledge his role in the evolution of this book. We would also like to thank John Crook, Brian Hazlett, Robert Hinde, and Peter Klopfer for their special contributions to our interest in socioecology.

A recurrent theme throughout the book is that animal societies can be better understood when different types of social relationships are clearly distinguished from one another. Recognition of the nature and biological importance of social relationships has grown substantially during the last decade as a result of long-term field studies. Nevertheless, funding for long-term studies is not easily obtained, because after several years of studying the same population it is often difficult to promise eye-catching results from another season's work. Instead, observations are typically built up slowly into a detailed description of the lives and relationships of a small set of individuals: an extra year's data often has trivial significance on its own, though it may be a critical part of the overall description of social variation. It takes insight and courage for granting agencies to support such studies, and it is therefore a pleasure to acknowledge some of the organizations that have not

Preface

only funded fieldwork but have been particularly prominent in maintaining long-term studies. These include the Harry Frank Guggenheim Foundation, L.S.B. Leakey Foundation, National Environmental Research Council (U.K.), National Geographic Society, National Institutes of Health, National Institutes of Mental Health, National Science Foundation (Psychobiology, Anthropology, and Population Biology Sections), New York Zoological Society, The Royal Society (U.K.), Royal Society Leverhulme Trust (U.K.), and the Science Research Council (U.K.). Many other organizations also provide critical support for field studies. Funding for the individual studies in this volume is acknowledged at the end of the book.

The preparation of this book was made easier by help from a variety of individuals and organizations. Mary Wiley of the American Society of Zoologists organized the Philadelphia symposium with care and efficiency. We are grateful to Nancy Rubenstein for bibliographic assistance, to Elizabeth Ross for drawings, and to Barbara DeLanoy for secretarial help. Joseph Manson kindly prepared the index. For further secretarial and administrative support we thank the Department of Biology, Princeton University, and the Department of Anthropology, University of Michigan, Ann Arbor. We also thank Judith May of Princeton University Press whose direction and perseverance helped transform our idea into reality. RWW acknowledges also extensive assistance from the Center for Advanced Study in the Behavioral Sciences at Stanford, 1983-1984. Finally, we are grateful to those who reviewed this book in manuscript. Many are acknowledged at the end of the volume, but there remains a substantial number of anonymous individuals who kindly contributed their time to evaluate chapters on particular taxonomic groups. That they should do so for a collective volume shows, as Darwin said in closing his *Journal* (1839), "how many truly kind people there are."

25 July 1985   D. I. Rubenstein
                                                    R. W. Wrangham

Contributors

———————————— • ————————————

S. J. ANDELMAN Department of Ecology and Behavioral Biology, University of Minnesota, 318 Church Street S.E., Minneapolis, Minnesota 55455, U.S.A.

K. B. ARMITAGE Department of Sytematics and Ecology, University of Kansas, Lawrence, Kansas 66045, U.S.A.

J. W. BRADBURY Department of Biology, University of California, San Diego, La Jolla, California 92093, U.S.A.

R.I.M. DUNBAR Department of Zoology, University of Liverpool, Liverpool, L69 3BX, U.K.

J. W. FITZPATRICK Division of Birds, Field Museum of Natural History, Chicago, Illinois 60605, U.S.A.

M. V. FLINN Society of Fellows, University of Michigan, Ann Arbor, Michigan 48109, U.S.A.

R. M. GIBSON Biology Department, University of California, Los Angeles, Los Angeles, California 90024, U.S.A.

L. M. GOSLING Coypu Research Laboratory, MAFF, Jupiter Road, Norwich, NRG 6SP, U.K.

P. J. JARMAN Department of Ecosystem Management, University of New England, Armidale, NSW 2351, Australia.

D. B. LANK Department of Biology, Queen's University, Kingston, Ontario K7L 1N6, Canada.

M. LEIGHTON Department of Anthropology, Harvard University, Cambridge, Massachusetts 02138, U.S.A.

B. S. LOW School of Natural Resources, University of Michigan, Ann Arbor, Michigan 48109, U.S.A.

F. MCKINNEY Bell Museum of Natural History, 10 Church Street S.E., and Department of Ecology and Behavioral Biology, University of Minnesota, 318 Church Street, S.W., Minneapolis, Minnesota 55455, U.S.A.

P. MOEHLMAN Wildlife Conservation International, New York Zoological Society, Bronx, New York 10460, U.S.A.

L. W. ORING Department of Biology, University of North Dakota, Grand Forks, North Dakota 58202, U.S.A.

C. PACKER Department of Ecology and Behavioral Biology, University of Minnesota, 318 Church Street S.E., Minneapolis, Minnesota 55455, U.S.A.

Contributors

M. PETRIE   Department of Biology, Open University, Milton Keynes, MK7 6AA, U.K.

S. K. ROBINSON   Illinois Natural History Survey, 607 East Peabody Drive, Champaign, Illinois 61820, U.S.A.

J. P. ROOD   Conservation and Research Center, Smithsonian Institution, Front Royal, Virginia 22630, U.S.A.

D. I. RUBENSTEIN   Biology Department, Princeton University, Princeton, New Jersey 08544, U.S.A.

C. J. SOUTHWELL   Department of Ecosystem Management, University of New England, Armidale, NSW 2351, Australia.

G. E. WOOLFENDEN   Department of Biology, University of South Florida, Tampa, Florida 33620, U.S.A.

R. W. WRANGHAM   Department of Anthropology, University of Michigan, Ann Arbor, Michigan 48109, U.S.A.

Ecological Aspects of
Social Evolution

1. Socioecology: Origins and Trends

———————————— • ————————————

DANIEL I. RUBENSTEIN

AND RICHARD W. WRANGHAM

UNDERSTANDING why species differ in their social behavior has fascinated biologists for years. Yet the search for explanations has never been more vigorous than it is now. In the past decade theories of the way evolutionary forces affect social behavior have been developed and tested in substantially greater detail than ever before. In the same period there has been a significant increase in our knowledge of the social behavior of wild animals. These developments have paved the way for two kinds of advances in our understanding of social behavior.

First, diverse patterns of behavior, from foraging to mating, are being understood in terms of their individual selective benefits. General rules are emerging that cut across different species and tie behavior to fundamental principles of evolutionary theory. Some rules are concerned with adaptations to the nonsocial environment, while others are related solely to the social strategies of conspecifics. Excellent reviews of these rules are available, particularly in Krebs and Davies' *Behavioural Ecology* (1984), where selected rules are applied to a wide variety of species. These examples show many elegant matches between theory and data, and thereby promise a remarkably tidy future for the analysis of social behavior.

Second, the adaptive significance of whole social systems is being dissected, analyzed, and understood in relation to environmental pressures. Remarkably, however, no recent book has systematically compared social systems as wholes. It is therefore often difficult to appreciate how different components of the system fit together, or why similar principles yield different results in different species. Nor is it yet clear what the major principles of grouping are in the vertebrates, or the extent to which they vary in different taxa. Accordingly, the present collection reviews the relationships between ecology and social behavior in a variety of birds and mammals. By comparing social ecology in different species groups, this book is intended to contribute to understanding the broad patterns of higher vertebrate social evolution.

The first pieces to the puzzle were discovered by John Crook over twenty-five years ago. Crook began in the late 1950s by comparing the behavior of different species of African weavers, birds that occupy a variety of habitats from primary rain forest to grassland savannah. He collected data on habitats, diets, nesting habits, and group sizes of approximately one hundred species, and found strong superficial correlations between ecology and social organization (Crook, 1964, 1970). This was exciting because it suggested that if the right variables were found, and if appropriate groups of species were compared, then the ways in which environments shape social systems could be identified. After a flurry of activity, the comparative method faltered. At this time social systems were described not in terms of the types and strength of social interactions, but in terms of the outcome of these interactions. Correlations between the outcomes, such as group size, dispersion pattern, mating system, and ecology, were of little use in showing how environments shaped social structure. What the comparative approach needed was a good theory. It got two: one from the social theorists and one from the ecologists.

The twin pillars of the modern theory of social behavior are the principles of individualistic reproductive maximization (G. C. Williams, 1966) and kin selection (W. D. Hamilton, 1964). With the development of these principles, animals came to be viewed as individuals who were armed with many behavioral options in their struggle for maximizing either their own reproduction, or that of their relatives. Extensions of the theory showed which options were better than others, in particular circumstances. By analyzing the behavior of individuals, the foundations for a comprehensive theory of social behavior were laid. Perhaps the most striking insight was that of Trivers (1972). He proposed a relationship between parental investment and sexual competition. He argued that when one sex invests more in the rearing of offspring than the other, members of the latter will compete for members of the former. The implication was simple: the sexes tend to invest their reproductive effort differently. When females invest heavily in parenting, males should invest heavily in mating, and vice versa.

The comparative method got an additional theoretical boost from the ecologists as well. Orians, Verner, and Willson analyzed polygamy and monogamy in red-winged blackbirds and proposed a general model for the evolution of polygyny (Orians, 1969). Like Trivers's theory, it was based on the idea that ecology influenced each sex differently. It suggested that when ecological conditions were just right, some males could defend large amounts of resources that females required. The model required that the habitat be divided into patches of vastly different qualities. When this occurred, many females could rear chicks in the best areas, whereas in the less rich areas, only one could succeed. In their model females, by attempting to maximize their reproduction, showed how they could induce the coexistence of both polygyny and monogamy in a population.

At this period, both social theorists and the ecologists were expanding their theories and attempting to account for the evolution of mating systems other than polygyny. A major breakthrough occured when Maynard Smith (1977) viewed the mating system problem as a game in which the optimal behavior of one parent depended on the behavior of the other. He argued that whether or not one parent should withhold parental care depends on how likely the other parent is to continue caring for the offspring. Since the parents' interests were similar, but not identical, Maynard Smith's idea was to search for a pair of strategies, one for the male and one for the female, which, when performed together, produced an 'evolutionarily stable strategy,' or ESS. Such a strategy would occur when it would not pay a male to depart from his strategy as long as the female continued to follow hers, and it would not pay the female to depart from her strategy as long as the male continued to follow his. The elegance of this approach was that all the pairwise options of: 1) both male and female care; 2) female care, male desert; 3) male care, female desert; and 4) both male and female desert, can be pitted against each other in a thought experiment and compared. Thus, the model offered the hope of determining the conditions that favored monogamy, polygyny, polyandry, and promiscuity, the four basic mating systems. Whether there is to be parental care and, if so, which parent is to provide it depends on the relative effectiveness of uni- or biparental care, the likelihood of a deserting partner finding an additional mate, and the extent to which a female's ability to provide care reduces her ability to produce future offspring. By estimating these parameters, the favored mating system can be predicted.

The model has many powerful features. First, it emphasizes sexual asymmetries, and focuses attention on a limited number of measurable variables that ultimately are associated with reproductive energetics. Second, it demonstrates that when selection favors only uniparental care, whether it is provided by the male or the female often depends on historical conditions and how they set the parameter values. This should provide powerful support to the comparative approach. The only problem with the model is its misuse by some in narrowing, rather than expanding the scope of the problem. Focusing on mating systems and their environmental determinants only provides a first step toward understanding a social system where social relationships other than those between males and females must be explained. Although a limited set of variables can adequately account for the evolution of each broad class of mating system, the problem will grow as explanations on a finer scale are required. The present reduction of dimension only becomes a problem if field workers limit their observations of natural systems to just this simple set of variables. If this were to happen, valuable insights would be lost.

While the social theorists were elaborating general rules for the evolution of the major mating systems, ecologists were expanding Orians' principle to account for the complex systems found in species other than blackbirds.

Emlen and Oring (1977) proposed that the potential for polygyny depended on both the spatial and temporal distribution of members of the limiting sex. For example, an even, scattered distribution of females makes it unlikely that one male will control more females than others. But when females live in groups, or as Orians suggested, the resources they need occur in patches, then some males will have the opportunity ultimately to control large numbers of females. Under these conditions the potential for polygyny is high. Whether it is realized depends on other factors, such as the degree to which females synchronize their reproductive behavior. If all females are receptive at exactly the same time, then males have little chance of mating with more than one female, and the polygyny potential is lost.

For the first time, a general theory of mating systems was emerging, but what it lacked was an understanding of how ecology shaped female reproductive interests. For Emlen and Oring, female distributions were given. But reliance on this assumption was eliminated by studies on antelope (Jarman, 1974) and bats (Bradbury and Vehrencamp, 1977a, b). Both studies showed that female distributions were finely tuned to the environment and depended on the needs of females to seek food and safety from predators. For antelope females, Jarman argued that diet was determined by body size; small females needed high-quality forage, whereas large females needed large quantities of low-quality food. And he showed that because high-quality items were more widely dispersed than low-quality items, smaller females were forced to live more solitary lives than larger females. As a consequence, he concluded that monogamy was the typical mating system of the smaller antelope, whereas polygyny, in its various guises, was to be the norm for the larger ones.

As Jarman's (1974) study shows, a framework is now in place for examining how ecology shapes certain intra- and intersexual relationships. Of course the framework will be strengthened as the ESS approach is applied to more types of competitive and cooperative interactions among social animals. But even in its simplest form, the model expects first, that female behavior—and this includes their social relationships—will be adapted primarily to meeting demands imposed by the physical environment. This is because the reproductive rate of females will normally be raised, or lowered, more predictably by their success in meeting these demands than by their success in other endeavors such as finding, or choosing among, mates. Second, it expects that male strategies are adapted primarily to competing for mating opportunities, because male fitness is more closely tied to mating success than to the acquisition of other resources. Taken together, they imply that to understand the ecological basis of social organization, it is imperative to understand separately the ecological and, as the ESS approach shows, social pressures operating on each sex.

Although only a beginning, the model contains some very basic ideas whose usefulness must be evaluated. With the recent proliferation of long-term field studies, there is sufficient data to apply these ideas to many different species with different ecologies and different phylogenetic histories. The collection of essays in this volume represents such an application, and one that succeeds in refining the old ideas and defining new problems.

We deliberately limited the scope of the analysis to birds and mammals, the main reason being that the complexity and individualistic nature of the social relationships exhibited by these groups poses the greatest challenge for the existing framework. The birds and mammals also exhibit the full range of standard mating systems, and comprise a sufficiently large number of different taxonomic groups. This has allowed us to maintain diversity, without spreading ourselves taxonomically so thin that no generalizations could be made. Moreover, for the first time, the long-term studies necessary for interpreting changes in social relationships of such long-lived animals are numerous enough to be evaluated.

The book is divided into two parts—Monogamous Variations and Polygynous Patterns. As is the case in the world at large, most of the avian studies are found in the former section and most of the mammalian studies are found in the latter. But as these studies show, similarities in the ecological bases underlying the evolution of the social systems of each type is surprisingly low.

The book begins with studies on the monagamous theme, in particular with Oring's study of facultative polyandry in spotted sandpipers. Of all the mating systems, polyandry is perhaps the least well understood. Few correlations with habitat type have been found, and when contrasted with monogamous systems, polyandry emerges from most studies as favoring females, but not males (Verhencamp and Bradbury, 1984). As our general framework suggests, such an analysis, if true, would certainly not be evolutionarily stable. Oring's analysis suggests that, at least for sandpipers, ecology plays an important role, and that both males and females benefit. According to Oring, sandpipers nest in areas where nest predation is extremely high, except on some islands. In these isolated places, the density of birds rises and because of intense female competition, some are excluded. Conversely, some secondary males are included. For males and females that breed the benefits are higher than they would be by breeding monogamously in less safe areas. Given that most polyandrous birds nest in open and defenseless areas, predation might be a critical variable for these other species as well.

Predation is clearly the major selective agent in the moorhen, a bird that usually mates monogamously, but occasionally exhibits polyandry. Petrie, in Chapter 3, has been able to show that nest failure resulting from predation places pressures on males to compete for territories comprising dense stands of aquatic vegetation. Yet females do not seem to choose mates on the basis of territory quality. Instead, they seem to choose males that are disproportionately fat. Petrie resolves this paradox by demonstrating that males perform most of the nest-guarding functions while females feed. With such a high likelihood of nest failure, it seems to be critical that a female stay well provisioned so she can lay replacement eggs, and that a male start the season with enormous fat reserves so he can incubate eggs laid late in the season. As in Oring's sandpipers, the threat of predation seems to establish a situation in which female behavior controls that of males. But in the moorhen the control is only partial, as males are simultaneously driven to compete for the safest territories. Nevertheless, in this monogamous society the division of posthatching parental labor is far from equal.

When it comes to examining the ecological adaptations underlying monogamy, few groups of animals provide a better basis for comparison than the canidae. All are monogamous, but the larger members of the family show polyandrous tendencies, whereas the smaller ones lean toward polygyny. In Chapter 4, Moehlman uses an insightful life history analysis to make ecological sense of this trend. She shows that larger species tend to invest disproportionately more in reproduction than smaller species, but they do so by making disproportionately larger litters containing relatively smaller young. Consequently, the young of large species need more postpartum care, and at a time when the mother has already invested heavily. Obviously, assistance is needed, but who should help? Moehlman shows that for large species, as typified by wild dogs, potential competition among many breeding females for the spoils of cooperative male hunting leads to enforced dispersal of all but one. As a result, a male-biased sex ratio sets the stage for polyandry. She also shows that virtually the converse occurs in the small species. Competition among females is much lower and the need for assistance in raising young can be met by the male. Using data on foxes, Moehlman shows that when food abundance increases, female densities increase, a condition that enables some males to mate polygynously. According to this line of reasoning, the species in the middle of the size range should be the most monogamous. Food is not as plentiful so the male will be necessary. As Moehlman's studies on jackals illustrate, helpers are the younger of previous litters, and members of both sexes contribute about equally to the rearing of their young siblings. But as they age, the threat of competition with their parents leads to their dispersal in search of their own breeding territories.

The ecological significance of helping in monogamous systems is also examined in the next chapter by Woolfenden and Fitzpatrick in their study of the Florida scrub jay. Their findings provide a striking contrast to the study of jackals, and provide a possible explanation as to why avian males are more philopatric than their mammalian counterparts. In the scrub jay, the breeding opportunities for newly maturing birds are extremely limited, because the small patches of suitable habitat are saturated. As a consequence, young of both sexes remain at home and help. But unlike the jackals, the fledging success of the parents remains constant after the addition of the first helper, and the behavior of helpers differs depending on sex. Like their adult counterparts, males are primarily responsible for finding food and bringing it to the mother and the young. Moreover, they are likely to remain as helpers longer and disperse shorter distances than females when they finally depart. Woolfenden and Fitzpatrick evaluate most of the standard explanations for these differences in behavior and philopatry and find them wanting. They suggest, instead, that male dominance, which ultimately results from disproportionate male aggressive defense of territories, induces female dispersal. However, since females must also aggressively insert themselves into territorial vacancies, it is not clear that they are less aggressive, or less effective. More work is needed to account for these exciting results. Some of the latest is presented in the next two chapters.

Even if severe habitat saturation favors that young stay at home and help parents defend territories and raise additional young, Leighton shows in Chapter 6 that other environmental features might forestall this outcome. Among hornbills, whether or not the young will have opportunities to disperse to vacant territories depends on body size. For some of the small African species, predation is so high that turnover among adults is rapid. For the larger Asian species, however, the opposite occurs, and the dispersal opportunities of the young are limited. But contrary to expectation, not all the large-bodied hornbills live in groups. Leighton argues that for some species the distribution of important fruit resources prevents economical exploitation once groups are composed of more than two individuals. Large size requires that more food be consumed per day. One solution would be to spend more time feeding and thus increase intake by visiting more patches. But this apparently is not how the largest hornbills behave. This remains an issue that when better understood is likely to bear its own fruit. Rather, hornbills reduce group size so that each patch can be mined more completely on a per capita basis. Thus, food limitation on one hand favors group living by reducing the dispersal options of young. On the other hand, it apparently encourages group dissolution.

Mongooses are some of the smallest carnivores and often behave in ways contrary to our usual stereotypes. In Chapter 7, Rood makes comparisons among a large number of species that differ in body size, diet, habitat, and activity period. Some strikingly broad conclusions result. Virtually all of the smallest species live in groups, feed on insects, and are diurnal, whereas the larger species are solitary, feed on small vertebrates, and are nocturnal. Rood argues that predation is the most important problem faced by mongooses; predatory attacks can be as high as one every six hours, with predator disturbances occurring at a rate of more than one per hour. Yet the nature of the food resource is the force that seems to constrain dispersion pattern. Whereas the small species can protect themselves by traveling in groups during the day because they feed on abundant insects, the large species can not solve their vulnerability problem in this way. Large individuals, by interfering with each other's capture of small vertebrates, are forced to forage alone, and thus at night. But a paradox emerges. If feeding competition among small mongooses is so low, why do dominant females supress the breeding activity of all other subordinate females, thus inducing them to become helpers? Reproductive suppression is common (see Moehlman, Chapter 2), and will be discussed in greater detail later in marmots (Armitage, Chapter 13) and in baboons (Dunbar, Chapter 14), but in general it usually occurs when females face stiff resource competition. Rood's analysis suggests an alternative answer. Suppression guarantees that the reproductive female will have vigilant helpers. And since survival rates of small groups are so low, reproductively suppressed females may simply be making the best of a bad job, waiting their turn to ascend to the top of a large, long-lived group. More data on the costs and benefits of the various options will be needed before Rood's intriguing hypothesis can be tested.

In Chapter 8, the last of this section, McKinney analyzes the social systems of a group of most unbirdlike birds, the dabbling ducks. Monogamy is the rule, but polygyny is common. Males do not help with parental care, new pair bonds form each year, and females are the more philopatric sex. McKinney shows that the exceptional nature of duck sociality is easier to understand by noting that the young are precocial, and are up and about soon after hatching. The need for biparental care is dramatically reduced, and, as McKinney argues, emancipates males in most species from performing parental duties. The few that stay defend feeding sites, but most leave and attempt mating with other already paired females, or try to get a head start reconditioning themselves for next year's reproductive challenge. But what is the rush? According to McKinney, females begin testing males early in the offseason for mate guarding ability. This is the only way they can reduce the threat of being subjected to forced copulations on the breeding grounds. Judging by the number of similar examples of mate guarding that appear in later chapters, the value of selecting males on the basis of their protective abilities seems to be high and widespread. For the unemancipated female, McKinney marshals strong comparative data showing that returning to a safe nest site favors not only the early testing of males but also female philopatry, early pairing, and an early return to the breeding ground.

As in the ducks, monogamy is uncommon in the blackbirds. Why is this so? This is just one of the many questions that Robinson tackles in Chapter 9 where he compares mating systems in blackbirds. By using the yellow-rumped cacique as an example, Robinson extracts a set of general rules that he then uses to examine the social systems of other blackbirds. He concludes that in caciques the major factor that controls female distributions and associations is predation. Sites that are safe from mammalian and avian predators are restricted to islands and females compete for access. By clustering and synchronizing nests in these areas, females markedly increase their chances of fledging young. Given that females form colonies, males compete for reproductive control of colonies, with the largest males acquiring rights to the safest ones. What is intriguing is that young males are larger than old males. At this point, it is not clear whether or not reproductive effort causes the age-specific decline in weight. What appears more certain, however, is Robinson's answer to the initial query: colonality usually leads to polygyny, and when it does not, the role of male parental assistance in the rearing of young turns out to be considerable. Blackbird polygyny rarely takes the forms of resource or harem defense polygyny. Again, the reason appears to be correlated with the colonial habit. Although females nest together, they must forage alone for insects. Thus, Robinson concludes that in order for males to assure their paternity, they are stuck guarding individual females. Thus, polygyny results from a series of matings. But if increasing paternity is certainly important, a new problem arises: Why do some blackbird males, such as red-wings, let their females leave the territory unguarded to forage?

In caciques, the threat of predation forces females to nest together, but the demands of feeding causes them to disperse. For Old World monkeys, the factors causing cohesive groups are still unknown (Wrangham, 1986). Predator pressure is a possible influence but whatever causes group-living, the nature of the food supply is different from that of caciques because it allows most monkeys to travel in permanent foraging groups. Some species of monkeys travel and forage in groups of about thirty individuals, whereas others are limited to feeding in parties of two. In Chapter 10, Anderman shows that group size has a striking effect on the mating system of Old World monkeys. Groups consisting of more than ten females contain many breeding males and exhibit a constant male-female ratio, whereas those consisting of five or fewer females only contain one adult male. At intermediate sizes, both systems occur. Her study examines the ecological determinants of these patterns, but it also raises some serious doubts as to the usefulness of typing species by mating system. She questions whether the findings that harem males in uni-male groups or dominant males in multi-male groups father most of the offspring, which are valid in a few well-documented studies, should serve as models for generalizing about social relationships and reproductive outcomes in both mating systems. Her comparative analysis shows that the degree to which females synchronize their reproductive activity can overturn the old generalizations. When females are highly synchronized, harem males and dominant males in multi-male groups have their reproductive success reduced because the lower-ranking males adopt novel reproductive strategies. These, in turn, lead to behavioral adjustments by the threatened males and suggest that a coevolutionary arms race is underway. Andelman's analysis shows that a wealth of information is still to be uncovered about male-male relationships, and that the solution will be tied to differences in the coordination of female behavior.

Most of the studies in this volume focus on one species and extract general rules linking ecology to social behavior, which are then applied to other closely related species. Another approach would be one that mimics that of Crook's study twenty-five years ago and involves comparing a large number of species and their social systems. But to succeed, it will have to focus on individuals and their reproductive interests. In Chapter 11, Flinn and Low perform such a comparison using the best-studied collection of social systems, those of humans! Human societies provide a wealth of data on relationships. For example, among female-female relationships, distinctions are often made between those of sister-sister, mother-daughter, aunt-niece, grandmother-granddaughter, cousin-cousin, and co-wives. In an ambitious analysis, Flinn and Low examine the ecological and economic settings of societies that span the four major mating systems and find some general relationships. For example, they show that, like most other animals, when ecological resources can be accumulated, polygyny results because some males control more resources than others. And when ecological resources can not be accumulated monogamy often results. But not always, because in humans males usually control a most important resource, females—mates, sisters, and daughters. Flinn and Low show that like all resources, the abundance, distribution, and unpredictability of females affect the strategy of control. But they also show that in human systems, male-male, and, to a lesser and more clandestine extent, female-female relationships influence the distribution of females, and thus the nature of marriage systems. Although we originally intended that this chapter teach biologists about humans, and anthropologists how biologists think, it has gone one step further. It teaches biologists to look at animals with the hand lens of anthropologists.

In addition to human societies, those of antelope are also well studied. In Chapter 12, Gosling examines these societies and shows that despite the myriad of ways in which females forage, move, group, defend themselves against predators, and raise young, there are only a few general mating responses shown by males. He argues that even before males can mate with females, they must establish both dominance over other males and a referent so that their status can easily be recognized by subordinates. Only then can males position themselves to maximize rates of copulation with females.

Gosling suggests that males can either follow females or sit and wait for females to arrive. The best strategy obviously depends on the density and movements of females. By analyzing numerous antelope studies, Gosling concludes that the "following" strategy is favored when food is evenly dispersed and of low density, or when females form groups, whereas the "sit-and-wait" strategy is favored when resources are of high quality and patchily distributed, or when foods needed to meet different seasonal requirements are contagiously distributed. What is perhaps most intriguing about Gosling's analysis is that he finds similarities in male strategies that previous workers have categorized as different. For example, male wanderings and harem holding are categorized as "following" strategies, and seasonal, year-long territories and lek territories are categorized as "sit and wait" strategies. By organizing antelope societies into these novel divisions, some startling insights in male-female relationships emerge.

Whereas antelope social organizations illustrate the role that female behavior plays in shaping male social interactions, the social systems of horses and zebras show that male-male interactions play just as important a role in shaping female social behavior. In Chapter 13, Rubenstein shows that even in the absence of predators, and despite the fact that feeding competition is heightened, unrelated female horses prefer living in cohesive groups. Only under severe conditions where grasslands are very patchy do females break up into temporary aggregations. He asks why should group cohesion be so prevalent, and suggests that the answer is related to male sexual behavior, which is so disruptive that it actually reduces female feeding rates. By associating in groups, and skewing the operational sex ratio, Rubenstein shows that only a few of the best males can defend the groups. By keeping other males away, the harassment levels females experience drop dramatically. Rubenstein also shows that in plain zebras, this strategy of employing males as shields has gone one step further. In these zebras, males without harems form their own cohesive groups. As a counterresponse, previously dispersed harems have come together and pairs of males jointly drive away competitors; alone neither male would have been successful. But perhaps the most fascinating question still remains: Who initiates the movements that bring the harems together? Are the males in control, directly protecting their reproductive interests, or are the females in control, manipulating their hired guns? Some insights into this problem emerge in the next two chapters on other harem living mammals.

The social relationships of another harem mammal, the yellow-bellied marmot, are analyzed in Chapter 14 and they are strikingly different from those of horses and zebras. Females form social groups of closely related individuals that persist for many generations as matrilines. Harems form when a single male attaches himself to one or more matrilines. Armitage's analysis shows that males always benefit reproductively by having a large harem, but that females do not. He shows that inter- and intramatriline competition leads to social suppression of reproduction by the younger females. Apparently parents are attempting to maximize their own reproduction without damaging too severely the long-term reproductive interests of their young. According to Armitage, suitable habitat is rare and since the risks of predation during dispersal are large, young females are induced to stay home and make the best of a bad job. Given that the nature of marmot harems and the ecological forces maintaining them are markedly different from those of horses and zebras, the usefulness of calling these social systems by the same name is questionable. Moreover, determining exactly why these offspring stay at home, compete with their mother, and breed with limited success, whereas those of jackals, hornbills, and scrub jays stay at home and help their parents, remains a fascinating problem.

In Chapter 15, Dunbar shows that gelada baboon society has two major levels: harems, which are stable one-male reproductive matrilines, and herds, which are unstable collections of harems. According to Dunbar, the need to reduce the risks of predation and the physiological side-effects of crowding lead to conflicting social pressures. On the one hand, predation favors the formation of large groups, which gelada are capable of doing because, like zebras, they are grazers. On the other hand, crowding leads to strife and physiological suppression of reproduction, much like the marmots. The social system that results is a compromise. Dunbar argues that related females form matrilineal harems to keep other unrelated females away and thus reduce the physiological effects of crowding. To support this conclusion Dunbar shows that harems only merge to form herds as a last resort when gelada find themselves in habitats where the ease of detecting and escaping from predators is extremely low. What is fascinating is that the compromise is only partially successful. Even within harems, low-ranking females suffer from reproductive suppression. As Dunbar shows, this tension among female kin leads to a diverse array of social strategies. Aging females form coalitions to hold on to their rank, low-ranking females attach themselves to "follower" males to form new groups, and females assess and alter their loyalty to their male depending on his effectiveness in interharem confrontations. Of particular importance is Dunbar's finding that rudimentary forms of these social relationships exist even in the phylogenetically distant forest-dwelling baboons that live in multi-male groups.

In Chapter 16, the limits of the ecological control of social organization are explored in Wrangham's study of chimpanzees and pygmy chimpanzees (bonobos). Although very closely related, field studies show that bonobo social organization is substantially different from that of chimps. Whereas female chimpanzees forage largely alone because of intense feeding competition, bonobo females form fairly stable subassociations. As for the males, they are often found in all-male coalitions in chimpanzees, but in bonobos they usually associate with the female subgroups. Wrangham suggests that the availability of a *secondary* or less preferred resource holds the key to understanding the differences. Although both species prefer fruit, bonobos regularly eat the ubiquitous terrestrial herbaceous vegetation (THV), just like their relatives, the gorilla. The drier habitats of chimps seemingly deny them access to this type of vegetation and thus prevent them from reducing their high levels of feeding competition. Wrangham concludes that bonobos have become so much like the gorilla that the formation of stable groupings has subsequently permitted the formation of protective relationships between males and females. But the transformation to the gorilla lifestyle is not complete. Wrangham cites a case where high-ranking males pursue a chimplike strategy of banding together and continuously patrolling the entire community of females. Clearly, further studies on this "missing link" will help clarify our understanding of the evolution of sociality in the great apes.

Of the polygnous mating systems, leks have been the most difficult to interpret. In Chapter 17, Gibson and Bradbury's analysis of sage grouse social relationships provides some important new insights. First, preliminary data show that female home ranges are extremely large and overlapping, and that females tend to use major traffic routes when moving from roosting to feeding area. All these features support the hot-spot model, which predicts that males will not follow females about in their large home ranges, but will sit and wait for them to pass by, especially if they predictably travel along certain routes. Then they examine male-female relationships on the lek, and produce a number of startling results. First, leks are not as stable as previous workers have thought since some males move display sites often. Second, male dominance is often relative, not absolute, as males generally win encounters only within their core area. Third, females do not necessarily choose central males, rather, chosen males become the center of attention. And fourth, females appear to choose males on the basis of strut rate. Interestingly, female-female competition, but not copying, seems to occur. How this competition affects the mating success of both sexes should prove to be fascinating.

When females exercise choice, they need not be limited to doing so on leks. While it is true that leks afford females the opportunity to shop around and make an informed choice, there are other ways by which the same amount of information can be extracted. In Chapter 18, Jarman and Southwell describe the novel information-gathering technique employed by the eastern grey kangaroo. When females become sexually receptive, which lasts about a week, they begin roaming widely, picking up male escorts. As time goes by, the train of males-in-waiting lengthens, but the largest male is always right behind the female. As Jarman and Southwell argue, the longer and more widely she moves the more likely she is to detect, and mate with, the largest and oldest male in the population. This seems to be important in kangaroos and may even be based on a unique marsupialian feature. Marsupials are indeterminant growers, so as they age they keep getting larger. Thus, if a male lives long enough, he will become the largest male and exert total dominance over all other males in the area. If estrus females have time enough to travel over the entire area, then the largest male could theoretically mate with all receptive females in the neighborhood. As Jarman and Southwell note, many of the ungulate (the kangaroo's eutherian counterparts) social relationships do not appear in kangaroo societies. The possibility that early phylogenetic divergence has led to such major differences in social responses to similar ecological features is fascinating, and should encourage even more comparative field studies.

If diet plays only a permissive role in shaping ape, equid, and gelada baboon societies, it plays the controlling role in shaping felid societies, especially that of the lion. As Packer notes in Chapter 19, the stealthy hunting habit of felids virtually requires a solitary lifestyle. So why are lions social? One explanation, that group hunting facilitates the capture of large prey, is not substantiated when Packer analyzes the average size of the prey in relation to female size. Leopards and cougars often prefer relatively larger prey than do lions. Another explanation is that group living occurs because hunting success is higher for pairs of lions than for solitary lions. By reanalyzing Schaller's data, Packer concludes that solitary hunters do best. He then shows that peculiarities of contemporary lion habitat make group living a "best of a bad job" situation. Lion society is based on a pride of genetically related females that share a common range, but travel in small open membership groups that regularly fuse and fission. Such a pattern of sociality is adaptive because in open habitats that are crowded, conspecifics can readily locate large carcasses. Thus, if lions are going to lose food to other lions, they might as well lose it to close kin who can at least augment inclusive fitness and might actually assist in driving strange lions away. Thus, kinship and mutualism seem to organize female relationships by reducing costs. Group hunting seems minor. Interestingly, for males the situation is somewhat different. Mutualism helps males take over prides and augment per capita reproductive success, regardless of the kinship relations among the helpers. Such major sexual differences in the social cements are intriguing and should encourage others to examine social relationships in this light.

As the examples described above show, many of the same behavioral pieces appear in many of the social puzzles. But they fall into place in different ways. For example, habitat limitation repeatedly induces juveniles to remain with their parents. Some juveniles, however, forgo breeding and help their parents raise young, whereas others attempt to breed and limit their own reproduction as well as that of their parents. Moreover, differences in which sex (male or female young) stays, and for how long, vary from study to study. Making sense of differences like these requires a detailed understanding of differences in ecology, demography, physiology, and phylogenetic histories of each species. Usually such an understanding is incomplete. In this volume, the authors have been encouraged to approach the problem of unravelling the links between ecology and behavior from a broad interdisciplinary perspective. Judging from the ideas, and the level to which they are developed, in the subsequent chapters, this approach has borne fruit. In a final chapter, we search for commonalities and specificities in the operation of each environmental pressure and extract a set of general rules that help shape the evolution of bird and mammal societies.

PART I

Monogamous Variations

2. Polyandry in Spotted Sandpipers: The Impact of Environment and Experience

———————————— • ————————————

LEWIS W. ORING

AND DAVID B. LANK

SANDPIPERS (family Scolopacidae, order Charadriiformes), have many aspects of reproductive biology in common. Most of the eighty-six species are migratory, and breed in the arctic or north temperate zone. Females typically lay clutches of four eggs in ground nests. The young have well-developed down, legs, and visual senses at hatch, and feed themselves on small invertebrates. Parental care involves brooding very small chicks, and protecting by vocal warning calls or by attacking or distracting predators.

In spite of these common features, members of this family exhibit a wide spectrum of mating and parental care systems. Monogamy with some degree of biparental care is most common, found in over fifty-five species; while the rest range from lek polygyny, with no male involvement in nesting (ruff, *Philomachus pugnax*; Andersen, 1948); to nonterritorial polyandry with no female participation after egg-laying (red-necked phalarope, *Phalaropus lobatus*; Hildén and Vuolanto, 1972); to systems in which both sexes are polygamous and each cares for a separate nest (Temminck's stint, *Calidris temminckii*; Hildén, 1975). This diversity provides the potential for an analysis of social systems in the context of variation in environmental features (Crook, 1965; P. J. Jarman, 1974; Pitelka et al., 1974).

Broad interspecific comparisons rest on understanding how ecological variables affect the behavior of individuals. Since 1973, we have studied the reproductive ecology of the spotted sandpiper (*Actitis macularia*) in order to evaluate the relationship between environment and social behavior. We recognized that the degree of polyandry was variable among populations (Oring and Knudson, 1972), but felt that an intense analysis of a single population, documenting interyear variation in environmental conditions, provided the best route to understanding social variability. Our study has examined breeding chronology (arrival on breeding grounds through migratory departure), spacing and territoriality, mating relationships, measurements of food availability, time and energy allocations, reproductive effort and success, patterns of parental care, predation pressure, philopatry of adults and offspring, and changes in density and sex ratio over four generations of sandpipers. We have found that variability in social behavior is due not only to changing environmental circumstances, but also to variation in the social environment created by differences in age and experience among the birds themselves. This chapter reviews how both kinds of variation influence the behavior of females and males in this species.

SPECIES

The spotted sandpiper and the common sandpiper (*Actitis hypoleucos*) of Eurasia comprise the modern genus *Actitis*. The close relationship of the two species is assumed on the basis of natal plumage (Jehl, 1968), cladistic analysis (Strauch, 1978), and displays (Oring, pers. obs.). The closest relatives of *Actitis* are the two species of tattlers (*Heteroscelus*) and Terek's sandpiper (*Xenus cinereus*). Ancestrally, *Actitis* was most likely derived from the *Tringa ochropus, T. glareola, T. solitaria* line of Tringinae. The primitive *Actitis* was probably boreal in distribution, whereas modern *Actitis* occupies a broader habitat and more southerly latitudinal range. It is

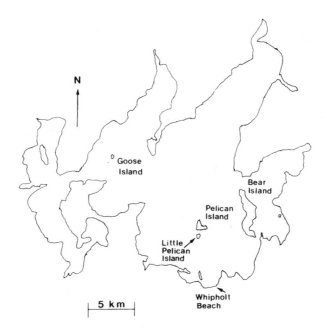

Fig. 2.1. Leech Lake, Minnesota. More than half of the spotted sandpipers breeding in the main basin of the lake are on Little Pelican Island. Small numbers of birds bred on Pelican, Bear, and Goose Islands, and at Whipholt Beach.

thus exposed to a longer period of primary productivity, which, in turn, provides the time for multiple breeding attempts (Hays, 1972; Oring and Knudson, 1972).

Spotted sandpipers breed in early successional habitat from the northern border of the boreal forest across North America, south to the central United States; and winter from the southern United States to northern Chile, Argentina, and Uruguay (American Ornithologists' Union, 1983). *Actitis hypoleucos* occupies similar habitat across a comparably wide Old World range.

A. macularia has a resource defense polyandry (Emlen and Oring, 1977) mating system in which females lay separate clutches with one or more males on all-purpose territories defended against other females. Normally, females acquire males sequentially; but simultaneous bonding to two males occurs occasionally (Oring and Maxson, 1978; Oring, unpublished). The incidence of polyandry has been found to vary from 1.17 to 1.86 males per female in three different populations (Hays, 1972; Oring and Knudson, 1972; Oring et al., 1983). Males also defend territories, and they provide most of the incubation and parental care.

Evidence of intrasexual selection of females is apparent in their morphology as well as behavior. Females have denser spot patterns than males, on average, especially posteriorly; and females are significantly larger than males in terms of both long measurements and weight ($p < 0.001$). Breeding ground weights of females (49.7 g), excluding laying and prelaying females, averaged 1.20 times the weight of males (41.3 g) (data from Little Pelican Island, Leech Lake, Minnesota).

APPROACH

From 1973 to 1984, we studied a color-marked population of Spotted sandpipers on Little Pelican Island (LPI), Leech Lake, Minnesota (Fig. 2.1) (Heidemann and Oring, 1976; Maxson, 1977; Maxson and Oring, 1978, 1980; Oring and Maxson, 1978; Oring and Lank, 1982, 1984; Oring et al., 1983; Lank et al., in press). The small size of the island (1.6 ha) and the birds' preference for open beaches allowed us to determine accurately the locations and timing of reproductive events from arrival to departure. No equally detailed breeding histories are available for any other polyandrous animal. The unique strength of this study lies in our having followed individuals year after year, determining precisely all mating combinations, space use, and elements of reproductive effort (RE) and success (RS). This, coupled with detailed analysis of food, and of time and energy allocation, allows a degree of analysis of factors influencing RS not possible with any other polyandrous animal.

The great detail of our study has dictated that we concentrate on a reasonably small number of birds. Our efforts were directed primarily at the eighteen to fifty-two birds breeding on LPI each year. This small sample size is our main weakness. Additional limitations include the following: 1) not all surviving chicks return to breed on LPI; 2) some adults, especially yearlings that fail to hatch chicks, emigrate to other locations; 3) a small proportion of our new breeders are probably older birds that first bred unsuccessfully elsewhere; and 4) we cannot follow our birds during the nonbreeding season.

An alternative approach for assessing lifetime RS involves computer simulation based upon yearly estimates of RS (Howard, 1983). The greatest advantage of this approach lies in the fact that it can be applied to species in which individuals cannot or will not be followed throughout their lives. In such cases, and where appropriate information about life history such as emigration-immigration exits and relevant demographic data are available, the simulation approach may allow meaningful comparisons of the life history strategies of the two sexes, or of different age classes. One weakness of this approach lies in the fact that it relies upon average circumstances to generate average results, rather than using specific circumstances to document a diversity of results. In higher vertebrates, where individual experience heavily affects decisions, there is no substitute for following individuals throughout their lives, monitoring their experiences, and relating experience to subsequent choices.

RESOURCE DEFENSE POLYANDRY:
AN EXAMPLE

As an introduction to the resource defense polyandrous system of spotted sandpipers, we review the nesting history of the northern tip of the island from 1974 to 1983. The example includes both exceptionally successful and faithful birds as well as temporary residents. Instances of mate and site fidelity and infidelity, natal philopatry, and interactions among close kin are also documented.

Female AR:B was a breeding bird on beach 3, on the south shore of the island (see Fig. 2.2; Table 2.1) when observations began in 1973. Her nest that year was unsuccessful. She returned in 1974 and established herself on the 9-10 point when this area's former owner failed to return. In four seasons at the point, AR:B had eight mates and laid forty-two eggs of which sixteen hatched and eleven chicks fledged. In her first three years as a breeder, however, none of her eggs hatched, and none of her four mates returned to the island. In 1976, both of AR:B's mates hatched eggs. One of these, a second-year male named BkPBk:APBk that had failed to hatch eggs the previous year, returned to breed with her again in 1977. AR:B had two

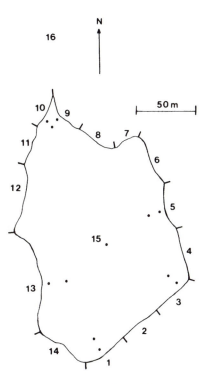

Fig. 2.2. Distribution of "territories" on Little Pelican Island. Space use is plotted in relation to fourteen "territories" laid out, beginning on the southwest corner of the island and proceeding counterclockwise, such that "territories" 14 and 1 are adjacent. Number 15 depicts territories that lack beaches, while 16 represents the site of a temporary island used for breeding in 1977. The locations of twelve food sampling sites are indicated by dots.

additional mates in adjacent beach 8 that year, and altogether fledged seven chicks from three broods. Two of her chicks with BkPBk:APBk (one male, one female) returned and bred on the island, as did four of her grandchildren.

In 1978, AR:B did not return, and the tip was taken over by two returning 1976 chicks: BWB:WA and her sister, dBR:AdB. In six seasons at the point, BWB:WA bred with six males, laid sixty-one eggs, thirty of which hatched, and fledged fifteen chicks. BWB:WA bred successfully as a yearling on an exposed sandbar adjacent to the the tip (beach 16, Fig. 2.2; Table 2.1). Her sister, who had not previously bred on LPI, occupied the western side of the tip in 1978 and 1979. In 1978, BWB:WA's mate of the previous year did not return, and she laid two clutches for AdG:dGW. Her sister laid two clutches with AR:B's old mate BkPBk:APBk, neither of which was

TABLE 2.1

Mating histories of the principal occupants of the north tip of Little Pelican Island, Minnesota, 1974–1983

		Primary pair					Other pertinent events			
Year	Female	(clutch #) ×	Male	(clutch #)	Terr #	Female	(clutch #) ×	Male	(clutch #)	Terr #
1973						AR:B	(1)	× Unbanded	(1)	3
1974	AR:B	(1)	× RA:YY	(1)	10					
	AR:B	(2)	× RA:YY	(2)	10					
1975	AR:B	(1)	× W:A	(1)	10	G:AG	(1)	× BkPBk:APBk	(1)	13
	AR:B	(2)	× W:A	(2)	10	A:RW	(2)	× BkPBk:APBk	(2)	13
	AR:B	(3)	× dGdG:dGA	(2)	10	A:RW	(3)	× BkPBk:APBk	(3)	14
	AR:B	(4)	× Y:RA	(2)	10					

Year	Female		Male		N
1976	AR:B	(1)	× BkPBk:APBk	(1)	10
	AR:B	(2)	× dGR:BA	(1)	10
1977	AR:B	(1)	× BkPBk:APBk	(1)	10
	AR:B	(2)	× BkPBk:APBk	(2)	10
	AR:B	(3)	× dGB:RA	(1)	8
	AR:B	(4)	× BA:RR	(2)	8
1978	BWB:WA	(1)	× AdG:dGW	(1)	10
	BWB:WA	(2)	× AdG:dGW	(2)	10
1979	BWB:WA	(1)	× BkPBk:APBk	(1)	9
	BWB:WA	(2)	× BkPBk:APBk	(2)	9
	BWB:WA	(3)	× GB:RA	(1)	9
1980	BWB:WG	(1)	× BkPBk:APBk	(1)	9
	BWB:WA	(2)	× dBA:YY	(1)	10
1981	BWB:WA	(1)	× BkPBk:APBk	(1)	10
	BWB:WA	(2)	× dBA:YY	(2)	10
	BWB:WA	(3)	× dGG:dBA	(1)	9
	BWB:WA	(4)	× dGG:dBA	(2)	9
1982	BWB:WA	(1)	× dBA:YY	(1)	10
	BWB:WA	(2)	× BkPBk:APBk	(1)	10*
	BWB:WA	(3)	× BkPBk:APBk	(2)	9
1983	BWB:WA	(1)	× dGYdG:MA	(1)	10
	BWB:WA	(2)	× BkPbK:APBk	(1)	10
	BWB:WA	(3)	× dBA:YY	(2)	10

Female		Male		N
dBR:AdB hatched				1–14
BWB:WA hatched				1–14
A:W	(3)	× AdG:dGW	(1)	10
BWB:WA	(1)	× AdG:BB	(1)	16
dBR:AdB	(1)	× BkPBk:APBk	(1)	10
dBR:AdB	(2)	× BkPBk:APBk	(2)	10
dBR:AdB	(1)	× BWB:RA	(1)	11
dBR:AdB	(2)	× AdG:dGW	(1)	10
dBR:AdB	(3)	× AdG:dGW	(2)	10
dBR:AdB	(4)	× AdG:dGW	(3)	10
MA:dGGdG	(1)	× dBA:YY	(1)	5
GR:AB	(1)	× AdG:dGW	(1)	10
BY:AdB	(1)	× AdG:dGW	(1)	10
BY:AdB	(2)	× AdG:dGW	(2)	10

NOTE: Underlined matings indicate that one one or more chicks hatched.
* Clutch experimentally removed

successful. In 1979, the sisters switched mates with each other, and both mated with additional males as well. The sister dBR:AdB failed to return in 1980. BWB:WA continued as a breeding bird through 1984, providing six clutches for BkPBk:APBk, five for male dBA:YY in a different territory on the island's north side (beach 8), and four for two other males. BWB:WA was simultaneously polyandrous with BkPBk:APBk and dBA:YY, and with BkPBk:APBk and AdG:dGW (Oring and Maxson, 1978; unpublished). BWB:WA had five chicks by three males return to breed; and she had five breeding grandchildren.

The complete nesting histories of the three males identified above are given in Table 2.1. To complete the picture of their offspring's nestings: BkPBk:APBk had four of his fourteen fledglings return and breed (two by AR:B and two by BWB:WA), and he had five breeding grandchildren. He was killed on his nest by a weasel in 1983. Two of dBA:YY's seven fledglings returned to breed, but no grandchildren, and none of AdG:dGW's ten fledglings bred locally. As a final note, in 1978, BkPBk:APBk adopted AdG:dGW's sole chick. Thus the chick was reared by its "aunt" and her mate.

As the histories above make clear, the structure of a spotted sandpiper population can be complex. We are still in the process of untangling affinities for individuals and places, searching for physical and behavioral qualities of successful and unsuccessful birds, and evaluating the role that chance events play in the development of individual life histories. In the remainder of this chapter we examine results to date, on the ways in which environmental variables and experience affect the breeding of spotted sandpipers.

ECOLOGICAL VARIABLES AFFECTING BREEDING

Effects of Food Abundance

Spotted sandpipers migrate north to exploit seasonally plentiful food supplies while breeding. Unlike arctic nesting waterfowl (Ryder, 1970; Ankney and MacInnes, 1978), shorebirds do not rely heavily on stored nutrients for meeting the costs of reproduction (Erckmann, 1983; Walters, 1984). Thus, food adequate for sustaining body condition, egg production, and growth of young must be obtained from the environment. The primary food of spotted sandpipers on LPI are adult midges (Chironomidae) and mayflies (Ephemeroptera) that emerge from the surrounding lake. We sampled arthropod abundance once a week with twelve cylindrical sticky traps (Rubbleke, 1976) located on six territories around the island, and exposed for forty-eight hours (Maxson and Oring, 1980; Fig. 2.2). Mean weekly trap catches for the island over all years are shown at the bottom of Fig. 2.3. The superabundance level of food indicated on the graph is the level at which

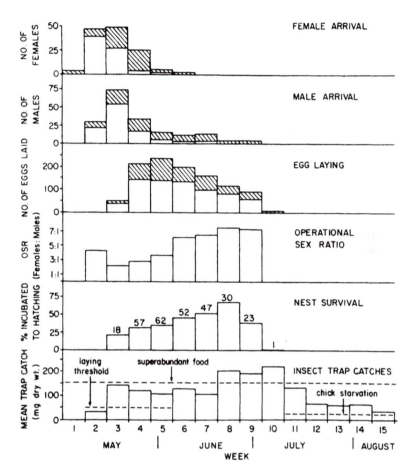

Fig. 2.3. Weekly distributions of spotted sandpiper arrival times, egg-laying, operational sex ratio, probability of nest hatch, and insect trap catches on Little Pelican Island. Shaded areas of arrival times and egg-laying refer to birds breeding on LPI for the first time, clear areas are experienced birds. OSR was calculated as weekly mean values of seven daily values based on all unpaired females and males seen on the island except incubating males. Nest survival rates are the percentage of nests laid during each week that were incubated through to hatching. Food is the mg dry weight of insects from all traps per week. Data for all variables are combined from 1975 to 1982; food values also include 1974. While summarizing over years eliminates interesting annual variation, these figures give an idea of the moderately long-term (two to three generations) situation faced by spotted sandpipers on LPI. Graph modified primarily from data in Lank et al. (in press).

non-incubating birds approached their minimal foraging times (Maxson and Oring, 1980, Fig. 11). The laying threshold was determined by correlating food levels with yearly variation in laying time. Chick starvation levels are based on one year's observations when food was extremely low during the hatching period (Lank et al., in press).

Comparing food levels with migratory arrival times (top two lines, Fig. 2.3) shows that females, especially experienced ones, usually arrive before food reaches the laying threshold, and before the time of male arrival, which coincides with increased food. Thus, they arrive before two essential reproductive resources—food and mates—are obtainable. Intrasexual competition for all-purpose territories, which provide females with both food and nest-sites, is responsible for the early arrival of females—an unusual pattern in migratory birds (Myers, 1981; Oring and Lank, 1982).

As a population, females lay substantial numbers of eggs over a six-week period ending abruptly the first week in July. Certain individuals have laid up to five clutches during this time; and the extended season is not due primarily to variability in laying times among females. Food is readily available throughout and after this period (Fig. 2.3). After laying started each year, there were no food effects on laying rates, interclutch intervals, or egg size (Lank et al., in press). Cessation while food was still abundant implies that other factors were responsible for the end of breeding. We suggest that chicks hatching from eggs laid after the first week in July would usually find insufficient food to sustain growth. Chick starvation was observed in one year when insect emergence was early throughout the season and ceased by mid-July, a time when food is usually still abundant (Fig. 2.3).

Maxson and Oring (1980, Table 8) found that the time allocated to various activities, e.g., foraging, varied with food abundance for both sexes at nearly all stages of the nesting cycle. The key ecological question is not only whether behavior varies under different conditions, but also how often various conditions are encountered. During the weeks when females lay (weeks 3–9, Fig. 2.3), food was sparse (below 50 mg) 28 percent of the time, abundant (50-150 mg) 30 percent of the time, and superabundant (above the 150 mg level) 42 percent of the time. Thus, three-quarters of the time, food abundance did not play a major role in determining the frequency of different behaviors.

Maxson and Oring (1980, 233–34) further concluded that food was not a major component of variation in territorial quality on LPI, and subsequent analyses confirmed this to be the case (Lank et al., in press). Food abundance varies among territories on the island, generally being highest on the downwind side (beach 5, to the east, Fig. 2.2), and lowest on the windward side (beaches 13 and 1, to the west), but overall, differences are not statistically significant. Territorial differences were not consistent from year to year, and the range among territorial means was minor compared to weekly variations on the island as a whole (see Lank et al., in press). Trap catches per territory were not correlated with male nesting densities (Lank, unpublished). These findings support the conclusion that after the first major emergence of the season, food is readily available to adults throughout LPI.

In summary, spotted sandpipers breeding on LPI have a sustained abundance of food for eight to nine weeks each summer. This is a longer period than that available to their more northerly breeding relatives (Lank et al., in press). These conditions are exploited by females, who lay up to five clutches per season. The primary difficulty for females lies not in producing eggs, but in obtaining mates to care for them.

Effects of Nest Predation

LPI is located about seven kilometers offshore. Its small size makes it unsuitable for year-round occupation by many of the spotted sandpiper's usual mammalian nest predators. In addition, over the past twelve years we have removed a mink (*Mustela vison*), a least weasel (*M. rixosa*), and a striped skunk (*Mephitis mephitis*) from the island. Despite this limited predator control, only 37 percent of the 1,289 eggs laid between 1973 and 1982 hatched. Nevertheless, LPI may be a relatively "safe site" for nesting sandpipers. Large areas of the mainland, and even nearby Pelican Island, provide suitable nesting habitat, yet are occupied by few if any breeders. From coastline surveys, we estimate that more than half of the breeding population of the main Leech Lake basin (Fig. 2.1) breeds on LPI. We suspect that mammalian nest and chick predation at other sites limits the density of breeding sandpipers. We have shown that this process works on LPI, where the level of breeding success strongly influences rates of philopatry and thus population densities (Oring et al., 1983; see below). Higher nesting density of males raises the potential for mate monopolization relative to low-density areas, but also increases competition among females for territories.

The most consistent source of nest failure on LPI from year to year is the deer mouse (*Peromyscus maniculatus*). The mice bite holes in the shells, but do not eat egg contents. Incubating birds detect the holes and usually abandon their nests (Maxson and Oring, 1978). Mice accounted for 29 percent of all egg loss, and about 18 percent of all eggs laid between 1973 and 1981 (Oring et al., 1983). In addition to mice, predation is caused by birds, especially common grackles (*Quiscalus quiscula*) and mustelids. Grackles are a problem every year; mustelids are absent most years. However, even a single mink or weasel can cause widespread nest destruction. In 1975 one mink caused total breeding failure.

Temporal Variation in Breeding

The probability that eggs laid in a particular week will hatch is shown in Fig. 2.3. Nests laid in weeks 3 and 4 had the lowest probability of hatching

(22 percent, 32 percent). The likelihood of success rose through week 8 (19-25 June, 72 percent). Nests laid after week 8 had a decreased chance of hatching, due primarily to increased mustelid predation in late June of some years.

Thus, early breeding is not "efficient" for females, in the sense of maximizing chicks fledged per egg produced, but it is the tactic favored by experienced birds of both sexes. Early breeding females, on average, do acquire more mates, lay more eggs, and fledge more young than later arriving birds, whereas, overall, early breeding males do no better than their later arriving counterparts.

Effects of Mate Availability

Worldwide, the sex ratio of spotted sandpipers is not strongly male-biased (Oring, in press). Nevertheless, at LPI we have witnessed a breeding sex ratio of 1.39 males per female; and in one year where the absolute ratio of all residents of the main Leech Lake basin was determined, it was 1.45 males per female. This skew reflects both the attractiveness of local circumstances to transient males, and the result of intrasexual exclusion of females from the area. While most females failing to establish territories left LPI, seven females remained in the area and were known not to breed.

Emlen and Oring (1977) defined the operational sex ratio as the number of sexually active females to males at any time. We computed the daily OSRs for LPI, including all birds each day, and considering incubating and brooding males as not sexually active (Lank, et al., in press). Weekly mean values averaged over seven years are shown in Fig. 2.3. The ratio is lowest during the influx of males in the third week of May, and rises thereafter through the end of the egg-laying period. Most of the rise is due to males becoming unavailable as they incubate. However, even in early May the ratio averages 2:1, reflecting the exclusion of some females from the breeding habitat by the territorial behavior of others.

The severe competition for mates implied by high OSR values late in the season has been measured in two ways. We compared the renesting probabilities of males and females after they lost or finished a clutch. Through week 5, both breeding males and females renested locally within a week at statistically similar rates, about 80 percent. While 60-80 percent of males continue to renest through week 8, however, the rate for females dropped steadily to 27 percent at that time (Lank et al., in press). Thus, the available mates were monopolized by a decreasing pool of females as the season progressed. Even among females that obtained mates, however, the time taken to do so became greater as the season progressed. Figure 2.4 compares mean intervals from completion of one nest to the start of laying with a new mate, and from destruction of a nest to relaying with the same mate. At the

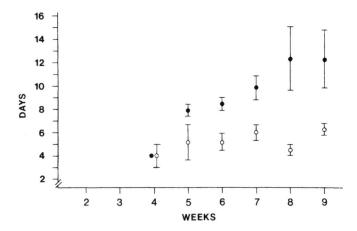

Fig. 2.4. Time taken by females to begin laying in a new nest (dots) with a new mate following completion of its previous nest, and (circles) with the same mate after nest loss.

start of the season, the new-mate and same-mate intervals were similar, despite the need to form new pair-bonds with males. Obviously, this occurred rapidly. Later, however, the new-mate interval became significantly longer, reflecting the change in osr. Since the same-mate line remained at four to six days, seasonal deterioration of the female's physiological condition can be rejected as the cause of longer interclutch intervals in late season.

EXPERIENCE

Experience in the Natal Year

Throughout the breeding period, families expand their home ranges into neutral or vacated territories, exposing the young to new places and individuals. By the time the young leave LPI they are familiar with much of its habitat and many of its residents. Once fledged, juveniles further expand their home ranges. We have seen young of the year resident on the mainland seven kilometers distant; and "foreign" juveniles become late season residents of LPI. Just how these experiences influence later territorial preference, and whether or not this early social experience plays a role in subsequent mate choice is unknown. When, where, and how parent-sibling and sibling-sibling bonds break is also poorly understood; and social interactions during migration and on the wintering grounds are not documented. Nevertheless, the fact that winter territories are primarily the domain of single individuals is known (S. M. Haig, unpublished). It is also known that yearlings arrive on the breeding grounds unpaired, indicating that behavior outside the breeding season does not play a primary role in determining mate choice. Below we examine philopatry and dispersal in a small cohort in which each individual was followed as a breeding yearling.

Philopatry and Dispersal of One Cohort

Although data on numbers of local chicks returning to breed on LPI are quite accurate, data on visitations by nonbreeding yearlings, and on distant breeding locations of local yearlings, are less accurate for most years. However, data on chicks fledged in 1982 are unusually complete, and provide insights as to the dispersal tendencies of young spotted sandpipers in general.

Only seven chicks fledged on LPI in 1982, and four of these returned to its vicinity in 1983. A search covering more than one hundred kilometers of nearby shoreline failed to find the three remaining chicks, and they were presumed dead. Two of the four returning birds were sisters and two brothers from unrelated families. The earliest arriving (17 May) and heaviest (55 g) sister laid two clutches for two separate males on LPI, while her later arriving (27 May) and lighter (47 g) sister bred monogamously about one kilometer away on Pelican Island. Similarly, the earliest arriving (18 May) and heaviest (44 g) male bred on LPI, while his later arriving (approx. 5 June) and lighter (40 g) brother bred on nearby Pelican Island (with the late arriving sister!). Both of the yearlings that bred on Pelican Island were also seen on LPI. The female tried to establish herself on LPI for two days before dispersing to Pelican Island; the male was seen on LPI twice while his mate was incubating. LPI may be a core breeding area from which subordinate or unsuccessful individuals disperse.

General Patterns of Philopatry

A total of fifty-five local chicks (through 1983) have been seen as adults on LPI—thirty-two females and twenty-three males. Of these, fifteen females and fifteen males bred, but females were significantly more likely to breed on LPI when one year old (Oring and Lank, 1982, 1984). Other local chicks are known to have bred on nearby Pelican Island and at Whipholt Beach (Fig. 2.1). In the final three years of a ten-year study, 31 percent, 40 percent, and 35 percent of the breeding females, but only 21 percent, 19 percent, and 17 percent of the breeding males were locally hatched (Oring and Lank, 1984).

The pattern that emerges is one where yearlings of both sexes home to the vicinity of their natality. Then, depending on their experiences as chicks and their assessment as yearlings of the overall breeding situation, they may or may not attempt to breed. Among those that attempt breeding, the time of arrival, their body condition, and the nature of the competition, all influence whether or not they succeed. Yearling females are much more likely than males to forsake breeding altogether due to the intensity of intrasexual competition. This tendency of some yearling females to forsake breeding is one of several reasons why the breeding population is male-biased.

Behavior of Yearlings

Yearling females arrive on LPI unpaired, an average of six days after experienced females, and at about the same time as experienced males (Oring and Lank, 1982; Fig. 2.3). In low-density populations such as Lake Itasca, Minnesota (Oring and Knudson, 1972) and Pelican Island, young females are essentially always able to establish a territory, and usually they can attract a late arriving male. A few may remain unpaired. In high-density populations such as LPI, yearling females face stiff competition, and some are unable to establish territories at all. Twice we have observed yearling females that were soundly defeated on LPI establish themselves as breeding birds on nearby Pelican Island. One badly beaten local yearling remained on LPI as an incubating "helper" for her former mate, and one remained in the vicinity as a nonterritorial, nonbreeding bird. Other yearlings that failed to establish themselves on LPI emigrated to unknown locations, and we have no way of knowing whether or not they bred.

Yearling males arrive long after the breeding of experienced birds is underway. The first of the inexperienced males pair with unmated females already defending territories. These females are almost always yearlings. Such males are able to settle with minimum aggression. If food is adequate these pairs may initiate egg-laying in as little as three or four days (Lank et al., in press). Yearling males arriving late into a saturated population may have to wait for a female to become available after her initial reproductive effort. Sooner or later, however, there are females that compete for these males. Primary males, disadvantaged by the loss of a female's help or subdivision of their territories, commonly attempt to exclude secondary or tertiary males. Females may then intercede and attack their own primary males in order to guarantee the settlement of additional mates. Thus, the behavior of the females themselves may enhance the male bias of a given breeding population.

Breeding Experience and Site Fidelity

First-time breeders are particularly sensitive to the success or failure of their reproductive efforts. In low-density areas, it is not uncommon for birds to move to opposite sides of their territories or even to emigrate in mid-season if eggs are lost. Twice we found such birds renesting two kilometers from their initial breeding site. On LPI, most immigrants remain within the population for the entire season regardless of whether or not nests are depredated or flooded. However, their return the following year is dependent primarily on whether or not they hatched one or more chicks (Oring and Lank, 1982, 1984; Oring et al., 1983). This response is stronger among females than males. Some yearlings emigrate after failure, but return to LPI in subsequent years, evidencing dual breeding site affinities. On the other hand, successful yearlings become locked into a single local population. The fact that young females respond to reproductive success more positively, in terms of showing site fidelity, is an additional factor influencing the sex ratio of a given population. The influence of age and success on return rates is explored further in Oring and Lank (1982, 1984) and in Oring et al. (1983).

Breeding Experience and Territorial Fidelity

There is a strong positive relationship between the territory a bird breeds in, in one year, and the one it occupies in subsequent years (Table 2.2). Exceptions to territorial fidelity include: 1) birds that breed in territories lacking adequate beach cover (mostly yearlings) may switch to territories possessing these attributes as they become available: 2) females occupying territories to which no males are attracted may move to positions closer to males (males appear to be attracted to territories with extensive beach areas and semi-open nest cover); and 3) yearlings whose reproductive efforts fail tend to change territories or breeding populations. Territorial switches may occur both within and between breeding seasons. There is no clear-cut sexual bias in preference for former territories: both sexes prefer to breed in former locations.

Breeding Experience and Mate Choice

Because of the great territorial fidelity of both males and females, demonstrating that birds prefer former mates, as opposed to territories, is difficult. If a female occupies her old territory and a former mate returns, separating his preference for that site from a preference for the female is practically impossible. Since females usually occupy territories before males arrive, their territorial fidelity is demonstrable, but mate preference is ambiguous. In a few instances, however, individuals have shown clear mate preference. Three such observations are detailed below.

1. Mate Tenacity by a Female. BR:YA was a yearling female in 1977. She laid a clutch for her yearling male (AB:dGdG) 23-28 May, and helped him incubate until 7 June. Then the nest was depredated and the female badly beaten by an older neighbor. BR:YA left LPI and that same day AB:dGdG paired on a new territory with RA:RR (he became her tertiary mate). Their clutch was completed 14 June, the female's third. She did not help incubate. On 16 June, BR:YA returned to LPI and began defending a new territory in front of AB:dGdG's nest; and from 20 June until hatching

TABLE 2.2
Territorial fidelity of spotted sandpipers on Little Pelican Island,
Minnesota

	Former territories	New territories	Probability
Females	75% (*n* = 52)	25% (*n* = 17)	< 0.01
Males	80% (*n* = 74)	20% (*n* = 19)	< 0.01

NOTE: Probabilities determined from a 2 × 2 contingency table with expected values determined, considering former territories to equal three territories—the precise former territory and the two adjacent to it. For this purpose, LPI is divided into fourteen "territories."

on 4 July, she regularly incubated the eggs of her former mate—ones she did not lay! BR:YA continued to defend this male's territory from intrusion by neighboring birds of both sexes through the third day posthatch and then disappeared for the year. In 1978, BR:YA returned to occupy the same territory she defended for AB:dGdG, and he became her second mate. BR:YA defended this same territory 1979-1983.

2. Mate Tenacity by a Male. In 1977, female RA:RR bred on LPI for the first time. She may have been two years old or older, as she was triandrous and provided essentially no parental care, both characteristics of older females. Her first mate was BA:RR, a yearling male. In 1978, RA:RR's leg was broken in a fight the day she arrived (13 May), as she attempted to regain her 1977 territory. Unable to defend a territory, RA:RR was seen only intermittently for the next few days. On 17 May, BA:RR returned. Rather than compete with paired males for their mates, or wait in a high-quality territory for a new female to arrive (five additional breeding females entered the population after this date), BA:RR paired with RA:RR in a densely wooded part of the island that lacked a beach. (Beaches are important for both adults and young as sources of food, because they allow foraging while keeping track of potential predators, and because they contain essential bathing sites). This year RA:RR was monogamous and shared regularly in incubation and care of the chicks. RA:RR's injured leg eventually fell off; and she was not seen again after 1978. BA:RR continued to breed on LPI in territories with good beaches (with the exception of 1980) through 1983 when he was killed by a weasel.

3. Mate Tenacity by a Female and a Male. Female —:RA and male dBA:dGdG bred on territory 13 in 1979 (Fig. 2.2). Four chicks were fledged by dBA:dGdG but his mate provided no parental care. Rather, she bred secondarily with male AM:YY on the adjacent territory 14. Together they

fledged two chicks. The success of dBA:dGdG on 13, and of his mate —:RA and her secondary mate AM:YY on 14, resulted in dBA:dGdG's being site-faithful to 13, but the other two birds to 14.

In 1980, all three of these birds returned to LPI. On 16 May, —:RA successfully claimed territory 14, and a yearling female, —:BWA, occupied 13. On 17 May, —:RA courted and appeared to pair with a new male, GRG:RA, while dBA:dGdG, who had just arrived, was chased off 13 by two other males. For the next six days, the following scene unfolded: beach 13 was successfully defended by a local yearling female —:BWA and male dBA:dGdG, and —:BWA courted dBA:dGdG vigorously. However, dBA:dGdG courted only —:RA, and attempted to attract her to 13 by repeated song flights at the border. On the other hand, —:RA successfully defended a portion of 14, and was favored there by AM:YY, her secondary mate of 1979. Not until 21 May did dBA:dGdG move into 14 long enough to copulate with —:RA, and not until 23 May did —:RA and dBA:dGdG occupy the same territory—14. In summary, —:RA retained a strong preference for her primary mate of 1979, dBA:dGdG, and for the territory (14) in which she personally participated in parental care; and both dBA:dGdG and AM:YY retained a preference for their old territories and old mate, —:RA. It took six days for the conflicting preferences to be sorted out. The conflict was eventually resolved by dBA:dGdG occupying the territory preferred by female —:RA. In other cases, we have seen the female move to a site preferred by the male.

Breeding Experience and Annual Reproductive Success

As a female gains experience, her chances for RS in future years increase. Females in their second breeding year obtain significantly more mates than yearlings, and females in their third year obtain significantly more mates and fledglings than younger females (Table 2.3; unpublished). The increase in a female's RS with experience results from a combination of several factors: 1) older females arrive earlier in the season, allowing them a longer time over which to breed; 2) older females are site-tenacious to territories where they have formerly bred, and are successful in their defense; and 3) older females are preferred by former mates. The fact that third-year breeding females do not arrive earlier than second-year females indicates that early arrival alone cannot account for the improved RS of old females. The factors mentioned above, combined with the nature of the social system of the species, allows females to capitalize on their experience to become polyandrous; and, on average, each additional mate a female obtains results in additional fledglings (Fig.2.5).

Males, too, may benefit from breeding experience. Older males return to LPI earlier and have a competitive edge in the occupation of former territo-

TABLE 2.3

Annual reproductive success of male and
female spotted sandpipers with differing amounts
of prior breeding experience on
Little Pelican Island, Minnesota

Previous experience	n	$\times$ no. chicks fledged	Probability
Females			
None	42	1.07	ns
One year	20	1.60	
Two years +	33	2.84	0.05*
Males			
None	49	1.20	ns
One year	23	1.22	ns
Two years +	38	1.55	

* $t = -2.11$, $df = 51$.

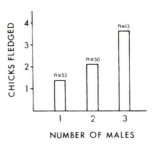

Fig. 2.5. Relationship between the number of males that females monopolize and the number of their chicks that fledged.

ries and the acquisition of former mates (Oring and Lank, 1982). However, the nature of spotted sandpiper parental care and the length of the incubation/brooding period precludes males from having more than one successful reproductive effort per year. In only one year did experienced males fledge more chicks than inexperienced males—a result of extensive late season predation by a weasel. Overall, there was no significant difference in the number of mates acquired, or young fledged by males with differing amounts of breeding experience (Table 2.3). Why is this the case? If egg loss were constant over the season, the early arriving, older males might be able to obtain, on average, a greater number of replacement clutches per season and, hence, reap greater RS. Greater RS is not achieved for several reasons: 1) on LPI during this study, egg loss was much greater early in the season than later, seriously depreciating the value of first clutches obtained by experienced males (Lank et al., in press); 2) the impact of years where early predation was heavy was amplified by the fact that, in some years, essentially all (e.g., 1976) or no (e.g., 1975) males succeeded (Oring et al., 1983).

Why then do experienced males precede young ones to the breeding ground at all? It might be related to such factors as shorter migratory distances, being in better migratory condition earlier in the season, and so on. It may also be that the seasonal pattern of predation recorded on LPI is not representative of the entire breeding range of the species and/or the era when this migratory pattern became genetically fixed. Finally, one might hypothesize that even though older males fledge no more chicks, the fact that their chicks hatch earlier, on average, might lead to their increased survival and higher rate of entrance into the breeding population (Perrins, 1970; Cooke et al., 1984). Fifty-five chicks have been seen and individually identified as adults in subsequent years. However, early hatching chicks were no more likely to be seen as adults or to enter the breeding population than later hatched ones.

Older females acquire more mates and fledge more young than younger females due to a combination of factors: early arrival, relatively good ability to defend preferred territories, and their being preferred by former mates and neighboring males. Males derive relatively little benefit from experience since there is little intermale variation in the number of mates acquired. These and other sex-related differences in breeding strategies are summarized below.

SEX-SPECIFIC STRATEGIES

Females

A female's RS increases on average with each mate acquired (Fig. 2.5); and the number of mates acquired increases through the first three years of life. Thus, female spotted sandpipers should prefer sites optimal for recruitment of males, and they should not take life-threatening risks, especially at an early age. Older female spotted sandpipers enhance their competitive ability through early arrival and acquisition of prime territories. As a female gains experience she may also become a favored mate; and the more males that prefer her, the more likely she is to be polyandrous. Females further augment their fitness through "bet-hedging"—vigorously pursuing alternate mating opportunities as they arrive, but, in their absence, investing in parental care.

Young females, lacking the competitive ability of their older counterparts, arrive and attempt to pair at times of minimum female-female conflict (Oring and Lank, 1982). As the operational sex ratio becomes strongly skewed in mid- to late-season (Lank et al., in press; Fig. 2.3), young females have little chance of acquiring secondary mates. At this time, yearling females contribute to their fitness primarily through the provision of parental care rather than the acquisition of additional males. Young females facing very difficult breeding circumstances may emigrate or forsake breeding altogether.

Why are young females less successful than older ones? Rubenstein (1982) illustrated the conditions under which young animals in moderate body condition may opt to delay reproduction. Curio (1983), too, challenged the traditional view that young birds were less productive due to environmental or social constraints, i.e., lack of experience. Rather, he proposed that young breeders with high future prospects of breeding refrain from expending maximal reproductive effort to forestall the associated risk of losing future reproduction (i.e., they may use restraint rather than constraint). Curio also suggested that constraint and restraint might be combined in such a way that parental constraint is used to gauge the extent of a reproductive effort.

In spotted sandpipers, some young females try hard and fail to obtain a satisfactory breeding situation; others appear to survey the field and give up with little overt effort. Proof of the degree to which the reduced RE and RS of young females is due to constraint versus restraint, however, must await the successful completion of female removal and/or male supplementation experiments.

Males

In contrast to females, male spotted sandpipers on LPI acquire no more mates and fledge no more chicks, on average, as they gain experience. Consequently, males show less age-related differences in their willingness to provide parental care, pursue potential mates, or enter conflict. Males invariably defend their territory against other males, and care for their eggs/young. They also show a willingness to fight in order to obtain their former mates and to keep their current mates from adopting a secondary mate. Males, in contrast to females, have a conservative lifestyle that remains the same throughout their breeding lives: to obtain a clutch of eggs and care for it, and if it is destroyed, get another. No male has two successful breedings per year, and normally no male forsakes breeding altogether. The lack of seasonal effects on fledging and return rates of chicks suggests that later-breeding males are not at a strong reproductive disadvantage. Were this not the case, selection would favor more synchronized breeding by males, making polyandry less likely (Emlen and Oring, 1977).

CONCLUDING REMARKS

Over time, alternate tendencies to disperse or be site tenacious have resulted in dense populations in some locales, sparse populations elsewhere, and no spotted sandpipers at all in some seemingly suitable places. In dense populations, intrasexual competition is acute and some individuals, primarily yearling females, are excluded. Since more females are excluded than males, the breeding sex ratio becomes skewed, and the incidence of polyandry is increased. This incidence is influenced further by such factors as a breeding synchrony and preference for particular breeding sites and for individuals. The degree to which differential mortality of the sexes further affects population sex ratios, and hence polyandry, is doubtless variable from time to time and place to place.

Nest loss plays a role in determining the incidence of polyandry since high loss leads to concentrations of birds in protected sites such as LPI. This, in turn, results in population saturation, female exclusion, and an increase in the incidence of polyandry. However, if one measures nest loss in these areas of high polyandry, the incidence may be low. The birds are there because that is where nest loss is low! On the other hand, in low-density populations, the incidence of nest loss may be high. Within a season, nest loss also plays a complex role in influencing the incidence of polygamy. If nests are lost early, i.e., during egg-laying, or the first days of incubation, the likely result is that pair members renest together. Thus, the chances that this particular female becomes polyandrous are diminished. On the other hand, where nest loss occurs after a male's initial mate has deserted him, his availability increases the likelihood that some other female will become polyandrous; and, at the same time, the male becomes polygynous.

Compared to the more conservative, monogamous members of the family Scolopacidae, spotted sandpipers tend to live relatively short lives, to breed first at a younger age, to be relatively dispersal-prone, and to have a greater female-female and year-year variance in reproductive effort (Oring et al., 1983; Oring and Lank, 1984). In this regard, spotted sandpipers are adapted for occupying highly disturbed, early successional habitat; and, relative to other sandpipers, they may be considered *r*-selected (Oring et al., 1983). Their social system and patterns of dispersal allow them to maximize reproduction at times and places where chances of success are strong; but to quickly adapt to disaster by dispersing to new areas, often ones occupied by other spotted sandpipers. Because of the ease with which spotted sandpipers can be monitored in high-density populations, the great behavioral lability they exhibit, and the power of using a sex-role reversed, polyandrous species to test hypotheses generated from monogamous and polygynous systems, the spotted sandpiper is an excellent subject for continued social system studies.

3. Reproductive Strategies of Male and Female Moorhens (*Gallinula chloropus*)

———————————— • ————————————

MARION PETRIE

HOW HAS the environment shaped the moorhen social system? A social system is essentially a description of two main aspects of animal existence: 1) the spacing pattern of the individuals (e.g., whether they form groups) and 2) the form of the mating system (e.g., whether individuals are monogamous or polygamous). These patterns are a consequence of "decisions" taken by individuals, for example, when an individual finds that the benefits of joining a group outweigh the costs. A social system is a description of the sum of the social behaviors of the individuals that compose it (Gosling and Petrie, 1981). It represents a level of organization on which natural selection does not act, since natural selection is thought to act on individuals or genes rather than at the level of the group or the species (W. D. Hamilton, 1964; G. C. Williams, 1966; Dawkins, 1976). Understanding the evolution of social systems must, therefore, start at the level of the individuals that compose them. To understand why a species exhibits a polygynous mating system, it is necessary to consider why it pays individual males to expend energy in competition for more than one female. Although it may seem obvious that by mating with more than one female a male may enhance his reproductive success, alternative means of expending reproductive effort can yield similar fitness increments. Males could instead invest energy in provisioning their offspring. This is especially true of birds, where males are capable of providing all of the postzygotic parental care.

In this chapter, I describe the social system of one species, the moorhen (*Gallinula chloropus*), and then analyze why this social system exists. I will consider why individuals of both sexes adopt particular reproductive strategies rather than theoretical alternatives. The social organization of the moorhen can be broadly summarized as follows. In the winter months the birds join small flocks where pair formation occurs. Birds leave these flocks to establish territories in the spring. The mating system is primarily monogamous although occasionally females form pair bonds sequentially with two males in one breeding season. It is not uncommon for more than one hen to lay in the same nest (Huxley and Wood, 1976), a phenomenon known as intraspecific brood parasitism or "nest dumping."

The questions considered here deal with both the spacing and mating behavior of moorhens. One set of questions considers why birds form flocks in the winter and are territorial in the summer; the other considers how and when the sexes apportion their reproductive effort into mating and parental components (sensu Alexander and Borgia, 1979) and why they adopt these strategies. Special attention will be paid to the role of the environment, if any, in shaping properties of the social system; where appropriate, the roles of other selective forces will also be discussed.

DESCRIPTION OF THE MOORHEN SOCIAL SYSTEM

The moorhen is a member of the Rallidae and in North America it is known as the common gallinule. Like most rails it lives in wetland habitats. It has a widespread distribution, including Africa, Europe, and America. The following descriptive account is based on a field study conducted in Norfolk, U.K., from December 1977 to September 1980. Large numbers of individuals were captured and marked (birds were caught in cage traps and marked with patagial wing tags and colored leg rings). The study area is part of a wider area of grazing marsh, a flat treeless habitat lower than the embanked river and crisscrossed by a network of man-made drainage ditches. The grass species on the marsh are a mixture of those sown as a fodder crop and wild grasses.

The moorhens nest along the edges of the ditches, in the marginal vegetation of reeds and sedges. Nests were found by searching the ditches and also by observation of breeding birds. The amount of cover at a nest site was measured by recording the distance at which the eggs in the nest first became visible to me. Nests situated in poor cover could be seen from farther away and therefore had high visibility values (see Petrie 1984 for details).

The moorhens on the ten-hectare study site formed small mixed-sex flocks (five to forty birds) between October and March (Fig. 3.1 shows the position of flocks on the area in winter 1978-79). Competitive interactions occurred while the birds were in flocks, ranging from one individual chasing another to fighting, with the birds pecking and striking at each other with their sharply clawed feet. Pair formation occurred in flocks before pairs left to establish territories. Moorhens frequently changed partners between breeding seasons. The proportion of marked birds that had different mates in successive seasons was 86 percent between 1978 and 1979 and 67 percent between 1979 and 1980. These changes were not always because one member of a pair died; in some cases the former mate was present elsewhere on the study area (Petrie, 1982).

Pairs of moorhens vigorously defended a length of ditch from intruders and neighbors during the breeding season (from March to September). Fig-

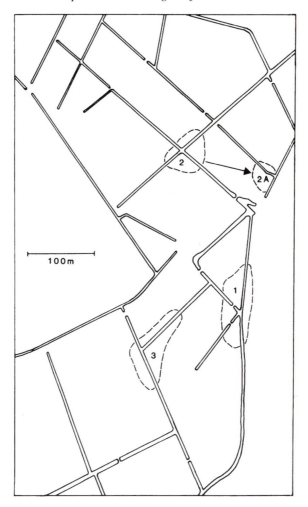

Fig. 3.1. Shows the position of the flocks on the study area in the 1979-1980 winter. The flock at site 2 moved to site 2A during the winter.

ure 3.2 illustrates the territories in one breeding season. Polyandrous trios in which females formed pair bonds with two males occurred at low frequency (e.g., two females out of eighteen in 1979). Such trios were established late in the season, and one was formed when a neighboring female deserted her mate. The female laid eggs in nests constructed by her first male and shared incubation with that male. Meanwhile, the second male oc-

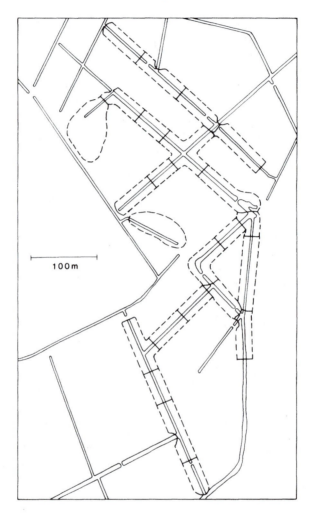

Fig. 3.2. A map of the territories on the study area for one breeding season (1978) illustrating their position in relation to drainage ditches. Territory size was determined from observations of boundary defense. Dotted lines represent approximate limits of area defense; exact measures were not possible at these points as there were no bordering territories.

casionally cooperated in defending the territory against neighbors but did not share any parental duties. Later in the season, the female, with the aid of her second male, initiated further nesting attempts.

There was considerable variation in the time at which pairs of moorhens left flocks to establish territories at the start of the breeding season. Some pairs established territories in March, while others did not do so until May (Petrie, 1982). Territory size was measured as the total length of ditch bank in the defended area. The mean length of ditch bank per territory ($n = 50$) was 149m. (SD = 51m). Both males and females took part in territory defense although males performed the majority (Petrie, 1984). The width of marginal vegetation along the ditches was measured three times in each ten-meter interval of ditch bank, and a mean value was calculated for each territory.

Birds constructed nests using fragments of the various monocotyledons growing in the drainage ditches. Males gathered nest material for females to arrange. Both birds shared incubation duties during the day, but the male performed all overnight incubation. The male's contribution to incubation amounts to 72 percent of the total (Siegfried and Frost, 1975). Both parents fed and tended the precocial young for a short time after hatching, but the exact contribution made by the sexes is unknown.

FEMALE REPRODUCTIVE STRATEGIES

Female moorhens expend the majority of their reproductive effort in two main ways, in intrasexual competition and in egg production.

Agonistic encounters between females were most common when they were in flocks. These encounters sometimes led to fighting, and marked females in flocks participated in more fights than marked males ($n_1 = 26$, $n_2 = 12$; binomial test, $p = 0.03$; combined observations from the 1978-79 and 1979-80 winters). These encounters typically occurred when a female approached a courting pair and tried to interrupt a courtship attempt by another female. Females initiated courtship more frequently than males ($n_1 = 87$, $n_2 = 12$; binomial test, $p = 1.0 \times 10^{-15}$; combined observations from the winters of 1978-79 and 1979-80). Courtship attempts were initiated by a bird approaching another and performing a characteristic neck arching display. The courtship sequence has been described by Wood (1974) and by Petrie (1982). Of eighty-seven courtship interactions initiated by a marked female, nineteen were interrupted by the aggressive behavior of another marked female (only marked birds were included and so this is a minimum estimate of the amount of competition for mates).

Since females court and compete for males in flocks where there is no shortage of males (flocks contain approximately equal numbers of males and females, with a small excess of males [Petrie, 1982, 1983a]), it seems likely that females expend energy in competition for access to scarce high-quality partners (Petrie, 1983b). If so, it can be predicted that those females that win fights should form pair bonds with the highest quality males. The best predictor of the outcome of female aggressive interactions in flocks was body weight. Body size, condition, and age were also considered but did not correlate with the proportion of agonistic encounters won by females. In two years a higher proportion of agonistic encounters were won by heavier females than by lighter females (Petrie, 1983a). Since heavy females win contests, the heaviest females should be paired to the highest quality males. But what is a high-quality male? Males perform 72 percent of the incubation (Siegfried and Frost, 1975) and so one hypothesis is that high-quality males are those that can incubate for long periods. Since incubation in the moorhen is energetically expensive and males lose 5–10 percent of their body weight during the breeding season, energy or nutrient reserves might be an important component of male quality. The heaviest females proved to form pair bonds with the males that were in best condition (condition in males was measured as the index of weight divided by the cube of the tarsus plus metatarsus length; this index was highly correlated with the weight of representative fat pads in dissected birds; see Petrie, 1983a for details) thus supporting the original hypothesis: male condition and female weight were positively correlated. If females select males that are in good condition, as these data suggest, then males without mates (and, since pairs establish territories, without territories) should be in poorer condition than those with mates; again the results obtained supported this prediction (Petrie, 1983a).

Of three factors that could affect the condition of males (age, size, and previous reproductive performance) size was found to be the most important. Smaller males were in significantly better condition than larger males, possibly because they have lower absolute food requirements. Comparing the smallest with the largest male moorhen, the larger bird has a 32 percent greater energy expenditure (Petrie, 1983a). The higher food requirement of larger birds may become critical when food or the time for feeding is limited. Moorhens appear to be food-limited in winter when there is a 12 percent decline in mean adult weight. Since the males in best condition were the smallest, the heaviest females tended to form pair bonds with the smallest males; female weight and male size were negatively correlated (Petrie, 1983a).

To summarize, it appears that females compete for access to high-quality partners since, presumably, there are fewer of these than competing females. A high-quality partner is a male with large fat reserves. This preference for good-condition males may explain why females do not pair for life but court and compete for mates in flocks each and every winter and commonly change partners between years. Male condition is likely to be variable, and thus unpredictable, from year to year.

Females can lay a large number of eggs in a season. The mean clutch size is between six and seven eggs. Predation rates are very high with up to 70 percent of nests lost in a season. Females may have to lay many eggs before they successfully rear a clutch and are capable of laying up to five replacement clutches in a season. If an early clutch is successful they lay again and rear a second clutch; occasionally moorhens rear three broods in the breeding season, which in the U.K. extends from March to August. Moorhen eggs on average weigh 23.5 g, which is 7.2 percent of the average female body weight of 326 g. In addition to laying eggs in their own nests some females parasitize the nests of other females, usually in the same laying period (Petrie, in prep.). Thus females make a major reproductive investment in egg production and sometimes lay thirty to forty eggs in a season.

MALE REPRODUCTIVE STRATEGIES

Although males in flocks do compete with each other there is no indication that this occurs in the context of courtship. Males do not usually take the initiative in courtship or compete for females. Male aggression, however, increases shortly before pairs leave to establish territories and males appear to be defending a space around themselves. There are no particular food items that they could be competing for, since at this time of year birds spend most of their time in flocks grazing the grass sward. Once pairs of birds establish territories, males become even more aggressive and they perform the majority of territory defense. Moorhen territories consist of a strip of open water along the center of each ditch and its vegetation-covered banks. These areas are vigorously defended against intruders. Residents also defend the well-defined boundaries of their territories against neighbors. Overt aggression is common in this context (Wood, 1974; Petrie, 1982). In ninety-two instances of territory defense observed from March to May 1979, sixty-six were initiated by male owners and only eight by female owners. The remaining eighteen cases involved defense by both pair members. Territory defense reached a peak in March and April, about a month before eggs are laid, showed a significant decline during May, and occurred at a relatively low frequency for the remainder of the breeding season. Male aggression therefore seems to be closely linked to the acquisition and defense of a length of drainage ditch and not to the defense of females. But why are males territorial?

Territoriality can be considered as a form of resource defense, which will occur when there are fewer resources than competitors. It would be favored by selection when the benefits of territory defense outweigh the costs. In the moorhen, territory defense coincides with the breeding season so it could be argued that the scarce resource defended is either directly or indirectly con-

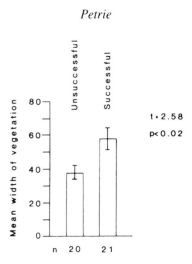

Fig. 3.3. A comparison of the mean
width (± standard error) of vegetation in
the territories of pairs that were unsuccessful in hatching any eggs. The sample
considers data from three breeding seasons (1978-1980).

cerned with reproduction. There are two main possible resources: 1) A food
supply necessary for egg production or for the growth of offspring; 2) suitable nesting areas that reduce the chance of predation on eggs and/or young.

The first possibility, that moorhens are defending a food supply, seems
unlikely since both parents and their young were regularly seen feeding
away from the defended area during the breeding season. Moorhens only
defend stretches of drainage ditches and the area immediately around them
(see Fig. 3.2) and, since the moorhens nearly always nest in the marginal
vegetation of the ditches, this suggests that the birds may be defending nest
sites. Moreover, there are four main lines of evidence for suspecting that the
limited resource defended is good cover for nest sites:

1) Birds that established territories first in the season had better hidden
nest sites (low visibility) (Petrie, 1982).

2) Male birds that competed successfully in winter flocks, the heaviest,
gained territories with better cover (Petrie, 1982). Heavier males were
larger birds, which had relatively less fat than smaller males. (There was no
relationship between success in agonistic encounters in flocks and the time
of establishing a territory).

3) The amount of vegetation in a territory was assessed, and successful
pairs (those who hatched at least one egg) had significantly more marginal
vegetation in their territories than unsuccessful pairs (Fig. 3.3). Thus good

cover for nests is likely to reduce the chance of predation. In the very open habitat of the grazing marshes the most important factor contributing to individual variance in reproductive success was whether nests survived to hatching; the majority of nests did not survive as predation rates were up to 70 percent.

4) Heavier males also obtained relatively larger territories and these contained more potential nest sites with good cover from predators (Petrie, 1984).

To summarize, males appear to compete with each other for territories with good cover nest sites, which reduce the chance of predation on eggs and young.

Apart from defending territories which contain resources that increase the likelihood of reproductive success, males contribute to the care of offspring in other ways. They select nest sites and collect nest material (Petrie, 1982); they make the major contribution to incubation which, in the moorhen, is energetically expensive (Siegfried and Frost, 1975); and, after hatching, they feed, brood, and protect the young. However, it is not known how their posthatching reproductive investment compares to that made by females.

DO THE FATTEST MALES ALSO OBTAIN THE BEST TERRITORIES?

As discussed earlier, females appear to pair selectively with fat, or good-condition males. But, as shown above, an important factor affecting reproductive success is the amount of cover in a territory. Do females also obtain the best territories by securing the fattest males? It is possible that males with good energy reserves can spend more energy on territory defense and could defend better territories. However, there were no significant correlations between the amount of vegetation that provides cover in a territory and the condition of the male owner. Thus, it appears that females, by selecting fat males, do not obtain preferential access to those territories with good cover.

WHY DOES MATE SELECTION OCCUR IN FLOCKS AND NOT AFTER TERRITORIES ARE ESTABLISHED BY MALES?

Why don't females choose the fattest males in the best-quality territories? Why do females apparently limit their options by selecting males in flocks? What is the advantage of selecting a mate in a flock before it has a chance of competing for the best nest sites?

The following are hypothetical advantages that could explain why females do not appear to take advantage of the full range of male attributes, their body condition and their territory quality:

TABLE 3.1

Comparison of the frequency of vigilant and
nonvigilant postures between birds that were not
isolated (usually members of a flock) and birds that
were isolated

	Vigilant	Nonvigilant
Not isolated	39	95
Isolated	39	55

NOTE: Isolated birds were defined as those whose nearest
neighbor was more than fifteen meters away. $\chi^2 = 3.77$, $p > 0.05$.

1) It is possible that pairs of birds could defend and hold onto more resources and this has led to the evolution of territory establishment by pairs.
If this was the case it could be predicted that the size or quality of the defended area is dependent upon the joint competitive ability (or the resource-
holding potential) of a pair. It was shown earlier that in both males and females the most important factor contributing to their competitive ability is
their body weight. So this prediction can be refined further: Is the size or the
quality of a territory dependent upon the combined weight of the male and
female owners? No significant correlation exists for either territory size and
the combined weight of the male and female ($r_s = 0.34$, $n = 20$), or the
mean width of the vegetation and the combined weight of the male and female ($r_s = 0.24$, $n = 20$). In fact, as shown earlier, territory size and quality is dependent upon male weight alone, which is consistent with the general observation that males perform the majority of territory defense.

2) Solitary male territorial moorhens may be at risk from predation and
this risk could be substantially reduced if pairs establish territories. It is
common in other species of birds, however, for males to establish a territory
that females subsequently select and, if females selectively paired with
males who had already established territories, this selection pressure could
outweigh the risks for males in being isolated. There is no direct evidence
that solitary individuals do suffer a higher risk of predation in the moorhen,
although in other bird species there is some evidence that they might, e.g.,
the wood pigeon (*Columba palumbus*) (Kenward, 1978). However, if this
was also the case for moorhens it might be expected that if individuals were
at risk they would spend significantly more time in a vigilant, head-up posture. Table 3.1 shows the amount of time isolated birds spend scanning for
predators in comparison with birds who had relatively close nearest neighbors. Although there is a tendency for isolated birds to spend more time in
vigilant posture, this difference is not quite statistically significant.

3) Those females that choose their mates in flocks obtain better quality males as a result of being able to sample more. It is not possible to compare the quality of mates of birds chosen in flocks with those chosen in territories, but it can be predicted that the quality of mate gained would be dependent upon the number of animals sampled, and, thus, on the size of flock where mate selection occurred. When the condition of males obtained by females that found their mates in a small flock (ca. ten birds) is compared with that secured by females attending a large flock (ca. twenty birds), no significant difference is found:

	Small flock (10)	Large flock (20)
Mean male condition	2.04 ($n = 12$)	1.98 ($n = 9$)
	$u = 50.5$, ns	

(The sample considers females that paired in flocks on the study area in 1979 and 1980.) However, this simple analysis considers only the absolute mate quality obtained, and earlier it was shown that the condition of mate secured is also dependent upon a female's competitive ability. It is therefore more meaningful to compare the quality of mate gained by a female taking into account her own ability to secure a mate. The following index of 'relative' mate quality was devised for this purpose:

Relative mate quality = Male condition rank / Female weight rank

Male condition and female weight were ranked to reduce the two measures to the same units. Males with a condition index (see Petrie, 1983a) value of 1.5×10^{-4} g.cm^{-3} were assigned rank 1 (the worst condition) and those of condition 2.5×10^{-4} g.cm^{-3} were assigned rank 100; females of weight 295 g (the lowest weight) were assigned rank 1 and those of weight 395 g were assigned rank 100; intermediate values were assigned corresponding ranks.

When the relative mate quality is compared for females obtaining mates in small and large flocks a significant difference is obtained. Females attending a large flock obtained a significantly better mate in relation to their own ability to acquire a mate than did those females obtaining a mate from a small flock:

	Small flock (10)	Large flock (20)
Relative mate quality	1.142	2.232
	$u = 30.5$, $p = 0.05$	

This fact could be an important determinant of why females select mates in flocks. By choosing from a pool of males, females may improve their chances of obtaining a high-quality mate. If males were dispersed in their

TABLE 3.2
Relationship between the quality of mate obtained by a female and the time at
which the pair left flock to establish territory

Correlation	n	Correlation coefficient (r_s)	Significance
Time of leaving flock vs. male condition	27	0.09	ns
Time of leaving flock vs. relative mate quality	21	−0.36	ns
Time of leaving flock vs. relative mate quality, excluding first-year birds	17	−0.61	$p < 0.02$

NOTE: Mate quality is measured as male condition and as a relative index accounting for the female's competitive ability.

territories, the costs to females of searching and competing for mates could be prohibitively high.

4) An evolutionary race between females for access to the best mates has led to selection of mates earlier in the year. If this was the case then it could be predicted that the females which form pairs earliest in the year would get the best mates. However, there was no relationship between the time at which pairs form and leave flocks and the condition of the male secured (Table 3.2). That is, females that left flocks with their mates earliest did not obtain the fattest males.

As in (3) above, the quality of mate obtained by a female also depends upon her own competitive ability. Therefore, it may be more meaningful to look at the relationship between the time at which pairs form and the relative mate quality secured by a female. Table 3.2 gives the relationship between the time at which females leave flocks and their relative mate quality and there is a nonsignificant tendency for females with relatively good quality mates to leave flocks earlier. However, if birds older than one year are considered, there is a significant negative correlation. First-year birds, excluded from the analysis, have never held territories and this factor may cause them to behave differently. First-year birds left flocks to establish territories either much earlier or later than the majority of older pairs (Petrie, 1982). By leaving a flock early, with a relatively good condition male, a female may avoid some of the competition for mates and thus manage to retain that male.

WHY DO FEMALES COMPETE FOR MATES AND NOT TERRITORIES?

The above considerations suggest that females go to flocks to select mates. This further suggests that the choice of male is more important to a female than the choice of a territory. Why? It may be that the advantages of possessing a good-quality male far outweigh the benefits of holding a good territory. What are the advantages of possessing a fat male, and how does this compare with having a good territory? Unfortunately the relative payoffs of possessing a fat male and possessing a good-quality territory cannot be simply computed in terms of the number of eggs successfully hatched by an individual female because of nest dumping. Brood parasitism occurs commonly in the moorhen. For instance, one estimate suggests that 34 percent of nests contain the eggs of a parasite (Huxley and Wood, 1976). As a result, it is difficult to assign eggs to hens with any precision. It may be that females which are relieved of some of the costs of parental care by possessing a good-quality male would be able to dump more eggs. No data are available to test this hypothesis, however.

Nevertheless, two points do suggest that fat males are more valuable than good territories. First, a possibility mentioned earlier, males with large fat reserves can spend relatively longer time incubating during the breeding season than thinner males. The number of days a male spends incubating in a season is correlated with his energy reserves at the start of the season (Petrie, 1983a). A male's fat reserves, and therefore his ability to incubate, limit a pair's reproductive performance; as a result, the number of clutches a pair can initiate in a season is also dependent upon the male's condition at the start of the season ($r_s = 0.73$, $p < 0.01$, $n = 14$) (Petrie, 1983a). An alternative explanation of these data is that the energy reserves of the female and her ability to lay eggs determine the number of nesting atempts. However, contrary to the expectations of this hypothesis, there was no significant correlation between a female's condition at the start of the season and the number of clutches started ($r_s = 0.35$, $n = 14$, ns) (Petrie, 1983a).

Secondly, good-quality nest sites may not be limiting to a female. Possessing a good-quality nest site has a fitness payoff, but all females may have opportunities to parasitize good-quality nests. If females have the option of parasitism as an additional strategy, partner quality may be critical in determining relative female reproductive success. However, whether or not owners allow their nests to be parasitized may also affect opportunities for parasitism and this factor should also be taken into account (see below).

WHY DO MALES COMPETE FOR TERRITORIES AND NOT FEMALES?

There are two circumstances that will favor competition among males for females. One is if females are in short supply, which could be a result of some males' securing more than one female; the other is if there is variance in female fecundity such that some males can gain high-quality partners. In the moorhen, females appear to be able to lay unlimited numbers of eggs (or at least a very large number) so that there is probably little variance in female fecundity (Petrie, 1983b). Also, the adult sex ratio is not significantly different from unity, and, since both parents contribute to parental care there is little opportunity for males to gain additional females. Thus the main conditions that could promote competition among males for mates are absent.

Given that males perform the majority of parental care, their reproductive effort goes mainly into incubation and rearing young. The cost to a male of replacing a clutch, up to the hatching stage, is high. This cost of possessing a poor cover nest site may have led to selection for intense competition among males for good cover territory sites. Thus good cover nest sites may be more important than fecund females as a limited resource for males. In addition, males could gain additional matings by securing a good-quality nest site. This might occur if parasitic females mate with a male owner before laying an egg in his nest. This last point is entirely speculative but it seems a reasonable hypothesis to explain why males allow their nests to be parasitized.

WHY DO MOORHENS FORM FLOCKS IN THE WINTER?

Males expend time and energy in establishing territories during March and April and show a weight loss during these months (Petrie, 1982). It is surprising that, at the end of the breeding season, they leave these hard-won areas to join flocks, with the accompanying cost of repeating the investment in territory establishment the following spring. Why do birds do this?

It may be that a territory, and the surrounding feeding area, simply do not provide enough resources to maintain a bird over the winter months and that birds leave their territories and aggregate on sites that offer a preferred food supply. This seems unlikely since the type of food available at flock sites does not differ from other similar sites where flocks do not form (Petrie, 1982). Flocks form at sites where there is tree cover or other features that offer protection from predators, but such sites are superabundant and only some have moorhen flocks. It seems more likely that moorhens gain some benefit from joining a group of other individuals.

The conclusions of preceding sections suggest that a female preference for groups of males could in turn exert a selection pressure on males to abandon their territories and form groups. An evolutionary scenario of this sort could explain why moorhens form flocks in the winter months. This suggestion is also supported by the fact that flocks form at the same sites year after year. These traditional mating areas could be analogous to leks, although there is no spatial reference for dominance and moorhens feed as well as select mates at these sites.

The two most widely cited selection pressures giving rise to group formation are feeding facilitation and predation. Could either of these alternative factors be important in the moorhen?

Moorhens in flocks spend the winter months grazing a close cropped sward. Many of the feeding facilitation hypotheses do not apply. One possible advantage is that the sustained cropping of a sward which occurs when animals feed in groups could increase the production of nitrogen-rich shoots. Barnacle geese (*Branta leucopsis*) are thought to benefit from group feeding in this way (Ydenberg and Prins, 1981). Similarly, grazing by groups of lesser snow geese (*Anser caerulescens caerulescens*) increased net primary production of marsh vegetation (Cargill and Jefferies, 1984).

The antipredator benefits of group living include a predator dilution advantage—the "selfish herd" effect (W. D. Hamilton, 1971)—and a vigilance advantage (for summaries see Bertram, 1978; Gosling and Petrie, 1981). All of these effects potentially apply to moorhens and they could well derive antipredator advantages from joining flocks, although it is difficult to judge whether these advantages are primarily responsible for the evolution of winter flock formation in this species.

DISCUSSION

The main aim of this discussion is to suggest why particular reproductive strategies have been adopted by male and female moorhens rather than some theoretical alternatives. These strategies result in the monogamous mating system typical of the species. The main determinant of a reproductive strategy, how an individual apportions its reproductive effort, will be considered separately for the sexes.

Males

To summarize, male moorhens appear to spend very little effort in acquiring mates. They expend their reproductive effort in 1) defending territories to acquire good cover nest sites and 2) in parental care; they perform the majority of both these tasks. This is an unusual strategy for males in the animal kingdom, where males more commonly expend the majority of their effort in acquiring matings. Over evolutionary time a male that spent its reproductive effort in caring for the young has been favored by natural selection over a male that competed for mates. What factors could increase the benefits of caring for the young and/or reduce the benefits of competing for mates, and are there any environmental correlates of these parameters?

Previous workers have considered the coevolutionary aspects of the division of parental care between the sexes, and have concluded that the pattern of parental care is determined in part by what the other sex does. For example, where a female cares for the young then it may pay a male to desert the female and search for more mates. However, whether a male opts to do this will depend entirely on the costs and benefits to males of deserting as opposed to caring. Both sexes could compete for matings and put little effort into parental care. There are species of fish where both sexes compete for mates and neither sex cares for the young (Reighord, 1920) and, in birds, some members of the Megapodiidae bury their eggs and can perform no postzygotic parental care (Frith, 1956); however, the exact form of the mating system in these species is unknown. Similarly, when one sex performs some parental care the other sex could also opt to perform an equivalent amount. Parental care is not an absolute necessity (see above examples) and there is no finite contribution required. Given this, it is not immediately obvious why male moorhens apportion their reproductive effort in the way that they do.

A factor that would operate to increase the benefits of caring for the young is one that will promote a high return for investment. Where the risk of predation is high and parental effort can provide a significant level of protection, parental care will probably evolve. This may have promoted parental care in moorhens. In a situation where it is more cost-effective for females to expend effort in egg production than to perform parental care there could be high benefits to a male's contributing parental care. Thus, although there is not a finite amount of care to be performed by one or other parent, lack of care by one sex will affect the cost/benefit function of caring in the other sex. But, if the benefits to a male of competing for mates are also high it may still not pay a male to care for the young. Factors that could operate to reduce the benefits to males of competing for mates are:

1) Where the benefits of additional matings are low. This may be the case in moorhens where successful rearing of the young is the main factor limiting reproductive success. There is no point in fertilizing more eggs if the chances of their survival are low without parental care.

2) Where the chances of gaining additional matings by allocating reproductive effort to searching for mates are low, because of the availability of mates both in space and time. If selection has favored care by both parents then the chances that additional females will be available for mating is low.

However, these arguments are all beset with cause-and-effect problems. For example, it could be argued that paternal care allows females to lay more eggs rather than the opposite. But what is clear is that the level of predation has had a major influence on the resulting strategy since it has favored parental care, multiple clutching, and territory defense of good cover nest sites.

One interesting question that could be asked about the consequences of the reproductive strategy adopted by male moorhens is why has resource defense polygyny not become a common pattern. This question arises since

the circumstances that favor this pattern have apparently been met according to the models of Verner and Willson (1966), Orians (1969), and Emlen
and Oring (1977). These models state that, where there is sufficient variability in the resource defended by males, polygyny will evolve, since it may
then pay a female to become a second mate of an already mated male rather
than to mate with an unmated individual holding a poorer territory. Male
moorhens defend an important resource that females need to ensure the survival of their offspring and there is variability in the amount of resources
defended by males, both as a result of environmental heterogeneity and also
as a result of variability in the resource-holding potential of males. Heavier
males who have a greater RHP can secure better and larger territories (Petrie,
1984). Why, therefore, is polygyny not seen in this population? The main
reasons may be:

1) Female moorhens prefer males that expend effort in parental care
rather than males that have expended effort in acquiring good territories.
The fact that females are more concerned with a male's parental capabilities
is illustrated by the fact that a) females court and compete for males that are
capable of investing large amounts of energy in parental care, and b) females select males prior to their establishing their territories probably in order to obtain the best of a choice of males. Why this female strategy may
have evolved will be considered in the next section.

2) Given that male parental effort is important to females, sharing a male
with another female might reduce a female's reproductive success to such
an extent that the benefits of good cover nest sites could not compensate,
i.e., it is more important to a female to have a male rearing her offspring
alone, even in a poor territory.

Polygyny may occur in situations where all available habitat is saturated,
since females could then be forced to make the best of a bad job by associating with an already mated male. D. W. Gibbons (pers. comm.) reports
cases of polygyny (and incidentally polyandry) in a different population of
moorhens where all the available breeding habitat was occupied by territories.

Females

Over evolutionary time a female that spent reproductive effort in competing for mates and egg production has been favored by natural selection
over a female that spent reproductive effort in caring for young. Thus in
cost/benefit terms, the fitness payoffs from egg production and competition
for mates outweighed those for contributing to parental care. The factor that
is most likely to have increased the benefits of competing for mates in the
moorhen is variance in mate quality. When there are fewer good-quality
mates than competitors, a good-quality mate becomes a scarce resource.

This will affect both the absolute and relative benefits of securing a mate (Petrie, 1983b). A female with a good-quality male has the absolute advantage of a male that can contribute a large amount of parental care (which will enable her to lay more successful clutches) and also a relative advantage in relation to other females without good-quality partners. The factor that has favored high egg production is probably the low survival chance of an egg. When predation rates are high, selection will favor those females capable of laying replacement clutches. The factor that may have reduced the benefits of caring for the young by females is probably the accompanying decline in egg production. This cost could become important when egg predation is high. However, there are no data to directly test this idea in the moorhen, although intuitively it seems likely that the high energetic demands of incubation and other forms of care would detract from the energy available for egg production.

Female choice of males that care for the young could be an important selective force in promoting paternal care, an argument first suggested by Trivers (1972). Males that did not contribute any help in caring for the young may be avoided by females and would thus fail to breed. This may have been the case in the moorhen since females show active preference for males that have large energy reserves and are thus capable of a greater expenditure in parental care.

A further question about female strategies, given that males already perform the majority of parental care, is why don't females completely desert a male and attempt to start a new clutch with a second male? In other words, why has simultaneous polyandry not evolved in this species? Occasionally a female will mate with two males sequentially in a season, sharing parental duties first with one male and then the other. Prior to mating with her second male she may simultaneously court two males present in her territory throughout the breeding season (Petrie, 1982). However, she does not have two clutches simultaneously cared for by the two males. It seems that some of the conditions for simultaneous polyandry are met, in that two males are available, but that the payoff to females in continuing to invest in one clutch is greater than deserting to initiate another. This may be because, although the female performs a relatively small amount of the parental duties, her limited investment has a high return. The main situation where her contribution is vital is during any absences by the male. Moorhens do not leave their nests unattended during incubation, and, when the male leaves the nest to feed for short periods, the female will tend to the eggs. Changeovers occur throughout the day and are usually on, or very close to, the nest. This behavior, I believe, is vital to the survival of the nest, since crows and other avian predators are quick to see and prey upon unattended eggs (pers. obs.). The alternative explanation is that females provide an energetic input in rearing their young to reduce the rate of depletion of the male's energy reserves so that the males can help to incubate subsequent clutches. However, this alternative explanation seems unlikely where a female already has a second male in her territory capable of incubating subsequent clutches. Thus, I would argue that predation pressure places a constraint on the exact form of the mating system that could develop in this species, since two parents are necessary for rearing the young, not for energetic reasons, but to provide adequate protection from predators.

Another possible strategy that females could successfully adopt is to expend more effort in parental care, especially since this can be a factor limiting reproductive success. A female's investment in incubation and care of the young is not fixed and could vary considerably between individuals, perhaps depending upon the chances of her initiating and rearing any replacement clutches. Thus a number of predictions could be made, which remain to be tested, about her parental contribution. For example, where a male is unlikely to be able to incubate a subsequent clutch a female may increase her parental investment in the current brood. One factor that could confound analyses of this sort is the importance of brood parasitism as an additional strategy. This could partly depend upon the number of available nests to parasitize.

The option of intraspecific brood parasitism could be important in determining which reproductive strategy both males and females adopt. For some females this strategy may be a real alternative to monogamously mating and they may lay all their eggs in this fashion. However, my most recent research suggests that it is an additional strategy for some females which also have nests of their own. D. W. Gibbons (pers. comm.) also found this pattern of egg dumping. If all else is equal, laying all one's eggs in one nest results in the same expected RS as scattering eggs among nests (Rubenstein, 1982; Andersson, 1984), but laying in many nests increases the likelihood that one offspring will survive to independence and is therefore an evolutionarily stable strategy. Rubenstein (1982) suggests that the advantages of dispersing eggs among nests increases as the intensity of predation increases. Where there are differences between territories, brood parasitism may have been favored by the selection pressure of predation, because if there is variability in the chances of nests surviving, it may pay a female to dump an egg in another nest if the nest is more likely to survive than her own.

Female moorhens show a preference for particular males. To carry out such a selection they clearly need to sample more than one male and the chance of obtaining a good male should theoretically increase with the number of males from which the choice is made. If this is the case females should be attracted to areas where there is more than one potential male.

This may be why moorhens form pairs in winter flocks, an argument that assumes a cost in sampling males dispersed over a wide area. This cost may not be simply the distance, but may also have something to do with the costs of not "knowing" any potential competitors and having to compete succes- sively for mates in contests where the outcome is less easy to assess. Brad- bury (1981) thought that the attraction of females to groups of males could explain the evolution of classical leks. Classical leks are those that offer no resources to visiting females except for the presence of males. In some neo- tropical frogs, males gather at ponds to attract females. These communally displaying groups are known as a chorus. In the frog *Physalaemus pustulous* where females are attracted to groups of males which vary in size from 40 to 450 individuals, the operational sex ratio (the proportion of females to males) was found to be higher in large choruses (Ryan et al., 1981). This could be because females find larger choruses more attractive. For whatever reason this could result in a higher frequency of matings by males in larger groups, which could provide the selective advantage for males to join cho- ruses.

Janetos (1980) has modeled the advantages of selecting the best-of-n males. He suggests that females should usually not compare more than five or six males before making a choice, since above this number the relative increase in benefits would be small. However, this model is probably not applicable to a competitive mate choice situation such as that in the moor- hen, where the number of potential available mates for an individual female is limited by its own competitive ability; females of relatively low status may thus have to sample a larger group than predicted by Janetos. Witten- berger (1983) also reviews the tactics that females may adopt when choos- ing males but, like Janetos, he does not consider the case where females mo- nopolize their mates after choosing them.

The findings that females prefer small fat males and that large body size affects success in intrasexual competition lead to speculation about selec- tion for male body size in this species. However, to be meaningful such con- siderations should include a discussion of all the selective advantages and disadvantages of small and large body size. For example, relationships may also exist between body size and incubation efficiency or the chance of pre- dation.

To conclude, I would argue that predation is the most important environ- mental factor shaping the social system of moorhens. The reproductive strategy of females can be ultimately attributed to high predation levels since it is this factor that has probably led to selection for the main repro- ductive efforts being spent in egg production and choice of caring males. It is also probably responsible for joint parental care, which results in monog- amy. In males it has led to the defense of scarce good cover nest sites. The additional strategy of brood parasitism is probably favored when there is high predation. However, female choice of male attributes is an important secondary factor affecting the social system of moorhens since this has resulted in competition among females and may have promoted paternal care. It may also be an important determinant of group formation in the winter months.

4. Ecology of Cooperation in Canids

———————————— • ————————————

PATRICIA D. MOEHLMAN

FAMILY CANIDAE is composed of approximately thirty-seven species with mean female body weight ranging from 1.5 kg (*Fennecus zerda*, fennec fox) to 31.1 kg (*Canis lupus*, timber wolf) (Clutton-Brock et al., 1976; Gittleman, 1984a). Among canids, the basic mating system is long-term monogamy, a system that is rare among mammals (< 3%; Kleiman, 1977). However, an examination of the social organization in the continuum of small to large canids reveals major trends in adult sex ratio, dispersal, mating systems, and neonate rearing systems (Table 4.1). Small canids (< 6.0 kg) tend to have an adult sex ratio skewed toward females, dispersal biased toward males, female helpers, and they exhibit a tendency toward polygyny. Medium-sized canids (6.0–13.0 kg) have equal sex ratios, equivalent sexual emigration, both sexes as helpers, and observations at present indicate that they are strictly monogamous. Large-sized canids, with the interesting exception of the maned wolf, have sex ratios skewed toward males, primarily female emigration, male helpers, and indications of polyandry (Table 4.2). Why do these suites of variables correlate to increasing body size? There is evidence that body weight has a causal relationship not only with metabolic rate in carnivores, but also with their life-history patterns (Millar, 1977, 1981; Bekoff et al., 1981; Gittleman, 1984a; Bekoff et al., 1984; Western, 1979; Eisenberg, 1981).

When log median neonate birth weight is regressed against log mean female body weight (Fig. 4.1) there is a strong correlation ($r^2 = 0.97$) and the slope is 0.76, indicating that as maternal weight increases, neonate weight increased as a three-quarters exponent of the mother's weight (final regression was run without major outliers; fennec fox, kit fox, and timber wolf). Therefore, large species in family Canidae will have smaller neonates relative to maternal body weight. Thus as body size increases in Canidae, neonates will be relatively smaller, and potentially have a longer period of dependency and require more postpartum investment. Smaller mothers will have more prepartum investment per neonate and larger mothers will have less prepartum investment in their individual offspring, and may require greater postpartum assistance.

When log mean litter size is regressed against log mean female body weight the slope is positive (0.33; $r^2 = 0.72$), and the correlation indicates that larger Canidae species have larger litter sizes (Fig. 4.2). To date, increasing litter size with increasing maternal weight has not been recorded

TABLE 4.1
Allometry, ecology, and social systems in the Canidae

Small canids (< 6.0 kg) Kit fox, Arctic fox, bat-eared fox, red fox	Medium canids (6.0–13 kg) Silverbacked jackal, golden jackal, coyote	Large canids (> 13 kg) Dhole, African hunting dog, maned wolf, timber wolf

Increasing mean female body weight

——————————————————→

I. *Absolute neonate size increases, but size relative to adult decreases.*
   Increasingly altricial offspring.
   Increasing dependency on adult investment.

II. *Litter size tends to increase.**
   Increasing need for paternal/alloparental investment.

III. *Sex ratio*

♀ Biased sex ratio	Equal sex ratio	*♂ Biased sex ratio
♂♂ Emigrate	Equal emigration	♀♀ Emigrate

IV. *Foraging behavior*

No cooperative hunting	Facultative cooperative hunting	*Obligatory cooperative hunting
most prey < than predator	prey often > than predator	most prey > than predator

V. *Mating system*

Monogamy + polygyny	Monogamy	Monogamy + polyandry

VI. *Cooperative breeding system (increase in ♂ investment relative to ♀ investment)*

Parental pair + ♀♀ helpers	Parental pair + ♂/♀ helpers	*Parental pair + primarily ♂♂ helpers

* Maned wolf is primary exception.

for any other mammalian family, and thus Canidae appears to be unique. Major outliers were removed (arctic fox, crab-eating fox, and maned wolf). Arctic foxes have large litter sizes compared to similar body weight canids. They often live in seasonally highly productive environments and appear to be relatively *r*-selected. Ecological parameters appear to have a major impact on their reproductive effort. In the Northwest Territory of Canada where major fluctuations in prey base (lemmings) occur and food can be very abundant during the whelping season, the mean litter size is 10.6 (Macpherson, 1969). By contrast, in the coastal areas of Iceland where food is patchily distributed but relatively consistent in availability, mean litter size is 4.0 (Hersteinsson, 1984). It is interesting to note that females have six to seven pairs of teats, while species of similar body weight have three to four pairs of teats (Table 4.2a). Little fieldwork has been done on crab-eating foxes, but one might predict that they may be food-limited during the whelping season. The maned wolf is a very unusual large canid. Since the

TABLE 4.2

Allometry, ecology, and social systems for eighteen species of Canidae

Species	FW	FB	NW	NR	LS	LR	PT	LW	G	GR	MS
Fennecus zerda	1.5	17.5	29.0	23–35	3.0	1–5	—	87.0	54.0	50–63	M
Vulpes macrotis	1.9	39.9	—		3.4	3–5	4	135.7	53.0	49–56	M, P
Alopex lagopus	3.0	37.0	75.0	60–90	7.0	4–21	6–7	525.0	53.5	51–60	M, P
Urocyon cinereoargentus	3.3	39.5	100.5	86–115	3.8	1–7	3–4	381.9	58.0	53–63	M
Vulpes vulpes	3.9	43.0	95.5	71–120	5.0	1–11	3–4	477.5	54.6	51–55	M, P
Otocyon megalotis	4.1	24.5	120.5	99–142	5.0	2–5	2	602.5	62.8	60–75	M, P
Nyctereutes procyonoides v.	4.7	28.5	110.0	105–115	4.0	2–5	—	440.0	62.0	61–63	M
Cerdocyon thous	6.0	40.5	140.0	120–160	3.1	1–6	—	434.0	56.0	52–59	M
Speothus venaticus	6.0	41.5	157.5	125–190	4.0	3–6	4	630.0	65.0	65–83	M?
Dusicyon culpaeus	6.7	51.0	168.0		5.0		—	840.0	57.5	—	—
Canis mesomelas	7.5	52.0	197.5	175–220	5.7	1–9	4	1,125.0	59.7	59–63	M
Canis adustus	8.3	53.5		—	5.4	1–7	2	—	63.0	57–70	M
Canis aureus	9.0	72.2			5.7	1–7	—	—	63.4	60–64	M
Canis latrans	10.9	84.5	254.5	225–284	5.4	1–12	4	1,374.3	61.5	58–63	M
Cuon alpinus	13.8	95.0	275.0	200–350	8.5	1–15	6–8	2,337.5	62.2	60–70	M, PA
Lycaon pictus	22.5	128.0	365.0	350–380	10.1	1–16	6–8	3,686.5	70.8	69–73	M, PA
Chrysocyon brachyurus	22.7	116.0	402.0	350–454	2.0	1–5	2 5	804.0	63.4	60–66	M
Canis lupus	31.1	130.0	400.0	350–450	6.0	1–14	5	2,400.0	63.0	58–66	M, PA

SOURCES: 1) Gittleman, 1984a, b; 2) Bekoff and Jamieson, 1975; 3) Ewer, 1973; 4) Gangloff, 1972; 5) J. C. McGrev 1979; 6) Egoscue, 1962; 7) Bekoff et al., 1981; 8) Hersteinsson and Macdonald, 1982; Hersteinsson, 1984; Macphersoi 1969; 9) Asdell, 1964; 10) Fritzell and Haroldson, 1982; 11) Trapp and Hallberg, 1975; 12) Goodall, pers. comm.; 1 Nel, 1978; Nel and Bester, 1983; Nel et al., 1984; 14) Lamprecht, 1979; 15) Smithers, 1983; 16) Macdonald 1979b, 198 1981; 17) Ables, 1975; 18) Schantz, 1981; 19) Storm and Ables, 1966; Storm et al., 1976; 20) Ikeda, 1983; 21) Eisenber 1981; 22) Brady, 1978, 1979; 23) Biben, 1981; 24) Kitchener 1971; 25) Ferguson et al., 1983; Ferguson, pers. comm 26) van der Merwe, 1953; 27) Rowe-Rowe, 1976, 1978; 28) Lombaard, 1971; 29) Moehlman 1979b, 1981, 1983; 3 Kingdon, 1977; 31) Seitz, 1959; 32) Macdonald, 1979a; 33) Lawick and Lawick-Goodall, 1970; 34) Wandrey, 1975; 3 Bekoff and Wells, 1980, 1982; 36) Berg and Chesness, 1978; 37) Camenzind, 1978; 38) Gier, 1975; 39) Kennelly, 197 40) Knowlton, 1972; 41) Bowen, 1978, 1981; 42) Nellis and Keith, 1976; 43) Andelt, 1982; 44) Sosnovskii, 1967; 4 S. A. Cohen, 1977; 46) Johnsingh, 1982; 47) Davidar, 1975; 48) Reich, 1981; 49) Frame et al., 1980, Frame and Fram 1977; 50) Malcolm, 1979; Malcolm and Marten, 1982; 51) Dekker, 1968; 52) Acosta, 1972; 53) Brady and Ditton, 197

Pleistocene, maned wolf food resources have been limited to small prey and it is the only large canid ($>$ 13 kg) that forages primarily on rodents and fruit (Dietz, 1984). These are small-sized food items that may impose energetic constraints on the parents and potential helpers, both in terms of food availability and the cost benefits of provisioning pups. Maned wolves have not been observed to hunt or breed cooperatively and have the smallest mean litter size (two) of any canid. Litter size for timber wolves also lies below the regression line. Timber wolves are mainly cooperative hunters in the winter but during whelping season they tend to feed on small-sized prey, solitary hunting prevails, and groups tend to disperse (Mech, 1970; Jordan et al., 1967; Peterson et al., 1984). Both Indian dholes and African hunting dogs lie well above the regression line. These two species have the largest observed litter sizes, six to seven pairs of teats, and from field studies are known to be obligatory cooperative hunters (Davidar, 1975; Johnsingh,

;S	GSR	HR	FB	CH	ASR	BSR	FE	H	PI	References
–	–	–	–		–	–	–	–	Y	1–4
⁇	2–3	–	O	NR	1:1 ↔ 1:2	1:2	–	–	Y	1, 5, 6
3.0	2–3	12.5	O	2	1:1 ↔ 1:2	1:1	♂	♀	Y	1, 2, 3, 7–9
–	–	<0.7	O	–	–	–	♂	–	–	1, 7, 9–11
⁇	2–6	0.1–20	O	NR	1:1 ↔ 1:5	–	♂	♀	Y	1, 7, 8, 9, 16–19
2.0	2–3	<1–2	I > O	NR	1:1 ↔ 1:2	–	–	♀	Y	1, 7, 12–15
–	–	0.01–0.03	O	–	–	1:1	–	–	Y	1, 7, 20
–	–	1–2	O	–	1:1	1:2	–	–	Y	1, 21–23
–	–	–	O	–	–	1:1.4	–	–	Y	1, 2, 7, 8, 24
–	–	–	–	–	–	–	–	–	–	1
2.9	2–5	2–22	O	2–3	1:1	–	♂ = ♀	♂ = ♀	Y	1, 7, 8, 15, 25–29
–	–	–	O/I	–	1:1	–	–	–	Y	1, 7, 15, 30
2.7	2–20	0.1–2.0	O	2–3	1:1	–	–	♂ = ♀	Y	1, 7, 8, 29, 31–34
3.4	2–7	4.5–80	O/C	2–7	1:1 ±	1:1 ±	= ±	♂ = ♀	Y	1, 2, 7, 21, 35–43
3.3	5–15	20–40	C > O	5–12	2:1 ↔ 7:1	♂ > ♀	–	♂ > ♀	Y	1, 7–9, 44–47
).0	2–40	500–2,000	C	2–40	2:1 ↔ 8:1	1:0.7	♀	♂♂ > ♀	Y	1, 7, 15, 48–51
2.0	–	22–30	O	NR	1:1	1:1	–	NR	–	1, 8, 9, 52–54
3.6	2–36	18–13,000	C > O	2–12	1:1 ↔ 3:1	1:1 ↔ 2:1	♀	♂♂ > ♀	Y	1, 7, 9, 15, 58–65

1) Dietz, 1984, pers. comm.; 55) Mech, 1970; 56) Rausch, 1967; 57) Harrington et al., 1982; 58) Fentress and Ryon,)82; 59) Harrington and Mech, 1982; 60) Harrington et al., 1983; 61) Jordan et al., 1967; 62) Keith, 1981; 63) Zimen,)76; 64) Van Ballenberghe and Mech, 1975; 65) Peterson et al., 1984.

EY: FW is mean female body weight (kg), FB is female brain weight (g), NW is median neonate weight (g), NR is ange of neonate weights (g), LS is mean litter size, LR is litter size range, PT is number of pairs of teats, LW is litter eight in grams (NW × LS), G is mode gestation (days), GR is gestation range (days), MS is mating system, M is onogamy, P is polygyny, PA is polyandry, GS is mean group size (adults), GSR is group size range, HR is home ange/territory (km²), FB is food base, O is omnivore, I is insectivore, C is carnivore, CH is cooperative hunting group ze, NR is not recorded, ASR is adult sex ratio, (+ ♂:♀ is more ♂ > ♀, – is less ♂ < ♀), BSR is birth sex ratio (♂:♀), ⁇ is farthest emigrants (♂ vs. ♀), H is helpers provisioning food and/or guarding pups, PI is paternal investment.

1982; Frame and Frame, 1977; Frame et al., 1980; Malcolm, 1979; Malcolm and Marten, 1982).

When log litter weight (median neonate weight × mean litter size) is regressed against log mean female body weight the slope is 1.14 ($r^2 = .89$, Fig. 4.3). The correlation indicates that as female body weight increases, average total litter weight is increasing at a slightly faster rate. Therefore, contrary to most mammals, as canid species get larger, litter weight increases absolutely as female weight increases, but the individual neonates are *smaller* and more dependent, and there are *more* of them. Presumably costs of raising litters are escalating and postpartum investment must correspondingly increase. Most of the species lying above the line have been documented by field research to exhibit cooperative breeding behavior (arctic fox, bat-eared fox, red fox, silverbacked jackal, coyote, Indian dhole, African hunting dog). A similar result has been obtained by Gittleman

Key to Figures 4.1–4.4
Fz = *Fennecus zerda*
Vm = *Vulpes macrotis*
Al = *Alopex lagopus*
Uc = *Urocyon
    cincereoargentus*
Om = *Otocyon megalotis*
Vv = *Vulpes vulpes*
Np = *Nyctereutes
    procyonoides*
Ct = *Cerdocyon thous*
Dc = *Dusicyon culpaeus*
Sv = *Speothus venaticus*
Cm = *Canis mesomelas*
Cs = *Canis adustus*
Ca = *Canis aureus*
Cl = *Canis latrans*
Cd = *Cuon alpinus*
Lp = *Lycaon pictus*
Cb = *Chrysocyon
    brachyurus*
Cw = *Canis lupus*
Circled letters indicate
that the species was
not utilized in the
computation of the
linear regression.

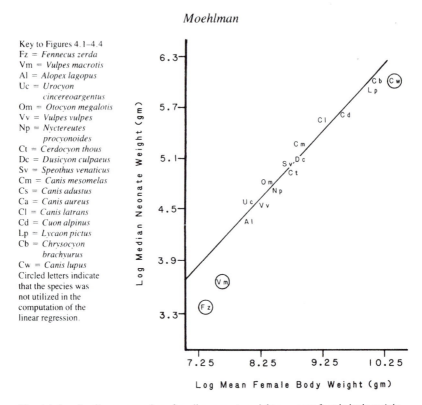

Fig. 4.1. Log-log linear regression of median neonate weight vs. mean female body weight. Slope = 0.76 ± 0.08 (at .05 level of significance), $r^2 = 0.97$.

(1984a, b) for carnivores in that communal species have heavier litter weights than maternal or biparental species. Of the species lying below the line only timber wolves are documented cooperative breeders. Once again maned wolves are outliers due to their very small mean littler size.

Regression of log modal gestation length versus log mean female body weight yields a slope of 0.07 ($r^2 = 0.58$, Fig. 4.4). This correlation is very weak but does suggest that with increasing canid body size gestation length is relatively shorter and neonates are smaller.

Examination of Table 4.2a reveals that neonate weights, litter size, and gestation periods are highly variable in Canidae. Thus it is with a certain amount of skepticism that these data were analyzed. However, these analyses do indicate fairly robust and interesting trends within Canidae. As canids get larger females will invest less prepartum per individual neonate and neonates will be relatively smaller and potentially more altricial. Concurrently, females will be producing more offspring such that their total prepartum investment in terms of litter weight remains proportional (slope = 1.14 ± 0.24) and may even increase with larger females. Female canids are

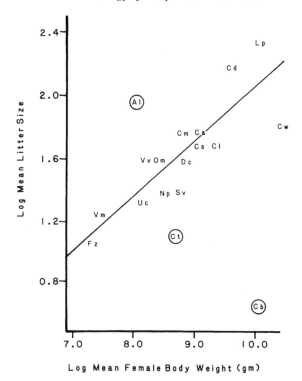

Fig. 4.2. Log-log linear regression of mean litter size vs. mean female body weight. Slope $= 0.33 \pm 0.12$ (at .05 level of significance), $r^2 = 0.72$.

investing heavily prepartum; larger females, in particular, may need assistance during gestation. It would be of interest to determine whether or not females are provisioned during gestation. Given the trend toward increasingly *larger* litters composed of proportionally *smaller* neonates, females will also require more postpartum assistance in provisioning and rearing offspring to the age of independence. The ability of females to produce larger litters will be dependent on predation pressure and food requirements. Canids whelp in dens and young typically spend most of their time in dens until they are weaned. Thus predation pressure is reduced and may allow canids to produce more young in a more altrical state with a longer period of development. However, the provisioning of increasingly large dependent litters does pose problems. Canids are also unusual among mammals in that most of them regurgitate food to their young. This behavior enables them to efficiently and safely bring larger amounts of food to the den.

Smaller canid females will require less male investment in their more pre-

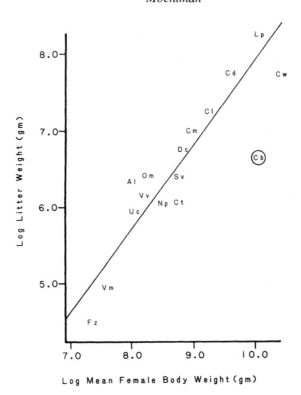

Fig. 4.3. Log-log linear regression of litter weight vs. mean female body weight. Slope = 1.14 ± 0.24 (at .05 level of significance), $r^2 = 0.89$.

cocial offspring and, as predicted by theory on parental investment and sexual selection, there will be less competition by females for males, and more of a tendency toward male dispersal, female helpers, and polygyny (Fisher, 1930; Trivers, 1972). Among smaller canids there is also a tendency for females to share reproductive costs by communal nursing (red foxes, bat-eared foxes). However, ecological parameters—in particular, food size and its spacial and temporal availability—can have major impact on spacing systems, breeding strategies, and reproduction. The arctic fox (*Alopex lagopus*), in areas where large fluctuations in prey base (lemmings) occur, has a mean litter size of 10.6 (Macpherson, 1969) but in coastal areas of Iceland where major food fluctuations do not occur the mean litter size is 4.0 (Hersteinsson, 1984). The variability in reproductive data such as litter size is an indication of the flexibility that canids are capable of in response to different ecological parameters.

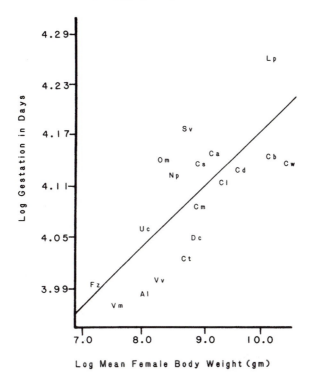

Fig. 4.4. Log-log linear regression of modal gestation length vs. mean female body weight. Slope = 0.07 ± 0.03 (at .05 level of significance), $r^2 = 0.58$.

At the other end of the continuum, large females require a lot of male investment in their more altricial offspring. A female cannot afford to share these contributions with other females and therefore competition for males can be intense. The best-documented species is the African hunting dog (Frame and Frame, 1977; Frame et al., 1980; Malcolm, 1979; Malcolm and Marten, 1982). Although the mating system is monogamous, polyandry has been observed (Lawick, 1973), there is a significant pup and adult sex ratio bias toward males, females emigrate, and males help (Fisher, 1930; Trivers, 1972). Females are occasionally mated by several males. There is some potential for multiple sires in a litter since this has been observed in *Canis familiaris* (F. Beach, pers. comm.). Cooperative hunting is obligatory, and concentrates on prey that is larger than the predator. Thus the cursorial hunting behavior of the canids coupled with the predation of larger animals by large canids (Gittleman, 1984a) and improved hunting success with larger

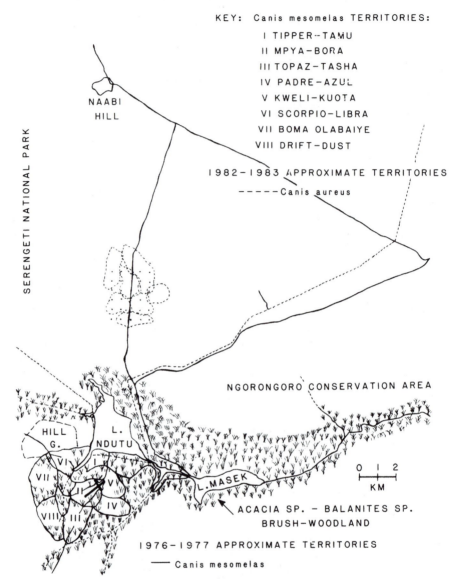

KEY: Canis mesomelas TERRITORIES:

I TIPPER--TAMU
II MPYA—BORA
III TOPAZ—TASHA
IV PADRE—AZUL
V KWELI—KUOTA
VI SCORPIO—LIBRA
VII BOMA OLABAIYE
VIII DRIFT—DUST

1982—1983 APPROXIMATE TERRITORIES

-----Canis aureus

SERENGETI NATIONAL PARK

NAABI HILL

NGORONGORO CONSERVATION AREA

HILL G.
L. NDUTU

L. MASEK

0 1 2
KM

ACACIA SP. — BALANITES SP. BRUSH—WOODLAND

1976—1977 APPROXIMATE TERRITORIES

——— Canis mesomelas

Fig. 4.5. Approximate jackal territories: Silver-backed jackals 1967-1977; Golden jackals 1982-1983.

72

packs (Kruuk, 1975) interacts with relatively high female investment needs to form stable patrilinear social groups that cooperatively rear the larger litters. The other large canids, Indian dholes and timber wolves, exhibit similar tendencies (Table 4.2a, b). The one exception, the maned wolf, as discussed above, is probably due to critical differences in food resources, i.e., prey are much smaller than the predator.

In the middle of the continuum, sex ratios are equal in terms of mating, social group composition, helping, and emigrating. Focusing on the ecology, helping, and social organization of two jackal species (*Canis mesomelas, C. aureus*), I will examine data on the need for postpartum assistance in rearing young, how individuals, according to sex and status, contribute, and how that interacts with type and availability of food resources in terms of pup survival, dispersal behavior, and cooperative breeding.

SPECIES AND ENVIRONMENT

Research on silverbacked and golden jackals began in July 1974 and at present has involved fifty-nine months of observations over a ten-year period. The two species live in adjacent but different habitat (Fig. 4.5). Silverbacked jackals were studied in the *Acacia sp., Balanites aegyptiaca* brush woodlands peripheral to Lake Ndutu, Serengeti Plain, Tanzania, where they typically whelped during the dry season (July to September), a period of high rodent abundance (Senzota, 1978, 1982). Golden jackals were observed in the adjacent short grasslands and whelped mainly during the wet season (December to March) when the migrating ungulates were in residence (Moehlman, 1979b, 1981, 1983; Macdonald and Moehlman, 1983).

MONOGAMOUS MATING SYSTEM

Both species form long-term monogamous bonds, some for as long as six to eight years, which may effectively mean that they pair for life. In silverbacked jackals no mate changes have been observed. In golden jackals one mate change was observed when the resident female was still alive. She was dominant to the intruding female but became more and more peripheral to the male until she disappeared. Among both species of jackal the form of monogamy is strongly obligatory (Kleiman, 1977) with little sexual dimorphism, an equal adult sex ratio, mating exclusivity, predominantly intrasexual aggression, similar behavioral roles and a high degree of synchrony of behavior including cooperative hunting and tandem marking. Jackal pairs have affiliative behavior including grooming each other, sharing food, and feeding and protecting sick or injured partners. They tend to rest and forage together. When individuals have been resting apart they locate each other with contact calls before starting to forage. Among silverbacked jackals, when the male of the pair disappeared during whelping season ($n = 2$) the female and pups all died within a week. The presence of both members of the pair may be essential to maintaining a territory during this critical period. In golden jackals, mate changes have been observed twice during the nonwhelping season and the resident individual succeeded in maintaining the territory (one male, one female).

TERRITORIAL SPACING SYSTEM

Both silverbacks and goldens are territorial throughout the year. Territories are maintained long term (five to eight years) with some changes in boundaries. A silverback pair's (Padre and Azul) boundary changed in size from approximately 4.5 km² (1974) to 2 km² (1976). Territory size decrease did not correspond to group size as the family was composed of three adults in 1974 and five adults in 1975 and may instead reflect an increase in food density (Moehlman, 1983). Golden jackal territories have also changed in size with the same resident pair (0.5–1.0 km²), but, again, variance has not correlated with group size.

Boundaries and the internal area of the territory were actively defended against intruding conspecifics. All observed territorial conflicts involved aggression between same sex individuals. The resident female threatens, attacks, and drives off intruding females without any assistance from her mate. The resident male will threaten and drive out of the territory any trespassing males. Occasionally the presence of a large carcass may attract individuals from adjacent territories in such numbers that it is impossible to drive them all away. However, these aggregations of trespassers are usually short term (< 4 hr). Residents leave their own territories to drink water and to scavenge. They will occasionally go on "exploratory trips" (e.g., 5 km) but are seldom off territory for more than a few hours. When territorial individuals leave their own area they do not scent mark.

Territories are maintained indirectly by scent-marking and vocalizations. Both members of a pair use a raised leg urination posture to scent mark grass tufts, bushes, and trees at nose height both within and at the edge of their territory. When foraging with their mates, territorial individuals mark at twice the rate as when they are foraging alone. When traveling together they mark the same spot in tandem. The presence of scent from both members of the pair may advertise to potential intruders that they are both actively in residence. Among silverbacked jackals single territorial females have not been able to maintain their area during the whelping season (Moehlman, 1983).

FEEDING ECOLOGY

Both species are omnivorous opportunists with varied diets ranging from fruit, to insects, reptiles, birds, rodents, hares (*Lepus sp.*) and actively predated adult Thomson's gazelle (*Gazella thomsoni*). During the dry season when silverbacked jackals are raising their pups, they primarily eat *Balanites aegyptiaca* fruit and a small diurnal rodent, *Arvicanthis niloticus* (~60 gm). *Balanites* seeds weigh 4.5–5.7 gm and have a high sugar content (38–40.3 percent sugar) (Watt and Breyer-Brankwijk, 1962). *Arvicanthis niloticus* is at the peak of its yearly cycle during the dry season, and in the study area was available in high density (13,130–32,080 rats/km^2 1977; Senzota 1978, 1982). Thus silverback jackal adults were feeding themselves and their pups from food resources that were abundant but available in small packages. An individual will typically travel six to eight kilometers in a foraging trip to hunt rodents and will stop beneath *Balanites aegyptiaca* trees to gather fruit. Although silverbacked jackals were never observed hunting Thomson's gazelle in the study area, from Wyman's (1967), Lamprecht's (1978a), and Ferguson et al.'s (1983) studies it is known that pairs of jackals are two to three times more successful at hunting gazelle fawns than individuals are.

Golden jackals raise their pups during the wet season when the migrating herds are resident on the short grass plains. During this time they forage successfully on gazelle fawns, hares, and lots of dung beetles (~35 per half hr). Similar to silverbacked jackals, golden jackal pairs are two to three times more successful than individuals at killing Thomson's gazelle fawns (Lamprecht, 1978a; Wyman, 1967). They also have access to wildebeest afterbirths and carcasses of zebra (*Equus burchelli*), wildebeest (*Connochaetes taurinus*), Grant's gazelle (*Gazella granti*), and Thomson's gazelle that were killed by a hyena clan living in the study area. Golden jackals have access to larger-sized food items during their whelping season. Golden jackal "pairs" feed their pups at a higher rate (0.524 feeds/hr, 63 hrs) than silverbacked "pairs" (0.318 feeds/hr, 232 hrs).

HELPERS

In both species some offspring of both sexes remain on their natal territory and help in the rearing of their younger brothers and sisters by feeding, protecting, grooming, and playing with them. They also feed their nursing mother and defend the territory. Both species are sexually mature at approximately ten or eleven months (Ferguson et al., 1983; Taryannikov 1976) and are delaying reproduction when they act as helpers from age eleven to eighteen months. Older helpers have been observed only twice among silverbacked jackals (two-yr.-old ♀: Moehlman, 1983; three-yr.-old ♂: Ferguson et al., 1983). Thus when year-old jackals remain with their parents they are presumably delaying reproduction for a year.

Jackal helpers are subordinate to their parents and may experience social suppression of endocrine function and reproductive behavior. Helpers did not raised-leg scent mark or exhibit sexual behavior and by visual estimate male helpers had smaller testes than their fathers. Parents rarely regurgitated to helpers but did share food (carcasses) with them and groomed them. Helpers did assist their same-sex parent in territorial defense against same-sex intruders. Helpers would trespass and forage on adjacent territories until they encountered a resident and were driven home. These exploratory trips may help a young jackal to assess the availability of unoccupied and suitable habitat and/or available mates. Studies of silverbacked jackal dispersal indicate that males and females do not disperse significantly different distances (Ferguson et al., 1983).

Among silverbacked jackals 24 percent of known surviving pups stayed and helped ($n = 20$) with a sex ratio of 1:1. Golden jackal pups tend to leave their natal territory for several months during their first year. During the dry season (July-November) when they are about seven + months old they can not be located on their natal territories in the short grass plains where visibility is good. During this season many transient golden jackals are seen in the brush woodlands. Golden jackal diet in the dry season consists mainly of dung beetle larvae and lizards and quite possibly there is not sufficient food available on the territory to support more individuals than the parental pair. Four young golden jackals were observed to emigrate temporarily and then return to help in the next breeding/wet season. Similar behavior has been observed in coyotes and also appears to be related to food availability (Bekoff and Wells, 1980). In contrast to silverbacked jackals (24 percent), 70 percent of known surviving golden pups were observed helping with the next year's litter with a sex ratio of 1:1 ($n = 16$). The higher proportion of goldens staying and helping versus dispersing may reflect a higher cost in obtaining a territory in the short grasslands where open visibility would make it more difficult to get established at the edge of a territory. Golden jackal territories are a square kilometer or less in size and form a tight mosaic.

PUPS

Pups are born in underground dens and during their first three weeks mothers spend a majority (90 percent) of their time in the den, which may alleviate problems of hypothermia. Pups of both species open their eyes at eight to eleven days (Seitz, 1959; Wandrey, 1975; van der Merwe, 1953) and most of their deciduous teeth have erupted by twenty-one days (Lombaard, 1971). Pups first emerge from the den at about three weeks of age and eat regurgitated food. They are weaned at eight to nine weeks. Pups stay within several hundred meters of the den until they are ten weeks old. By fourteen weeks they are well coordinated, no longer use a den, and start to forage with the adults. At this age, silverbacked jackals are approximately half of adult body weight (Lombaard, 1971). The first fourteen weeks constitute the period of time within which the pups are the most dependent on parents and helpers for food and protection. All observed mortality occurred within this initial fourteen-week period. Thus, pups reaching the age of fourteen weeks and starting to forage could be equated with fledging in birds. Pups, although they do most of their own foraging, continue to receive food from parents and helpers until they are about eight months old. As helpers (11–18 mos) they will be fed infrequently by their parents.

COOPERATIVE BREEDING SYSTEM

Although the female parent bears heavy reproduction costs in terms of gestation, lactation, and regurgitation to pups, she is assisted by her mate and older offspring (helpers) when they are present. In both species, when a pair without helpers raises a litter they must divide their activity between staying at the den guarding the pups and foraging for food. Typically, adults do all their resting near the den such that when they are not guarding they are foraging. The presence of a single adult constitutes protection for the pups. Adults vocally both warn the pups (who run into the den) and threaten the potential predator. Jackals are adept at biting 55 kg hyenas (*Crocuta crocuta*) in the rump and chasing them away from the den.

During the first fourteen weeks in families with no helpers silverbacked and golden pups were left unguarded about 20 percent of the time (295 hrs) By contrast, when helpers were present they made significant contributions to pup guarding. With the addition of one helper, pups were left alone 16 percent (silverbacked jackals) and 2 percent (golden jackals) of the time (299 hrs). With the addition of two to three helpers, pups were left alone 8-1.5 percent of the time (176 hrs) (Fig. 4.6). The presence of one helper in silverbacked jackal families led to a significant decline in the amount of time the parental pair spent at the den, allowing more time for foraging. Thus the presence of helpers may indirectly improve the provisioning of pups by allowing the parents to hunt as a pair and hence increase their hunting success rate on larger prey (Wyman, 1967; Lamprecht, 1978a). It may also allow the parental pair to defend, retain, and exploit carcasses killed or scavenged more successfully (Moehlman, 1983; Lamprecht, 1978a). However, in silverbacks this may also reflect the data for families with one helper ($n = 2$) having been taken during weeks ten to fourteen (postweaning). With two helpers in silverbacked jackal families, the parental pair spends more time at the den, which may indicate that they can spend more time resting with more adults provisioning the pups. However, data for the family with two

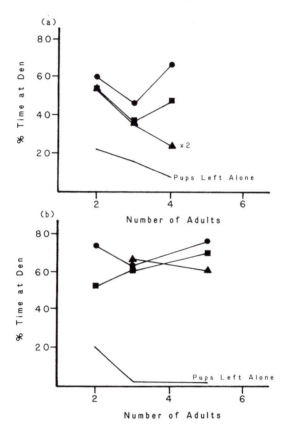

Fig. 4.6. Mean pupsitting time by jackal families
with varying numbers of helpers during weeks 3 to
14 of pup rearing. (a) Silverbacked jackals ($n = 7$:
472 hrs of observations) and (b) Golden jackals ($n =$
7: 298 hrs of observations).

helpers were taken during weeks five to nine and may also reflect that the
female spends more time at the den while lactating. With two helpers, each
individual helper spends significantly less time guarding/resting at the den.

The presence of one helper in golden jackal families led to a decline in
the amount of time that the lactating mother (all golden data includes nursing pups) spent at the den; and the father increased the amount of time spent
guarding/resting. With three helpers the parents spent more time foraging
(similar to silverbacks). With three helpers, each individual spent a similar
amount of time guarding/resting at the den compared to one-helper families.
In golden jackals the presence of helpers had less effect on parental pair
trade-offs between guarding and foraging. All individuals spent more time

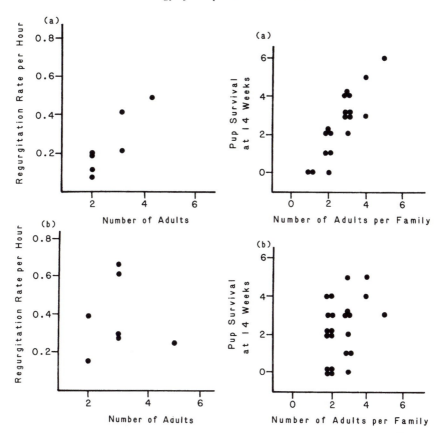

Fig. 4.7. Mean hourly rate of regurgitations by jackal families with varying numbers of helpers during weeks three to fourteen of pup rearing. (a) Silverbacked jackals ($n = 7$; 472 hrs of observations; $r_s = 0.90$; $p = < 0.01$) and (b) Golden jackals ($n = 7$; 298 hrs of observations; $r_s = 0.04$, ns).

Fig. 4.8. Relationship between number of adults per family and pup survival at fourteen weeks of age in (a) *Canis mesomelas* ($r_s = 0.89$, $p = < 0.01$) and (b) *C. aureaus* ($r_s = 0.36$, $p = < 0.05$).

at the den than silverbacks, and with the addition of helpers, pups were rarely left unguarded. The higher percentage of time spent at dens may reflect less time needed for foraging in order for adults to feed themselves and the pups.

Helpers regularly contribute food to young pups and the lactating mother. In silverbacked litters, detailed observations indicate that significantly higher ($r_s = 0.90$, $p < 0.01$) rates of regurgitations per hour occur in families with more adults (i.e., those with helpers) (Fig. 4.7). Food resources for silverbacked jackals during the whelping season consist primarily of abundant but small-sized food items (e.g., ~60 gm rodents and ~5gm fruits). The energetics of foraging for many small food packets appear to limit adult abilities to provision themselves and the pups (maned wolves presumably are limited by similar energetic contraints). In families where pups outnumbered the adults, they were rough-coated, weak-legged, appeared malnourished, and died ($n = 2$) or disappeared until the adult:pup ratio approached 1:1. Pairs on average raised 1.3 pup and with the addition of helpers, mean pup survival at fourteen weeks increased (e.g., 3 adults:3.25 pups, 4 adults:4.0 pups, 5 adults:6.0 pups). Pup survival at fourteen weeks of age correlates significantly with the number of adults (parents + helpers) in the family (Fig. 4.8:$r_s = .89, p < 0.01$). For ten litters in which number of pups at emergence from the den (three weeks) were known, number of adults also correlated significantly with pup survivorship from three to fourteen weeks ($r_s = .88, p < 0.01$; Moehlman, 1983). Thus there is a significant correlation with little variability between a) number of adults in a family and regurgitation rate and b) number of adults in a family and pup survival at fourteen weeks.

Examination of relative contributions by members of a pair raising pups on their own and pairs with helpers reveals that the mother consistently contributes the most feedings (regurgitations + nursings) (Fig. 4.9). However, with the addition of helpers her total contributions decline from 76 percent (no helpers) to about 40 percent. Fathers increase their feeding contributions slightly. Helpers on average contribute 33 percent (one helper) to 25 percent (two helpers). When one helper is present there is no significant difference between male helpers (0.04–0.14 r/h) and female helpers (0.6 r/h) in regurgitation rate per hour and/or percent time spent pup sitting ($\male\male$: 24–72%, $\female$: 72%). When two helpers are present in a family there is no significant difference between males (0.05–0.10 r/h) and females (0.05 r/h) in regurgitation rates and/or percent time spent pupsitting ($\male\male$: 7–94%, $\female\female$: 32%).

In golden jackal families with the addition of helpers there was an upward trend in the regurgitation rate but data were variable and there was no significant correlation (Fig.4.7). However, provisioning rates in golden jackals are consistently higher than those for silverbacks. In fact, golden pairs feed their pups at a higher rate than silverback families with one helper (goldens: two adults, $n = 2, \bar{x} = 0.52$ feeds/hr; silverbacks: three adults, $n = 2, \bar{x} = 0.36$ feeds/hr). Nursing rates tend to be higher in goldens ($n = 7, \bar{x} = 0.23$) than silverbacks ($n = 6, \bar{x} = 0.15$). Food resources for golden jackals during the whelping season consist of small items like dung beetles, but also hares and Thomson's gazelle fawns and adults. They also scavenge on carcasses of Thomson's gazelle, Grant's gazelle, wildebeest, and zebra. There is a readily available abundant food supply courtesy of the wildebeest/zebra/gazelle migration to the short grass plains during the rainy season. Thus in

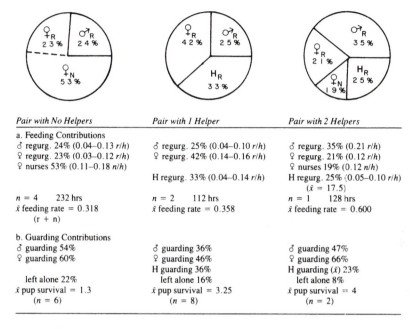

Pair with No Helpers	Pair with 1 Helper	Pair with 2 Helpers
a. Feeding Contributions		
♂ regurg. 24% (0.04–0.13 *r/h*)	♂ regurg. 25% (0.04–0.10 *r/h*)	♂ regurg. 35% (0.21 *r/h*)
♀ regurg. 23% (0.03–0.12 *r/h*)	♀ regurg. 42% (0.14–0.16 *r/h*)	♀ regurg. 21% (0.12 *r/h*)
♀ nurses 53% (0.11–0.18 *n/h*)		♀ nurses 19% (0.12 *n/h*)
	H regurg. 33% (0.04–0.14 *r/h*)	H regurg. 25% (0.05–0.10 *r/h*)
		($\bar{x}$ = 17.5)
n = 4 232 hrs	*n* = 2 112 hrs	*n* = 1 128 hrs
$\bar{x}$ feeding rate = 0.318	$\bar{x}$ feeding rate = 0.358	$\bar{x}$ feeding rate = 0.600
(r + n)		
b. Guarding Contributions		
♂ guarding 54%	♂ guarding 36%	♂ guarding 47%
♀ guarding 60%	♀ guarding 46%	♀ guarding 66%
	H guarding 36%	H guarding ($\bar{x}$) 23%
left alone 22%	left alone 16%	left alone 8%
$\bar{x}$ pup survival = 1.3	$\bar{x}$ pup survival = 3.25	$\bar{x}$ pup survival = 4
(*n* = 6)	(*n* = 8)	(*n* = 2)

Fig. 4.9. Relative feeding and guarding contributions of adults according to status and sex in silverbacked jackals. Average feeding rates include both regurgitations and nurses.

this study area, golden jackals are not food-limited during the whelping season unless the rains fail (in 1982 one pair was observed to have pups in a 100 km² area). Pup survival does correlate significantly with the presence of helpers (Fig. 4.8 r_s = .36, $p < 0.05$) but the data have more variability and the correlation is much weaker than for silverbacks. Among golden jackals the addition of helpers is less critical in terms of provisioning pups. Golden jackal pups are better provisioned and guarded than silverbacked pups, but fewer survive at fourteen weeks of age. Although the rains help provide food in abundance, they also repeatedly flood dens and pups die of exposure and illness. This density-independent factor negates the contributions of parents and helpers.

Among golden jackals the mother consistently contributes the most feedings to pups (Fig. 4.10). With the addition of helpers, her total food contributions to pups show no real change (66 → 58 → 61%), but she received an increasing rate of regurgitations (0.02 → 0.05 → 0.17 *r/h*) from other adults in the family. Within the parental pair the male tends to regurgitate to pups more often than the female, but she makes major contributions by nursing the pups. With the addition of helpers there is a decline in the father's share of provisioning. Helpers on average contibute 14 percent (one helper) to 29

81

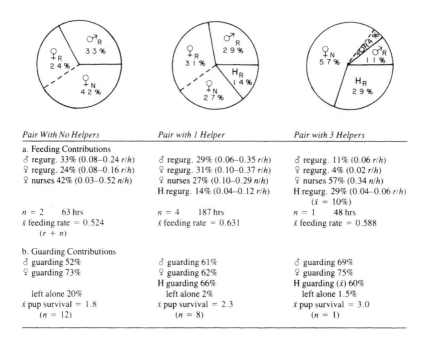

Pair With No Helpers	Pair with 1 Helper	Pair with 3 Helpers
a. Feeding Contributions		
♂ regurg. 33% (0.08–0.24 r/h)	♂ regurg. 29% (0.06–0.35 r/h)	♂ regurg. 11% (0.06 r/h)
♀ regurg. 24% (0.08–0.16 r/h)	♀ regurg. 31% (0.10–0.37 r/h)	♀ regurg. 4% (0.02 r/h)
♀ nurses 42% (0.03–0.52 n/h)	♀ nurses 27% (0.10–0.29 n/h)	♀ nurses 57% (0.34 n/h)
	H regurg. 14% (0.04–0.12 r/h)	H regurg. 29% (0.04–0.06 r/h)
		($\bar{x}$ = 10%)
$n = 2$ 63 hrs	$n = 4$ 187 hrs	$n = 1$ 48 hrs
$\bar{x}$ feeding rate = 0.524	$\bar{x}$ feeding rate = 0.631	$\bar{x}$ feeding rate = 0.588
($r + n$)		
b. Guarding Contributions		
♂ guarding 52%	♂ guarding 61%	♂ guarding 69%
♀ guarding 73%	♀ guarding 62%	♀ guarding 75%
	H guarding 66%	H guarding ($\bar{x}$) 60%
left alone 20%	left alone 2%	left alone 1.5%
$\bar{x}$ pup survival = 1.8	$\bar{x}$ pup survival = 2.3	$\bar{x}$ pup survival = 3.0
($n = 12$)	($n = 8$)	($n = 1$)

Fig. 4.10. Relative feeding and guarding contributions of adults according to status and sex in golden jackals. Average feeding rates include both regurgitations and nurses.

percent (three helpers) within families. Their rates of regurgitation are similar to those of silverbacked jackal helpers, but their contributions tend to be a smaller percentage of the total number of feeds. When one helper is present there is no significant difference between males (0.04–0.10 r/h) and females (0.10 r/h)in regurgitation rates and/or percent time spent pupsitting (♂♂: 48–85%, ♀: 91%). When three helpers are present in a family there is no significant difference between male (0.06 r/h) and females (0.04–0.06 r/h) in regurgitation rates and/or pupsitting (♂: 54%, ♀♀: 60–67%) although there is a tendency for females to spend more time pupsitting at the den.

Cooperative breeding in both species of jackal involves an equal sex ratio in reproductives (long-term pair bonds) and an equal sex ratio in helpers (Table 4.3). Although the mother is solely responsible for gestation and lactation she receives significant assistance in the feeding and guarding of the pups. Other than gestation and lactation there is symmetry of sexual roles among jackals. Both members of the mated pair mark and defend their territory. Both parents regurgitate to the pups at similar rates. Both parents guard and defend the pups at similar rates (♂♂ slightly > ♀♀), and both groom, play, and otherwise socialize with the pups. Helpers are also present

TABLE 4.3

Sex ratios in silverbacked and golden jackals in the Serengeti

	Age in weeks	Silverbacked $\male:\female$	Golden $\male:\female$
Emerge from den	~3	1:0.9 ($n = 34$)	1:0.8 ($n = 33$)
Surviving at	~14	1:1.1 ($n = 49$)	1:0.8 ($n = 44$)
Helpers	~52	1:1 ($n = 20$)	1:1 ($n = 16$)
Reproductive	52+	1:1 ($n = 42$)	1:1 ($n = 46$)

in equal sex ratios and also display symmetry of roles in defense of the territory, and feeding, guarding, and socializing with the pups. Helpers make significant contributions to pups and the presence of helpers correlates significantly with pup survival at fourteen weeks (silverbacked jackals $p <$ 0.01, golden jackal $p < 0.05$). The size, spatial, and temporal availability of food affects the magnitude of helper contributions such that in silverbacked jackals additional adults (helpers) in the family correlate with significantly higher regurgitation rates. Survival of silverbacked jackal pups appears to be food-limited. By contrast, although golden jackal helpers contribute a substantial percentage of regurgitations to pups they have less effect on the feeding rate. Due to the abundance of large food items during the wet season golden pups are normally not food-limited. However, density-independent factors such as rain can negate parental and helper contributions.

DISPERSAL

At eleven or twelve months of age a young jackal has the option of staying on its natal territory for an extra six to eight months and helping to rear its full sibs (on average $r = 0.5$) or dispersing and attempting to acquire a mate and a territory and to rear a litter of its own pups (Moehlman, 1979b; Emlen, 1982a, b). Data on success at emigrating, mating, establishing a territory, and rearing pups at one year of age versus two years of age are needed to fully answer questions concerning the cost/benefits of helping versus dispersing. However, existing data enable one to interpret differences in two closely related and socially similar species inhabiting areas with different vegetation and food resources.

Among silverback jackals a small percentage (24 percent) of surviving pups stay and help. The ecological conditions in terms of food resources are relatively stable. The rodent (*Arvicanthis niloticus*) density varies from 3,130 to 7,790 rats/km² during the wet season to 13,130 to 32,080 rats/km² during the dry season (Senzota, 1978, 1982). Even during the low point of the rodent cycle there is an abundant biomass available to silverbacked jackals. During the wet season the influx of the migrating herds occurs. Thus adult silverbacked jackals appear to have an abundant food supply. Resource availability and dispersion conditions (Macdonald, 1983) are favorable for the retention of offspring on the natal territory. No bottleneck in terms of food availability occurs on silverback territories. The food resources situation that makes it possible for young adults to stay also presumably makes it easier for them to disperse. The brush woodland habitat with its poorer visibility conditions would allow individuals initially to establish themselves at the edges of existing territories. However, as Emlen (1982a, b) has delineated in his ecological constraints model, even if breeding openings exist, the cost of rearing young may be prohibitive. It is difficult to raise pups since pairs on average raise 1.3 pups ($n = 6$) and even experienced pairs lose whole litters. Jackals that stay and help derive important benefits. They acquire extended experience on familiar terrain, and may improve their survivorship and the quality of their future parental care. Since helpers are full sibs, by delaying their reproduction for one year they potentially increase their inclusive fitness on average by 1.74 + pups × 1/2; [$n = 17$ litters: on average a pair raised 1.3 pups ($n = 6$), pair + 1 helper raised 3.3 pups ($n = 8$), pair + 2 helpers raised 4 pups ($n = 2$), and pair + 3 helpers raised 6.0 pups ($n = 1$) (Moehlman, 1983; Macdonald and Moehlman, 1983)]. But even with the constraints on rearing offspring, most one-year-old silverbacks appear to opt for dispersing and attempting reproduction at an earlier age.

By contrast 70 percent of known surviving golden jackal pups were observed helping with the next year's litter. During the dry season when they are seven+ months old there is a food ''bottleneck'' and young of the year leave their natal territory. They return in the wet season when food is abundant. This ''high'' rate of return by eleven-to-twelve-month-old golden jackals may reflect a high golden jackal density with a saturated habitat in which it is difficult to disperse successfully, mate, and acquire a territory. One-year-old goldens are simply forced to stay. Current golden jackal data indicate that pairs more than adequately provison their litters and the major cause of mortality is a density-independent factor. Thus for golden jackal dispersal, the major ecological constraint appears to be the difficulty in aquiring a territory. Golden jackals ($n = 2$) have also been observed on their natal territories in years when their parents did not have another litter. This is perhaps another indication of the relative benefits of staying on the home territory and the difficulty of dispersing. Golden jackals, similarly to silverbacked jackals, potentially derive the benefits of extended experience on familiar terrain in terms of improved survivorship and quality of future parental care. They also derive inclusive fitness benefits but not as significantly as silverbacked jackals.

Staying and helping versus dispersing in both jackal species is affected by ecological constraints of food type and spatial and temporal dispersion throughout the year, the availability of suitable territories, and the probability of successfully raising pups. Male and female reproductive strategies benefit from the retention of helpers since their presence correlates significantly with higher pup survival rate and hence parental reproductive success. Mothers with helpers may also improve their own future reproductive success since helpers alleviate maternal nutritive stress during the lactation period.

Selection for monogamy and long-term pair bonding may reflect in these jackals both physiological and ecological constraints. Jackals have a relatively large litter size and there is a long period of infant dependency (Kleiman and Eisenberg, 1973; Kleiman, 1977). Silverbacked jackal pairs in the Ndutu study area are omnivorous and utilize small-sized abundant food resources that are energetically costly to collect. Paternal investment is critical to pup survival and if a male were to divide his care between several litters, then possibly no pups would survive and both male and female reproductive success would suffer. Domestic dogs (*Canis familiaris*) can produce litters with multiple sires (Beach, pers. comm.). Thus maintenance of mating exclusivity may be important in all *Canis* species so that the monogamous male does not invest in offspring that he did not sire. Long-term pair bonds are also critical for territorial maintenance. When one member of the pair dies/disappears during whelping season, the remaining adult cannot sustain the pups and control the territory. Golden jackals in the adjacent short grass plains utilize larger-sized abundant food resources. Faculative cooperative hunting and defense of carcasses play more of a role in their foraging behavior. Pairs on average raise 1.8 pups ($n = 12$) but variability is high (0-4) and pup survival does not appear to be limited by food provisioning. Paternal investment is still important in terms of provisioning and guarding. However, the nature of the food resource availability might enable a male to provision successfully several litters and the pair bond does not appear to be as strong as the silverbacked jackal pair bond. Polygyny and polyandry have not been observed in jackals, but with their median body weight and variable foraging strategies they are capable of flexibility in spacing, group size, and, potentially, mating and rearing systems. Larger stable territorial groups of golden jackals were observed in Israel and Macdonald (1979a) suggested that these large social groups were possible due to the large clumped food resource that was economically defendable (Bradbury and Vehrencamp, 1977a, b; Emlen and Oring, 1977). However, no information on the mating and breeding system was available from this study.

Thus in two jackal species differences in habitat and food resource size, type, and spatial and temporal availability correlated with pup rearing and survival, dispersal behavior, pair bonds, and cooperative breeding. Within Canidae, body scaling analyses can provide important clues to constraints

85

on mating systems and cooperative breeding, but it is equally important to examine these patterns with regard to the best field studies available. Variability in food packet size and temporal and spatial availability of food resources can affect spacing systems and breeding strategies. Canids can exhibit great intraspecific variability in terms of food resources utilized and home range/territory size (Table 4.2b). The importance of resource dispersion to the philopatry/dispersal, group size, and social organization of solitary foraging carnivores has been discussed by Kruuk (1978b), Macdonald (1983), and Schantz (1981). Increased foraging success and prey retention has been considered critical to the evolution of sociality in the larger carnivores (Kleiman and Eisenberg, 1973; Kruuk, 1975; Lamprecht, 1981). Within Canidae the allometry of neonate weight and litter size to maternal body weight appears to impose critical requirements for the successful rearing of young in terms of investment by adults other than the mother. However, the availability of food and the potential for nutritive input to the mother and pups can have dramatic effects on this equation, for example, arctic foxes and maned wolves. Within species food availability is also correlated with major reproductive changes. One of the most impressive examples is the arctic fox, which has mean litter sizes of 10.6 in Northwest Territories, Canada, where it feeds on lemmings (Macpherson, 1969) and mean litter sizes of 4.0 along the coast of Iceland where they scavenge on carrion washed ashore (Hersteinsson, 1984).

Medium-sized canids may have the most flexibility in terms of altering their social behavior and organization. In coyote field studies home range size can vary by a factor of 2 to 3 and have no correlation with group size (Camenzind, 1978; Bekoff and Wells, 1980, 1982) while in another study area group size can be strongly correlated with home range size (Bowen, 1978, 1981). Sex ratios can be 1:1 or skewed toward males or females (Berg and Chesness, 1978; Bowen, 1978; Bekoff and Wells, 1980). Males tend to emigrate farther in one population and females in another (Berg and Chesness, 1978; Nellis and Keith, 1976). Mated pairs can hold territories together in one population (Bekoff and Wells, 1982) and in another, according to radio tracking data, males can hold a territory that contains several female territories within it (Berg and Chesness, 1978). But more information is needed on the social and reproductive status of individuals. In coyotes, sex ratio, emigration, and spacing systems may be equal, or skewed toward males *or* females. Such data, though incomplete, are intriguing and emphasize how much food availability can affect medium-sized canid spacing systems and possibly mating and rearing systems.

Thus allometry can provide important insights to the origins of social organizations in Canidae. However, it will be field studies that reveal how ecological constraints determine spacing systems and breeding strategies within these limitations.

5. Sexual Asymmetries in the Life History of the Florida Scrub Jay

———————————— • ————————————

GLEN E. WOOLFENDEN

AND JOHN W. FITZPATRICK

FLORIDA scrub jays (*Aphelocoma c. coerulescens*) breed as monogamous pairs. These pairs mate for life, and dwell in large all-purpose territories which they defend the year round. As breeders, males and females experience equal survivorship, and the same appears to be true for fledglings during the first twelve months of life. As young yearlings all jays, male and female, remain in their natal territory where they help the resident breeders (usually their parents) raise a new brood of offspring. Despite these fundamental similarities between the sexes in survivorship and mating strategy, substantial differences exist between the sexes in dispersal and reproductive behavior. These sexual asymmetries in dispersal behavior parallel those found in most birds (e.g., Greenwood, 1980), and their persistence in a population where predispersal and postdispersal demography are so symmetrical between the sexes is doubly surprising. In this chapter, after describing the social milieu of the Florida scrub jay, we describe these sexual differences and discuss them in the context of the jays' overall social system. Our data and musings are based on complete life histories of hundreds of individually marked jays in a population we have studied intensively since 1969. Our speculations about the evolution of certain of these differences, especially regarding dispersal, may apply to a range of animals much more diverse than American jays. We focus on the general case of monogamous birds that defend all-purpose territories.

STUDY SITE AND METHODS

Fieldwork has been carried out from 1969 through the present at the Archbold Biological Station in Highlands County, Florida. The station property consists of a 1,600 ha preserve near the southern terminus of the sandy Lake Wales Ridge, about three-quarters down the center of the Florida peninsula. Since the onset of the study we have followed the lives of 683 individual jays from fledging to death. Jays that died before reaching their first birthday constitute most of this sample, but it also includes 189 yearlings and 80 who fledged and became breeders in our study tract. For over a decade the size of the tract, within which essentially all scrub jays are individually marked, has been about 400 ha (1.5 square miles).

Each year we find all nests and color-band all young and all immigrants. Immigrant yearlings can be identified as to age class, which adds slightly to certain samples of known-age breeders. Acceptable habitat for the jays extends for many kilometers in several directions beyond the study tract, and we regularly search this surrounding scrub for marked dispersers from our tract.

Demographic data are derived from monthly censuses of all jays in the study tract. Territory boundaries are mapped in detail on aerial photographs each spring during the nesting season. The jays are tame to humans, and this allows all phases of our work to be accomplished with unusual accuracy. Extensive analyses of the demographic data gathered during the first decade of our study, and elaboration of our views on the evolution of cooperative breeding, appear in a recent book (Woolfenden and Fitzpatrick, 1984).

THE HABITAT

The single most important ecological feature about Florida scrub jays is their restriction to a rare, patchy, and vanishing habitat that is sharply bounded by unusable habitats (Woolfenden and Fitzpatrick, 1984; Fitzpatrick and Woolfenden, in press). The jays are entirely confined to the xeric oak scrub of central and coastal peninsular Florida. Florida scrub vegetation grows only on excessively drained sandy soil, mostly ancient or modern sand dunes. It is dominated by a few species of stunted, woody perennials, chiefly oaks (e.g., *Quercus inopina*, *Q. geminata*, *Q. myrtifolia*). Widely scattered pines (*Pinus elliotti*, *P. clausa*) interrupt a vegetation profile that rarely exceeds two meters in height. This desertlike habitat, replete with scattered cactus (*Opuntia compressa*), is a relic of earlier, drier times in the Pleistocene when a broad belt of xeric vegetation, now largely confined to southwestern North America, extended eastward to the Atlantic Ocean. In Florida, scrub remains only as scattered, often small, isolated patches (see maps in Jackson, 1973 and J. A. Cox, 1983). Today the nearest scrub jays to those in Florida live 1,600 kilometers to the west, in central Texas.

In contrast to southwestern North America, usable suboptimal habitat for scrub jays in Florida is rare to nonexistent: oak scrub is sharply bounded at its margins by a variety of very different, unsuitable habitats. We have shown (Fitzpatrick and Woolfenden, in press) that even in marginally habitable oak scrub, reproduction and survival are too low to support a stable population, principally because of increased predation. Presumably for similar reasons, scrub jays in Florida shun all forms of forest and woodland.

which in eastern North America are the domain of a widespread competitor, the blue jay (*Cyanocitta cristata*). Blue jays, notorious nest robbers, rarely venture into the oak scrub, where they are dominated by scrub jays. No jay species lives in the other major habitat types found in Florida, the treeless grasslands and the vast freshwater marshes.

Florida scrub jays eat mostly insects and acorns, though small vertebrates (e.g., tree frogs and lizards) and some other plant foods (e.g., small berries) are taken when encountered. Insects are gleaned from low shrubs and palmettos and from the sandy ground surface, on which the jays often stand while foraging. Bark-gleaning, digging, and aerial hawking are rare foraging activities. Acorns appear to be a necessary staple all year. They are taken directly from the abundant oak shrubs from August through December each year. They also are cached during these months, making them available almost throughout the year. Insects are scarcest on cool days in mid-winter and during spring droughts, and are most abundant during the warm, wet summers. Cached acorns supplement animal food most heavily during times of insect scarcity, especially in late winter months. Year-to-year variation in insect availability seems high, but acorns seem abundant every year (these features are currently under detailed study).

THE SOCIAL MILIEU

Numerous aspects of our population data suggest that acceptable scrub is constantly saturated with breeding Florida scrub jays, resulting in a surplus of potential breeders (Woolfenden and Fitzpatrick, 1984). With the exception of a brief population crash in 1980, breeder density within the study tract between 1970 and 1983 has been extremely stable. In all it has varied between 5.8 and 8.0 breeders per 40 ha (cv = 10 percent); eliminating the unusual year (see Woolfenden and Fitzpatrick, 1984, for details) yields twelve years in which breeding density varied from 6.3 to 8.0 (cv = 8 percent). These minor fluctuations took place during years when the total population of nonbreeders (helpers) varied from 3.1 to 7.1 individuals per 40 ha (cv = 28 percent) at breeding times. During these same years, helpers two years old or older (i.e., clearly at breeding age) varied between 1.0 and 2.8 per 40 ha (cv = 30 percent). These figures show that breeder density is substantially less variable than total population density, indicating that a roughly constant number of breeding slots exists and that any vacancy is filled as soon as it appears. Even in years of lowest density of older nonbreeders, one exists for every three to four pairs of jays. The ratio can get as high as one potential breeder for every pair of actual breeders.

Nearly constant habitat saturation within the breeding population results in a strong inverse correlation between breeder density and average territory size. The correlation ($r = .81$; $p < .01$) occurs for the simple reason that essentially all usable scrub is contained and defended within existing jay territories every year.

Florida scrub jays defend their territories all year, and virtually all activities take place within these boundaries. The territories are relatively large for the size of the bird (Schoener, 1968), and much larger than scrub jay territories in western North America (e.g., Atwood, 1980). Based on measurements of 221 territories from 1971 through 1979, territories range in size from 1.2 ha to 20.6 ha with a mean of 9.0 ha. Pairs with helpers have significantly larger territories than those without (10.4 versus 7.2 ha, respectively; Woolfenden and Fitzpatrick, 1984). Normally, once a jay becomes established as a breeder it lives the rest of its life in essentially the same piece of scrub. The size of its territory fluctuates as family members grow up and depart, and as neighboring families gain or lose members. Sometimes, families with older male helpers actively usurp space from their neighbors. Despite these fluctuations, however, core areas of prime scrub habitat are defended by each breeding pair for many years in succession, and often are passed between generations (Woolfenden and Fitzpatrick, 1978).

The social world of the Florida scrub jay is crowded and competition for space is manifested almost daily as groups encounter intruding jays at or near their territorial boundaries. During peak times of dispersal (mid-fall and early spring) encounters are most frequent, occurring many times per day in each territory. Any loss of a jay from a breeding pair (usually from predation) is detected within a day or two by other jays of both sexes in the neighborhood, who begin to intrude into the weakened territory. The ensuing increase in loud vocalizations and conspicuous aerial displays alerts distant jays that an unsettled area exists, even if it is many territories away. Such unsettled areas often become a focus for visits by many jays from a wide neighborhood, and particularly by nonbreeding jays of the same sex as the missing breeder. In such a manner, vacancies are discovered and filled.

Life-time monogamy typifies Florida scrub jays. More than 95 percent of the breeders we have followed remained with the same mate until one member of the pair died. Widowed jays soon pair with a replacement mate from outside the territory. The rare divorces seem to be caused by injury or illness preventing a mate from functioning normally, which causes the healthy mate to begin pairing with a healthy, immigrant replacement. A few divorced jays who did recover wandered furtively between territories before either disappearing or locating a new mate. No jay has remained a wanderer for more than a few months.

The nesting season lasts from mid-March into June, peaking during April. The percentage of nesting pairs with helpers fluctuates through the years, and depends mainly on the average fledging success within the population during the previous breeding season. From 1970 to 1983, pairs with helpers varied from 36 to 78 percent of all pairs, averaging 54 percent. Most helpers (64 percent) are one year old, having hatched the preceding spring. Two-year-olds make up an additional 24 percent, and the remaining 12 percent vary from three to seven years old.

The social system of scrub jays in western North America is similar to that of the Florida population in that both are sedentary, monogamous, territorial, and confined to oak-dominated habitats (Verbeek, 1973; Atwood, 1980; Ritter, 1983). However, a profound and potentially revealing difference exists: all available evidence indicates that the western populations entirely lack helpers. Our suspicion, briefly summarized, is that ecological crowding is less intense in the vast, climatically and vegetationally variable habitats acceptable to scrub jays in western North America. We surmise that habitat-crowding is a correlate to delayed breeding and cooperation in *Aphelocoma*, and that intrademic competition for space is comparatively low among scrub jay populations in the west. Nonbreeders who disperse early and attempt to breed on their own succeed frequently enough to be selectively favored in such a situation (Woolfenden and Fitzpatrick, 1984, contains lengthy discussions and a model of this hypothesis).

HELPERS AND REPRODUCTION

Of special interest in analyzing cooperative breeding is the issue of whether helpers actually do increase the reproductive output of the breeders they assist. Florida scrub jay helpers join the breeding pair in a variety of activities, including territorial defense, predator detection and mobbing, alarm-calling, and the feeding of nestlings and fledglings (Woolfenden and Fitzpatrick, 1984). A resulting average increase in the production of genetically related sibs theoretically could selectively favor nonbreeders who postpone or even sacrifice their own breeding attempts in order to afford this help (W. D. Hamilton, 1964). In monogamous, diploid species such as birds, this form of kin selection would affect males and females identically, and might even lend some genetic "inertia" to any divergence between the sexes in their propensities to help or to disperse.

Determining whether helpers really do help is difficult. Some investigators suggest that the hypothesis can be adequately tested only through removal experiments. Brown and Brown (1981b) used removal experiments to show that helpers apparently do improve reproduction in grey-crowned babblers (*Pomatostomus temporalis*). Field tests such as these present their own risks, however. For example, the socially disruptive effects of sudden removals from experimental groups cannot be measured or controlled. Disruption itself might produce biases in the reproductive performances of the groups under study, and these biases might vary according to the individual histories and constitutions of the different groups.

We have elected not to perform removal experiments in our Florida scrub jay population, because we are attempting to follow the long-term, natural course of a wild population. Fortunately, however, each year our population includes some breeders with helpers and some without. Over many such years, various natural comparisons between the two samples allow us to test the hypothesis that helpers increase the reproductivie success of the breeders they assist. Based on fourteen consecutive years in which fledging success was measured among natural pairs with and without helpers, we have determined that helpers do help.

Fledgling production by Florida scrub jays is highly variable. Between 1970 and 1983 it varied from 0.9 to 2.9 fledglings per pair per season. Nevertheless, during each of the fourteen years pairs with helpers fledged more young ($\bar{x} = 2.36$) than did pairs without helpers ($\bar{x} = 1.58$). Nest success, defined as the proportion of nests fledging at least one young, was higher among pairs with helpers in twelve of the fourteen years. The exceptions were the two years when nest predation was unusually low, causing overall fledging success to be highest (2.9 and 2.6 fledglings per pair). These and other data suggest that the presence of helpers improves reproductive success principally by reducing the probability of nest failure. This issue is examined and documented in detail in Woolfenden and Fitzpatrick (1984).

Nest failure is almost always caused by predation, probably most often by snakes, bobcats, raccoons, hawks, owls, and other jays. Starvation of young jays is essentially absent, and certainly does not account for disappearances of whole broods from a nest. Furthermore, we have found that eggs (which cannot starve) have a significantly higher probability of reaching hatching age when the group includes one or more helpers as well as the breeding pair. We conclude that helpers help by reducing nest predation, presumably through improved predator detection and increased intensity of predator mobbing.

Many factors confound the seemingly direct relationship between presence of helpers and nesting success. For example, the presence of helpers correlates with fledging success only for nests initiated early in the season. Helpers seem to be equally attentive during later attempts, so their contributions may be swamped by a heavy increase in predator activity later in the nesting season. Whatever the reason, overall nest success plummets from 80 percent for nests initiated in March to 25 percent for those initiated in late May.

Many details of the effects of helpers through the season require further study, but the principal result is clear: Florida scrub jay helpers do help breeders to raise an average of 50 percent more young than do breeders without helpers. However, fledgling production remains constant with increasing *number* of helpers beyond one. The increase in inclusive fitness gained by a single helper is therefore diluted when more than one helper is present. On the average, any one helper is sufficient to raise the level of production of offspring, so that extra helpers would appear to add little to their own inclusive fitness by helping. We have found no evidence, however, that this dilution results in any change in dispersal or helping propensities among helpers belonging to larger family units. Fitzpatrick and Woolfenden (in press) discuss the demographic conditions that favor delaying dispersal and helping even in the absence of kinship benefits thereby gained.

SEXUAL SYMMETRIES IN BEHAVIOR AND DEMOGRAPHY

At age one year virtually all Florida scrub jays, male and female, function as helpers. These jays do not appear to be physiologically incapable of breeding, however, as several yearlings have paired and nested, two of them successfully (one male, one female). These exceptions constitute only 3 percent of a sample of over 200 yearlings. We determine the sex of nonbreeding jays by behavior. One distinct trait is a vocalization known as the "hiccup" call, which is given only by females. Dominance interactions (Woolfenden and Fitzpatrick, 1977) also are useful in determining sex. Between 1971 and 1983 we observed 210 jays who reached age one year; 96 were males, 98 were females, and 16 were of undetermined sex when they disappeared. Even if all 16 unknowns were of one sex, for which we can think of no good reason, this sex ratio would not differ significantly from parity. We assume that the primary sex ratio is even, so it appears that the sexes suffer equal mortality during their first twelve months of life.

The sex ratio among breeders always is even, because monogamy is the rule. Furthermore—and extremely important to the demography of nonbreeders—the death rates of male and female breeders appear to be identical. As of October 1983 we had lost 166 breeders: 84 (51 percent) were males, 82 (49 percent) were females. As a partially independent measure of breeder survivorship by sex, we calculated the mean duration of *known-age* jays as breeders; it was similar for the two sexes: 3.14 breeding seasons for males, 3.73 for females (ranging from one to eleven years). During this time breeders had an average of 1.58 mates: 1.52 for males, 1.64 for females.

SEXUAL ASYMMETRIES IN BEHAVIOR AND DEMOGRAPHY

Males and females thus far seem to face symmetrical social and ecological pressures. Neither sex is limiting, because they are strictly monogamous and sex ratios and death rates are similar between sexes among both juveniles and breeders. A nonbreeding surplus population of both sexes always exists, and they stand to gain equally from any propensity to help produce more sibs than could be raised without them. Yet, as we have shown elsewhere (Woolfenden and Fitzpatrick, 1978, 1984), great differences exist between the demographic regimes of helper males and females. Predominant among these differences is the greater mobility and dispersal tendency exhibited by females, a pattern that typifies virtually all bird species (e.g., Greenwood, 1980). We suspect that the origin of these asymmetries may lie in two primary behavioral differences that seem to be primitive among monogamous passerine birds. These are 1) division of labor during nesting, and 2) male dominance and female subordinance. We emphasize these differences here and in the discussion, because ultimately they may underlie all the demographic asymmetries as well.

Breeders

At the nest, the female breeder performs all incubating, brooding, and shading. The male breeder provides much of the food consumed by his mate during egg production, incubation, and brooding of small young. The male breeder also provides most of the food consumed by the nestlings, especially in the absence of helpers. Even late in the nesting cycle, when the young have improved at thermoregulation, the female breeder typically attends the nest while the male breeder forages for food. Presumably, she is on guard against potential nest predators. Male breeders with active helpers reduce their feeding trips to the nest, but not their nest attendance. They shift to more frequent sentinel duty on exposed perches near and above the nest. As sentinels, jays watch for predators and monitor the territory for potential conspecific intruders.

Males appear to be more active than females as defenders of the territory. They are the first to respond to territorial intrusions (Woolfenden and Fitzpatrick, 1977) and their boundary displays more frequently include overt aggression and even fights. Females participate actively in territorial defense, but their role often is only vocal; their distinctive, sex-specific "hiccup" call is used almost exclusively during territorial displays. This call announces the presence of intruders and appears to act as support for the males during their chases. Either sex will chase intruders of the opposite sex upon occasion, but because of male dominance, unassisted females usually are reluctant to chase males. Some reluctance by males chasing females exists, but is less evident.

Male Florida scrub jays dominate females regardless of family status. Thus, male helpers are able to dominate female breeders (usually their mothers) as well as female helpers, for example, in competition over a food item (Woolfenden and Fitzpatrick, 1977). Dominance behavior is not obvious during the first several weeks of a jay's life, but is well developed by the time of postjuvenal molt, at about age six months, when adult plumage is attained (Bancroft and Woolfenden, 1982). Male breeders are dominant over all other members of the group.

Helpers

Behavioral differences similar to those among breeders are apparent among male and female helpers, as if the division of breeding activities arises even before a breeding position is obtained. Male helpers, especially older ones, make more feeding visits to the nest than do females, and they deliver more total food on individual visits. In general, male helpers deliver two to four times as much food as do female helpers (Stallcup and Woolfenden, 1978). Older female helpers often visit the nest without food. Sometimes they even attempt to join the breeding female on the nest, although the breeder resists these attempts. We have no evidence that a helper female ever has laid an egg in the nest of a breeder.

As mentioned above, male helpers dominate all females in their group, although, as with breeding males, this dominance is manifested only rarely and usually subtly. Male and female helpers appear to be equally active at detecting and mobbing potential nest predators. Males, especially older ones, are slightly more aggressive in territorial disputes, but neither sex defends as actively as do breeders (Woolfenden and Fitzpatrick, 1977).

Dispersal

Even extreme habitat saturation would present little problem to a dispersing individual if breeding vacancies were to appear with sufficient frequency. Therefore, a true measure of population pressure upon nonbreeders is the relationship between breeder death rate and the rate of input of potential breeders who will fill those vacancies. In Florida scrub jays annual breeder mortality is variable, but has averaged 19 percent (SD $= 11.8$, range from 7 to 45 percent) between 1969 and 1983. Most deaths apparently result from predation. During the same period, pairs produced an average of 2.0 fledglings annually, of which 0.7 survived to become yearlings. This yields an average of 0.35 yearlings per individual breeder. But death rates of these same breeders average only 0.19 annually. These "birth" and death rates tell the overall story of breeding-space competition: on average, only about half the potential breeders ($.19/.35 = .54$) can fill a vacancy each year (Fitzpatrick and Woolfenden, in press).

Given such competition for breeding vacancies, successful dispersal becomes a game of 1) *surviving* during a potentially long nonbreeding period, 2) *locating or creating* a breeding slot, and 3) *winning* the slot once it is located. In this crowded regime, where livable habitat is entirely defended by territorial jays, survival is maximized by remaining with the tolerant family group on familiar ground. Clearly, however, permanent philopatry

TABLE 5.1
Routes to breeding among male and female
Florida scrub jays with known dispersal histories

	% Total	$\bar{x}$ dispersal distance $(m)^a$	SD
Males ($n = 61$)			
Inherit __.	10	102	± 94
Bud	38	222	± 85
Mate replacement	49	$536,638^b$	$\pm 298,631$
De novo	3	525	—
Females ($n = 48$)			
"Bud-mate"^{c}	29	665	± 413
Mate replacement	67	765	± 428
De novo	4	330	—

^{a} Dispersal distances measured from site of last nest helped to site of first nest constructed as a breeder.
^{b} Shorter of two mean dispersal distances excludes one male whose exceptionally long dispersal (3,600 m; see Fig. 5.1) apparently resulted in breeding, presumably through mate replacement, but was never confirmed.
^{c} Females neither inherit nor bud; those that *paired* with males who inherited or budded their breeding territory ($n = 14$) are grouped as "bud-mates." Sample size for females is smaller than for males because most immigrants, whose dispersal histories are unknown, are females.

would limit the opportunity to discover breeding vacancies within the population, especially ones that would not be incestuous. (Our evidence for strong incest-avoidance is discussed below.)

Sexual asymmetries among helpers become most apparent at the dispersal stage, when older nonbreeders (i.e., yearlings or older) begin attempting to establish themselves as breeders. Below, we briefly describe the various routes they take during this process, but the results can be summarized in a few words: males remain longer as helpers, disperse over shorter distances, and often begin breeding on ground once contained in their natal territory; females leave home more frequently as nonbreeders, their permanent dispersal comes sooner, and they end up farther from home compared to males. Mean ages at first breeding reflect these differences in the timing of dispersal: 3.8 years for males versus 2.4 for females. Increased mobility results in greatly increased female mortality during this dispersal phase.

We distinguish four pathways to breeding: 1) direct territorial inheritance, 2) territorial budding, 3) mate replacement, and 4) establishing a territory *de novo* between existing ones. Table 5.1 lists the frequencies with which sixty-one males followed these four routes in our population. These sixty-one males represent the sample of jays whose entire "dispersal history" we know, from early nonbreeding (almost all were banded as nestlings) to the time they began breeding within the study population. Excluded are jays of unknown age or whose helping or early breeding history contains uncertainties in our records. Because of differences in dispersal behavior, we have fewer records for females than for males. Females virtually never inherit natal ground, but of course they do pair outside their own family with males that inherit. Of forty-eight females with adequately documented histories, fourteen (29 percent) paired with male heirs to territory. The discrepancy between this percentage and that for male heirs themselves $(10 + 38 = 48$ percent) is accounted for by the arrival of long-distance dispersers from outside the study tract, who seem more frequently to pair with males that are budding than with widowed males.

1) Direct inheritance of the entire natal territory is rare even among males (10 percent), because it requires a combination of rare events. When a breeding male dies, even in a territory containing a helper male of breeding age, the widowed female usually holds her breeding position by pairing with a new male from outside the territory. Resident helpers remain helpers or else disperse. In the six clear cases of full inheritance, both breeders died essentially simultaneously $(N = 3)$, the helper male was unrelated to the breeding female and paired with her $(N = 2)$, or the helper succeeded in "expelling" his mother and pairing with an immigrant $(N = 1)$.

2) Territorial budding (Woolfenden and Fitzpatrick, 1978) typically is preceded by an increase in the size of the territory where the male is a helper. This territorial growth usually is preceded by an increase in family size, combined with a decrease or a breeder's death in a neighboring family. A resident helper—usually the dominant one—begins defending a segment of the expanded territory with the help of a female from outside his family. They become sole occupants of the new territory, while the male's kin remain in the remainder of the family territory, now reduced in size. Twenty-three (38 percent) of the sixty-one males became breeders in this fashion (Table 5.1).

3) About half of the male jays (and presumably an equal fraction of the females) become breeders by replacing lost breeders away from home. Males from many territories frequently attempt to pair with a widowed female, and we have circumstantial evidence that the process often includes active fighting. Females appear more furtive, perhaps because they frequently are much farther from home than are males when attempting to form a pair bond (see below).

4) Establishing a territory *de novo* away from home and between existing territories is a rare event, having occurred only twice in thirteen years within our tract. One of these was notably unsuccessful: the territory appeared to be tiny (1.7 ha), and the pair deserted their nestlings shortly after hatching. Both returned to their respective families as helpers, leaving their young to die.

Jays searching for a place to breed normally visit many territories before gaining residency in one, although we have not yet quantified this phenomenon. Temporary forays out from the natal territory begin during the first autumn of life, when loose bands of five to ten juveniles from several families can be seen wandering among many territories in the neighborhood of their home territory. These intruders are tolerated by territorial residents until postjuvenal molt nears completion, when the juveniles are virtually indistinguishable from adults. A second period of juvenile wandering occurs near the onset of the breeding season in March. Both in autumn and spring these forays are short, seemingly undirected, and end with the juveniles returning home. As nesting season progresses the spring wanderers shift attention to helping at the nest in their natal territory.

Forays by yearling helpers become regular again after the nesting season, especially among females. Perhaps the neighborhood is now familiar to them as a result of their occasional wanderings as juveniles. Now, however, the forays are not random. Nonbreeding jays appear to seek out places of unrest, which frequently ensues from the death of a breeder. The increased trespassing brings on heightened territorial defense by remaining residents and neighbors of both sexes, and this in turn seems to attract more jays. While not nearly as violent, these periods of unrest share many characteristics of the so-called "power struggles" of acorn woodpeckers (*Melanerpes formicivorus*; Koenig and Pitelka, 1981).

Numerous preliminary forays characterize Florida scrub jay dispersal, and these appear to be more numerous, of longer duration, and covering wider areas among females compared to males. As yet we have no independent measure of how far nonbreeding individuals wander from home during these forays. An estimate can be made, however, based on the distribution of distances in "successful" forays, namely, completed dispersal movements. Table 5.1 summarizes the distances moved by the jays included in the preceding analysis. Figure 5.1 includes a different and slightly expanded sample, showing the dispersal distances for 119 jays whom we followed from place of last helping to place of first breeding. Several of these were discovered as breeders outside the study tract, hence could not be included in Table 5.1.

Figure 5.1 illustrates the difference between overall male and female dispersal distances (median and mean distances shown by arrows). Virtually all male dispersal occurs within half a kilometer of the last nest helped ($\bar{x}$ = 456 m) while many females travel over a kilometer between natal and breeding site ($\bar{x}$ = 1180 m). This difference is highly significant (median test, $\chi^2 = 23.6, p < .001$).

Figure 5.2 diagrams the dispersal distances listed in Table 5.1, and illustrates the magnitude of the sex difference in terms of territories potentially investigated during dispersal forays. An average-sized Florida scrub jay territory, if circular, would have a radius of about 170 m. Among the six males

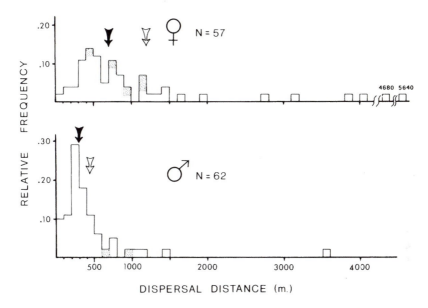

Fig. 5.1. Frequency distribution of dispersal distances for male and female Florida scrub jays, measured between nest last helped and first nest of first season as a breeder. Median (solid arrows) and mean (open arrows) dispersal distances are shown for each sex. Shaded portions of histogram represent known minimum dispersal distances for known-age immigrants whom we know had not bred elsewhere before entering the study tract.

who gained breeding status through direct inheritance, average distance "dispersed" was 102 m, which is about equal to the normal distance between successive nests within a continuing pair's territory (Woolfenden and Fitzpatrick, 1984). Males who became breeders through territorial budding moved an average of 222 m, or just outside the normal radius of a territory (but recall that budding is most frequent after major territorial growth). Males who replaced lost breeders moved only slightly less than did the females within our known-history sample (Table 5.1). However, this measure of female dispersal is highly biased toward the shorter distances because it excludes the sample of dispersers that left our study tract. When these are included, as in Figure 5.1, the female dispersal-radius (outer circle in Figure 5.2) clearly extends well beyond that of males.

As shown schematically in Figure 5.2, females can be thought of as searching through a neighborhood that contains about five times as many territories as that of males. Of course, the circles in Figure 5.2 represent only mean distances, and the actual effective neighborhood sampled by nonbreeders is better estimated by multiplying the frequencies of successful

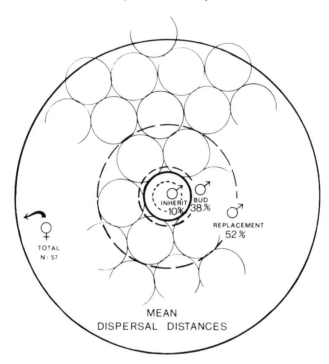

Fig. 5.2. Mean dispersal distances for three categories of males (dashed circles) and for all females combined (solid outer circle), expressed as concentric circles around the center of a natal territory (heavy, solid circle in center). In this schematic diagram, territories all are average size, 340 m across, and hexagonally packed (smaller solid circles) to illustrate different average neighborhood sizes of males versus females. Percentages shown for three male categories show the proportion of the total that follow each dispersal route (*de novo* is included under replacement).

dispersals at each distance away from home by the number of territories distributed at those distances. We made such a calculation, using our actual distributions of dispersal distances and assuming a uniform habitat of hexagonally packed territories with diameters of 340 m. Our results: males' average ''dispersal neighborhood'' contains approximately 14.3 territories, while that of females contains 73.5, a fivefold difference.

Helper Mortality

Calculating death rates of helpers is complicated by the few helpers that become breeders outside our study tract. We find many of these breeders eventually, but we do not find all. Rather than counting the deaths directly, then, we calculate how many *must* have died, given the known, stable rate of appearance of breeding vacancies, the directly observed proportions of

TABLE 5.2
Survivorship of male and female yearlings during dispersal and early breeding ages

	Age in years						
	1	2	3	4	5	6	7
Males	1.00	0.78	0.65	0.54	0.44	0.36	0.29
Females	1.00	0.65	0.51	0.42	0.34	0.28	0.23

NOTE: Synthetic survivorship figures (= calculated proportions surviving to age x), based on calculations of male and female mortalities as helpers (which differ) and as breeders (.19 for both sexes) and on observed proportions of each sex as breeders at each age class (see Woolfenden and Fitzpatrick, 1984).

each age class filling them, and the observed ratio of nonbreeders at each age class to the number of breeding vacancies (calculations described in detail in Woolfenden and Fitzpatrick, 1984). For samples of 106 yearling and 31 two-year-old helper females (1970–1982) the death rates were 43 and 23 percent, respectively. In a sample of 94 yearling and 40 two-year-old helper males, the rates—calculated in exactly the same way—were 20 and 18 percent. These different rates of mortality result in a sex ratio among older helpers that is highly biased toward males (Woolfenden and Fitzpatrick, 1984).

We attribute the substantial difference in yearling death rates to their different dispersal strategies. Females begin to leave their natal territory regularly at age fourteen months, sometimes disappearing for days on end (though usually returning each night). Conservatively calculated from data in Woolfenden and Fitzpatrick (1984) we estimate that older female helpers spend at least fourfold, and perhaps as much as tenfold, more time away from their territories than do helper males. During forays, females travel farther away than do males. Little doubt exists that these wanderings expose nonbreeding females to a variety of dangers that are less threatening when these jays are on familiar ground among familiar jays. We suspect that increased rates of predation account for most of the added mortality, as jays away from home are not familiar with the best local hiding places and evening roost sites, and they frequently spend much of their time around noisy, unsettled areas easily detected by potential predators.

Table 5.2 summarizes the differences in overall survivorship as they appear in the life tables of male and female Florida scrub jays after age one year. The cost for earlier and longer-distance dispersal is a greater risk of death before breeding, resulting in a drop in survivorship among females aged two and three years. The benefits, we can only assume, lie in their being able to breed earlier (if they survive) and farther from any potential for incentuous matings, even accidental ones, *given that their male relatives tend to breed close to home.* In the Discussion, we speculate as to why this feature of males has evolved.

Incest Avoidance

We define incest as an attempt by close relatives to breed with one another. The most common potential examples would be parent-offspring or sib-sib matings, in which the breeders share a degree of relatedness of .5, and half-sib, aunt-nephew, or uncle-niece matings, each representing degrees of relatedness of .25. Incestuous matings such as these are extremely rare in our Florida scrub jay population. In fifteen years we have witnessed only two matings between individuals as closely related as .25. One (aunt-nephew) was a chance occurrence in which typical dispersal and mate replacement brought together two relatives who had never lived in the same family unit. The other (mother-son) was a full-fledged case of incest that lasted only a few weeks, before the mother was killed while incubating a clutch of fertile eggs. This case is described in detail in Woolfenden and Fitzpatrick (1984).

Estimating the number of expected cases of incest if pairing were random with regard to relatedness is extremely difficult. We shall not attempt such an analysis here. Instead, we present the following observations as evidence that Florida scrub jays actively avoid incestuous matings: 1) Despite the regular occurrence of opposite-sexed siblings of breeding age (helpers < age 2 years) in the same territory during a breeding season ($n = 22$, 1971–1983), never have we seen siblings even seriously court one another, much less attempt to pair and breed. Opportunities for siblings to pair have included several cases where both parents died in the same year, leaving the territory to their offspring. In two cases both male and female remained and became breeders, but each in its own half of the natal territory paired with unrelated jays from outside the group. 2) We have witnessed thirty-six cases in which an older helper was present in its natal territory at the time that its natural parent of the opposite sex was widowed; in only one case, the mother-son pairing discussed above, did pairing occur between the widow and offspring. At least six times, pairing has occurred under similar circumstances when the surviving breeder was a stepparent of the helper rather than its natural parent. 3) Mate replacement in an adjoining territory inhabited by a widowed close relative (accidental incest) has occurred only once (aunt-nephew case mentioned above), and never between full or half-siblings despite many opportunities for such a mating.

DISCUSSION

Symmetries and Asymmetries

In demographic terms, the breeding regime that Florida scrub jay helpers strive to join is sexually symmetrical, because monogamy combined with identical death rates of breeders provide equal room for males and females as breeders. Furthermore, breeding opportunities are chronically limited within the rare and patchy habitat of this population. Breeder death rates are low, and juveniles of both sexes survive to breeding age at a rate that exceeds their potential for recruitment as breeders.

In behavioral terms, sexual asymmetries do exist among breeders. Division of labor is well developed, especially during the nesting season. Females tend the nest while males provide most of the food, defend the territory more vigorously, and spend more time on exposed perches as sentinels. At all times males are dominant over females and they take the more aggressive role in territorial defense.

Both demographic and behavioral asymmetries exist between the sexes as helpers. Compared to females, male Florida scrub jays help more, actively; they deliver more food to the nest; they defend the territory more vigorously; they help for more years; they leave the home territory less frequently as helpers; they have lower mortality during this stage; they disperse over shorter distances; and they are far more likely to inherit some or all of their natal territory when they become breeders. These differences arise during a period of their lives that is sandwiched between the demographically symmetrical stages of juvenile and breeder. Below, we discuss some hypotheses for the origins of these asymmetries among helpers. We begin with a discussion of the behavioral asymmetries between breeders, however, because these may underlie the whole story.

Division of Labor and Male Dominance

It appears that virtually all corvids are monogamous (Goodwin, 1976) despite having a range of social systems varying from solitary pairs to highly cooperative social units (e.g., Verbeek, 1973; J. L. Brown, 1974). Indeed, we still have little reason to question Lack's (1968) assertion that almost all *birds* are monogamous. As emphasized by Emlen and Oring (1977), shared parental care is a major aspect of avian monogamy, and presumably represents a selective agent enforcing it. Shared parental care often is accompanied by division of labor between the sexes, probably for reasons of increased efficiency through specialization.

We emphasize that the term "division of labor" implies that members of the pair have shared goals, and that the activities required to achieve these goals are divided between them. Paramount among the goals of paired Florida scrub jays is the production of successful offspring in a difficult and crowded environment, in which combined parental effort clearly is essential. We suspect that division of labor is most common among species in which the monogamous pair bond is a permanent one, especially if it lasts many breeding seasons. This condition favors long-term investments in the mate as well as the offspring. Division of labor is probably most prevalent in situations where permanent bonded pairs live on permanent, all-purpose territories that require vigilance and strong defense the year round.

As the egg producer, a paired female incurs a substantial energy drain during the reproductive period (see, e.g., Ricklefs, 1974). Garnering of energy by females may be doubly important in populations, such as the Florida scrub jay, where nest predation is frequent and replacement clutches must be laid soon thereafter. Oviparity, and the associated need to conserve energy, therefore introduces a major asymmetry into the male-female relationship in birds. Given that division of labor in some form is advantageous, then the female should assume the role of nest attendant. This role allows her to acquire energy by receiving food from her mate, while conserving energy through reduced participation in both food-gathering and territory defense. Reduced activity during egg formation also reduces the risk of injury to the female or the developing egg. It follows that the males perform the energetically demanding tasks while females perform the equally necessary tasks of incubation, brooding, shading, and guarding the nest. Indeed, such is the pattern by which nesting labor is divided in virtually all monogamous birds that defend all-purpose territories.

We suggest that male dominance can arise as a side effect of this division of labor. Resource defense, especially across a large territory, and possibly guarding against cuckoldry, presumably select for increased size, strength, and aggressiveness. These features will be favored in either sex that takes over the role as primary defender. Dominance by the stronger sex is a natural, if quite incidental, result.

Asymmetries in Helping Behavior

In an excellent recent paper, Koenig et al. (1983) list six possible conditions that could explain sexual differences in the amount and kind of help provided by helpers. Briefly stated, asymmetries could arise under any of the following conditions: 1) one sex is more likely to inherit all or part of its helping territory (Stallcup and Woolfenden, 1978; Woolfenden and Fitzpatrick, 1978); 2) one sex is more likely to breed in an adjacent territory, near its close relatives (Greenwood and Harvey, 1976; Woolfenden and Fitzpatrick, 1984); 3) one sex is more likely to disperse with siblings as a cooperative unit (Ligon and Ligon, 1978); 4) one sex is more likely to die before breeding and therefore will profit more from the indirect component of fitness by raising extra sibs (A. B. Clark in Koenig et al., 1983); 5) one sex is more likely to profit more as a breeder from the experience gained as a helper (Lancaster, 1971; Hrdy, 1977); 6) division of labor among the helpers results in equal investment by each sex but involving different roles. To these we add another hypothesis: 7) different behavior patterns among helpers represent genetic predispositions toward different activities, reflecting the different roles of the sexes as breeders.

Interestingly, Koenig et al. (1983) found that gross asymmetries in amount and nature of helping behavior performed by the two sexes ran in precisely the reverse direction from the Florida scrub jay. Female acorn woodpeckers are the more active helpers, for reasons still unclear to those authors and to us. They suggest that a "complex of ecological and genetic factors" is responsible, including conditions 4, 5, and 6 listed above. Unfortunately, adequate data on helper behavior, separated by sex, are scarce for other species. In general, males tend to be the more "helpful" sex, and in many species they are the only sex to remain regularly as helpers (reviewed by J. L. Brown, 1978).

Conditions 1 and 2 seem most applicable to the Florida scrub jay case. As we have suggested (Woolfenden and Fizpatrick, 1978), male helpers are more likely to inherit the home territory or to breed adjacent to it, and therefore stand to gain more from helping raise extra sibs than do females. However, we also stress that the difference in performance by helpers exactly matches the different behavior patterns that will typify them later as breeders. At the very least, this complicates any attempt to interpret behavioral differences among helpers as different adaptive strategies of the two sexes. As helpers and as breeders, males are more aggressive territory defenders and more active providers at the nest, while females seem intent upon watching or brooding over the nest, although they do bring in substantial amounts of food as well. We see no plausible way to test the idea at present, but we entertain the possibility that the different amounts and styles of help between the sexes are little more than byproducts of extreme differences in nesting behavior characterizing male and female breeders. Viewed in this way, the helping behavior is selectively rather neutral from the helpers' standpoint, even more so than the differential need for "practice" implied by condition 5 above. We doubt that the slight difference between body sizes of males and females selects for their different amounts and styles of helping behavior, but we cannot rule out this possibility at present.

Dispersal Asymmetries

As emphasized by Greenwood (1980) and others, sex biases in dispersal are the rule among vertebrates. Among birds, with few exceptions, females disperse farther and in many cases earlier than do males. Therefore, any explanation we put forward here regarding the Florida scrub jay case must be viewed in the larger context of birds in general. Indeed, we agree with R. Mumme (pers. comm.) that female-biased dispersal probably is a primitive condition among birds, and that once established it may be difficult to reverse. Reversal is inhibited because the only intermediate conditions are that both sexes become philopatric (leading to close inbreeding) or that both become long-distance dispersers. The latter condition would require the formerly philopatric sex to forfeit the advantages to philopatry in favor of riskier dispersal to less familiar surroundings. We find such a scenario implausible. In short, philopatry is advantageous, but only one sex can enjoy it

(Shields, 1982). Assuming, then, that one sex or the other must be a long-distance disperser (to avoid incest), social systems may be forced to evolve around this primitive condition instead of ever evolving away from it.

We agree with Shields (1982) that incest can be genetically disadvantageous while inbreeding may not be. We define incest as breeding between members of an immediate family, where mates share a degree of relatedness on the order of .25 or greater. The distinction between this phenomenon and elevated levels of inbreeding at the population level is important. Moore and Ali (1984) argue that general inbreeding-avoidance is not sufficient as an explanation for sex-biased dispersal, but they do not directly address the question of incest. Indeed, they do state (p. 108) that "dispersal in some species probably does function to avoid inbreeding." Among many cooperative-breeding species, including the Florida scrub jay, inheritance of breeding space is frequent (Selander, 1964; J. L. Brown, 1974; Woolfenden and Fitzpatrick, 1978). This leads to extreme philopatry, which often is sex-biased for reasons outlined above. In such cases, we submit that sex-biased dispersal among otherwise extremely sedentary individuals *can* result from selection to avoid incest (also Zahavi, 1974; Greenwood and Harvey, 1976; Koenig et al., 1983). Florida scrub jays are extremely sedentary. Most females disperse only a few territories away (Fig. 5.1), so that effective population sizes are quite small. We have shown (Woolfenden and Fitzpatrick, 1984) that this does lead to slightly elevated levels of populational inbreeding. However, sex-biased dispersal, combined with absolute avoidance of pairing within the functional family unit, causes incest to be extremely rare.

Sex-biased dispersal may confer advantages, but why are females the dispersers? Four hypotheses have been advanced: 1) genetic asymmetry between the sexes (Whitney, 1976), 2) female–female competition (Koenig et al., 1983), 3) resource defense by males (or females, in mammals; Greenwood and Harvey, 1976; Greenwood, 1980), and 4) male dominance, causing females to disperse (Gauthreaux, 1978).

We concur with Greenwood (1980) and others that the bearer of the homogametic versus heterogametic sex chromosomes has little bearing on the issue because ecological conditions easily could reverse a predisposition toward philopatry stemming only from being homogametic. A good case is made for female–female competition being a factor in acorn woodpeckers (Koenig et al., 1983), but this is predicated on males joining their parents or sibs as breeders in a peculiar, polygamous mating system. Florida scrub jays are strongly monogamous, with both sexes having equal opportunities to become breeders. The same is true for most other known cooperative-breeders, including all corvids (and most woodpeckers), hence female–female competition is unlikely to be generally applicable as an explanation for sex-biased dispersal systems.

Greenwood (1980) proposes that sex biases in dispersal result from the type of mating system characterizing birds and mammals. In birds, males tend to defend resources in a fixed territory, which selects for philopatry among males and (by default) more extensive dispersal by females, both to locate good territories and to avoid close inbreeding. Among mammals, which are generally polygynous, females are more tied to home ranges while males are free to disperse more widely to garner additional mates. Florida scrub jays—and most other monogamous, territorial birds—seem to fit Greenwood's avian model. Indeed, male Florida scrub jays spend one to three years helping before breeding on their own. Females, which disperse earlier, are less active defenders. However, Greenwood leaves unanswered a fundamental question: why are *males* the resource defenders rather than females? We suggest that the explanation is related to Gauthreaux's (1978) hypothesis, discounted by Greenwood (1980, p. 1149) but in our view closely connected to Greenwood's own model.

Gauthreaux argues that female-biased dispersal in birds is mediated through behavioral dominance by males over them, allowing males to win nearby or favored territories and forcing females to disperse farther to avoid inbreeding or locate suitable resources. Like Greenwood, Gauthreaux leaves unanswered the question of how males become the dominant sex in most birds. Furthermore, as argued by Koenig et al. (1983), "behavioral dominance is likely to shift, depending on ecological factors affecting the sexes," yet female-biased dispersal remains the rule. However, as we have argued above, at least in the case of monogamous species that defend territories males will generally become the more aggressive sex, and often larger, as a result of female oviparity combined with selection favoring division of labor between the sexes. Intersexual dominance follows as a secondary result of these common features, and might even serve as a proximate factor enforcing female-biased dispersal in some species. (We have no evidence that males actively expel females from the natal territory in Florida scrub jays, so it appears not to be such a proximate factor in this case.)

We suspect that both male resource defense *and* male dominance arise together in monogamous, territorial birds, and their effects go hand in hand in promoting dispersal systems that remain female-biased across most bird groups. The fundamental asymmetry, which has always been true even among the primitive birds, is extreme anisogamy. In birds this condition, combined with selection for territoriality and extensive parental care, ultimately may have selected for male philopatry as a general trait. The mediating influence in such a scenario is division of labor, leading to male resource defense and male dominance, thereby favoring males as the sex to stay near home.

6. Hornbill Social Dispersion: Variations
on a Monogamous Theme

———————————————— • ————————————————

MARK LEIGHTON

ALL HORNBILLS for which there is information mate monogamously (Kemp, 1979), but in some species a dominant, breeding pair lives in territorial, "cooperatively breeding" groups with auxillary, nonbreeding helpers-at-the nest (as defined by Emlen and Vehrencamp, 1983; Emlen, 1984). A rich body of long-term empirical studies have been instrumental in integrating behavioral, demographic, and ecological factors into a theory of cooperative breeding in birds (J. L. Brown, 1978; Emlen, 1982a, 1982b; Emlen and Vehrencamp, 1983; Gaston, 1978; Koenig and Pitelka, 1981; Ricklefs, 1975). Exemplary studies include those of J. L. Brown (1974) on Mexican jays, Woolfenden and Fitzpatrick (1978; Chapter 5) on Florida scrub jays, Koenig and Pitelka (1981) on acorn woodpeckers, and the Ligons (Ligon, 1981) on green woodhoopoes (this African woodland species is closely related to hornbills). Studies of hornbills cannot match these in relating behavioral strategies to components of inclusive fitness (e.g., Emlen and Vehrencamp, 1983; Koenig, 1981; Woolfenden and Fitzpatrick, 1984). Therefore, in this chapter I have focused on how ecological factors constrain cooperative breeding. I draw on studies of tropical rain forest hornbills in Borneo (Leighton, 1982; Leighton and Leighton, 1983) and of African savanna hornbills (Kemp, 1976a; Kemp and Kemp, 1980) to address the question of why some hornbill species typically live in pairs while others occur as cooperatively breeding groups.

Animal social dispersion, the spatial arrangement of individuals of different age and sex classes into groups of various sizes and age/sex compositions (Bradbury and Vehrencamp, 1977a, b), is proximally the result of both a) demographic processes such as age- and sex-specific mortality and fecundity (J. Cohen, 1971; Pulliam and Caraco, 1984) and b) decisions made by individuals to immigrate into or emigrate from certain groups or to permit others to enter or remain in a social group. The primary assumption of behavioral ecologists is that these decisions can be related to particular social, demographic, and ecological factors in accordance with inclusive fitness theory (Krebs and Davies, 1978; Pulliam and Caraco, 1984). If so, differences among species in group size and composition can be understood in terms of how the fecundity and mortality of individuals of different age, sex, and status and their kin would vary in groups of different size and composition (Emlen and Oring, 1977; Bertram, 1978; Wittenberger, 1980b; Wrangham, 1980).

By comparing a set of related taxa that are similar in biology but vary in patterns of social dispersion (Altmann, 1974; Clutton-Brock and Harvey, 1984; Jarman, 1974, 1982), one can generate or test hypotheses linking grouping patterns to specific demographic, behavioral, and ecological variables (Wittenberger, 1980b). Hornbills are good subjects for such a comparative analysis because sympatric hornbill species vary in whether cooperative groups occur, but are very similar in morphology and ecology. I will adopt a comparative approach to examine the relationship between foraging and group size in territorial hornbills (cf. Bradbury and Vehrencamp, 1977a, b).

Unlike theories of group living in other vertebrates, which stress the positive benefits of sociality (e.g., foraging efficiency or predator avoidance), ecological models of cooperative breeding postulate that social groups form merely because immatures lack opportunities to breed themselves and so accumulate in groups (J. L. Brown, 1978; Emlen, 1982a, b; Emlen and Vehrencamp, 1983; Gaston, 1978; Koenig, 1981; Koenig and Pitelka, 1981). Koenig (1981) analyzed how reproductive success varied with group size in sixteen species of cooperatively breeding birds and found that the modal group size usually exceeded the size(s) that fledged the greatest number of offspring. In twelve cases, per capita fitness was not greater in groups larger than a single pair. He concluded that in most species, birds were forced to live in groups larger than the optimal group size because of a shortage of territorial openings. Koenig and Pitelka (1981) proposed that cooperative groups would form when a species has specialized habitat requirements, so that there are limited opportunities for mating birds to breed and/or survive in suboptimal habitats.

Ricklefs (1975), however, has argued that juveniles may remain in their natal territories even when a gradual habitat gradient exists. Of crucial importance is the relative turnover of territorial adults and the productivity of surviving immatures that might form a pool of potential helpers. Selection could favor juveniles' staying and helping in the parental group (or joining a group of unrelated birds) (Emlen and Vehrencamp, 1983; Ligon, 1981) when faced with a gradual habitat gradient because suboptimal habitat, though plentiful, would be full due to high productivity and/or low mortality in the population.

In species in which cooperative breeding societies develop, ecological conditions must not only induce juveniles to stay, but also prevent adults from expelling the young. Although it is recognized that trade-offs are involved in such situations (Emlen and Vehrencamp, 1983), few suggestions have been offered about what the ecological constraints might be. An underlying assumption of the ecological models of cooperative breeding cited above is that juveniles either join a territorial pair group or try to breed on their own. In fact a third option is available, as demonstrated by some hornbills: juveniles can either live solitarily within the territories of pairs or form flocks in which survival but not breeding opportunity is the primary attraction (see below). Alternatively, parents may exclude immatures from staying within their territory even though the demographic conditions outlined in other models were precisely fulfilled. Hence the ecological constraint models for territorial cooperative groups developed by J. L. Brown (1978), Emlen (1982a, 1984), Gaston (1978), and Koenig and Pitelka (1981) outline some necessary conditions for nonbreeders to prefer helping rather than dispersing, but neglect to indicate the ecological conditions necessary for the resident parents to tolerate nonbreeders that remain to form cooperative breeding groups. The model of territory group size recently advanced by J. L. Brown (1982) explicitly points out the importance of identifying the relationship between resource depletion and group size in territorial species. The intended contribution of the comparative analysis of hornbills offered here is to address how cooperative breeding may be constrained by the reduced efficiency of group foraging, even though other sufficient demographic and ecological conditions hold.

I first describe relevant background on hornbill breeding biology and present the comparative baseline information on social dispersion and ecology for several hornbill species. I then proceed to consider the grouping options available to different classes of individuals. I finish by summarizing how variation in group size in Bornean hornbills might be related to fruit resources (the bulk of these data are available elsewhere; Leighton, 1982, in prep.).

BACKGROUND

Hornbills comprise an ancient family (Bucerotidae) of large-bodied birds of the Old World tropics and subtropics in the order Coraciiformes, most closely related to the hoopoes (Upupidae) and wood hoopoes (Phoeniculidae) (Feduccia, 1980). Unlike related hole-nesting coraciiforms and piciforms, all forty-five species of hornbills (except the two *Bucorvus* species) share a peculiar nesthole sealing habit (Kemp and Kemp, 1980). After entering the nesthole to begin egg-laying, the female plasters mud around the edge of the hole, leaving only a thin crack through which she and her nestlings obtain food. The female remains ensconced in this hole throughout the egg-laying and part or all of the prolonged nestling period, which lasts for one to four months (Kemp, 1971, 1979; the period is longer in larger hornbill species).

The hornbill nesthole-sealing habit predisposes the female to mate monogamously. Because the female and nestlings depend entirely on provisioning, the burden of food-gathering on the male constrains his ability to provision two nests simultaneously (as hypothesized by Wittenberger and Tilson, 1980), which is supported by the low observed hunting rates in Bornean hornbills I observed (see below and Leighton, 1982). Sequential monogamy by either sex is limited by a long period of provisioning of fledglings and, additionally, for males by the likelihood that the first-breeding female could emerge from her nesthole and disrupt subsequent nesting attempts by subordinate females within her territory.

Morphological, dietary, and grouping characteristics of the sympatric African savanna hornbill species (Kemp, 1976a; Kemp and Kemp, 1980) and seven sympatric species of Bornean tropical rain forest (Leighton, 1982; Leighton and Leighton, 1983; in prep.) are summarized in Table 6.1. For biogeographical distributions, see Sanft (1960). The phylogenetic affinities follow Kemp (1979). *Tockus* are small-bodied forms, argued by Kemp to share primitive hornbill anatomical and behavioral characteristics, and are perhaps only distantly related to the rest of the family. The other African genus, *Bucorvus* is related to the large-bodied Asian *Buceros* and *Rhinoplax* rather than to other African genera. *Rhyticeros* is most closely related to these two Asian genera. The other three Asian genera of Table 6.1 (*Berenicornis, Anorrhinus,* and *Anthracoceros*) are smaller-bodied and most closely related to one another.

Diets and Seasonal Patterns of Grouping and Spacing

The habitat of the African species differs dramatically in physical structure, seasonality, and available food resources from Bornean rain forest. Nonetheless, the species-specific dichotomy between territorial pairs and groups occurs in both habitats (Table 6.1). Pairs of *Tockus* maintain small territories (means are 10 ha for *erythrorhynchus*, 17 ha for *flavirostris*, and 63 ha for *nasutus*) around their nestholes during the breeding season, but join nonterritorial foraging flocks at other times. Yearlings may breed in the first year, but not all pairs breed (Kemp, 1976a). *Bucorvus* breed as cooperative groups that maintain year-round territories of about a hundred square kilometers. Females tend to disperse from their natal groups, and these solitary females (plus one observed group of three females) are tolerated within the group territories. Groups consist of a breeding pair without other adult females (or, less commonly, with one subordinate helper: in 21 percent of twenty-eight groups), but commonly with more than one adult male helper (two to three male helpers in 57 percent of twenty-eight groups), and one to three immatures.

Five of the Bornean species maintain year-round territories, but only against conspecifics. The two largest species occur as pairs, at times with one dependent juvenile, in relatively large territories (*Buceros* ca. 2–3 km²; *Rhinoplax* ca. 7–8 km²). The three smaller species all have territories of slightly more than one square kilometer; their relative population densities are proportional to their differences in group size (Leighton, 1982). *Anthracoceros* live in pairs with up to two juveniles, *Berenicornis* pairs are accompanied by one to three helpers in cooperative breeding groups (*n* = 3 groups), and *Anorrhinus* breeding pairs live in cooperative breeding groups with four to six helpers, composed approximately equally of males and females (*n* = 3 groups), plus up to three juveniles. *Rhyticeros* adults occur as pairs that maintain a nesthole during the breeding season (no evidence of territoriality) but otherwise join flocks of fluctuating composition and size, with pairs frequently leaving one flock and joining another in the course of a day. Juvenile and subadult *Rhyticeros* and *Buceros* occur in separate flocks from adults. Although both the African and Bornean data are from a limited area, the patterns of social dispersion shown in Table 6.1 appear to be typical for each species (Kemp, 1979; Kemp and Kemp, 1975).

Bornean hornbills combine visits to fruit trees with bouts of hunting animal prey (Table 6.1). Fruits are either figs (*Ficus* spp.) or two types of fruit (note b to Table 6.1) with lipid-rich flesh (Leighton and Leighton, 1983). Except for *Rhinoplax*, hornbill species eat all three types of food (Leighton, 1982). Prey are widely scattered small vertebrates and large arthropods. African hornbills are more carnivorous: *Tockus* eat some fruit when seasonally available, but *Bucorvus* is entirely carnivorous. *T. nastus* hunts arboreally but the other three African species hunt on the ground.

Food availability fluctuates seasonally for all species (Kemp, 1976b; Kemp and Kemp, 1980; Leighton and Leighton, 1983), and corresponds to changes in breeding, spacing, and grouping patterns. Breeding is apparently limited to periods of high food abundance. In Africa, animal prey become abundant after the onset of the wet season, when breeding by all species took place (Kemp, 1976b). In unusually dry years, fewer pairs of all species nested (Kemp, 1976b). In the Bornean study, all hornbills nested and successfully fledged young only during a five-month period of high fruit availability in the first year (Leighton, 1982; Leighton and Leighton, 1983). No species was known to nest during the second year, even though some pairs were without dependent young.

The incidence of territoriality among these species corresponds closely to diet. The nonterritorial *Rhyticeros* species do less hunting of animal prey and consume more lipid-rich, drupaceous fruits (e.g., Lauraceae, Burseraceae), which are more seasonal in availability relative to the diet of other Bornean hornbills (Leighton and Leighton, 1983). At fruit trees, *Rhyticeros* species tend to be subordinate in competitive interactions with the territorial species (Leighton, 1982). Outside the breeding season, instead of competing with the territorial species for access to the rare fruit patches, *Rhyticeros* instead are either nomadic or maintain huge home ranges aggregating where local concentrations of fruit trees occur (Leighton and Leighton, 1983). Unlike the territorial loud calls of the sedentary species, *Rhyticeros* loud calls seem to elicit assembling by other pairs or flocks, which then travel together. Flocks fly high above the canopy, often directly for five to ten kilometers, presumably commuting between roost sites (Kemp and Kemp, 1975; pers. obs.) and various fruit-rich areas of forest. This complex of flocking, calling, foraging, and diet characteristics is consistent with the hypothesis that these birds parasitize or share information about the locations of local patchy resources (Ward and Zahavi, 1973; Krebs and Davies, 1984).

Year-round territoriality, in contrast, is associated with access to stable food supplies outside of the resource-rich breeding season. Interestingly, territoriality among hornbills is linked to special foraging uses of their large, laterally-flattened bills. Although one important function of the flat side of the bill is to plaster mud around the nesthole (Kemp, 1971), the bill is also used as a lever, enabling hornbills to feed on peculiar resources. *Bucorvus* uses its bill to dig and turn over objects while hunting on the ground, especially during the food-poor dry season. Kemp and Kemp (1980) proposed that this ability allows *Bucorvus* to maintain year-round territories. By contrast, species of *Tockus* do not use the bill in this manner, and maintain only seasonal breeding territories, which they abandon during the (food-poor) dry season to forage as flocks in more mesic habitat (Kemp, 1976a). Bornean species use their bills as levers both to flake off bark, thereby exposing hidden animal prey (especially *Buceros*, pers. obs.) and to pry open the splitting husks of their preferred fruits to extract seeds covered with lipid-rich flesh (Leighton, 1982). The seeds of these fruits are specialized for dispersal by hornbills (Leighton, in prep.), and become particularly important for cooperative breeding species during fruit-poor times (Leighton and Leighton, 1983). In keeping with its "aberrant" sharp, short bill, *Rhinoplax vigil* does not visit these fruit trees, but relies almost entirely on figs for the fruit component on its diet (Table 6.1).

The territorial groups of *Anorrhinus* and *Berenicornis* were larger during fruit-rich seasons. The most closely studied of *Anorrhinus* foraged as two subgroups of fluctuating composition when fruit trees were sparse. Although the two subgroups shared and defended the same territory, agonistic interactions characterized many of the encounters between them. In *Berenicornis* the helpers were not seen to travel with the dominant breeding pair during the fruit-poor time. Shifts to smaller group size during the fruit-poor times was a general phenomenon for these species: three *Anorrhinus* and two *Berenicornis* groups simultaneously changed in the same manner.

METHODS

Studies of African *Tockus* hornbills (Kemp, 1976a) were conducted during parts of three years, and of *Bucorvus* (Kemp and Kemp, 1980) for five months with a recensus of group composition three years later. These studies concentrated on the breeding season. Research priorities were the quantitative description of food resources, predators, territorial behavior, breeding biology, and demography. Observations were directed toward sampling twelve to seventy pairs or groups for each species, rather than intensive studies of individual birds.

My studies of the Bornean hornbills were specifically designed to examine relationships between ecology and social organization. To relate diet and foraging behavior to food resources, sampling was limited to those individuals occupying a three-square-kilometer area at the Mentoko Research Camp, located in lowland evergreen rain forest within the Kutai National Park, East Kalimantan, Indonesia, during twenty-four consecutive months of study (Leighton, 1982). Portions of the territories of two to four groups of all five territorial species occurred within this site. The individuals of all territorial species could be identified from permanent variations in mottling of casque, bill, and/or orbital skin colors, or casque shape. Independent observations of group size and composition, activities, and diet were collected by censuses along a grid of about thirty kilometers of trails, and by following pairs or groups of birds. One group each of *Berenicornis* and *Anorrhinus* became habituated to human observers and could be followed closely without disrupting the birds' normal activities. Other species were followed by stealth. Over two thousand hours were spent watching fruit trees visited by hornbills to record the sequence of visitation by different species, feeding group sizes, and competitive interactions at these concentrated food sources. Descriptions of social behavior were collected opportunistically during all types of observation.

Densities, distributions, and fruiting phenologies of tree and liana species producing fruit eaten by hornbills were measured monthly within thirty 0.5 ha vegetation plots and along trails (Leighton and Leighton, 1983). Estimates of crop size (numbers of fruit ripening on a tree), and rates of fruit ripening were scored for individual plants. Fruits were collected from most of the important species, measured for metric and qualitative traits, then dissected and preserved for later chemical analysis. The variable of patch size was measured both as the crown volume (in m^3) within which ripe fruit was available on a plant and, more relevant to this study, as the grams dry wt of ripe fruit flesh available (i.e., the product of the estimated number of ripe fruit simultaneously available within a patch and the average dry wt of flesh per fruit) (Leighton, 1982).

HORNBILL SOCIETIES AND SOCIAL RELATIONSHIPS

Relationships among Adult Females

In both of the nonterritorial *Rhyticeros* species adults occurred as mated pairs, which associate in flocks of fluctuating composition outside the breeding season. Females often travel, perch, feed, and call near other mated females, although mated pairs maintain their spatial integrity within the flock. No agonistic interactions were seen among females.

In the other five Bornean hornbill species and the African *Bucorvus*, however, dominant adult females remain dispersed through territorial behavior (Kemp, 1976a; Kemp and Kemp, 1980; Leighton, 1982). Adults of both sexes gave loud calls several times a day, which often elicited or were a response to calls by neighboring pairs or groups. Protracted territorial exchanges frequently ensued in which calls were exchanged and the participating pairs or groups shadowed one another's changes in position along their common boundary (Leighton, 1982). Females were active participants with their mates in enforcing spacing, indicating that a female directly benefits by excluding other females from her range.

In the cooperatively breeding species of *Bucorvus*, *Berenicornis*, and *Anorrhinus*, only a single dominant female bred. In some *Bucorvus* groups, a second subordinate adult female was tolerated. *Anorrhinus* groups contained two to three female helpers ($n = 3$ groups). In both Bornean species the breeding female of the group, but not the male(s), vigorously chased away extragroup, "strange" females. In *Berenicornis*, this was observed three times during the fruit-poor period when groups had been reduced to pairs. Once during the same period, when the breeding *Anorrhinus* female was traveling with two adult males and a juvenile female, she chased, grappled with, and repeatedly stabbed a strange juvenile female that had been accompanying the two subordinate adult females of the group. I mistakenly disrupted the interaction prematurely and the juvenile managed to escape.

Relationships among Males

Relationships among males parallel those seen among females in the Bornean species. In *Rhyticeros* flocks, males associate with their mates but tolerate other males.

Territory-holding males of the other species participated in calling displays with their mates, and with other group members in the cooperatively breeding birds. *Bucorvus* groups are male-biased: because solitary females (but not males) are common (Kemp and Kemp, 1980), it is likely that only females disperse from their natal groups. Adult males have amiable relationships with one another. Similarly, *Anorrhinus* males showed little intrasexual aggression and cooperated in calling and preening. A solitary subadult male sometimes joined the adult male of one *Anthracoceros* pair in giving territorial calls and was tolerated completely by the adult: it is not known if the subadult was the resident's son.

Relationships between Adult Males and Females

Adults may be lifelong mates in the territorial species. No changes in adult composition in any of the territorial pairs or groups of any of the species were noted over the two-year Bornean study. Adults suffer higher mortality among the African *Tockus* species, and some took new mates in successive years.

Relationships among individuals of the *Anorrhinus* groups were complex. During most of the study, males and females freely intermingled with one another at perches where they preened, regurgitated seeds, and gave territorial calls. However, when the main study group foraged as two independent subgroups, their compositions changed several times. Though these subgroups were at times mutually hostile, individuals within each group gave territorial displays and thus appeared to participate in common defense. During one period of several days, the three adult females traveled separately from the three adult males plus a juvenile female. Later, the dominant pair traveled with a second adult male and a juvenile female, and repeatedly chased and displaced the birds of the other subgroup from fruit trees.

In *Rhyticeros undulatus* and *Buceros*, flocks of (mostly) immatures ranged over large areas (probably hundreds of square kilometers, judging from the rapidity and infrequency with which they passed through the study site). Courtship feeding (described in Kemp, 1976a) within these flocks is common. Because some of the flock's subadults associate closely in male-female pairs, it is likely that pair formation takes place here.

Relationships between Adults and Immatures

Parental care is extended in hornbills. Juveniles in all Bornean species continued to receive food in response to their begging calls for at least six months after fledging (e.g., *Anorrhinus, Anthracoceros, Berenicornis*, and *Buceros*). For the first several months, birds acquire all food from parents and helpers. All older birds in the focal groups of the two Bornean cooperative breeders fed the fledgling(s). In *Anthracoceros*, each parent traveled separately with and procured food for one of the two fledglings: all four birds traveled or perched together only occasionally.

Dispersal of immatures takes different forms. *Tockus* juveniles mature rapidly and may breed the following year (Kemp, 1976a). In cooperative breeders only one sex typically disperses. As noted above, *Bucorvus* females but rarely males disperse from their natal groups. In the single known case for *Anorrhinus*, the male left his natal group at five months after fledging, while the female remained within the group for at least twelve more months (when the study ended). *Rhyticeros* juveniles travel with their parents for at least six months after fledging, then join juvenile flocks.

In one case each of *Buceros* and *Anthracoceros*, a subadult was sighted repeatedly over periods of several days living within the range of a territorial pair. The *Buceros* subadult may have been the same individual seen traveling with this pair (presumably its parents) when the study was initiated. Presumably it left when a flock of juveniles temporarily used this area. In the *Anthracoceros* case, a subadult male was consistently seen close to the territorial pair over an eighteen-month period. Thus, at least some juveniles may continue to range within their natal territories, though seldom traveling with their parents. However, nearly all *Buceros* subadults and nonterritorial adults are in flocks that range over large areas through the territories held by adult pairs. When these flocks entered the study site, territory owners made no attempt to repel them—perhaps an impossible task, given that these flocks traveled and foraged as very loose aggregations. Subgroups were typically separated by hundreds of meters, but nonetheless maintained contact by frequent, loud, single-note calls. Flocks of immature *Anthracoceros* and *Rhinoplax* may also form, but these flocks are small and uncommon (Kemp and Kemp, 1975; pers. obs.).

To summarize, in territorial species a single pair breeds. Females either prevent other adult females from entering the territory or such females are tolerated as helpers. Immatures are either retained in the group, or disperse and live alone (in some cases tolerated within the territories maintained by adults), or join juvenile flocks.

ECOLOGICAL PRESSURES AND PATTERNS OF SOCIALITY

Grouping Options for Adult Males, Adult Females, and Immatures

A consideration of the options regarding grouping for male and female hornbills indicates common solutions for maximizing fitness, all of which are based ultimately on the nest-sealing habit (Kemp, 1971) and its enforced monogamy. Female confinement may limit nestling production indirectly, because her mate must feed her and her offspring. Clutch size usually exceeds fledgling number, with the supernumerary eggs either not being incubated or their nestlings dying from sibling competition or parental neglect: presumably they are produced to hedge the bets of the female (Kemp, 1971, 1976a). The larger hornbills (> 1.5 kg) invariably have a single fledgling; smaller-bodied Bornean species (*Anorrhinus* and *Anthracoceros*) fledge two young (pers. obs.); and *Tockus* species fledge three to five young. The advantage of nesthole-sealing has been postulated to derive from low rates of female and nestling predation (Kemp, 1971). Accordingly, Kemp (1971, 1976a) found that 90–91 percent of twenty-two to seventy-three nests in each *Tockus* species produced fledglings (and only a small percentage of the failures could be ascribed to nest predation). By contrast, 29–42 percent of the nests of those similar-sized, hole-nesting but not sealing sympatric (not hornbill) species failed. Recently, J. Kalina (pers. comm.) observed frequent intraspecific nest disruption in a dense population of *Bycanistes subcylindricus*, a West African forest hornbill. Pairs without nestholes evicted nesting females and their eggs or nestlings, usurping the nesthole; females were particularly aggressive. In this species, then, the significant risk to nest failure may come from conspecifics. The nest-sealing habit does allow extended periods for nestling growth before fledging, but it constrains other parameters of the reproductive rate for males and females, including the number of fledglings per clutch and the rate of clutch production.

Breeders Benefit from Helpers

Potential routes for enhancing female reproductive success are: 1) to increase her own survivorship, via lower maternal investment per clutch and higher residual reproductive value (Cody, 1966); 2) to increase offspring survivorship or quality; and 3) to produce more frequent clutches. Fitness could be enhanced in all three ways by the addition of nonbreeding helpers that assist in provisioning the female and her offspring. There is evidence that all three processes operate. Nonbreeders have been observed to feed the nesting female and nestlings in *Bucorvus* (Kemp and Kemp, 1980) and *Anorrhinus* (Madge, 1969); in both these species and *Berenicornis* (pers. obs.), helpers fed fledglings. Female survivorship may be improved also because her burden of parenting is reduced (Ricklefs, 1975). Offspring survivorship can be improved by the higher feeding rates provided to nestlings and fledglings that helpers could provide (J. L. Brown, 1978; Woolfenden and Fitzpatrick, 1978). Fledgling success was relatively high (range of 60–84 percent) in the three *Tockus* species, but the major source of mortality was starvation (Kemp, 1976a,b). More generally, the restriction of breeding to the food-rich seasons in all species suggests that food limits reproductive rate. Madge (1969) noted that a group of *Anorrhinus* was able to raise two sets of fledglings in quick succession apparently because helpers did most of the feeding. One of my three *Anorrhinus* groups did likewise. No other hornbills attempted a second clutch. Because Bornean hornbills restrict breeding to relatively infrequent, apparently supra-annual fruiting seasons, double-clutching can significantly augment the fitness of breeding pairs.

The benefits of helpers similarly accrue to the breeding male as well as to the female. By reducing his burden of parental care, the breeding male may thereby enhance his own condition (Ricklefs, 1975) and his future reproductive success.

Besides the presumed advantages of helpers in feeding offspring and females, helpers participate fully in territorial calling and encounters with neighboring groups. That the maintenance of a territory is an important activity is suggested by the high rate at which Bornean species interact with the neighboring groups (several times a day) (e.g., see J. L. Brown, 1982) and the zeal with which they respond to the calls of neighbors (Leighton, pers. obs.). In Bornean species, it is likely that territories protect food resources, and, in particular, fruit trees, rather than nest sites. Because territories are very large (1–8 km²) nestholes would not seem to be in limited supply. The nestholes of different Bornean hornbills occurred in large trees, apparently in similar sites (Madge, 1969; pers. obs.): Kemp noted that *Tockus* species exchanged nestholes from year to year. Several pairs of non-territorial *Rhyticeros* species nested successfully within the study site. Because these species are behaviorally subordinate to all territorial species except *Anthracoceros* (Leighton, 1982), it seems unlikely that territory sizes were based on a mosaic of a limited supply of available nestholes.

In contrast, encounters between territorial groups or pairs most often occur at or near important fruit trees near borders. Competition is intense at these trees because a) hornbill species (except *Rhinoplax*) prefer the same kinds of fruits (Leighton, 1982, in prep.), and b) these preferred fruit trees are rare during most periods. Thus, a given tree is often visited by several species (Leighton, 1982) and also by conspecific neighboring groups.

Two advantages of foraging in large groups have been revealed from observations of *Anorrhinus*. First, intraspecific dominance at disputed fruit trees was related to group size. The focal group, comprised of seven adults and subadults was driven from fruit trees near their territorial border whenever the "West" group of ten claimed a particular tree, but the focal group in turn could displace the "South" group of six birds. Therefore. one advantage to large groups may be in obtaining priority of access to preferred food patches (as proposed for primates by Wrangham [1980] and for stingless bees by Johnson and Hubbell [1975]). These intraspecific hornbill interactions apply only to fruit patches near borders. Groups rarely penetrated deep into their neighbors' territories and upon discovery withdrew.

The second advantage applies to relatively small species in interspecific competition and is probably less important. *Anorrhinus* was able to improve its position in the interspecific dominance hierarchy for access to commonly used fruit trees by group foraging (Leighton, 1982, in prep.). Synchronized aggressive attacks by two or three birds successfully drove the much larger-bodied *Rhyticeros* and *Berenicornis* and similar-sized *Anthracoceros* individuals from fruit trees.

In summary, breeding pairs of territorial hornbills are hypothesized to benefit from helpers in two important ways. First, the nesting habit of the female and the long period of fledgling dependence creates pressure on a breeding pair that can be alleviated by helpers. Second, holding a territory is necessary for breeding, and the quality of the territory is likely to be linked (via female and offspring survivorship) to maintaining access to preferred fruit trees (i.e., the benefit of reducing resource uncertainty within the territory, as proposed by Gaston, 1978). I would argue that these benefits apply plausibly to all breeding pairs of the territorial Bornean hornbill species.

One cost of living with reproductively mature helpers is the possibility of cuckoldry for a male, and for the females, of sharing paternal investment with a second breeding female. This problem has been solved to some extent by the dominant, breeding male green woodhoopoe through mate guarding (Ligon, 1981). Similarly, the breeding pair of *Anorrhinus* traveled separately from the rest of the group prior to nesting. There are no observations of subordinate females attempting to breed in cooperative hornbill groups, such as occurs in Mexican jays (J. L. Brown, 1974). This apparently favorable option for males and for subordinate females may be uncommon or absent if, for instance, the dominant female can disrupt such attempts (perhaps by later attacking the fledglings after she emerges from her nesthole).

Lack of helpers in some territorial hornbill species must then be due to differences in the dispersal strategies of juveniles or in limits to group size.

Dispersal Strategies of Immatures

The optimal dispersal strategy of hornbills as they approach reproductive maturity should depend on the combined probabilities of 1) survivorship, 2) acquiring a territory and successfully breeding there, and 3) enhancing the reproductive success of close relatives, thereby enhancing their own fitness indirectly (J. L. Brown, 1978: Emlen, 1984). Whether survivorship is higher by remaining within the natal territory or by dispersing, and, further, by traveling and foraging alone or in association with others (either with the parents or in a flock) depends on the mortality risks such birds face. There would seem to be no a priori rationale for endorsing either case for hornbills. Staying within the natal territory offers the potential advantages of 1) intimate knowledge of a home range and hence more efficient access to food resources and safe roosting sites and 2) more tolerant competitors for food (parents and siblings). Even so, resources may be more abundant elsewhere. Foraging in subadult flocks might enhance survivorship for some hornbills over that experienced by solitary dispersers. Flocks of subadult *Buceros* freely forage within territories of breeding pairs, who do not attempt to repel them. In this way they have access to large areas of forest and can take advantage of local superabundances of fruit, which can periodically be limited to a single territory.

Kemp (1976a) has shown that the small-bodied adult *Tockus* suffers substantial predation from raptors, so that there is relatively high turnover of territories. Also, because territories are not maintained year-round, first-year birds may fare relatively well in competition for territories; Kemp (1976a) showed that territories and nestholes were not limiting in the populations he studied. Under these conditions, the juvenile option of searching for available territories becomes more attractive than the helping alternative.

In the larger-bodied territorial Bornean species and *Bucorvus*, the options for immatures are more limited. In all species, the habitat is saturated with contiguous territories occupied by breeding adults (whose survivorship is high) plus a large pool of nonbreeders (also with high survivorship). Under such conditions, the option of remaining within the parental territory is attractive relative to the option of dispersing (J. L. Brown 1974, 1978; Emlen, 1982b; Ricklefs, 1975). First, eventual success in competing might be predicated upon maximizing survivorship during a prolonged preadult tenure in the natal group. A refuge is provided from which nonbreeding offspring can make periodic forays to search for territories, but within which survivorship might be high (J. L. Brown, 1974; Woolfenden and Fitzpatrick, 1978). Perhaps more importantly, the relative inclusive fitness payoff from helping to raise siblings and enhancing maternal survivorship (upon which access to the refuge of the parental territory may depend) is also high (J. L. Brown, 1978; Gaston, 1978; Emlen and Vehrencamp, 1983).

The method by which birds compete for territories is important to consider in analyzing patterns of dispersal. If a cooperating pair is both necessary and sufficient to challenge successfully for territorial occupancy, then juveniles may fare better by joining flocks and seeking a mate than by staying in their natal group and helping. This hypothetical strategy should be seen, however, as a consequence of the prior condition that territory-holders are pairs. If instead territories were held by cooperative groups, then coalitions may be required for successful usurpation (e.g., Koenig and Pitelka, 1981; Ligon, 1981).

Constraints on Group Size in Bornean Hornbills

From considering the benefits to the breeding pair of having helpers and to immatures of being helpers (or at least of remaining within the parental territory), we would expect all territorial Bornean hornbills to feature cooperative breeding groups, yet they do not. Two hypotheses, both unrelated to ecological explanations, can be dismissed. First, if extraterritorial birds were unavailable to form a pool of helpers, either because the balance of age-specific mortality and fecundity rates favored immatures competing for territories, or because the habitat was not saturated with territories, then all subadults and adults would breed. This cannot be the case with the Bornean species, since nonbreeding, nonterritorial adults are common in *Anthracoceros* and *Buceros*. Potential helpers in these species occur in flocks or as solitary birds.

Second, there do not appear to be phylogenetic constraints to evolving either a) tolerance of older juveniles or subadults by territorial pairs or b) helping behavior among nonbreeding birds in which cooperative groups do not occur. While collecting hornbill skins in Borneo, Hose (in Shelford, 1899) unwittingly performed an important experiment. He shot the male *Buceros* provisioning a female and nestlings. Surprisingly, other birds began to provision the trapped birds, indicating that helping by nonbreeding birds occurs in this species. The likely candidates for these helpers are the solitary juveniles/subadults tolerated within the adult territories. Under appropriate ecological conditions, then, it is sometimes possible for cooperative groups to form in territorial species normally lacking helpers.

Alternative explanations for the lack of helpers and for limits on group size must involve ecological factors. Several sorts of ecological constraints can be dismissed as highly implausible. There is no evidence to suggest that Bornean territorial species are differentially susceptible to predator or parasite risk or that such risks vary directly with group size. No predatory attacks were observed, and the fauna appears to lack potential diurnal predators on hornbills. No behavioral reactions were witnessed that could be construed as antipredatory. In fact, the activity during which birds seem most susceptible to predation is the hunting of animal prey, because hunting often occurs in foliage dense enough to obscure predator attack. Yet these birds hunt independently in widely dispersed pairs or groups, and are thus poorly positioned to benefit from mutual vigilance and alarm cries. Although hornbills may be susceptible to nocturnal predators (small felids, civets) while roosting, the relative advantages of either clustering or spacing out to elude nocturnal predators cannot indirectly restrict diurnal group size, since the birds could merely reassociate in the morning. Finally, there appears to be no relationship between vulnerability to predators or parasites and group size: two species of essentially identical body mass and ecology, *Anorrhinus* and *Anthracoceros*, are at opposite extremes of the range of group sizes.

Hypotheses Invoking a Foraging Constraint on Group Size

An alternative explanation for the differences in group size among the territorial Bornean species would be provided if species vary in diet, and if the different food resources affect the efficiencies of foraging in different-sized groups by influencing the ease with which food is found, consumed, or defended. This is the classic food-based explanation for the comparative analysis of social organization (Jarman, 1974; Bradbury and Vehrencamp, 1977a, b; Krebs and Davies, 1978). A priori this approach seems plausible because fitness in hornbills may be strongly tied to feeding success, which underlies territoriality.

Examination of hornbill hunting behavior suggests that if diets vary

TABLE 6.1

Body mass, diet, and adult dispersion of the contrasted hornbill species

Species	Body mass[a] (g)	Diet[b]	Adult dispersion Breeding season	Adult dispersion Nonbreeding season	Dispersion of nonbreeding adults and subadults
Bornean rain forest					
Anorrhinus galeritus	900	CLR > DLR > figs ≫ animal	Terr. groups (7–10)	same (3–5)	terr. groups
Berenicornis comatus	1,700	‚‚	Terr. groups (4–5)	same (2)	terr. groups/solitary
Anthracoceros malayanus	1,000	‚‚	Terr. pairs	same	solitary/(flocks also?)
Buceros rhinoceros	2,600	‚‚	Terr. pairs	same	flocks/(solitary also?)
Rhinoplax vigil	3,100	Figs > animal	Terr. pairs	same	‚‚
Rhyticeros corrugatus	1,600	DLR > CLR > figs ≫ animal	Nonterr. pairs	flocks (2–22)	flocks (of immatures)
R. undulatus	2,500	‚‚	Nonterr. pairs	flocks (2–14)	‚‚
South African savanna					
Tockus erythrorhynchus	200	Animal > fruit	Terr. pairs	flocks	?
T. flavirostris	200	‚‚	‚‚	‚‚	?
T. nasutus	200	‚‚	‚‚	‚‚	?
Bucorvus leadbeateri	4,000	Animal	Terr. groups (3–8)	same	male in terr. groups/ females solitary

From Kemp, 1979 (there is little or no sexual dimorphism).
Food types are arranged in order of estimated relative importance by dry weight. Types are: CLR = capsular fruits (i.e., husks split open when ripe) with lipid-rich flesh; DLR = huskless fruits (mostly drupes) with lipid-rich flesh; Figs = *Ficus* fruit; Animal = small vertebrates or large arthropods.

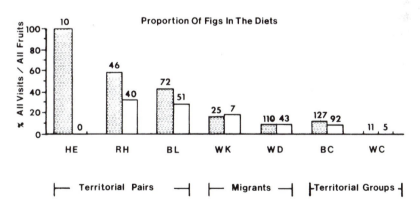

Fig. 6.1. Percentage of visits to figs (= fruiting *Ficus* plants) out of total visits to all fruit patches. Textured bars are from spot observations of fruit-eating recorded while walking census trails. Open bars are from following pairs or groups of birds. *N*'s are numbers of observed independent visits by feeding groups to fruit patches. Species codes are: HE = *Rhinoplax*, RH = *Buceros*, BL = *Anthracoceros*, WK = *Rhyticeros corrugatus*, WD = *R. undulatus*, BC = *Anorrhinus*, and WC = *Berenicornis* (see Table 6.1).

123

among hornbills such that foraging efficiency changes with group size, then the component of the diet responsible for this must be fruit rather than animal food. All five territorial species are omnivorous. Data from activities at first contact were used to calculate relative time spent hunting as a percentage of total time hunting plus fruit-eating. *Buceros* spent 50 percent of their feeding time hunting ($n = 68$ total feeding first contacts), *Rhinoplax* 50 percent ($n = 14$), *Berenicornis* 64 percent ($n = 17$), *Anorrhinus* 24 percent ($n = 104$), and *Anthracoceros* 25 percent ($n = 72$) (Leighton, 1982). *Buceros* spent significantly more time hunting than did *Anthracoceros* or *Anorrhinus* ($p < .01$; sample sizes for the other two species are too small for comparison). Note that hunting time per se could not explain grouping difference because these percentages bear no relationship to group size: *Anthracoceros* live in pairs, but *Anorrhinus* occur in large groups, yet the two species hunt similar amounts of time. All four species (*Rhinoplax* being too infrequently observed) catch prey at very low rates (e.g., 38 prey items per approximately 700 bird-hours of observation in *Anorrhinus*). By contrast, fruit is relatively rapidly harvested within trees and makes a much greater contribution to the diet by weight (Leighton, 1982). The regularity and persistence of hunting behavior, despite its low rate of return, may be due to the complementarity of the nutrients provided by animal prey versus fruit flesh, that is, protein versus energy (Leighton, 1982).

Differences in the Use of Fruit Resources

Differences in use of fruit resources could explain variation in group size if, for instance, hornbills that forage in pairs utilize small patches of fruit (Jarman, 1974; Bradbury and Vehrencamp, 1976a, b; Leighton and Leighton, 1981). To examine this possibility, frugivory by hornbills was compared to determine a) if differences in fruit preferences exist (as measured by a selectivity index that compares relative frequencies of visitation and availability of fruit species) and b) if the relative importance (by time spent feeding) of fruits varied. The results are merely summarized here (for details see Leighton, 1982).

First, hornbills show extraordinarily high overlap in their preferences for fruits. All but one of the seven species select a combination of large-seeded fruits with lipid-rich flesh and figs (Table 6.1). *Rhinoplax* fed on only figs, but no strong differences could be seen among the other four territorial species in their use of the lipid-rich fruits. The species show similar preferences among the various fruits, and feed in the same places and manner within the crown (Leighton, 1982).

Second, they generally preferred the lipid-rich fruits over figs, but varied in their relative use of figs. This seems to provide the sole correlation between group size and diet: there is an inverse relationship between fig use and group size (Fig. 6.1). This relationship could be explanatory if fig

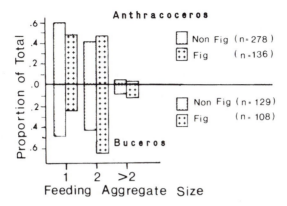

Fig. 6.2. Relative proportion of all observations of *Buceros* and *Anthracoceros* feeding in either fig or in nonfig fruit patches in which the number of simultaneously feeding birds was one (either the male or female), two (the pair), or three (plus a juvenile). *N*'s are independent observations from censuses and follows of birds combined.

patches are smaller (i.e., provide fewer fruits) than are the patches of lipid-rich fruits. Then hornbills that rely more heavily on figs might be constrained to feed in smaller groups because feeding competition within groups would be more intense, exerting pressure toward a reduction in group size.

Few arguments of this form go beyond such demonstrated associations as seen in Fig. 6.1 (Jarman, 1974; Krebs and Davies, 1978; Wrangham, 1980) to provide measures of the resource variables (Leighton and Leighton, 1981). In this case, however, the evidence indicates that fig patches are in fact larger than patches of lipid-rich fruits. First, individual fig plants ripen their fruit synchronously (Leighton, pers. obs.), and are renowned for the large patches of fruit they provide (Janzen, 1979). This generalization was confirmed by comparing direct estimates of fruit availability within the two types of patches (Leighton, 1982, in prep.).

Second, we can test this hypothesis directly by examining if the pairs of *Buceros* and *Anthracoceros* tend to feed more commonly together in lipid-rich fruit trees than in fig trees, as predicted by this hypothesis. As anticipated from the ripening pattern of fig crops outlined above, the pairs more frequently feed together in figs (Fig. 6.2; $p < .05$, χ^2), refuting the hypothesis that group size is limited by the small size of fig patches.

Scaling Patch Size to Hornbill Size and Competitive Behavior

To summarize, these differences in diet do not provide an explanation for the fact that *Buceros* and *Anthracoceros* live as pairs, whereas *Anorrhinus* and *Berenicornis* live in groups. A more careful analysis of the problem, however, taking into account the implications of body size and of interspecific competitive interactions between these territorial species at their preferred patches of lipid-rich fruits, has provided evidence for more refined hypotheses that relate hornbill foraging to differences in group size.

The hypotheses are related to the special conditions of frugivory in these hornbill species. They all prefer to eat the lipid-rich fruit pulp produced by a limited subset of rain forest trees. Outside the restricted times of fruit superabundance (when breeding occurs), the fruits are primarily dehiscent capsules with large seeds dispersed almost entirely by hornbills (Leighton and Leighton, 1983; Leighton, in prep.). These trees have the important properties that a) large fruit patches of any of the species are rare, so that often groups and pairs of all territorial species are drawn to the same trees; and b) fruit ripens asynchronously within a crop, so that a limited number of fruits are available each day. These large fruit trees are foci of interspecific competition, especially early in the morning when they hold the most ripe fruit.

Data from early morning watches of fruit trees were analyzed to examine if species vary in their times of first arrival for their morning feed. The time from 0550 to 0700 hours was divided into three periods. For each species, the percentage of first arrivals that occurred during the earliest (0550–0610) period was 40 percent for *Anorrhinus* ($n = 192$ arrivals), 29 percent for *Berenicornis* ($n = 106$), 14 percent for *Anthracoceros* ($n = 205$), and 9 percent for *Buceros* ($n = 19$) (Leighton, 1982). Thus, the group-living species tend to more often arrive earlier than those living in pairs ($p < .01$, $p < .05$ for *Buceros*). Analysis of interspecific displacements from fruiting trees indicated that the competitive hierarchy is *Buceros* > *Anorrhinus* > *Berenicornis* > *Anthracoceros*. The latter species was displaced twenty-five times by others (including nine times by the similar-size *Anorrhinus*). Coordinated attacks by several *Anorrhinus* were even successful in driving off the larger *Berenicornis*. The net result is that when several species use the same fruit tree, as is often the case, the largest patches are usurped by the competitively superior species. In essence then, *Anthracoceros* group size may be constrained because its available preferred patches have been reduced to smaller sizes by the competitive dominants.

In addition to competitive processes, a second process is hypothesized to operate in this system, which also would have the effect of changing the measured patch size to match the organism's perspective more closely. The scaling of daily energy requirements to body mass raised to the three-quarter power (Mace and Harvey, 1983) suggests that large hornbills require more energy per unit time. Because of the similarities between hornbills in activity budgets (Leighton, 1982) and general physiology, this scaling rule is very likely to hold. They must, therefore, acquire this larger ration through

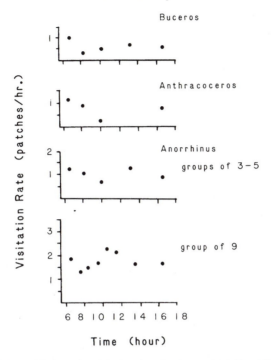

Fig. 6.3. Numbers of fruit patches visited per hour by three territorial hornbills. A patch was tallied if any member of a pair, or group, visited it. Observations are placed into hourly classes according to the midpoint of the sample. Each mean is computed from observing a minimum of four groups per interval to a maximum of twenty per interval.

feeding more in each patch and/or by visiting more patches. Figure 6.3 shows that among the territorial species the large pair-forming *Buceros* does not visit patches at a faster rate than either the smaller pair-forming *Anthracoceros*, or the smaller group-living *Anorrhinus*. Since only about half of the members of any *Anorrhinus* group feed during a patch visit, the per capita expected visitation rate is approximately equal for all three species. Consequently, a larger hornbill must consume more fruit per visit in order to meet its greater metabolic requirements (or put another way, a given patch should provide fewer meals for larger birds than for smaller ones). This argument therefore predicts an inverse relationship between body mass and group size among the territorial species that overlap extensively in their fruit preferences. Consistent with the prediction (Fig. 6.4), the largest species, *Buceros*, occurs in pairs, the mid-sized *Berenicornis* occurs in small groups, and the small *Anorrhinus* lives in large groups (and, interestingly,

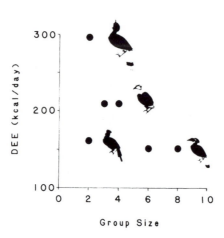

Fig. 6.4. Relationship between daily existence energy (DEE) and group size (numbers of sub-adults and adults) for the four territorial species in Borneo that overlap extensively in fruit preferences. DEE is estimated as $1.7 \times$ BMR (basal metabolic rate), from the formula for non-passerines, BMR $= 86.4M^{0.734}$, where M = body mass (Aschoff and Pohl, 1970). Solid circles correspond to observed group sizes. Hornbill sketches are of *Buceros* (top), *Berenicornis* (middle), *Anthracoceros* (lower left), and *Anorrhinus* (lower right), and are placed to the right of corresponding data points.

breaks into smaller groups during fruit-poor times). The anomalously small groups of *Anthracoceros* follow from its competitive displacement to use smaller patches than the species superior in exploitative and interference competition.

CONCLUSIONS

The picture I have painted of hornbill social systems is one in which subtle demographic and ecological factors importantly influence whether helpers will be found associated with breeding pairs. In the case of *Tockus* hornbills, relatively rapid maturation of offspring and their early participation in breeding may be due to the vulnerability of adults to predators, and to the availability of breeding territories. Both factors encourage early dispersal and breeding by offspring. In the case of the territorial Bornean hornbills, neither of these factors operates. Instead, available territories are saturated with breeding pairs of great longevity, making helping an attractive option. The option is constrained in some species because group living may lower foraging efficiency by intensifying feeding competition.

This treatment emphasizes the complexity of factors influencing the benefits and costs of group living, and argues that relatively slight changes in the balance can result in sociality or relative solitariness, with cascading constraints on the possible patterns of social behavior and mating systems. This study has more general significance for the way we should view food resource variables such as patch size. Actual measures of the variables have to be weighted by behavioral and morphological characteristics of the organism if general principles of how resources influence social dispersion are to emerge.

Further Studies

Three types of studies might be undertaken to enlarge upon or revise the interpretation of hornbill social systems described here. First, long-term studies of selected species using banded individuals would identify the relatedness among members of cooperative breeding groups and also of the solitary immatures that use the territories of *Anthracoceros* and *Bucorvus*. Combined with detailed observations of social relations among group members of species like *Anorrhinus*, enigmatic features of coalition formation could then be interpreted.

Second, further observations or experiments could test some of the interpretations offered here. The patch-size arguments postulated for territorial Bornean hornbills are susceptible to experimental invalidation through manipulating patch sizes to examine short-term changes in foraging group size. The hypothesis for the lack of helpers in *Tockus* might be amenable to artificial restriction of nesting opportunities in order to create a pool of nonbreeding potential helpers. Complementing such experimental approaches, which are not without their logistic difficulties, would be observations of populations of these same African or the Bornean species in which patterns of group size were different. Alternative hypotheses about the underlying resource base or demographic factors could then be erected and tested by measurement of the relevant variables.

Third, further comparative studies of other hornbill species, or of the neotropical toucans and aracaris (Ramphastidae), which to some degree are ecologically convergent with hornbills, would be enlightening. African rain forests contain several sympatric hornbills with more frugivorous diets than the savanna species; their social systems might be predicted to conform to the constraints postulated as important for Bornean species. Similarly, studies of the convergent (but distantly related) omnivorous rain forest ramphastids would shed light on the generality of the relationships between various demographic and ecological factors postulated here. Some relationships might be fundamentally different for toucans since they are of much smaller body sizes than forest hornbills, they are hole-nesting but not nesthole-sealing, and they may not be tied to the peculiar types of capsular fruits that importantly influence relationships among the sympatric Bornean hornbills. The relative importance of predation versus foraging on survivorship, and of interspecific competition at fruit trees on group size, may then alter the costs and benefits of the decisions made by adults and juveniles that affect group size. The several different behavioral, demographic, and ecological variables which have been identified here must each be examined by comparative studies in order to exclude alternative explanations of social dispersion.

7. Ecology and Social Evolution
in the Mongooses

———————————— • ————————————

JON P. ROOD

THE MONGOOSES comprise a large and relatively unstudied carnivore group showing considerable diversity in ecology and social organization. They occur in habitats from rain forest to semi-desert, may be either diurnal or nocturnal, and feed on a variety of prey from insects to rats and birds. Most of the thirty-six species of mongooses are solitary, but eight species in four genera travel and den in cohesive groups which usually contain several adult males and females. What ecological factors have been responsible for differences in the social systems of the mongooses and how has sociality evolved in the group? This chapter summarizes available field data and describes some associations between ecology and behavior. It also attempts to identify the selective pressures that have been important in the evolution of mongoose social systems, and suggests a route to sociality in the group.

In searching for correlations among ecology, behavior, and social organization, I have been limited by a lack of field data in most species. Ewer (1973) remarked on the absence of field studies on mongoose social organization, and, while the situation is somewhat better today, only a few species have been studied in their natural habitats. This chapter represents an attempt to correlate gross ecological traits with social systems in this group and I hope may stimulate additional field work. Continuous field observations are often difficult because mongooses tend to be shy and secretive, disappearing into cover at the slightest disturbance. However, with the advent of telemetry and night scopes, even the shyer, nocturnal species can now be located and observed.

Mongooses have traditionally been placed in the subfamilies Herpestinae and Galidiinae of the Viverridae (e.g., Simpson, 1945; Ewer, 1973) but recent work indicates that they form a distinct family, the Herpestidae (Wozencraft, 1982), containing seventeen genera and thirty-six species. *Galidia, Galidictis, Mungotictis*, and *Salanoia* are endemic to Madagascar, and all other genera are represented in Africa. One genus (*Herpestes*) is widely distributed in Asia and *H. auropunctatus* has been introduced into the West Indies and the Hawaiian and several other Pacific islands.

The dentition of mongooses closely resembles that of the ancestral Miacid carnivores (Butler, 1946). *Herpestes*, the most widely distributed genus, has the longest fossil history, extending back to the Miocene (Barry,

TABLE 7.1

Mongoose social structures and ecological traits

Genus	Social unit	Habitat	Activity cycle	Diet	Weight (kg)	No. species	References
Galidiinae							
Galidia	Pairs?	F	D	V	<1	1	Albignac, 1973; Larkin and Roberts, 1979; Walker, 1983
Mungotictis	Pairs?	S	D	I	<1	1	Albignac, 1976; Walker, 1983
Galidictis	Pairs?	F	N	V	<1	1	Albignac, 1973; Walker, 1983
Salanoia	Pairs?	F	D	I	<1	1	Albignac, 1973; Walker, 1983
Herpestinae							
Herpestes	Solitary	F, S	D or N	V	1–4	14	Ewer, 1973; Kingdon, 1977; Gorman, 1979; Walker, 1983
Atilax	Solitary	F, S*	N	V, C	3–4	1	Kingdon, 1977; Rosevear, 1974
Ichneumia	Solitary	S	N	I	3–4	1	Kingdon, 1977; Waser, 1980; Waser and Waser, 1985
Rhynchogale	Solitary	S	N	I	2–3	1	Dorst and Dandelot, 1970; Kingdon, 1977; Walker, 1983
Bdeogale	Solitary	F, S	N	I	1–3	3	Kingdon, 1977; Rosevear, 1974; Walker, 1983
Mungotinae							
Helogale	Groups	S	D	I	<1	2	Kingdon, 1977; Rood, 1983a
Dologale	?	S	D?	I?	<1	1	Dorst and Dandelot, 1970; Kingdon, 1977; Walker, 1983
Liberiictis	Solitary?	F	D?	I?	2–3	1	Rosevear, 1974; Schlitter, 1974; Walker, 1983
Crossarchus	Groups	F	D	I	1–2	3	Ewer, 1973; Kingdon, 1977; Rosevear, 1974
Mungos	Groups	S	D	I	1–2	2	Kingdon, 1977; Rood, 1975; Rosevear, 1974
Cynictis	Pairs?	S	D	I	<1	1	Lynch, 1980; Smithers, 1971; Walker, 1983
Paracynictis	Solitary	S	N	I	1–2	1	Smithers, 1971; Walker, 1983
Suricata	Groups	S	D	I	<1	1	Ewer, 1973; Lynch, 1980; Smithers, 1971

KEY: F = forest; S = savanna; D = diurnal; N = nocturnal; V = primarily small vertebrates; I = primarily insects; C = crustacea; *near water.
NOTE: Genera are grouped according to phylogenetic relationships. Number of species based on Wozencraft in Honacki et al., 1982.

1983). The mongooses probably migrated from Europe to Africa in the early Miocene and radiated within Africa during the Miocene, Pliocene, and Pleistocene (Savage, 1978). According to Petter (1969), a single herpestine ancestor gave rise to all genera of African mongooses. Savage (1978) states that *Herpestes* probably migrated from Africa to Asia in the late Pleistocene (but see Barry, 1983, for an alternative view). Wozencraft (pers. comm.) divides the Herpestidae on morphological grounds into three subfamilies and six groups, each group sharing a common ancestor. He considers the "Herpestine group" (*Herpestes, Atilax*) and the four genera of Malagasy mongooses to be the most primitive on the basis of tooth pattern; the Malagasy group split off from the Herpestine group very early in the evolution of the Herpestidae.

METHODS

Comparative Procedures

In Table 7.1, I have grouped the seventeen mongoose genera according to phylogenetic relationships (Wozencraft, pers. comm.) and given data on number of species, body weight, social unit, habitat, activity cycle, and diet for each. All species for which I could find data were then classified as: 1) large (adult weight of > 2 kg) or small (adult weight of < 2 kg); 2) primarily diurnal or primarily nocturnal, and 3) primarily insectivorous or primarily a small vertebrate feeder. This classification gave seven combinations, for example, small, diurnal, insectivorous (Table 7.2).

Several problems were encountered in preparing these tables, primarily because of the lack of reliable field data for most species of mongooses. Two species could not be classified because of insufficient data and are not included in Table 7.2. *Dologale dybowskii* is closely related to the group-living *Helogale*—indeed Wozencraft (pers. comm.) considers them probably congeneric—and is very likely to share ecological traits with this genus. Field observations are not available and group-living in *Dologale* cannot presently be confirmed. *Liberiictis kuhni* is known only from two whole specimens and ten skulls collected in forest habitats in northeastern Liberia (Schlitter, 1974). Present limited evidence suggests that it may be large, diurnal, insectivorous, and solitary (Schlitter, 1974; Walker, 1983). If so it is the only mongoose that combines these traits.

Since there are no thorough studies of food habits for most species of mongooses the classification of diet was sometimes difficult. In addition, mongooses are opportunistic feeders. Most will take a variety of small available prey, both vertebrate and invertebrate, and the relative proportions of dietary items frequently vary in different areas and habitats. For example, the *Herpestes* species possess a well-developed carnassial shear and hunting methods adapted to catching small vertebrates, especially rodents. These

133

TABLE 7.2
Relationship between ecological categories and social structure

Ecological category	Number of species			
	Groups	Pairs?	Solitary	Total
Small, diurnal, insectivorous	8	3	0	11
Crossarchus alexandri, C. ansorgei, C. obscurus,				
Cynictis penicillata, Helogale hirtula, H. parvula,				
Mungos gambianus, M. mungo, Mungotictis				
decemlineata, Salanoia concolor, Suricata				
suricatta				
Small, diurnal, vertebrate feeders	0	1	6	7
Galidia elegans, Herpestes auropunctatus,				
H. edwardsi, H. fuscus, H. javanicus,				
H. pulverulentus, H. sanguineus				
Large, nocturnal, insectivorous	0	0	4	4
Bdeogale jacksoni, B. nigripes, Ichneumia				
albicauda, Rhynchogale melleri				
Small, nocturnal, vertebrate feeders	0	1	3	4
Galidictis fasciata, Herpestes brachyurus,				
H. hosei, H. smithii				
Small, nocturnal, insectivorous	0	0	2	2
Bdeogale crassicauda, Paracynictis selousi				
Large, nocturnal, vertebrate feeders	0	0	4	4
Atilax paludinosus, Herpestes naso,				
H. semitorquatus, H. urva				
Large, diurnal, vertebrate feeders	0	0	2	2
H. ichneumon, H. vitticollis				

are preferred prey when available (Prater, 1948; Ewer, 1973; Rood and Waser, 1978; Gorman, 1979; D. Nellis and Everard, 1983), and *Herpestes* are here considered primarily small vertebrate feeders although in areas where these are scarce invertebrates may predominate in the diet (e.g., La Rivers, 1948; Gorman, 1975).

Defining the social unit of some species is also difficult. Species that normally forage and den in groups containing more than one adult male and female are termed group-living, while those that normally forage and den alone or in temporary groups consisting of a mother and her dependent offspring are termed solitary. Most mongooses are unambiguously either solitary or group-living, but some do not fall into these categories. For example, the four Malagasy mongooses and *Cynictis penicillata* from southern Africa are listed as living in ''pairs?'' in Table 7.1, since the male is an important part of the social unit and they are often seen in pairs or family groups. However, they frequently forage alone and might be classed as solitary under some definitions, cf. Waser and Jones (1983). Observations on *Cynictis* and *Mungotictis* suggest that in these forms individuals may be members of social units that are more complex than a simple family unit. *Cynictis* dens communally (Ewer, 1973), and Earle (1981) reported colonies of about eight individuals including a dominant male and female. Albignac (1976) used radiotracking to determine that twenty-two *Mungotictis* living on 300 hectares were divided into two stable social units, which sometimes engaged in agonistic encounters in areas of range overlap. Individuals foraged singly, in pairs, maternal family parties, and all-male groups.

Field Methods

Fieldwork on the dwarf mongoose has been carried out at the Serengeti Research Institute, Serengeti National Park, Tanzania, from 1974 to the present. During the study 478 dwarf mongooses have been live-trapped and freeze-marked (see Rood and Nellis, 1980) for individual recognition. Full-time observations were made in the first three years of the study and I have subsequently visited the area for six weeks to eight months each year. The main study area is located in open *Acacia, Commiphora* woodland containing a high density of *Macrotermes* termite mounds, which are used by the mongooses as dens. The study area was enlarged to twenty-five square kilometers in 1980 and presently includes the ranges of thirteen dwarf mongoose packs. Each year at the start of the birth season (October-November), all dwarf mongooses on the study area are censused, and young of the year and immigrants are live-trapped and marked. In addition, suitable mongoose habitat surrounding the area is searched for emigrants.

Banded mongooses were studied for over two years in Queen Elizabeth Park, Uganda, from 1971 until 1973, and subsequently in the Serengeti National Park until the present. A total of 147 individual banded mongooses have been trapped and marked during the study. However, because freeze-marking does not work well on this species it has not yet been possible to obtain long-term data on the life histories of more than a few individuals. Mongooses marked with Nyanzol dye can be recognized for about four months and combinations of ear notches provide long-term identification if the animal can be retrapped. All animals in a pack using the area surrounding the Serengeti Research Institute have been individually recognizable for more than three consecutive years.

Most observations of the social interactions of dwarf and banded mongooses were made when the animals were at dens in the early morning or late afternoon. However, observations were also made opportunistically throughout the day, sometimes for entire days. I normally observed the mongooses from a Land Rover using 10 × binoculars. Helper behavior was recorded during all-day watches at breeding dens.

TABLE 7.3
Representative mongoose social systems

	Spatial distribution	Mating system	Care of young	Dispersal	References
Small Indian mongoose (*Herpestes auropunctatus*) Small, diurnal, vertebrate feeder	Solitary. Wide overlap in home ranges of all sex and age groups	Polygamous	By mother only	Young disperse at puberty when 4–5 months old	Gorman, 1979; D. Nellis and Everard, 1983
White-tailed mongoose (*Ichneumia albicauda*) Large, nocturnal, insectivorous	Solitary. Exclusive adult male ranges. Male-female ranges overlap widely. Some adult females share ranges	Polygamous	By mother only	Young forage independently of mother by 9 months but may continue to use natal range. Young males and some females later disperse, but some females remain in mother's range	Waser and Waser 1985
Dwarf mongoose (*Helogale parvula*) Small, diurnal, insectivorous	Social groups containing several adult males and females. Home ranges frequently overlap	Monogamous. Each group contains a dominant breeding pair	All group members help care for young	Many emigrate when 1–3 years old, but some of each sex may stay and breed in natal groups. Natal philopatry more common in females	Rood, 1978, 1980, 1983a, 1985
Banded mongoose (*Mungos mungo*) Small, diurnal, insectivorous	Social groups containing several adult males and females. Home ranges frequently overlap	Polygamous. Groups contain several breeding males and females	All group members help care for young	Some young males and females emigrate; some females remain and breed in natal groups	Neal, 1970; Rood, 1974, 1975, pers. obs.

Social Evolution in Mongooses

Information on mongoose social systems is most complete for the dwarf mongoose. I here summarize this material and make some comparisons with the group-living banded mongoose and with some solitary species. Parameters used to compare the social systems are spatial organization, mating system, care of the young, and dispersal. A brief comparison of these parameters in the two group-living and two solitary mongooses is given in Table 7.3.

The Dwarf Mongoose

This species lives in multimale, multifemale groups (packs) which average about eight individuals but may contain more than twenty. Group ranges frequently overlap, but each range contains an area of exclusive use (Rood, 1983a). Spacing is mediated by frequent scent marking at dens, by avoidance of large packs by smaller ones, and by occasional brief skirmishes in areas of range overlap. These aggressive interactions commonly involve chasing and vigorous anal marking of objects; physical contact with biting occasionally occurs. The encounters usually end with one or both packs moving toward the centers of their ranges.

Young at breeding dens are aggressively protected from conspecific packs. On three occasions a pack of six dwarf mongooses was observed to repulse a larger pack of sixteen which was attempting to enter a den containing young (Rood, 1983a). All members of the smaller pack participated in chasing the intruders in these encounters. Aggression usually occurs between animals of the same sex but, during the excitement of intergroup encounters, attacks may also be made on mongooses of opposite sex. However, transients that encounter packs are only chased by same-sex individuals.

Within a pack affiliative behavior is common between adult males and females and between adults and juveniles. In one pack consisting of three adult males and three adult females, 94 percent of the 200 grooming interactions recorded were heterosexual and 71 percent involved mutual grooming (Rood, 1983a). Either sex may initiate a bout of grooming. In most packs the alpha male grooms preferentially with the alpha female while the beta male grooms all females with approximately equal frequency.

Since dwarf mongooses feed mainly on insects, which they find individually, food competition is uncommon. Occasionally an individual may be displaced from a foraging site, and if a mongoose opportunistically catches a large prey item such as a mouse, others may give chase and attempt to expropriate the prize. Similarly, Macdonald (pers. comm.) reports almost no food competition in another insectivorous group-living mongoose, the meerkat (*Suricata suricatta*). In the dwarf mongoose, overt aggression (primarily chasing) occurs among males competing for females during mating periods and is regulated by a dominance hierarchy.

Each pack contains a dominant breeding pair, the alpha male and female, who are usually the oldest individuals, and are the parents of offspring raised by the group (Rood, 1980). The alpha female is able to displace all other mongooses at food sources but in the wild she seldom does so. In an artificial feeding situation where there is room for only one animal at the food she displaces others at will (Bostian, pers. comm.; Rood, pers. obs.). The alpha female is usually the first mongoose to emerge from the den in the morning and the first to leave the den to forage. When she comes into estrus the alpha male stays near her and chases any subordinate adult males who attempt to approach. Subordinate adult females come into estrus synchronously with the alpha female and frequently solicit mounting from the alpha male. Toward the end of the mating period—which continues for several days—he may mate with them, and the beta male may then copulate with the alpha female (Rood, 1980). Subordinate males also sometimes mate with subordinate females, but typically only the alpha female conceives. On the few occasions when more than one female becomes pregnant, the resulting number of young born is not larger than a single female's average litter size, suggesting that subordinate females lose their litters (Rood, 1980). Rasa (pers. comm.) reports that all young born to nine captive subordinate females were killed directly after birth; on one occasion she observed the alpha female committing the infanticide.

All pack members cooperate in raising the young. One or more mongooses remain at the den to guard them while the rest of the pack forages. These babysitters change frequently throughout the day so that all have some time to forage (Rood, 1978). Helpers also aid by bringing insects to the young mongooses and by transporting them in frequent den changes. When the young are old enough to forage with the pack, adults continue to feed them by digging up insects and allowing the young to take them; young mongooses continue to receive adult protection and each stays close to an adult of either sex during foraging and traveling.

In two large packs the principal helpers were found to be subordinate adult females and yearlings (Rood, 1978). Observed values for babysitting were higher than expected for subordinate adult and yearling females, near the expected value for yearling males, and lower than expected for other age and sex classes and the alpha pair. Yearling males and females did most of the feeding. Unrelated immigrants are frequently good helpers. In a pack of six mongooses consisting of a dominant breeding pair, an immigrant male and female, and a natal male and female, the main helpers were the immigrants and natal animals; the parents of the young did little babysitting and were not observed to bring food to the young (Rood, 1983a).

The presence of helpers allows the breeders more foraging time and assures that the young will be guarded at the den during the mating period, which commences approximately two weeks after parturition. Except for nursing, the alpha female provides little parental care, but the alpha male plays an important guarding role in scanning for predators from high points on or near the den (Rood, 1983a). In very small packs alpha males help by babysitting. For example, in a pack of three mongooses, the alpha male was the main babysitter: the following breeding season, when the pack consisted of five mongooses, including an immigrant female, the alpha male rarely stayed alone with the young. The immigrant female spent more time at the den and more alone with the young than any other pack member (Rood, 1978).

Large dwarf mongoose packs raise more young than small packs, indicating that the helpers do help (Rood, in prep.). Pack size at the start of the breeding season was positively correlated with number of young raised to the start of the next breeding season ($r_s = 0.55$, $p < .001$, $n = 99$ packs). Forty-seven packs in which the number of individuals exceeded the median (i.e., packs of eight or more mongooses) raised an average of 3.5 young per year while fifty-two packs containing less than eight mongooses raised an average of only 1.3 young. The mean number of young raised per year increased steadily to 4.3 at a pack size of ten and then leveled out. Young raised per capita (0.43) was also highest at a pack size of ten.

Adult mortality, estimated from the disappearance rate of mongooses that are unlikely to emigrate, was also significantly higher in small than in large packs ($\chi^2 = 15.3$, $df = 1$, $p < 0.001$; Rood, in prep.). In packs containing less than eight individuals, 45 percent of the breeders and subordinate adults more than four years of age disappeared per year, while only 23 percent of these adults disappeared from packs containing eight or more mongooses.

In dwarf mongooses, both males and females maximize their reproductive potential by becoming alpha individuals in a large pack. A mongoose can adopt a conservative tactic, remain in its natal pack, and accede to alpha status on the deaths of older same-sex individuals. Alternatively, it can leave the natal pack and seek breeding opportunities elsewhere. Both males and females may adopt either strategy. Mongooses that emigrate usually leave alone, or with another pack member of the same sex, when between one and three years of age (Rood, 1983a, 1985). Emigrants attain breeding status by joining opposite sexed emigrants to form new packs, or by joining existing packs (see Table 7.4). Some mongooses attain immediate alpha status by joining packs in which breeders have died, but others must wait, sometimes for several years, until older same-sex individuals are gone. Mongooses that transfer increase their reproductive potential by moving from packs with more older same-sex individuals to packs with fewer (Rood, 1983a, 1985).

Group size is of critical importance to mongoose fitness, and the lifetime

TABLE 7.4
How sixty-five dwarf mongooses attained alpha status

	Males	Females	Total
Joined existing pack	9 (30%)	11 (31%)	20 (31%)
Formed pack	11 (37%)	9 (26%)	20 (31%)
Takeover	7 (23%)	1 (3%)	8 (12%)
Remained in natal pack	3 (10%)	14 (40%)	17 (26%)
Total	30	35	65

reproductive success of breeders is considerably higher in large packs than in small ones (Rood, in prep.). Large packs maintain their identity in a defined home range for long periods (some have persisted throughout the study), while small packs are likely to fail. Of fifteen packs that failed during the study, thirteen were below the median size when last observed, and eleven contained only one adult female. If the alpha female in a small pack dies, the remaining pack members may emigrate, resulting in the disappearance of the pack from the area (Rood, 1985). Areas of apparently ideal habitat where packs exist for long periods but eventually fail may remain unoccupied for many years—in this, dwarf mongooses differ from many species of cooperatively breeding birds (e.g., Woolfenden and Fitzpatrick, Chapter 5 this volume) in which all suitable habitat is saturated. For a mongoose, it may often be more beneficial to remain in or join a large pack and delay reproduction than to form a small pack in which breeding status can be immediately attained.

Dwarf mongoose males and females show many similarities in reproductive strategies. However, females—as in many other mammals—are less mobile in the following ways:

1) Females are more likely to remain in their natal pack than males. Of 126 females and 96 males which survived their year of birth, 40 percent of the females, but only 14 percent of the males, were still in their natal packs as three-year-olds. In consequence, many alpha females are natal animals, whereas most alpha males are immigrants. In breeders of known life history, 40 percent of the females, but only 10 percent of the males, were still in the pack of their birth (Table 7.4).

2) One or two females sometimes move in and establish a home range in an unoccupied area. Such females may remain alone for over a year before being joined by a male, thereby forming a new pack (Rood, 1985). Males usually keep on the move in searching for breeding openings and do not stay long in unoccupied areas.

3) If, as infrequently occurs, a pack disintegrates leaving only the alpha female, she remains in her range and waits for a male to join her (Rood, 1985). In a similar situation, alpha males leave the area.

There are also sexual differences in the amount of aggression occurring during intergroup transfer. Females have not been observed to join packs containing more than one adult female and the process of transfer has been peaceful. Indeed, in one case the alpha female actively groomed a young immigrant who had just contacted her pack. Males, however, sometimes attempt to join packs containing two same-sex adults. They often encounter considerable aggression, most frequently from the beta male, and they may be injured in these attacks (Rood, 1983a). Prospective male immigrants may trail a pack for over a month before they are allowed to share the communal den. When the number of potential immigrants exceeds the number of resident males, groups of intruder males are sometimes able to drive resident males from their ranges and join the pack females. Seven of thirty males of known life history (23 percent) attained alpha status as a result of male takeovers (Table 7.4). This suggests that alpha males, especially those in groups that do not contain other adult males, face a considerable risk of being ousted. The only recorded female takeover occurred following the death of the breeding female, when a yearling female was forced out by two older intruders. Successful female breeders maintained their status for life.

The Banded Mongoose

Banded mongoose packs are usually considerably larger than dwarf mongoose packs. Pack size averages fifteen but large packs may contain over thirty individuals. As in the dwarf mongoose, home ranges frequently overlap and aggressive intergroup encounters are regulated by a dominance hierarchy dependent on pack size. Small packs attempt to avoid those of larger size but they are often seen and chased. Banded mongoose intergroup encounters typically differ from those of dwarfs in being of longer duration and involving higher-intensity aggression. Components normally include aggressive chasing, biting and hair-pulling, loud screeching churr vocalizations, and frequent and vigorous anal marking of objects and other pack members. Often the animals became dispersed over a wide area with most of the chasing occurring between individuals or small groups (Rood, 1975).

During high-intensity aggressive interactions the mongooses frequently appear oblivious to external stimuli. On these occasions I could approach to within a few meters and once four hyenas were attracted and watched a fight at a distance of about thirty meters. Conflicts between packs of similar size may continue for over an hour. Often they terminate with both packs retreating toward the centers of their ranges.

Adult males take a leading role in aggressive encounters and most of the chasing and fighting occurs between males. A group of males will often chase a rival male group into a termite mound and keep them there for long periods, swarming over and marking the mound and lunging at the holes with dorsal and tail hairs bristling. Groups of males occasionally leave their pack ranges and invade the ranges of neighboring packs. These intruders are vigorously attacked by the adult males of packs they encounter.

Observations on marked packs of banded mongooses suggest that, as in the dwarf mongoose, there is a dominant pair who are the oldest individuals. The oldest female is the first to emerge in the morning, and she leads the pack from the den to forage. This species differs from the dwarf mongoose in that packs contain several breeding males and females. The females come into estrus synchronously and produce multiple litters totaling up to sixteen offspring. Reproduction is synchronized within but not between packs, indicating that social stimuli are important in triggering estrus in this species (Rood, 1975).

As in the dwarf mongoose, the young are raised communally and most or all pack members babysit and bring insects to them. In observations on two packs at breeding dens, particular adult males (superhelpers) gave disproportionate amounts of aid. In one pack, observed for 154 hours, an adult male spent 57 percent of his time at the den with the young while figures for the other fourteen pack members varied from 26 percent to 38 percent. The male superhelper usually stayed alone with the young when the pack first left the den in the morning and did not forage until they returned several hours later. This male was of unknown relationship to the young and had emigrated by the following breeding season. At that time another adult male (a natal animal) acted as superhelper, remaining with the young on fourteen of twenty-seven mornings—no other individual stayed on more than three mornings (Rood, pers. obs.). D. W. Macdonald (pers. comm.) reports that in the group-living *Suricata suricatta*, nonbreeding males and females may stay with the young for twelve hours at a time.

Only preliminary data are presently available on dispersal and breeding strategies in the banded mongoose. Some females breed in their natal packs and probably spend their entire lives there. In one pack in which all individuals have been distinguished for over three years, three of the four breeding females are natal animals. It is not yet known whether males sometimes remain and breed in their natal packs. All five adult males in the above-mentioned pack are immigrants, having joined between January and December 1983, while I was absent from the study area. During this period, the five resident males (the dominant male of unknown origin, an immigrant, and three natal animals) all disappeared, suggesting that a male takeover may have occurred. In *Suricata suricatta*, Macdonald (pers. comm.) observed several intruder males attack and drive out the resident males of a pack. The intruders were later seen associating with the pack females on several occasions.

Solitary Mongooses

The two solitary species listed in Table 7.3 are polygamous and show no paternal care. Home ranges of the small Indian mongoose (*Herpestes auropunctatus*) are reported to overlap widely although Gorman (1979) points out that his data were collected over a long period of time and that at any one time individual ranges may not overlap. Waser and Waser (1985) found that in the white-tailed mongoose (*Ichneumia albicauda*) degree of range overlap depended on population density. At low densities the mongooses showed a pattern characterizing many solitary mammals with a mosaic of exclusive or nearly exclusive female ranges overlapping a separate mosaic of larger male ranges, while at higher densities several females shared nearly congruent home ranges, forming "clans." Range-sharing probably results from natal philopatry in females (Waser and Waser, 1985). Clan members forage independently and usually sleep separately, but two adult females were found sharing a den on several occasions. Like members of a clan of European badgers, *Meles meles* (Kruuk, 1978a), the mongooses recognize and tolerate individuals within their clan, probably scent mark a common boundary, and avoid the area used by neighboring clans (Waser and Waser, 1985).

The wide range overlap reported in these two solitary mongooses indicates that individuals have many opportunities to interact socially with conspecifics. While undoubtedly solitary in the sense that they are usually found alone (Waser and Jones, 1983), these and other solitary mongooses show gregarious tendencies. In the Serengeti National Park, Tanzania, pairs of *Herpestes sanguineus* sometimes forage together, and Kingdon (1977) reported that a family group of this species consisting of a male, female, and one young lived in a drain near his house for four months and occasionally hunted together. Family parties of *H. pulverulentus* are also reported to den together (Ewer, 1973).

The Egyptian mongoose (*Herpestes ichneumon*) may be the most social member of the genus. In East Africa this species forages alone or in family groups (Kingdon, 1977; Rood, pers. obs.), but in other areas there are numerous records of larger groups (see references in Kingdon, 1977). Ben-Yaacov and Yom-Tov (1982) studied Egyptian mongooses inhabiting the levees between fish ponds on a kibbutz in Israel. Food availability was high due to the occurrence of dead fish, dead chicks from a hatchery, and a garbage dump. Most observations were made on two semi-tame "families" occupying an area of 900 square meters which were provisioned from a vehicle. The families consisted of an adult male, two to three adult females, and their young, and they exhibited division of labor. The young mongooses suckled from any mother in the family and family members cooperated in grooming, babysitting, feeding, and playing with the young. Such complex behavior resembles that of the group-living mongooses and may have resulted, at least partly, from the unusual habitat limiting dispersal, and the abundant, artificial food resources available to the mongooses.

TABLE 7.5
Ecological parameters of some mongoose species

	Location	Main food items (listed in order of importance)	Home range (km²) Males	Females	Group	Population Density (No./km²)	References
Solitary Mongooses							
Herpestes auropunctatus	Fiji	Rodents, toads, grasshoppers, dragonflies	0.39	0.22		50.0	Gorman, 1979 D. Nellis (pers. comm.)
Herpestes auropunctatus	West Indies	Rodents, birds, lizards, Orthoptera	0.04	0.02		350.0	D. Nellis and Everard, 1983
Herpestes sanguineus	Serengeti, Tanzania	Rodents, snakes	0.50	0.25			Rood and Waser, 1978
Ichneumia albicauda	Serengeti	Beetles, crickets, ants, termites	0.97	0.64		3.7	Waser and Waser, 1985; Waser, 1980
Atilax paludinosus	Natal, South Africa	Crabs, small mammals, birds, frogs					Rowe-Rowe, 1977
Pair?-living Mongooses							
Cynictis penicillata	South Africa	Termites, grasshoppers, beetles, moths					Lynch, 1980.
Mungotictis decemlineata	Madagascar	Beetle larvae, millipedes, snails, amphibians			1.5	7.3	Albignac, 1976
Group-living Mongooses							
Helogale parvula	Sangere River, Serengeti	Beetles, termites, insect larvae, grasshoppers			0.34	30.9	Rood, 1983a, pers. obs.
Helogale parvula	Kirawira, Serengeti	Beetles, insect larvae, rodents, termites			1.6	2.9	Rood, 1983a, pers. obs.
Mungos mungo	Short grass plains, Serengeti	Beetles, centipedes, crickets			15.0	1.0	Rood, 1984, pers. obs.
Mungos mungo	Queen Elizabeth Park, Uganda	Millipedes, beetles, ants, termites			0.8	17.4	Rood, 1975
Suricata suricatta	South Africa	Beetles, moths, termites, grasshoppers					Lynch, 1980

144

Social Evolution in Mongooses

How do gregarious and solitary mongooses differ in ecology? Table 7.2 shows that group-living mongooses are all small, diurnal, and insectivorous. No solitary species combines these traits, but three pair?-living forms (*Cynictis, Mungotictis, and Salanoia*) do. Except for *Crossarchus*, the group-living mongooses all live in open habitats, but this is also true of many solitary species (Table 7.1). While the group-living species share ecological traits the solitary forms show great variability, much of it due to adaptive radiation in the widely distributed genus *Herpestes*. The fourteen species all feed primarily on small vertebrates, but they vary in preferred habitat, body size, and activity cycle (Table 7.2). Six species are small, diurnal, vertebrate feeders. The remaining solitary mongooses, which can be classified *Ichneumia, Rhynchogale, Bdeogale, Paracynictis*, and *Atilax*, are all nocturnal, and, with the exception of *Atilax*, insectivorous.

Studies of the food habits of individual mongoose species indicate the importance of beetles in the diet of most insectivorous forms (Table 7.5). The large herds of African ungulates attract an abundance of dung beetles and these are an important food source for all the group-living mongooses that have been studied.

Home range size tends to be larger in group-living than solitary mongooses but intraspecific variation is great as a consequence of differing habitats, food dispersion, and population densities (Table 7.5). In open habitats where food is sparsely distributed, mongoose densities are low and home ranges large. For example, on the open Serengeti short grass plains where food and cover are comparatively scarce, banded mongooses live at densities of about one square kilometer, use ranges of approximately fifteen square kilometers, and travel up to ten kilometers per day. On Mweya Peninsula, in Queen Elizabeth Park, Uganda, food resources (particularly millipedes and dung beetles) and cover are abundant. Banded mongoose densities are high (ca. 17 / km^2), ranges average 0.8 square kilometers, and the packs move approximately 2.3 kilometers per day. Similarly, dwarf mongoose densities are considerably lower, and ranges larger in the more open Kirawira habitat in comparison with the *Acacia, Commiphora* woodland, Sangere River area (Table 7.5).

The highest population densities occur in the small Indian mongoose (*Herpestes auropunctatus*) where it has been introduced on islands of the Caribbean. Nellis (pers. comm.) believes that 350 animals per square kilometer is a reasonable figure for large areas of suitable habitat, and this degree of overcrowding has forced some animals to spread into less preferred areas (Nellis and Everard, 1983). This adaptable mongoose has thrived on island habitats in the virtual absence of predators and competitors in spite of persecution by man.

SELECTION PRESSURES

Food Abundance and Dispersion

The dispersion and abundance of resources (especially food) are fundamental to the spacing and structure of carnivore societies, and certain patterns in the availability of food may allow group living (Macdonald, 1983). What might these patterns be in small carnivores? Why are the group-living mongooses all insectivorous while species that specialize on small vertebrates are solitary?

Ewer (1973) states that "the more predacious mongooses are solitary since, with the killers of small prey, one hunter tends to interfere with another," and Gorman (1979) observes that "social groups, in constant vocal communication, are not likely to be conducive to the successful exploitation of small vertebrates." Mutual interference could help to explain why small, group-living carnivores might find it difficult to catch small vertebrates. In contrast, many insects and other invertebrates on which mongooses feed are slow-moving inhabitants of the soil litter and are less likely than small vertebrates to escape from group foragers. Such food resources need only to be "harvested."

Gregariousness has been argued to have evolved in many carnivores as an adaptation allowing the more efficient exploitation of patchy food resources, e.g., large ungulates or concentrations of earthworms (references given in Waser, 1981). In the Serengeti National Park, Waser (1980, 1981) found no evidence of clumping in the distribution of most insects important as prey for small nocturnal African carnivores, but he did find a high prey renewal rate. Invertebrate prey removed experimentally from sample quadrants reached 67 percent of initial levels within twenty-four hours. Waser (1981) suggests that where prey are not distributed in patches large enough to be shared by multiple foragers, rapid prey renewal is a necessary precondition for the evolution of sociality. In comparison with insects, small vertebrates are less abundant and have a slower rate of renewal, characteristics that promote food competition among small carnivores and increase the cost of sharing with neighbors. Foraging packs of mongooses spread out and dig for insects or snap them up from the surface with little apparent competition. The capture of a vertebrate such as a mouse or small snake disrupts the foraging, as individuals chase one another trying to obtain the prize. And the disturbance can attract predators. It therefore seems that the fairly even distribution, abundance, and rapidly renewable nature of insects versus small vertebrates as food resources allows group-living to evolve in insectivorous small carnivores. The correlation between insectivory and mongoose sociality results primarily from the availability and abundance of insects as a food resource that can be harvested with minimal competition.

Predation

Insectivory *allows* group formation in the mongooses, but what is the pressure that causes them to group? Protection from predators is widely recognized as one of the primary reasons why animals live in groups (e.g., Treisman, 1975; Wilson, 1975; Rubenstein, 1978), and I believe predation has been the main selective pressure promoting and maintaining cohesive social groups in the mongooses. Small, diurnal, insectivorous carnivores can detect and deal with predators more effectively in a group than when alone.

As in most field studies of small mammals, predation on mongooses has only rarely been observed in the field. However, the frequency of predator attacks, and the complexity of mongoose antipredator behavior, argue that predation has been an important force in the evolution of mongoose group life.

When emigration seemed unlikely, mortality through predation was the probable cause of disappearance of Serengeti mongooses. Diseased animals were not observed on the study area and individuals that disappeared usually appeared in good condition when last seen. I observed only two successful predator attacks at dens, both by black-backed jackals (*Canis mesomelas*). The victims, an adult dwarf and a young banded mongoose, had become slightly separated from their packs and were attempting to return to a den in the late afternoon when they were captured. On another occasion a banded mongoose was seen eating a freshly killed month-old dwarf mongoose, which it had apparently caught in or on the breeding den.

Mongooses are most vulnerable when foraging and most predation probably occurs at this time. In the Taru Desert, Kenya, dwarf mongooses form mutualistic foraging communities with hornbills, *Tockus deckeni* and *T. flavirostris* (Rasa 1983a), and these conspicuous birds facilitate following the mongooses for entire days. In over five hundred hours of observation, Rasa recorded approximately one raptor attack every six hours and her animals were disturbed by raptors more than once an hour. Of the eighty-seven observed attacks, eight (9 percent) were successful. Young dwarf mongooses are particularly at risk, and six of the victims observed by Rasa were juveniles five to fourteen weeks old. In the Serengeti study, less than half (46 percent) of 148 juveniles counted when less than two months old were still present at one year (Rood, 1983a). Most (78 percent) of the young that disappeared did so before they were six months old.

Both juvenile and adult mortality is significantly higher in small packs than large ones (see above), probably because of the greater vulnerability of mongooses in small packs to predation. The following incident illustrates the difficulty small packs have in protecting young mongooses from predators: ". . . a group containing three adults and three young approximately six weeks old was attacked practically simultaneously by three pale chanting goshawks (*Melierax poliopterus*) which captured all three young within 56 s. Since each adult was, at the time, accompanying a single youngster while foraging, no guard was 'on duty' and the raptors were not noticed until they had almost completed their stoops'' (Rasa 1983b).

Group-living mongooses have evolved several types of antipredator behavior. Dwarf mongooses rely on vigilance and alarm calls, while the larger banded mongooses frequently harass predators. Dwarf mongoose packs stay near cover when foraging and there are usually one or more individuals scanning the surroundings from termite mounds or other elevated areas. Alpha males are particularly active in this sentry role. If a predator is detected, loud penetrating alarm calls cause all group members to run to cover.

In the Serengeti National Park, dwarf mongooses do not occur on the open short grass plains, probably because of the lack of cover, but groups of banded mongooses are able to use this habitat. Available shelter consists mainly of widely dispersed springhare (*Pedetes capensis*) holes or kopjes (rock outcrops). Banded mongooses run—often for several hundred meters—to the nearest cover if a predator such as a lion approaches, but large packs respond aggressively to predators up to the size of a jackal. All pack members typically form a tight bunch and frequently stand up while approaching the predator, giving the appearance of a single large animal. This causes the predator to retreat and it is frequently chased. Adult males normally take the lead in these displays. Banded mongooses also bunch and threaten stationary raptors, causing them to leave the area. In one instance harassment, led by the dominant male, resulted in the rescue of an adult pack member that had been captured by a martial eagle (Rood, 1983b). Dwarf mongooses do not bunch and harass potential predators, probably because their smaller individual and group size would render such behavior ineffective and even dangerous.

The protection from predators of young and vulnerable mongooses is an important function of the social groups. Alloparental behavior has been recorded in all group-living mongooses which have been studied and is common in many carnivore societies (Macdonald, 1983).

DISCUSSION

Wozencraft (pers. comm.) places all the group-living mongooses in the subfamily Mungotinae (Table 7.1), suggesting that sociality may have evolved only once in the Herpestidae. A likely route to sociality is through the mated pair and family. Pair living may have evolved from a solitary, polygamous ancestor in which females were dispersed at low densities making it difficult for a male to include more than one within his range. By remaining with a single female and guarding her from rival males he could prevent cuckoldry, assuring that all offspring raised would be his. The male could assist the female in guarding and feeding the young, and both mongooses would benefit from their increased ability to detect and respond to predators. The male might also help the female to defend resources. However, in the case of food, it seems unlikely that resource defense would be an important reason for pairs to remain together. It is probable that an ancestral social mongoose would have fed primarily on insects and other invertebrates as do all living members of the Mungotinae. And, as argued above, insectivory does not promote competition in small carnivores.

Once pair-living evolved, it is not difficult to conceive how a *Helogale*-like society might occur. Predation pressure would select for pairs whose offspring remained in the family group and helped to raise their younger siblings. Young adults who remained with their parents would probably have been reproductively suppressed but the cost in delayed reproduction may have been outweighed by kin selection benefits and increased survival—with possibilities for later reproductive opportunities—deriving from the protection offered by group life. Once groups of extended families formed there would be pressure for individuals to leave their natal groups and join others in which the breeding opportunities were greater.

In attempting to trace the evolutionary origins of group life it may be useful to consider the method of pack formation in the group-living *Helogale* and the costs and benefits of immigration to members of breeding pairs. Packs sometimes form when single or related pairs of females leave their natal ranges, establish residence in an unoccupied area, and are joined by one or more males. A high proportion of newly formed groups fail but successful groups grow through recruitment and immigration and may continue for many years. Members of newly formed pairs should (and do) encourage opposite sexed immigrants as potential breeding partners and helpers and as extra antipredator sentries. While an immigrant male might be a threat to a breeding male's status, in fact single male immigrants have always joined at a subordinate level. Immigrant males also benefit the breeding male by lessening the possibility that he will lose his position in a male takeover. Since natal philopatry is common in females, dominant individuals might be expected to repulse prospective immigrant females because they are likely to become breeders if the alpha female dies, thereby depriving her daughters of obtaining a breeding opening. However, breeding females have been seen apparently welcoming female immigrants; perhaps the advantages of increasing group size in a small vulnerable pack outweigh the possible costs.

Waser and Waser (1985) present an alternative route to mongoose sociality. They suggest that the group-living mongooses have evolved from a form with an *Ichneumia*-like social structure characterized by shared female ranges resulting from natal philopatry. Some selective factor such as predation pressure may have favored group foraging in female clans, and males may then have joined these groups because of the higher reproductive payoff accompanying defense of multiple rather than solitary females, or as a consequence of the same selective forces favoring group living among females.

However, mongoose species in which groups of females forage separately from males do not exist, although this type of social system is found in one small- to medium-sized carnivore, the coati, *Nasua narica* (Kaufmann, 1962). Adult males are an integral part of the group in all gregarious mongooses. In *Helogale*, social bonds, as measured by the frequency of allogrooming and allomarking, are strongest between adults of opposite sex, further suggesting that the formation and maintenance of the pair bond has been fundamental in the evolution of mongoose social life. Also, some less social species commonly den and forage in pairs and family groups—and in *Cynictis* and *Mungotictis* more complex social units may occur. *Cynictis* appears to den communally but often forages alone; it frequently uses hearing in detection of prey (pers. obs.) and this may favor solitary hunting. *Cynictis* might be transitional between more solitary and gregarious forms. Long-term studies on this and others of the so-called pair-living mongooses are greatly needed. Information on the stability of the pair bond, the genetic relatedness of associated animals, and the extent if any of paternal and alloparental care of the young would be of particular interest.

Among small carnivores, only the Herpestidae have produced forms that forage and den in groups containing several adult males and females. Neither the closely related Viverridae nor the Mustelidae—the carnivore family with the greatest diversity of species—have evolved life in cohesive groups. Why has group living not evolved in other taxa? The gregarious mongooses are all small, diurnal, terrestrial, and insectivorous. No other small carnivore combines these traits. The viverrids are primarily nocturnal and arboreal and these characteristics may make it difficult for individuals in a group to maintain contact (Ewer, 1973). The viverrinae and many mustelids are adapted for killing rodents and other small vertebrate prey, a diet that promotes feeding competition and does not favor gregariousness.

In the most social of the mustelids, the European badger, several adult males and females share a territory and sleep in a communal den (Kruuk, 1978a). Badgers are nocturnal and forage and travel individually, not in cohesive groups; they feed primarily on patchily distributed earthworms whose biomass and productivity is vastly in excess of what the badger population is estimated to consume (Kruuk, 1978b). As in mongooses, sociality in badgers has been facilitated by the utilization of an abundant food resource which enables individuals to share a range with a minimum of feeding competition. However, unlike mongooses, badgers do not cooperate in antipredator situations, and they inhabit a saturated habitat where there is probably a high penalty on leaving the area because all niches are occupied and residents are ferociously territorial. Badger groups occur when young badgers are allowed by their parents to remain in their natal area—sometimes for several years—until a breeding opening in a neighboring clan becomes available (Kruuk, pers. comm.).

The route to sociality suggested here for the group-living mongooses resembles that in the Canidae (see Kleiman and Eisenberg, 1973) in being based on long-term affiliations between a pair and matured offspring. However, the selective pressures that have shaped the societies in these two families have been different. Cooperative hunting is often advanced as the principal reason many large carnivores, including canids, live in groups (Mech, 1970; Kruuk, 1972; Schaller, 1972; Kleiman and Eisenberg, 1973), but see Lamprecht, 1981; Mills, 1982; Packer, Chapter 19 this volume for alternative explanations in some species. Many large carnivores are vulnerable to predators primarily, or even only, when young and predation may not have been a primary factor in the evolution of their groups. In contrast mongooses are vulnerable to a wide variety of predators thoughout their lives.

The evolution of sociality in the African mongooses may have been expedited by the abundance of shelters constructed by other animals that provided suitable cover for groups of small mongooses and allowed communal denning. Banded and dwarf mongooses do not dig their own dens and typically use termite mounds, which are large enough to provide shelter for twenty or more individuals. The ventilation shaft openings are frequently just large enough for a banded mongoose-sized animal to squeeze through, and the mounds provide ideal bolt holes when predators are detected. While meerkats (*Suricata suricatta*) are capable of excavating their own burrows, their small size allows them to use those dug by ground squirrels, *Xerus inauris* (Lynch, 1980). Little is known of the denning habits of *Crossarchus*, but according to Kingdon (1977), *C. alexandri* shelters in termitaries and hollow trees.

A shortage of suitable dens may limit communal denning and hence group size in the white-tailed mongoose (Waser, 1981) and other large, solitary forms. Waser and Waser (1985) found that most *Ichneumia* dens were inactive *Odontotermes* termite mounds, often at the base of trees, in which the mongooses had enlarged the ventilation shafts. Several were so shallow that the mongoose was visible from outside, suggesting that *Ichneumia*'s ability to excavate more suitable dens was limited and that good sites were in short supply.

The radiation of large carnivores in the Felidae and Canidae was promoted by the explosive Pliocene radiation of the Artiodactyla, which in turn accompanied the appearance of extensive areas of grassland during the late Miocene and Pliocene periods (Kleiman and Eisenberg, 1973; Sunquist, 1981). The presence of these ungulates inhabiting open terrain encouraged sociality in some large carnivores by promoting cooperative group hunting. And it may also have allowed sociality in small carnivores by creating the conditions for an abundant food supply. A chain of events that links large ungulates and insectivorous small carnivores exists in many parts of Africa today. For example, when the great herds of wildebeest, zebra, and gazelle migrate from the short grass plains of the Serengeti National Park, Tanzania, at the end of the rainy season, they leave behind large quantities of dung. The abundant dung beetles attracted by this rich resource allow banded mongooses to forage and travel in groups throughout the ensuing dry season when many Serengeti animals are forced to leave the plains. Such an abundant food resource may have facilitated group formation in mongooses ancestral to the group-living forms existing today.

8. Ecological Factors Influencing the Social Systems of Migratory Dabbling Ducks

———————————— • ————————————

FRANK MCKINNEY

DABBLING DUCKS (Anatini) pose several intriguing problems for behavioral ecologists. Why do they differ from most birds in that females, rather than males, are philopatric? Why are they basically monogamous, given that males do not incubate and play no part in brood care in most species? Why does pairing occur in winter, often months before breeding begins, and why are new pair bonds usually formed each year? Why do males of certain species defend breeding territories? In the light of current views on sex-biased dispersal (Greenwood, 1980), mating system evolution (Emlen and Oring, 1977), and pair-bond duration (Rowley, 1983), these characteristics call for evolutionary explanations in terms of phylogenetic, ecological, and behavioral factors.

Several authors have discussed factors that could have led to male-biased dispersal, monogamous mating systems, and winter pairing in waterfowl (Anatidae) in general (Lack, 1968; Wittenberger and Tilson, 1980; Greenwood, 1980, 1983; Oring, 1982). However, there are important variations in morphology and life history within this large family of birds that need to be kept in mind. For example, the factors promoting monogamy in swans and large geese (in which families remain intact for six to nine months and pair bonds tend to be lifelong) may well be different from those promoting monogamy in migratory dabbling ducks (in which parental care lasts only about 1.5 months and new pair bonds typically form each year). Furthermore, even within the Anatini, there may be important differences between the mating systems of sedentary or nomadic species (McKinney, 1985) and those of migratory species. In this broad review, I will focus on key sources of selection that appear to have shaped the social behavior of migratory (primarily Northern Hemisphere) species and their relationship to life-history characteristics distinctive to this group.

KEY FEATURES OF DABBLING DUCK BIOLOGY

Dabbling ducks (including thirty-seven species of *Anas* within the tribe Anatini) are among the most flexible and adaptable members of the family Anatidae. (For reviews of dabbling duck biology see Weller, 1964; Bellrose, 1976; Cramp and Simmons, 1977; Johnsgard, 1978.) Like most waterfowl, they are strong fliers. Twenty-two species are wholly or partly migratory (Johnsgard, 1978). In contrast to ducks that feed by diving (pochards, stiff-tails, sea ducks) the elongated body shape and relatively small feet of dabbling ducks equip them well for walking as well as for swimming on the water surface where they obtain most of their food (Raikow, 1973). Consequently nest sites are not necessarily adjacent to wetlands, and females are capable of walking up to several kilometers to lead their newly hatched ducklings to water. Nest sites are usually on the ground, incubation is by females only, and the primary antipredator strategy of nesting females is crypticity. Nevertheless incubating females are very vulnerable to mammalian predators, and studies of remains at fox dens have shown that many are taken on the nest (Sargeant 1972; D. H. Johnson and Sargeant, 1977; Sargeant et al., 1984). Also, the eggs and ducklings of dabbling ducks are often subject to heavy predation (e.g., 50 percent or higher losses of clutches) and females of most species commonly lay replacement clutches (Bellrose, 1976).

Although ducklings are precocial and feed themselves, they do need to be brooded to protect them from cold, they must be led to good feeding areas, and they need to be warned about and protected from predators (e.g., by parental warning calls, escorting to cover, and distraction displays). Parental care is entirely by females in migratory Northern Hemisphere dabbling ducks and, even in certain Southern Hemisphere species in which the male also stays with the brood, it is the female who leads and broods the ducklings. Females remain with their ducklings for forty to sixty days, usually until they can fly (Evans et al., 1952; Oring, 1964a; Ball et al., 1975).

Although migratory dabbling ducks have radiated to occupy diverse wetland habitats and to exploit varied food sources (primarily aquatic plants and invertebrates; Thomas, 1980, 1982; Sugden, 1973; White and James, 1978; Eadie et al., 1979; Swanson et al., 1979; Weller, 1972, 1975; Nudds and Bowlby, 1984), each species has considerable dietary flexibility. This permits seasonal changes in diet, for example in response to temporary flooding (Krapu, 1974) or when birds move from wintering to breeding ranges (Wishart, 1983a). Diet can also change in response to changes in energetic and nutritional needs. Especially significant in this respect is the switch from an omnivorous or vegetarian diet to mainly invertebrate foods by females during the period of egg production (Krapu, 1974, 1979, 1981; Swanson et al., 1979).

Ducks lay large eggs with large yolks and thus eggs are costly to produce. J. R. King (1973) estimated that as much as 52-70 percent of a female's daily energy intake goes toward egg production. Krapu's (1974, 1981) studies in North Dakota have shown that migrant pairs of northern pintail (*Anas acuta*) and mallard (*A. platyrhynchos*) are heavy and fat when they arrive on the breeding grounds. Lipid reserves built up before arrival are used dur-

TABLE 8.1

Homing rates of dabbling ducks banded on the breeding grounds in North America

| | Females | | | | Males | | | | |
| | Adult | | Juvenile | | Adult | | Juvenile | | |
	No. banded	% returned	No. banded	% returned	No. banded	% returned	No. banded	% returned	References
Mallard	15	13	20+	5			13+	0	1
	24	42	122+	6					2
	113	46	140	5					3
North American black duck	89	25	289+	2					2
Northern pintail	44	39	115+	13			132+	2	1
Gadwall	16	37	8+	12			9+	0	1
	52	29							4
	33	63	47	6	242	9*	25	4	5
							17+	0	5
Northern shoveler	19	42	12+	8			12+	0	1
	20	15	116	3	19	11*	134	1	6
Blue-winged teal	58	14	30+	0			19+	0	1
	16	0	200+	0					7
American wigeon	21	43							8

SOURCES: 1) Sowls, 1955; 2) Coulter and Miller, 1968; 3) Doty and Lee, 1974; 4) J. M. Gates, 1962; 5) Blohm, 1978, 1979; 6) Poston, 1974; 7) McHenry, 1971; 8) Wishart, 1983b.

+ Captive-reared ducklings, released as juveniles.

* Twenty-one of twenty-three adult male gadwalls and two adult male northern shovelers that returned were unpaired.

ing the early part of the breeding season, while females are producing the first clutch of eggs and males engage in mate-defense and energetically costly pursuit flights. Females obtain most of the protein required for egg production by feeding on invertebrates, which is extremely time-consuming (Dwyer et al., 1979). Krapu (1981) believes that, in the case of the mallard, protein intake limits clutch size, and lipid reserves provide the energy resources necessary for the female to acquire the protein needed for the initial clutch. Thereafter, lipid reserves are depleted and the ability of females to produce renest clutches depends on availability of food on the breeding area. During incubation, females lose weight as they use up lipid reserves and have reduced time available for feeding (Folk et al., 1966; H. J. Harris, 1970; Krapu, 1981; Gatti, 1983).

Like other waterfowl, dabbling ducks have an annual, postbreeding wing-molt, during which the flight feathers are molted simultaneously and the birds are flightless for three or four weeks while the new feathers are growing. An annual flightless period occurs in only a few other bird groups, mostly aquatic types, such as loons and grebes, that can afford the risks involved (Woolfenden, 1967). The main advantages of a simultaneous rather than a gradual wing-molt are probably related to seasonal partitioning of major time and energy demands in the annual cycle (J. R. King, 1974). In migratory dabbling ducks, activities that require full flying ability coincide with the periods in the year when energy demands are great (migration, breeding, and wintering phases) (Fig. 8.1). The scheduling of the flightless period after the breeding season is probably favored (at least in the Northern Hemisphere) by the high predictability of abundant food and warm weather in the summer months.

Flightless waterfowl are restricted in their mobility and are especially vulnerable to predators (Oring, 1964b; Gilmer et al., 1977). Replacement of the flight feathers is energetically demanding and requires a nutrient-rich diet (Owen, 1980). In many species, there are well-marked seasonal movements to traditional molting places where food and cover are in close proximity (Salomonsen, 1968). Dabbling ducks favor large marshes that provide safety from drought, a rich reliable food supply, emergent vegetation for escape cover, and seclusion from disturbance (Hochbaum, 1944). In many species, however, it is primarily males and those females that have been unsuccessful in their breeding attempts that move to these molting marshes; females that raise broods tend to stay in the breeding area to molt.

In summary, for females of migratory dabbling ducks, reproduction entails a) high risks of predation while at the nest and when escorting ducklings; b) high energetic demands of egg production calling for increased time spent feeding; c) frequent need to produce replacement clutches; d) loss of body weight during incubation associated with reduced time available for feeding. During the annual flightless period, dabbling ducks have special needs for a rich, localized food supply and habitat providing security from predators; females who raise broods may have less ready access to these requirements than males. Also, in some species, the ability of females to reproduce successfully can be influenced by their body condition on arrival at the breeding grounds, which in turn may depend on conditions on the wintering grounds. Therefore the costs and risks of reproduction are inevitably high for females, and males may be expected to behave in ways that tend to reduce these costs and risks for their mates.

FIDELITY TO THE BREEDING AREA

Studies by Sowls (1955) and others (Table 8.1) have documented strong tendencies for female mallard, northern shoveler (*A. clypeata*), North American black duck (*A. rubripes*), gadwall (*A. strepera*), and American wigeon to return to breed in the same area in successive seasons, as long as

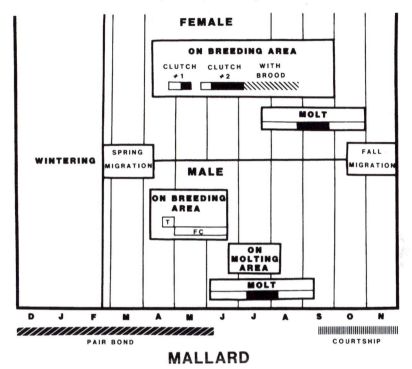

MALLARD

Fig. 8.1. The timing of seasonal activities in a pair of mallards. In this hypothetical example, the female lost her first clutch to a predator soon after incubation had started but the replacement clutch hatched successfully. On the breeding area, this male was territorial (T) initially, then made forced copulation (FC) attempts on other females. The periods of body feather molt are indicated in white, wing-molt in black. The representative chronologies for male and female are based on data on North American migratory populations summarized by Bellrose (1976).

they survive and the habitat remains suitable. However, this "homing tendency" is weak in female blue-winged teal (*A. discors*) and greatly influenced by habitat conditions on the breeding grounds in the northern pintail. Although Sowls (1955) showed that pintails home well when conditions are suitable for breeding, during periods of severe drought in the North American prairies this species can be nomadic and pioneering (G. S. Hochbaum and Bossenmaier, 1972) and many fly north to the arctic (R. I. Smith, 1970; Calverley and Boag, 1977; Derksen and Eldridge, 1980).

Studies of two species in which females show strong homing tendencies (gadwall and northern shoveler) have revealed that males rarely return to the same breeding area in successive years unless they fail to acquire a mate during the winter (Table 8.1). Natal philopatry (return of yearlings to the area of their birth) has been documented also in several migratory *Anas* species, and again the tendency is stronger in females than males (Table 8.1).

Dabbling ducks of both sexes are capable of breeding in their first year (Bellrose, 1976), but yearlings weigh less, have reduced lipid reserves in spring, lay smaller clutches, and lag behind older birds both in pairing and laying (Krapu and Doty, 1979; Dean and Skead, 1979; Wishart, 1983a, b). Postfledging mortality rates in the first year of life are high in migratory populations subject to hunting (Boyd, 1962; Cramp and Simmons, 1977), but annual survival is much higher among adults (e.g., 62% for male, 54% for female North American mallards according to Anderson, 1975). Maximum ages for wild mallards are thought to be sixteen to twenty years (Anderson, 1975), and an exceptional individual can live as long as twenty-nine years (Clapp et al., 1982). In summary, birds that survive beyond the first year have many advantages over yearlings, and presumably they stand a better chance of evading predators and producing offspring because of their experience in breeding, molting, and wintering areas.

Female dabbling ducks take the lead in selecting: a) the breeding area; b) the nest-site; c) the places visited by the brood (H. A. Hochbaum, 1944; Sowls, 1955; Poston, 1974; Wishart, 1983a). They are dependent on the area selected as the breeding home range for the food resources they need to produce eggs and to sustain them during incubation; their choice of a safe nest-site can determine their own survival and that of their eggs, and, by their decisions on where to lead their ducklings, they must play a major role in brood survival. In making all of these decisions, females stand to benefit by being in familiar terrain, and it is not surprising that fidelity to the breeding area is strongly developed in female dabbling ducks.

WHY ARE DABBLING DUCKS BASICALLY MONOGAMOUS?

All dabbling ducks form pair bonds and monogamy is the general rule. This poses an interesting problem since only females incubate and in most species males do not take part in care of the young. Why then do males restrict their reproductive efforts by practicing monogamy?

The most popular hypothesis is that male-biased sex ratios produce such strong competition among males for access to females that each male is forced to defend a single female (Lack, 1968; Wittenberger and Tilson, 1980). For example, in eight North American species of *Anas*, males comprise 53-63 percent of fall populations (Bellrose, 1976) and similar male-biased ratios have been reported for several Southern Hemisphere species (Caithness and Pengelly, 1973; Dean and Skead, 1977). The main factor producing these imbalances is generally thought to be heavy mortality of females during the breeding season.

Two aspects of the phenomenon of male-male competition are relevant here. Does a male have any opportunity to breed if he does not obtain a mate, and does a male need to guard one female to ensure that his sperm will compete successfully with sperm from other males? Both factors are probably important in dabbling ducks.

Several studies have shown that some male dabbling ducks (even old, experienced individuals) fail to obtain mates in certain years and arrive on the breeding grounds unpaired (e.g., Poston, 1974; Dwyer, 1974; Blohm, 1978; Humburg et al., 1978; Seymour and Titman, 1978). Some of these males succeed in pairing with renesting females whose pair bonds have broken, but others remain unpaired for the whole breeding season. In the species studied to date, unpaired males apparently continue to court females rather than switching to a promiscuous strategy. Therefore, failure to obtain a mate may well mean that a male has little chance of breeding that season.

Mate-guarding is generally strongly developed in male dabbling ducks, but there are specific variations in the vigor and duration of male defense (Titman and Seymour, 1981; McKinney et al., 1983). As well as closely escorting and remaining watchful, paired males actively threaten other males that show interest in the female and attack males that attempt to copulate with her. Females can be subjected to forced extrapair copulations that lead to sperm competition and, if they occur during the period before and during egg-laying, these inseminations can result in the fertilization of eggs. In species such as the mallard that show moderately strong development of mate-defense behavior, males follow their mates more consistently and closely during the female's fertile period (Goodburn, 1984). Therefore defense of a male's genetic paternity appears to be an important aspect of his monogamous tendencies.

Mate-guarding by males also provides benefits to female waterfowl. By minimizing disturbance by predators and rival males, a male can enhance his mate's feeding efficiency. Such feeding enhancement has been demonstrated in eider ducks (*Somateria mollissima*) (Ashcroft, 1976; Spurr and Milne, 1976), shelducks (*Tadorna tadorna*) (I. J. Patterson, 1977), swans (*Cygnus columbianus bewickii*) (D. K. Scott, 1980), geese (*Branta canadensis*) (McLandress and Raveling, 1981), and goldeneyes (*Bucephala clangula*) (Afton and Sayler, 1982). Time budget studies indicate that mateguarding occurs in many species of dabbling ducks during the period of egg production (Dwyer, 1975; Seymour and Titman, 1978; Afton, 1979; G. R. Stewart and Titman, 1980; Titman, 1981). Males of some dabbling duck species also defend breeding territories within which their mates feed during the incubation period.

Therefore, males stand to benefit from defending individual females by ensuring that they will have an opportunity to breed and by protecting their genetic paternity. Females also benefit by having a male consort during their fertile periods because this coincides with the period when they must spend extra time feeding for egg production, and male vigilance and protective behavior can enhance their feeding efficiency. If males are contributing essential support for their mate's breeding effort, perhaps females really control dabbling duck mating systems by forcing males to be monogamous? Some light can be thrown on this question by examining the ways in which males engage in secondary reproductive strategies.

SECONDARY MALE REPRODUCTIVE STRATEGIES

As well as maintaining a pair bond with one female, male dabbling ducks attempt to copulate forcibly with other females. This behavior has been recorded in twenty-one species of *Anas* and, at least in well-studied species such as the mallard and northern pintail, forced copulation appears to be a secondary reproductive strategy of paired males (see McKinney et al., 1983 for review of this and other hypotheses). Experiments with captive mallards have shown that females can store sperm for up to seventeen days (Elder and Weller, 1954) and that eggs can be fertilized by sperm delivered during forced copulations (Burns et al., 1980). Forced copulation attempts are directed mainly at females during their fertile periods. As well as defending their mates and attempting to prevent forced copulations, paired males have also been seen to force copulation on their own mates after the latter have been assaulted. Such behavior suggests an antidote insemination strategy. Therefore some (perhaps many) dabbling duck species appear to have "mixed male reproductive strategies" of the type predicted by Trivers (1972): monogamy as the primary strategy, promiscuity as a secondary strategy.

Similar mixed strategies have been reported for bank swallows (*Riparia riparia*) (Beecher and Beecher, 1979), snow geese (*Anser caerulescens*) (Mineau and Cooke, 1979), cattle egrets (*Bubulcus ibis*) (Fujioka and Yamagishi, 1981), and lesser scaup (*Aythya affinis*) (Afton, 1985). Extrapair copulations have been observed in primarily monogamous species of birds in many families but the forced nature of these copulations is especially obvious in waterfowl (Gladstone, 1979; Ford 1983; McKinney et al., 1984). One factor that may have favored the evolution of forced copulation strategies in waterfowl is the presence of an intromittent organ (phallus). This structure is probably primarily an adaptation to aquatic life; most anatids copulate while the pair is swimming on the water surface.

As well as attempting forced copulations, paired males of several species have been observed to direct courtship displays toward other females. This behavior occurs during the early stages of pair-formation while bonds are still tentative (e.g., Wishart, 1983a), and may continue into the breeding season in species such as the green-winged teal (*A. crecca*) (McKinney and Stolen, 1982). In the mallard, although many pair bonds remain intact or reform for renesting attempts within the same breeding season, mate switches sometimes occur (Humburg et al., 1978; Ohde et al., 1983), producing serial monogamy. Therefore, extrapair courtship is probably associated with tendencies for paired males to assess alternative mates.

Opportunities for males to hold more than one mate simultaneously are rare in migratory dabbling ducks because of male-biased sex ratios. Wishart's (1983a) observations on pair formatiton in the American wigeon (*A. americana*) suggest that persistent courtship and aggression among rival males prevent bigamous relationships from developing. In species where females have a strong tendency to return to the same place to breed, it is difficult to imagine how polygynous bonds could persist after the birds leave their wintering areas in spring. Each male can follow only one female back to her breeding home range.

The importance of these factors in preventing polygyny in migratory dabbling ducks is supported by a study in which the sex ratio was experimentally altered. Ohde et al. (1983) showed that some wild mallard males were capable of pairing simultaneously with two females when males were removed from, and extra females released into, a local breeding population in Iowa. Bigamous behavior has also been documented in captives of three Southern Hemisphere *Anas* species (Cape teal *A. capensis*, white-cheeked pintail *A. bahamensis*, and speckled teal *A. flavirostris*) (Stolen and McKinney, 1983; McKinney and Bruggers, 1983; McKinney, 1985), and perhaps such behavior occurs at times in wild populations of these species where it is favored by local conditions. In some parts of their range these species may have extended and/or irregular breeding seasons that could produce unusual operational sex ratios. Some sedentary populations offer more opportunities for extrapair liaisons (e.g., African black ducks *A. sparsa*; McKinney et al., 1978), and some may nest in small colonies (e.g., speckled teal in nestholes of monk parakeets *Myopsitta monarcha*; Weller, 1967) where the proximity of females could facilitate bigamy. These possibilities remain to be investigated in the field.

Therefore, male dabbling ducks can diverge from strict monogamy in at least three ways: by seeking forced extrapair copulations, by switching mates during a breeding season (serial monogamy), and by attempting to hold two mates simultaneously (bigamy). Forced copulations have been observed in many species and, at least in migratory species, this appears to be the main way in which paired males can supplement the reproductive effort they invest in their mates. However, promiscuous activities inevitably compete with mate-guarding and mate-support activities. The ways in which males resolve these conflicts can be instructive in revealing the importance of male mate-support roles in certain species.

THE TIMING OF PAIR FORMATION

In the Northern Hemisphere, pair formation in migratory dabbling ducks occurs during the fall, winter, and/or spring months while the birds are in flocks away from the breeding grounds. The chronology of pairing varies greatly among the seven best-studied North American species (Paulus, 1983; Hepp and Hair, 1983, 1984); for example, gadwall pair early in the winter (October-November) while green-winged teal delay pairing until February-March. This means that most pairs of these two species form eight and three months respectively before the next breeding season begins. Why does pairing occur so early?

Observations on courtship and pairing behavior in urban mallards (Weidmann, 1956; Lebret, 1961) and other *Anas* species (Johnsgard, 1960a, b, 1965; von de Wall, 1965; McKinney et al., 1978; Soutiere et al., 1972; Wishart, 1983a, b; Hepp and Hair, 1983) combined with analytical and experimental work with captives (Weidmann and Darley, 1971; McKinney, 1975; Laurie-Ahlberg and McKinney, 1979; Dervieux and Tamisier, 1979; Cheng et al., 1978, 1979; Standen, 1980; Bossema and Kruijt, 1982; Kruijt et al., 1982; D. M. Williams, 1982, 1983) indicate that pair formation includes the following steps:

1. Early in the pairing season, some individuals of both sexes show interest in one another by associating and performing displays (usually while swimming). Male attention-getting displays are highly directional, specifying a particular target female. Responsive females stimulate male courtship by giving calls and swimming jerkily (nod-swimming). "Social courtship" results when several males court one female. Courted females indicate preferences for certain males by swimming after them while threatening other males (a ritualized performance called "inciting"). A favored male responds by moving ahead of the female, showing her the back of his head as though "leading" her away from the group. Many early liaisons formed in this way are tentative and temporary, both sexes frequently switching attention to different individuals.

2. As the pairing season advances, some birds become firmly paired: mates are almost always together, their daily activities are synchronized, and both repel rival males. However, as the proportion of unpaired females in the population dwindles, courtship and competition among males increase in intensity. There is evidence for North American black ducks (Stotts and Davis, 1960), gadwall (Blohm, 1982), and American wigeon (Wishart, 1983b) that older males obtain mates earlier than yearlings.

3. When all females are paired, the remaining unpaired males continue to try to secure mates by courting paired females. Courtship and the testing of bonds continue during spring migration and after arrival on the breeding

grounds. Some unpaired males may eventually pair with renesting females whose pair bonds have already broken, but others remain unpaired (Humburg et al., 1978; Blohm, 1978).

As Lack (1968) pointed out, this pattern suggests that pairing early is advantageous for males because of the male-biased sex ratios in duck populations. Males that delay may fail to obtain a mate and forfeit an opportunity to breed that year. However, although competition between males for mates may be a key factor promoting male courtship during winter, bonds cannot form unless females respond. Are there advantages to females in pairing early?

A major consequence of pairing three to eight months before breeding begins is that individuals have a long time in which to test potential mates and make final decisions with care. Several studies indicate that males as well as females are selective in their choice of mate (reviewed by D. M. Williams, 1983). Males direct their courtship to specific females and females make their selection from those males that actively court them (Kruijt et al., 1982). Mate-choice experiments on captive mallards suggest that females favor high-ranking males with undamaged plumage (D. M. Williams, 1982, 1983).

There are several ways in which females could be assessing a male's competence as a breeding partner. Copulations occur during winter, starting months before males are producing sperm (Höhn, 1947). Although males may initiate precopulatory signals, females determine whether and when mounting will occur by assuming the prone posture. Therefore, females have the opportunity to test male competence in achieving intromission as part of the process of mate assessment. Wishart (1983a, b) suggests that female American wigeon are assessing males on the basis of their performance in mate-guarding and pair-bond maintenance during winter, because these characteristics are likely to reflect their efficiency as breeding partners. There is a rich field for further research on mate-selection processes in these ducks that form pairs away from the breeding grounds.

In addition to advantages of early pairing associated with mate choice, dabbling ducks may benefit from being paired while they are on the wintering grounds. The gadwall is one of the earliest species to pair (October-November) but one of the latest to begin breeding (May-June) and Paulus (1983) has suggested that early pairing is advantageous to individuals in enhancing their access to food resources on the wintering grounds. In contrast to most Northern Hemisphere *Anas*, gadwall feed primarily on leafy vegetation and algae, relatively poor quality foods requiring a high investment of time spent feeding (Paulus, 1982). Paulus (1983) has shown that paired birds dominate unpaired birds, giving the former greater access to preferred feeding areas. Unpaired male gadwall spent more time in locomotion than paired birds (Paulus, 1984), probably in part because of movements related to food search. Hepp and Hair (1984) present further evidence that paired birds are dominant in wintering flocks of dabbling ducks.

Benefits of having paired status while on the wintering grounds are likely to be especially significant for females. The constant presence of a vigilant male escort can enhance a female's access to localized food sources (as in gadwall), protect her from disturbance by courting males, and give her warning of surprise attack by predators. Advantages of these kinds in enhancing the feeding efficiency of females have been demonstrated in a number of waterfowl but until very recently most attention has been given to the period of egg production in the case of dabbling ducks.

Excepting the gadwall discussed above, specific differences in pairing chronology among *Anas* species remain to be explained. A delay in pairing until later in winter could be related to age or body condition, as in the case of yearling males that weigh less and lag behind older individuals in assumption of breeding plumage (e.g., American wigeon; Wishart, 1983b). Also, energetic constraints related to body size, dietary needs, habitats, and climatic conditions could have important influences on pairing chronology. The costs of courtship and mate-guarding may be too high for some species during the early part of the winter (e.g., green-winged teal; Tamisier, 1972). Cost-benefit analysis for different species is another fruitful field for future research.

In summary, males of migratory dabbling ducks are expected to benefit by securing a mate early in the pairing season because of male-biased sex ratios in wintering populations. Both sexes, but especially females, probably benefit from having several months in which to test prospective mates and make their final decisions on breeding partners. Females stand to benefit from having a mate while on the wintering grounds because the male's vigilance and protection allow her to feed more efficiently, and paired birds dominate single birds giving them greater access to preferred feeding sites. The pattern of pairing chronology observed in each species is likely to be influenced especially by energetic and social requirements associated with specific differences in feeding ecology and winter habitats. Within each species, key factors promoting pair formation are different for males and females. Although males compete, often strenuously, to monopolize individual females, females also play active roles in mate selection.

WHY DO MALES DESERT THEIR MATES AND PAIR ANEW EACH YEAR?

Advantages of retaining, or reuniting with, the same mate have been demonstrated in several long-lived bird species (reviewed by Rowley, 1983) and, in view of the benefits that dabbling ducks appear to derive from being paired, it is necessary to explain why a pattern of re-pairing each year has evolved in this group.

The annual breakup of pair bonds in migratory Northern Hemisphere species is well documented. In the fourteen holarctic species concerned, females are observed alone with their broods (Bellrose, 1976; Cramp and Simmons, 1977; Gilmer et al., 1977). Since males of many species leave the breeding grounds weeks ahead of brood-tending females, mates could repair only if individuals return to the same wintering areas. There is evidence that several sea ducks do renew their pair bonds in successive seasons by meeting on traditional wintering areas (Savard, 1985), but this has not been documented in dabbling ducks. One study of marked mallards demonstrated that mates can reunite (probably before leaving the breeding area) and return to breed in the same area in subsequent years (Dwyer et al., 1973), but this is thought to be a rare occurrence.

Several hypotheses have been proposed to account for male desertion. C. G. Sibley (1957) and Kear (1970) suggested that the male's presence may be disadvantageous in betraying the nest-site. At least for dabbling ducks, this is not a very convincing idea because males rarely go near the nest, and in any case it would not apply to such species as the northern shoveler in which males remain conspicuously on territory throughout incubation (Poston, 1974; Seymour, 1974). The possibility that males leave to reduce competition for food with the female and ducklings (Selander, 1966; Salomonsen, 1968) is difficult to envisage as a product of individual selection. Dabbling duck broods are highly mobile, their food resources are not usually economically defendable, and it is unlikely that an individual male's departure could benefit his own family in respect to the food available to them. Moreover, there is evidence that males of two species (blue-winged teal and northern shoveler) may not leave the breeding area after pair bonds break (McHenry, 1971; DuBowy, 1985).

As pointed out by Bellrose et al. (1961), early departure from the breeding area has several potential advantages for males that are not open to brood-tending females. In particular, males have the option of moving to a molting site that provides a rich, assured food supply and greater safety from predators than they would have if they stayed on the breeding areas (Oring, 1964a, b; Salomonsen, 1968; Fredrickson and Drobney, 1979; Prince and Gordon, in press). Males can begin wing-molt a month or more before brood-tending females, which generally delay molt until the ducklings fledge (Oring, 1964a; Gilmer et al., 1977). Thus males have more time to recover from the energetic costs of renewing the flight feathers, to accumulate reserves for migration to the wintering areas, and (especially in species that pair early) to attain body condition that permits expenditure of time and energy in courtship and competition for mates.

In summary, in Northern Hemisphere migratory species of dabbling ducks with short, regular breeding seasons pair bonds break annually because males desert their mates while the latter are incubating. Likely benefits of desertion for males are ability to move to favorable areas for the wing-molt, and early completion of the wing-molt in preparation for migration to the wintering areas and subsequent competition for mates for the next breeding season. Opportunities for mates to reunite for subsequent breeding seasons are limited because of early departure of males from the breeding grounds and presumed difficulties of finding old mates on wintering areas. In general, evidence on the ultimate factors causing male desertion and the rarity of re-pairing is circumstantial and further studies of marked pairs on the breeding grounds are needed.

WHY ARE SOME SPECIES TERRITORIAL?

Most dabbling ducks engage in aerial pursuits at certain stages of the breeding cycle. These pursuits are related to at least four different phenomena: pair formation, forced copulation, mate defense, and territory defense (Dzubin, 1969; Titman and Seymour, 1981; McKinney et al., 1983). Flights associated with pair formation and forced copulation usually involve several (or many) males pursuing a single female, but they can be distinguished by the behavior of the participants. If pair formation is the objective, the males give courtship calls and postures, and the flight proceeds at a leisurely pace; if the males' goal is forced copulation, they are silent, the pursuit is rapid, and the female makes energetic attempts to escape. Mate defense is indicated when chases are closely associated with rivalry between males for mates or copulations. Territory defense is indicated when a paired male chases conspecific pairs as well as single birds, if his pursuits are brief and he returns to his starting point, and if he initiates chases whether his mate is present or not.

Typically a territorial male's "waiting area" (Dzubin, 1955) (also called the "core area," Poston 1974 or "activity center," Dwyer, 1974) is on a wetland (a pond, cluster of ponds, or stretch of shoreline) where he waits for his mate while she is at the nest. Most of his pursuit flights start from here, but in highly territorial species encounters involving threatening and fighting also occur on boundary zones between neighboring territories.

The occurrence, degree, and duration of territorial aggression vary greatly among dabbling ducks and there are several hypotheses on the factors selecting for this behavior (reviewed by Wishart, 1983b). The main proposals are that territoriality a) ensures a male's exclusive access to his mate; b) ensures protection of a food source; or c) reduces nest predation. Underlying recent discussions of these ideas are the assumptions that territory defense entails costs and risks to males, that these are offset by benefits to males, and that attention should be given to the economic defendability of resources crucial to reproduction (J. L. Brown, 1964; Davies, 1978).

There are two aspects to the idea that territoriality ensures a male's exclusive access to his mate. First, by defending a territory a male might be able to reduce the risk of other males copulating with his mate. By spreading out, breeding pairs tend to reduce the number of males in the vicinity of each female, and by driving out trespassing males each territory-owner could minimize the risk of being cuckolded. Male intolerance toward conspecific males during the prelaying, laying, and at least the early incubation phases have been noted in many *Anas* species, especially during the peak period of mate-guarding (e.g., Titman, 1983 for mallard). However, territorial males usually direct most of their pursuits at the females of intruding pairs, which suggests that the objective is to expel pairs (e.g., Dwyer, 1974 for gadwall).

Secondly, establishment of the male's "waiting area" as a rendezvous place to which his mate flies when she leaves the nest could be important in allowing the male to monitor the progress of his mate's nesting attempt. If the clutch is taken by predators, the male's continued use of the core area will favor maintenance of the pair bond for renest attempts. However, it seems that this result could be achieved by site tenacity without defense of the area.

If territoriality provides protection for food resources, the benefits are likely to be greatest for the female during egg production and, in some species, during incubation. In migratory species of *Anas* there is no evidence that food for the ducklings is important. Territories are not defended during the brood-rearing period and broods are so mobile that they often move away from the nesting territory.

In one species, the northern shoveler, territory defense usually continues throughout the incubation period (Seymour, 1974; Poston, 1974) and there is evidence that females are dependent on food obtained on the territory to enable them to survive. Afton (1979) has shown that females feed most of the time they are off the nest during incubation, the proportion of time spent feeding increases as incubation proceeds, and in one instance a female deserted her nest after her mate was killed, apparently because harassment by a new territorial male prevented her from feeding. Afton (1980) attributes the female shoveler's dependence on food resources during incubation to the small body size of this species. Male blue-winged teal (Stewart and Titman, 1980) and American wigeon (Wishart, 1983a, b) are also territorial late into incubation, apparently with similar benefits for females.

The view that the main function of the chasing activities of male dabbling ducks on the breeding grounds is to disperse nests and reduce the risk of predation on eggs was formerly widely held (Hammond and Mann, 1956; Tinbergen, 1957; R. I. Smith, 1968; McKinney, 1965). But this hypothesis has not been critically tested and recent studies suggest that high-quality

protective cover and security from mammalian predators are more important to females in influencing the choice of the site than is the distance from
other duck nests. High densities of nests have been found on many predator-
free islands (Duebbert et al., 1983; Lokemoen et al., 1984) and by planting
high-quality nesting cover and controlling predators Duebbert and Lokemoen (1980) were able to produce high concentrations of nests on experimental plots in farmland areas. While chasing activities could be influencing nest dispersion under certain circumstances, the effects may be
incidental. Perhaps male chases can be explained more simply in terms of
mate-guarding and protection of a feeding area for the female rather than as
products of selection for a nest-spacing mechanism.

Many factors probably influence the incidence and characteristics of territorial behavior in each species as well as the individual variations observed
within species. Specific differences in body size, diet, and feeding methods
must be of fundamental importance in determining the home range requirements of breeding birds (Nudds and Ankney, 1982). The spatial and temporal distribution of food resources and the stability of preferred wetland
types evidently influence the mobility of breeding pairs and the defendability of feeding areas. For example, in the northern pintail, breeding pairs exploit dispersed and ephemeral wetlands and territoriality is probably not an
option for such species (Derrickson, 1978). Within a species, males tend to
hold their territories longer in the early part of the breeding season than they
do late in the season (Bellrose, 1976). Presumably these variations result
from individual decisions by males under the influence of conflicting interests and varying environmental conditions.

In summary, while male interests may be served directly in minimizing
the risk of cuckoldry and in helping to maintain the pair bond for renest attempts, these benefits are insufficient to account for the behavior of the most
highly territorial species. In the latter, territory defense is more likely an important component of indirect male parental investment through protection
of a secure feeding area for the female during egg production and, to varying degrees, incubation.

DISCUSSION

The studies reviewed here have shown that for female dabbling ducks reproduction entails high risks of predation, high costs of egg production, and
energetic stress during incubation. One important consequence, as Greenwood (1980, 1983) pointed out for Anatidae in general, is that return to a
familiar breeding place has great advantages for females, and these advantages provide a convincing explanation for the occurrence of female philopatry in this group of birds. Another consequence, less widely recognized, is
that female dabbling ducks receive support from their mates in the form of

vigilance and protection from disturbance while they feed. Therefore, although males are emancipated from incubation, and in most species broodrearing, their indirect parental investment may be crucial for female reproductive success.

The requirements of breeding females for male support appear to be key factors promoting primarily monogamous mating systems in dabbling ducks. In species where females require male protection during critical periods of the annual cycle (in wintering flocks, while producing eggs, during incubation), male contributions cannot be shared and polygyny is rarely an option open to males. Monogamy is also favored because of benefits to males associated with male-biased sex ratios. By maintaining a pair bond with a single female, a male can be assured of a breeding partner, and by attending her closely he can protect his genetic paternity. However, although the interests of females and males in maintaining exclusive pair bonds do coincide closely at certain times (notably during the female's fertile period while she is forming eggs) the female's requirements appear to be of primary importance in influencing the timing of pairing (e.g., in gadwall) and the duration of pair bonds and male territory defense (e.g., in northern shoveler).

In migratory dabbling ducks with short regular breeding seasons, the partitioning of the annual cycle into three distinct phases—breeding, molting, and wintering—has profound influences on relationships between males and females. Within each species, food and habitat requirements differ at each stage, and in some cases they are different for each sex. The resulting conflicts of interest between males and females are fundamental to the pattern of seasonal pair bonding in which each sex can influence the behavior of the other in various ways.

In winter, females are the limiting sex and they appear to be largely in control of the process of pairing. By their responses to male courtship and their expressions of mate preferences, they probably have the last word in deciding which male they will lead back to the breeding grounds. For males, courtship is not merely a matter of monopolizing one female by keeping other males away from her. They must also practice salesmanship to gain female acceptance through the vigor of their displays, the constancy of their attention, and the demonstration of their competence as effective escorts. There is no evidence that a male can force an unwilling female to pair with him.

After the pair arrives on the breeding grounds, the female remains largely in control of the activities of the pair during selection of the breeding area and nest-site. Male mate-guarding and enhancement of the female's feeding efficiency are especially important during the prelaying and laying phases. Once the female begins to incubate, her need for mate support may or may not persist. In some species females appear to be very dependent on it, but in others they seem able to do without it. In any event, this is the stage at which males can take the initiative in reducing their mate support, attempting forced copulations, and eventually deserting.

Unlike swans and geese, in which biparental brood care is the rule and families remain intact throughout the following winter, male dabbling ducks have the option of deserting their mates during the incubation period. It is possible that some bonds remain intact or re-form, especially in sedentary populations of certain tropical and Southern Hemisphere species, but in migratory species of *Anas*, male desertion seems to be usual. In view of the time and energy required to obtain mates and the probability that well-adjusted partners would benefit by maintaining their bonds from year to year, the occurrence of male desertion needs to be explained.

Two factors favoring male desertion in migratory dabbling ducks are: 1) absence of any further opportunities to breed during the current season and 2) advantages of proceeding with the molt. In Northern Hemisphere species with short breeding seasons, fertilizable females are no longer available after the last renest clutches are laid and unsuccessful females leave the breeding areas. At the same time, males may benefit in several ways by proceeding with the molt. In contrast to their mates, who must remain on the breeding area until their ducklings fledge, males may be able to move to a favorable molting site that provides greater safety and a more assured food supply than they would have by remaining on the breeding area. They can also save time by going into wing-molt early and building up their body condition in preparation for fall migration, the stresses of winter, and the costly process of competing for mates.

It is unfortunate that by labeling mating systems according to their basis in "resource-defense" or "mate-defense" many authors have decided to classify migratory dabbling ducks in the "mate-defense" category. This label can be misleading because it tends to imply that monogamy is imposed by males on females. In fact, the question of which sex "controls" these mating systems is not yet settled. Probably the manipulative power changes hands at certain key stages in the annual cycle, and even within the migratory *Anas* species there can be important differences between species.

SUMMARY

Evidence on the high costs of reproduction for females, male-biased sex ratios, competition among males for mates, advantages of early pairing, female philopatry, male mate-support, requirements of birds in wing-molt, and benefits of male desertion is substantial but the causal links between these phenomena are inferred rather than proven. If the interpretation proposed here is correct, however, it suggests that females have a controlling influence on pairing and on the behavior of males in the early part of the breeding season but as the female's need for male mate-support wanes, the male can determine if and when the bond is to be broken. Use of the term "mate-defense monogamy" in categorizing the mating system of this group should not carry the overly simple implication that monogamy is imposed on females by the monopolizing activities of males.

PART II

Polygynous Patterns

9. The Evolution of Social Behavior and Mating Systems in the Blackbirds (Icterinae)

———————————— • ————————————

SCOTT K. ROBINSON

THE SPATIAL and temporal distribution of reproductive females and the extent of male parental care needed are increasingly regarded as the primary determinants of avian mating systems (Orians, 1969; Emlen and Oring, 1977; Oring, 1982). Ecological factors that promote clumping of females, such as predation and spatially variable food resources, favor the evolution of polygyny because males have the opportunity to defend more than one female (Emlen and Oring, 1977). Studies of the subfamily Icterinae, which includes New World orioles and blackbirds, have been very important in developing and testing this model (e.g., Orians, 1980). Species in which females nest far apart tend to be monogamous, while those with clustered nests tend to be polygynous (Selander, 1958, 1972; Orians, 1972, 1980; Lowther, 1975), though there are exceptions (Lowther, 1975). Comparative studies, however, reveal a fundamental dichotomy in the mating systems of both monogamous and polygynous species. In some species, males defend females directly, while in others, males defend resources used by females. This raises an interesting question: How can males be assured of their paternity if they do not guard females during the egg-laying period?

In this chapter I first present an analysis of the determinants of the mating system of the yellow-rumped cacique (*Cacicus cela*), a blackbird in which males sequentially defend females. I then use the insights gained from this study to examine the social behavior and mating systems of other Icterinae, including species in which males defend resources. I conclude with a discussion of the factors that are likely to favor female defense and those that favor resource defense.

DETERMINANTS OF POLYGYNY IN THE YELLOW-RUMPED CACIQUE

Emlen and Oring (1977) proposed five determinants of mating systems in animals (Fig. 9.1). In this section I examine how each of these factors may influence the mating system of the yellow-rumped cacique. This section summarizes the results of a study of the social behavior of the yellow-rumped cacique in a remote site in Amazonia, the Cocha Cashu Biological

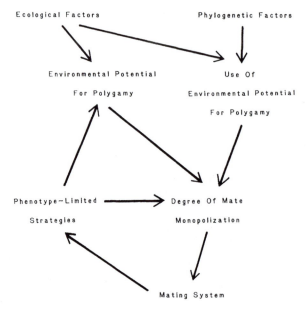

Fig. 9.1. General scheme of the determinants of mating systems. From Emlen and Oring (1977) with modifications proposed by Rubenstein (pers. comm.).

Station in the Manu National Park of southeastern Peru. During the five years of the study, I marked over 750 individuals and followed the fates of 1,120 nests in two nearby populations.

Phylogenetic Factors

A species can be restricted in the ways it responds to ecological pressures by its phylogenetic past, but such restrictions are less likely to be important in the Icterinae, a family that shows great plasticity in the evolution of social and mating behavior (Orians, 1972, 1980; Lowther, 1975). Mating systems vary within and between genera, and even within species (Orians, 1972, 1980). Within the genus *Cacicus*, for example, there are colonial polygynous species (*C. cela* and *C. haemorrhous*; Feekes, 1981), and monogamous territorial species (*C. holosericus* and *C. uropygialis*; Skutch, 1954, 1972). There is also considerable variation in the mating systems of the approximately twenty-seven colonial species, which are chiefly in four genera (*Psarocolius, Caccus, Quiscalus,* and *Agelaius*). In addition to this diversity of mating systems, the ninety-six species of the Icterinae (Blake, 1968) live in a wide diversity of habitats and have a remarkable variety of spacing patterns, diets and foraging tactics (W. J. Beecher 1951; Skutch, 1954; Selander, 1958, 1972; Bent, 1958; Orians, 1972, 1980; Lowther, 1975). The Icterinae are therefore analogous to the Old World Ploceinae (Crook, 1964) in their degree of social plasticity when faced with different environments (Selander, 1972). Perhaps because of their evolutionary plasticity, comparative studies of the Icterinae have been very important in the development and empirical testing of mating system theory (Orians, 1980).

Ecological Factors Affecting Spatial Distribution

Emlen and Oring (1977) argue that the spatial and temporal distribution of resources determines the extent to which multiple mates are energetically defendable, or the "environmental potential for polygamy" (Fig. 9.1). In most birds, the environmental potential for polygamy is strongly influenced by the spatial and temporal distribution of nests, which in turn affects the distribution of females during the breeding season (Lack, 1968; Wittenberger, 1979; Oring, 1982). In the yellow-rumped cacique females nesting in synchronous clusters on islands and around wasp nests experience greater fledging success than females nesting in other kinds of sites (Fig. 9.2). Islands and wasp nests, which are extremely scarce, offer protection against mammalian predators such as brown Capuchin monkeys (*Cebus apella*) which destroy many colonies in sites that are less isolated from the surrounding forest (Robinson, 1985a). Clustered, synchronous nests within colonies are more successful because they are defended by more individuals, which mob avian predators such as toucans and caracaras (Robinson, 1985a). The likelihood that an attack by black caracaras (*Daphius ater*) or Cuvier's toucans (*Ramphastos cuvieri*) will be successful declines significantly as the number of caciques mobbing them increases (Robinson, 1985a). Clustered nests that contain a mixture of active and abandoned nests also provide protection against great black-hawks (*Buteogallus urubitinga*), which often give up an attack if the first few nests they search are empty. Large clusters may act as "mazes" that confuse predators. Selection to avoid nest predation therefore favors nests that are clustered in space and time.

It seems less likely that the spatial distribution of food resources favors clumping of nests in the yellow-rumped cacique. According to Horn's (1968) model, blackbirds with spatially unpredictable, ephemeral food resources can minimize travel time and improve their chances of locating food by nesting colonially in the geometric center of potential food patches. Species with evenly distributed, predictable resources can minimize travel time by defending solitary territories in which nests are placed in the center. Caciques feed their young primarily on arthropods (chiefly Orthopterans, family Tettigoniidae, and larval Lepidoptera) gleaned from the foliage of trees,

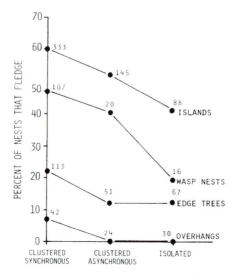

Fig. 9.2. Estimated fledging success of nests in different degrees of clumping in space and time within different types of colony sites on Cocha Cashu, 1979-1983. Nests are arranged in order of *decreasing* local clustering in space and time. "Clustered" nests are located within 1 m of at least two active nests. Isolated nests are located at least 1 m from the nearest active nest. Synchronous nests are those finished within 1 m of at least two other nests within a five-day period. Asynchronous nests are those finished within 1 m of none or one other nest within a five-day period. Synchrony as used here is therefore a measure of local synchrony. Reprinted from Robinson 1985c.

vines, and shrubs (Robinson, 1985b). Such foliage-dwelling insects are generally regarded as an evenly distributed, predictable, and hence defendable resource (J. L. Brown, 1964; Horn, 1968). Therefore, coloniality is unlikely to enhance foraging efficiency in the cacique. Females feed alone or in groups of two or three individuals. If anything, coloniality should decrease foraging efficiency by increasing competition for food near the colony, thus necessitating longer flights to feeding areas that have not been depleted of their resources. Foraging considerations may thus represent a cost rather than a benefit of coloniality for the cacique.

Environmental Potential for Polygamy

As a result of the spatial clumping of cacique nests, there is a high "environmental potential for polygamy" in the yellow-rumped cacique. Males that could defend the best colony sites would have the potential to mate with many females. Indeed, the evolution of polygyny in caciques may have followed a path such as that proposed by Verner and Willson (1966) and Orians

(1969). To avoid mammalian predators, females may have initially aggregated in the territories of males that contained islands or wasp nests. The advantages of nesting on islands or around wasp nests (see Fig. 9.2) may have been sufficient to overcome the disadvantages of reduced parental care from territorial males, which would have to divide their attention among more than one female. The further advantages of social defense against avian predators, which can attack any colony site, would favor increased clustering of nests within habitats, even if it meant forgoing male parental care altogether. S. A. Altmann et al. (1977) hypothesized that cooperative interactions such as mobbing in the yellow-rumped cacique would favor the evolution of polygyny. This scenario depends critically upon the capacity of females to feed young without help from males, which I discuss next.

Ability to Capitalize on "Environmental Potential for Polygamy"

Trivers (1972) and Emlen and Oring (1977) hypothesize that the ability to capitalize on the "environmental potential for polygamy" depends upon the extent of parental care required to raise young. This, in turn, can be affected by both ecological and phylogenetic factors (Fig. 9.1). In the cacique, females raise young unaided by males, except for defense against some avian predators. Two factors appear to make this possible. First, females greatly accelerate their insect-foraging rate when feeding nestlings (Robinson, 1984). Females fly two to four times more often when searching for arthropods to feed their nestlings than they do when foraging for self-maintenance (Robinson, 1984). During the nestling period, females fly between branches as rapidly as many birds less than a third of their weight (Robinson and Holmes, 1982; Robinson, pers. obs.). Such energetically expensive foraging may only be possible because females feed themselves on abundant, easily obtainable fruit and nectar. Even with this high-speed foraging, 80 percent ($n = 25$) of all females fledge only one of two nestlings. Females also lose weight steadily during the nestling period (Robinson, 1984) in spite of the availability of fruit and nectar. These data suggest that females are able to raise young without male parental care, but only at considerable cost.

Second, the deep, enclosed, pouchlike nests of the cacique may provide enough protection from predators to allow females to be absent for extended periods. Many species with open, cuplike nests are extremely vulnerable to "hit-and-run" predators, conspecific marauders, and climatic extremes (reviewed in Haartman, 1969). Such nests require nearly constant attendance by at least one parent. The enclosed, pouchlike nests of the yellow-rumped cacique, however, are very difficult for predators to enter or tear open and should provide protection against rain and cold. Pouchlike nests represent a phylogenetic factor that enhances the ability of caciques to capitalize on the "environmental potential for polygamy," since most arboreal-nesting Icterinae have pouchlike nests including many monogamous species (Skutch, 1954, 1972; Bent, 1958; Koepcke, 1972). Pouchlike nests may therefore be a preadaptation that enhances the potential for polygamy in caciques. Haartman (1969), in fact, found that polygyny in European birds is correlated with enclosed nesting.

In summary, the availability of abundant fruit and nectar, the capacity of individual females to accelerate their insect-foraging rate, and the protection afforded by enclosed nests all make it possible for females to incubate and feed young alone with no help from males. This in turn makes it possible for males to seek more than one mate since their time is not taken up in raising young. Male caciques are thus able to capitalize on the "environmental potential for polygamy."

Degree of Monopolization of Females

Males gather at colony sites where females are building nests and fight among themselves to establish dominance, which determines which males gain access to females, both at and away from the colony (Robinson, 1985c). When a new male arrives at a colony, it engages in a series of aggressive interactions with males that are already residents of that colony. Encounters between comparably ranked males sometimes escalate to aerial grappling fights and prolonged "shouting" matches (Robinson, 1985d). Each aggressive interaction results in a clear winner and loser. The winner can and does supplant the loser from anywhere in a colony. From a table of wins and losses for each male, an approximately linear dominance hierarchy can be constructed (Robinson, 1985c). Males move freely among colonies, gathering wherever there are nest-building females. There is no indication that males play an active role in "herding" females together in colony sites. Females nest together, but feed separately or in small flocks away from the colony in the forest. Since copulations take place away from the colony, males must consort and guard each female during the entire egg-laying period in order to be assured of paternity. Consorting, therefore, limits the extent to which any one male can monopolize the matings in a colony. For example, if five females lay eggs synchronously, they will be consorted by five different males. Therefore, the clumped nesting of females does not *necessarily* create a situation in which males can economically defend more than one female.

Instead, the temporal distribution of females allows a few males to monopolize most females. During the long nesting season, breeding is almost continuous. Females build from two to seven nests per season; most females that lose nests to predators start renesting almost immediately. Successful females often renest later in the season once their young become independent. The number of nests completed in the study area during the main part

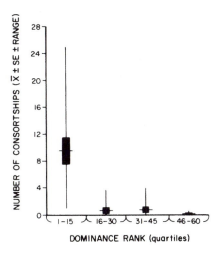

Fig. 9.3. This figure plots the mean number ± SE and range of the number of females consorted by males in different quartiles of the sixty-male hierarchy in 1981 (each quartile consists of fifteen males). Each female counts as a single consortship lasting three days during the period from four to two days prior to the onset of incubation, which covers the period when both eggs are most likely to be fertilized. If a female was consorted by several males, I assigned each male a fraction of a consortship based on the proportion of the three-day period during which it consorted the female.

of the 1981 season ranged from a high of seventy-four in August to a low of thirty-six in December (Robinson, 1985a). In addition, a few females built nests in July (six completed) and January (thirteen completed). Likewise, within colonies, or groups within colonies, many females nest asynchronously (Robinson, 1985e). Top-ranked males therefore have the opportunity to consort many females sequentially. Males can move to new clusters within colonies or switch to new colonies when the local supply of egg-laying females runs out. As a further result of this asynchronous nesting, caciques have a very high "operational sex ratio" (OSR), which Emlen and Oring (1977) define as the number of males per fertilizable female in a population at any one time. In most colony sites the OSR ranges from three to twelve males per female, being highest in the best-protected colony sites (Robinson, 1985b). This indicates that males in the top quarter of the hierarchy have a high potential to mate with most females in the population, which indeed proves to be the case (Fig. 9.3). Males from the top quarter of the hierarchy consort over 80 percent of all females, with each male averaging ten females per season. In summary, asynchronous nesting of females makes it possible for a quarter of the males in the population to monopolize over three-quarters of the fertilizable females.

The Mating System of the Cacique

Emlen and Oring (1977) distinguish among three kinds of polygyny: 1) resource defense, 2) female (or harem) defense, and 3) male dominance. In resource defense polygyny, males defend resources critical to female reproductive success. In female defense polygyny, males defend females, which are clustered together. In male dominance polygyny, males sort themselves out by their positions of dominance and females choose to mate with the highest-ranking males. The mating system of the cacique does not fit any of these categories. As in resource defense polygyny, dominant males defend sections of colonies where there are clusters of nest-building females. However, mating takes place away from the colony, which means that males must consort and guard females wherever they go, as in female defense polygyny. But, because females do not feed together, males can guard only one female at a time, and therefore do not defend harems. The males that gather at colonies establish a dominance hierarchy, as in male dominance polygyny, but dominance determines which males consort females rather than which males are chosen by females.

Rather, the mating system of the cacique could best be characterized as "dominance defense polygyny," which combines all three of the above mating systems. Males gather wherever there are females, establish a dominance hierarchy, and the most dominant males consort individual females throughout their period of receptivity. This mating system is characteristic of many herbivorous mammals, such as bison, giraffes, African buffalo, and elephants, in which males and females are aggregated, but each individual female is highly mobile during her receptive period and must therefore be consorted by males (Jarman, 1983). This may be the first time that such a mating system has been described for a bird. Jarman (1983) in fact stated that such "complex" mating systems are only found in mammals.

Phenotype-Limited Strategies

Wrangham and Rubenstein (Chapter 20) argue that intrasexual competition among individuals of different phenotypes can have a major impact on the "environmental potential for polygamy" and the degree to which males can monopolize females (Fig. 9.1). The existing social pattern determines which phenotypic traits enhance reproductive success. Typically, individuals with different phenotypes adopt different strategies for obtaining the resources that limit reproductive success. Females compete for access to the best nest sites, that is, those that are safest from predators. Larger females tend to win aggressive encounters and therefore control the best nest-sites both within and between colonies (Fig. 9.4). Long-term residents tolerate each other and crowd together in densely interwoven groups in the best positions within colonies (Robinson, 1985e). As a result, small newcomer females are forced to nest asynchronously or in poor colony sites. As I argued

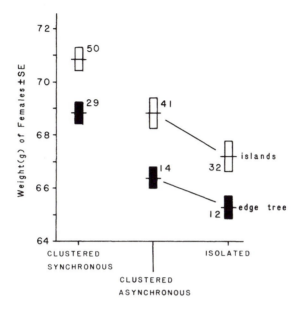

Fig. 9.4. Prenesting weights of females that established nests in different positions in two different kinds of colony sites. Weights are from birds captured within two hours of dawn during a period extending from two weeks prior to the date on which a nest was begun through the middle of nest-building. After this period, females begin to gain weight in preparation for egg-laying. Note that females in the better-protected colony site (islands) and in the better-protected positions within colonies (clustered, synchronous nests—see Fig. 9.2) are heavier than those in more vulnerable sites. Definitions of clustered, isolated, synchronous, and asynchronous as in Fig. 9.2. Lines connect distributions that do not differ significantly by one-way ANOVA. All other distributions differ at the .01 level, including distributions of weights of females in the same positions in different colony sites (e.g., weights of females in clustered, synchronous nests on islands are significantly heavier than those in clustered synchronous nests in edge trees). I do not have enough weights of females nesting near wasp nests and on overhanging branches to compare them with islands or edge trees.

above, asynchronous nesting increases the opportunities for a few males to monopolize most matings. However, having nests scattered among several colony sites decreases spatial clustering of nests and therefore reduces the degree to which females can be monopolized, though many males switch colonies regularly. Therefore, intrasexual competition for nest-sites has essentially the opposite effect of nest predation, which selects for synchronous clusters of nests in a few well-protected sites.

Dominance, which determines access to females, is positively correlated with weight in males (Fig. 9.5). Heavy males tend to win the grappling bouts that determine dominance and therefore consort with most females (Robinson, 1985c). Smaller males adopt ''satellite'' tactics, that is, they

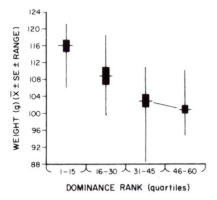

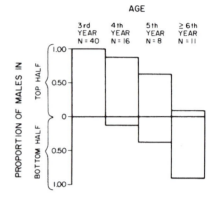

Fig. 9.5. Mean weight ± SE and range of males in different quartiles of the sixty-male dominance hierarchy of 1981. Line connects distributions that do not differ significantly by a one-way ANOVA. All other distributions differ at the .01 level. Weights are from males captured within two hours of dawn from August to December.

Fig. 9.6. Proportion of males of known ages that are in the top and bottom halves of the hierarchy. Data are from two years, 1981 and 1982. This figure does not include males of unknown ages, unless they are known to be at least in their sixth year. Note that nearly all older males are in the bottom half of the hierarchy.

consort females before they are taken over by dominant males and they occasionally double up with more-dominant males that are consorting females. Therefore, heavy males have the potential to monopolize most females. Weight, however, declines with age (Robinson, 1985c), which limits the number of years during which any one male is dominant. Males in the top half of the hierarchy lose an average of 4.6 g per year (Robinson, 1985c), which is sufficient to cause the decline in dominance shown in Fig. 9.6. Likewise, dominance only lasts for part of a season (Robinson 1985c). The nesting season of the cacique may be too long for any one male to maintain its rank and consort females for the entire period. This conforms with Emlen and Oring's prediction that extreme asynchrony may reduce the extent to which males can monopolize females.

The stresses of sequential consorting, costs incurred only by the dominants, may be responsible for the brevity of dominance. During the entire consort period, dominant males must defend their females against satellite males, which are present at the colony and in the forest. It seems likely that males would lose condition during the consort period since they have little time to forage. This, in turn, may make them less likely to win fights with newly arrived young males that have delayed their entrance into the hierarchy until later in the season. Thus, even though satellite males are excluded from direct access to females, they may have a profound effect on the duration of dominance in high-ranking males.

In summary, phenotype-limited strategies of males and females have often contradictory effects on the extent to which males can monopolize females. Competition among females for the best nest-sites tends to cause many females to nest asynchronously, which increases the potential for males to monopolize them. However, it also causes females to spread out over several colony sites, which increases the number of males that consort females. Similarly, heavy dominant males can consort with many females, but changes in weight with age limit the period of time during which males can obtain exclusive consortships.

FACTORS FAVORING NEST-CLUSTERING IN OTHER ICTERINAE

In this section I test the usefulness of the model developed above for the yellow-rumped cacique by using it to account for patterns of sociality in other Icterinae. In the yellow-rumped cacique, the nest distribution appears to be determined by the scarcity of safe nest-sites and the advantages of group defense rather than by foraging considerations. For this reason, I emphasize the possible role of nest predation in accounting for nest distribution. When possible, however, I will use existing data on foraging to evaluate the completeness of the model.

The tendency of marsh-dwelling blackbirds to nest in clusters has generally been attributed to rich, but patchy resources (reviewed in Orians, 1980). There is some evidence, however, that blackbirds may nest in marshes to escape nest predation. Robertson (1972), for example, found that nests in upland fields suffered nearly twice as much predation as those in marshes. Orians (1961) hypothesized that coloniality in tricolored blackbirds, which nest in colonies of up to 200,000 individuals (Lack and Emlen, 1939), may have resulted from a scarcity of marshes relative to the carrying capacity of widespread feeding areas. Tricolored blackbirds forage primarily in agricultural land surrounding marshes, so it is unlikely that the food resources of marshes cause them to aggregate their nests. Marshes, like islands, provide at least some protection against arboreal and terrestrial mammals, though there are few data on the kinds of predators that attack nests in most blackbirds. Social foraging may also be an important factor promoting nest clustering in tricolored blackbirds (Orians, 1961).

The clustering of nests within marshes observed in most blackbirds (reviewed in Orians, 1980) may also serve to increase protection against predators. Female red-winged blackbirds, for example, tend to cluster their nests in territories that have the best cover for hiding nests from predators (Lenington, 1980), though patchiness of food resources may also play a role. Picman (1981) found that female red-winged blackbirds that nest in clusters suffer reduced loss of eggs to marsh wrens, which often puncture eggs in nests of other species. Female red-wings apparently cooperate in defense against marsh wrens. Marsh wrens also attack the nests of yellow-headed blackbirds (Verner, 1975), which may favor clustering of nests, though Willson (1966) argued that female yellow-headed blackbirds aggregate in territories with the best food resources. In the yellow-hooded blackbird, females cluster their nests within marshes and cooperate in chasing shiny cowbirds, a brood parasite, out of colonies (Wiley and Wiley, 1980). Spatially isolated and asynchronous nests of yellow-hooded blackbirds tend to suffer much higher percentages of brood parasitism than clustered, synchronous nests. These studies indicate that marsh-nesting blackbirds may derive the same benefits from coloniality as the yellow-rumped cacique.

Yasukawa and Searcy (1982), however, found that female red-winged blackbirds show no tendency to cluster their nests within high-quality territories. Indeed, female red-wings are territorial in some areas (reviewed in Searcy and Yasukawa, 1983). This suggests the interesting possibility that female red-wings may vary their nest dispersion in response to the kinds of predators that are present in each marsh. In areas without marsh wrens, females may scatter their nests to minimize food competition near their nests. It would be very interesting to compare the spatial distribution of nests in marshes with and without marsh wrens. Overall, however, it is difficult to determine whether food or predation is the primary determinant of nest distribution using the available data.

Like the yellow-rumped cacique, Brewer's blackbirds are restricted in their colony site requirements. In central Washington, colonies tend to be located near the geometric center of feeding areas, but only in areas of extensive sagebrush with nearby bluffs or trees which males use as sentinel perches (Horn, 1968). Males and females mob predators, but are ineffective at deterring mammalian predators. As a result of these conflicting selective pressures, there is no optimal distribution of nests within colonies as there appears to be in the cacique. Horn (1968) and Furrer (1975) found no significant net differences in the intensity of predation in different positions within colonies. Rather, it seems more likely that coloniality in Brewer's blackbirds has evolved as a means of exploiting a patchy, unpredictable food resource as hypothesized by Horn (1968). Brewer's blackbirds forage socially and eat many aquatic insects, which are unpredictable in space and time (Orians and Horn, 1969).

COMPARISON OF MATING SYSTEMS IN THE ICTERINAE

A "broad-brush" comparison of social and mating systems in the published studies of Icterinae provides general support for Emlen and Oring's model

(Fig. 9.1). Species in which females nest together tend to be either polygynous or "promiscuous" (Table 9.1). In these species, males can presumably either defend females directly, or defend the areas in which females nest. Lowther (1975) also found strong correlations between coloniality and "nonmonogamous" mating systems in the Icterinae.

Coloniality is also strongly correlated with extreme sexual dimorphism in size in the Icterinae (Selander, 1958, 1972; Lowther, 1975). Species in which females nest far apart, on the other hand, tend to be monogamous (Lowther, 1975), again, presumably because males cannot monopolize more than one female when nests are scattered. Further support for Emlen and Oring's model comes from the orchard oriole, which is monogamous in most areas, but may be polygynous in areas where nest trees are scarce, and females nest colonially (references in Bent, 1958). Many other species with variable nest dispersions show varying degrees of polygyny (Table 9.1). This view contrasts somewhat with Orians' (1972) review of the mating systems of the Icterinae. Orians emphasized the importance of diet rather than nest distribution and hypothesized that frugivores such as cacique and oropendola species may have clutch sizes limited by the amount of protein females can mobilize for eggs. In this case, females could easily find food for the young without assistance from the male, which would reduce the need for male parental care. However, at least in southeastern Peru, yellow-rumped caciques and russet-backed oropendolas primarily feed their young arthropods (Robinson, 1985b). It is likely that males could be of considerable assistance, especially in the cacique, in which females greatly accelerate their insect-foraging rate when feeding nestlings and seldom fledge more than one young from their clutch of two eggs (Robinson, 1984). Females therefore appear to pay a high price by forgoing male parental care.

There are, however, some considerable differences in the mating systems of species with comparable spatial distributions. For example, some colonial blackbirds are monogamous rather than polygynous (Table 9.1). Likewise, males of some polygynous blackbirds defend territories rather than defending females directly as in the cacique. There are also examples of convergent mating systems in species with very different social systems, such as the brood parasitic brown-headed cowbird and the colonial yellow-shouldered blackbird (Table 9.1). In this section I compare the mating system of the cacique with other mating systems that have been documented in the Icterinae. The purpose of this comparison is to look for ecological similarities between species with the same mating system, and ecological differences between species with different mating systems.

In addition to the mating system of the yellow-rumped cacique, there are at least four other kinds of mating systems found in the Icterinae. Below I discuss each separately.

TABLE 9.1
Nest distribution, habitat, and mating systems in the Icterinae

Nest distribution	Species	Nesting habitat	Mating system	References
Colonial	Tricolored blackbird (*Agelaius tricolor*)	Marshes	Resource defense polygyny	Orians, 1961
	Yellow-hooded blackbird (*A. icterocephalus*)	Marshes	Resource defense polygyny	Wiley and Wiley, 1980
	Brewer's blackbird (*Euphagus cyanocephalus*)	Sagebrush/potholes	Female defense monogamy; female defense polygyny	Horn, 1968, 1970; L. Williams, 1952
	Common grackle (*Quiscalus quiscula*)	Isolated trees in open areas	Monogamy	Bent, 1958; Wiley, 1976
	Boat-tailed grackle (*Q. major*)	Isolated trees in open areas	Resource defense polygyny	Selander and Giller, 1961
	Great-tailed grackle (*Q. mexicanus*)	Isolated trees in open areas	Resource defense polygyny	Kok, 1972
	Yellow-rumped cacique (*Cacicus cela*)	Isolated trees near forest	Dominance defense polygyny	This study
	Red-rumped cacique (*C. haemorrhous*)	Isolated trees near forest	"Polygyny or promiscuity"	Feekes, 1981
	Montezuma oropendola (*Psarocolius montezuma*)	Isolated trees near forest	"Polygyny or promiscuity"	Skutch, 1954
	Crested oropendola (*P. decumanus*)	Isolated trees near forest	Harem defense polygyny(?)	Tashian, 1957; Drury, 1962
	Russet-backed oropendola (*P. angustifrons*)	Isolated trees near forest	Harem defense polygyny(?)	Robinson, pers. obs.
	Chestnut-headed oropendola (*P. wagleri*)	Isolated trees near forest	"Promiscuity" or female defense polygyny	Smith, 1983; Chapman, 1928

	Species	Habitat	Mating system	Reference
Aggregated	Red-winged blackbird (*Agelaius phoeniceus*)	Marshes	Resource defense polygyny	Various refs.
	Yellow-shouldered blackbird (*A. xanthomus*)	Mangrove islands	Female defense monogamy	Post, 1981
	Yellow-headed blackbird (*Xanthocephalus xanthocephalus*)	Marshes	Resource defense polygyny	Willson, 1966
Variable	Bobolink (*Dolichonyx oryzivorus*)	Fields	Resource defense polygyny	Wittenberger, 1980, Martin, 1974
	Yellow-winged blackbird (*Agelaius thilius*)	Marshes	Monogamy (nonterritorial?)	Orians, 1980
	Orchard oriole (*Icterus spurius*)	Woodlands	Monogamy or polygyny (when nests aggregated)	Bent, 1958
	Northern oriole (*I. galbula*)	Woodlands	Monogamy (regardless of nest dispersion)	Pleasants, 1979
	Brown-and-yellow marshbird (*Pseudoleistes virescens*)	Marshes	Monogamy (cooperative) (nonterritorial)	Orians et al., 1977
Dispersed	Eastern meadowlark (*Sturnella magna*)	Fields	Monogamy or polygyny	Lanyon, 1957
	Western meadowlark (*S. neglecta*)	Fields	Monogamy or polygyny	Lanyon, 1957
	Lesser red-breasted meadowlark (*S. defilippi*)	Grasslands	Monogamy or polygyny	M. Gochfeld, pers. comm.
	Long-tailed meadowlark (*S. loyca*)	Farms, fields	Monogamy or polygyny	M. Gochfeld, pers. comm.
	Pale-eyed blackbird (*Agelaius xanthophthalmus*)	Marsh	Monogamy or polygyny	Robinson, pers. obs.

189

TABLE 9.1 Continued.

Nest distribution	Species	Nesting habitat	Mating system	References
	Melodious blackbird (Dives dives)	Second growth shrubs	Monogamy	Orians, 1983
	Rusty blackbird (Euphagus carolinus)	Lakes, bogs	Monogamy	Bent, 1958
	Austral blackbird (Curaeus curaeus)	Woodland	Monogamy (cooperative)	Orians et al., 1977
	Most orioles (24 spp) (Icterus spp.)	Woodland forest	Monogamy (at least 13 of 24 species)	Bent, 1958; Skutch, 1954; Robinson, pers. obs.; Various refs.
	Scarlet-headed blackbird (Amblyramphus holosericus)	Marsh	Monogamy	Orians, 1980
	Scarlet-rumped cacique (Cacicus uropygialis)	Forest	Monogamy	Skutch, 1972
	Yellow-billed cacique (C. holosericus)	Dense shrubs	Monogamy	Skutch, 1954
Brood Parasite	Brown-headed cowbird (Molothrus ater)	Variable—usually woodlands near fields	Female defense monogamy	Ankney and Scott, 1982
	Giant cowbird (Scaphidura oryzivora)	Oropendola and cacique colonies near open areas	Female defense polygyny(?) or promiscuity	Robinson, pers. obs.

NOTE: Mating system follows classification scheme of Emlen and Oring (1977) when applicable. "Colonial" species are those with nests located within a few meters of each other. Species with "aggregated" nests are those in which nests occur at high densities in appropriate habitats. Species with "variable" distributions usually have widely spaced nests but aggregate their nest in some habitats. "Dispersed" nests are widely scattered, usually not within sight of each other.

1. Female Defense Monogamy

Brewer's blackbirds, common grackles, and yellow-shouldered black-birds have social organizations similar to that of the cacique, but males consort with only one female. Males of all three species consort and guard females during the nest-building and egg-laying period. In Brewer's black-birds, pairs form in nonbreeding flocks, so that when colonies are established, each female already has a mate (Horn, 1970). Soliciting females, however, attract all nearby males, so males must guard females in order to assure paternity. Sometimes, many females from the same colony solicit simultaneously, a behavior Horn interpreted as a means by which females guarantee the fidelity of their mates. Males help feed young and act as sentinels during incubation. In common grackles, most competition among males occurs at the time of colony establishment (Bent, 1958; Wiley, 1976). After pairs become established, competition is reduced, but males continue to consort. In yellow-shouldered blackbirds, each male defends a small section of a colony tree and consorts and guards whichever female chooses to nest in its territory (Post, 1981).

Why do males consort only one female in these species? Post (1981) argued that male parental care may be essential in yellow-shouldered black-birds because of the long distances (up to 2 km) between feeding areas and colony sites. Even if females accelerated their foraging rates, the travel time required for long flights would necessitate extended periods away from the colony. Males feed incubating females and nestlings, guard nests when females are away, and participate in mobbing predators. Brewer's blackbirds and common grackles also feed far from colony sites and males participate in guarding nests against predators (Horn, 1968; Bent, 1958). All three species have open, cuplike nests, which may require more guarding than the enclosed nests of caciques. Therefore, the importance of male parental care in these species may limit the extent to which males can capitalize on the environmental potential for polygamy. Tricolored blackbirds, which also have open-cup nests and feed far from the colony, may be monogamous because males are needed to help raise young.

The social organization of the brown-headed cowbird, a brood parasite, is also very similar to that of the yellow-rumped cacique, in spite of the considerable differences in the reproductive biology of the two species. Studies of radio-tagged cowbirds have shown that females commute daily between areas where they search for nests to parasitize and feeding areas (Dufty, 1982; Rothstein et al., 1984). Females spend their mornings searching for nests in territories which they defend against other females (Ankney and Scott, 1982). In the afternoons, females feed in flocks in cattle yards, pastures, or lawns (Rothstein et al., 1984). This pattern is the reverse of that of female caciques, which nest together, but feed separately. Nevertheless, in both caciques and cowbirds, females are clustered at some times of the day and scattered at others.

As in caciques, male cowbirds consort and guard females at feeding areas and when females search for nests. Males gather at feeding sites and form dominance hierarchies (Darley, 1978, 1982; Rothstein et al., 1980, 1984). Dominant males are more likely than subordinates to obtain consortships (Darley 1978, 1982) and may also copulate more often than subordinates (West et al., 1981). The OSR of males is about 1.5 males to 1 female in many populations (reviewed in Ankney and Scott, 1982), so roughly one-third of all males are excluded from forming consortships. Dominance and consorting are correlated with weight in males, and young males actually experience greater consorting success than older males (Ankney and Scott, 1982). Small, subordinate males adopt "satellite" or sneaky tactics, and at least occasionally copulate with females when dominant consort males are absent (P. F. Elliot, 1980; Darley, 1982). Male caciques also form weight-correlated dominance hierarchies that determine which males consort and which adopt subordinate tactics.

The major difference between the two mating systems is that male cowbirds consort only one female, while dominant male caciques sequentially consort many females. Each female cowbird lays twenty to forty eggs during the eight-week breeding season when all of the other females in a population are also egg-laying (D. M. Scott and Ankney, 1980). In comparison, each female cacique is only receptive for a four-to-five-day period and lays only two eggs per clutch. By consorting a single female, a male cowbird could possibly father as many as forty eggs. A male cacique would have to consort twenty females to father an equal number of eggs. Therefore, it may pay for a male cowbird to consort a single female continuously for an entire breeding season (Wittenberger and Tilson, 1980; Ankney and Scott, 1982).

The variable spatial distribution of females may determine the mating system of the brown-headed cowbird and the yellow-rumped cacique. In both species, females gather together for part of the day, but are widely scattered for the rest of the day. A few males could control access to females when they are clumped, but not when they are scattered. Therefore, in order to be assured of paternity, males must consort and guard females individually, rather than as a group. Unfortunately, there are few data on where and when cowbirds copulate (Ankney and Scott, 1982). The spatial distribution of females may also vary geographically (Rothstein et al., 1984) in which case the mating system may also vary, as P. F. Elliot (1980) suggested.

2. Harem Defense Polygyny in Oropendolas?

Oropendolas (*Psarocolius* spp.) are similar in many aspects of their nesting ecology to yellow-rumped caciques, with which they often share colony trees (Koepcke, 1972; Robinson, 1985b). Oropendolas nest colonially in groups of two to fifty nests in sites where they cannot be reached by mammalian predators (Chapman, 1928; Tashian, 1957; Drury, 1962; Skutch, 1954; N. G. Smith, 1968). Oropendolas also nest around wasp nests, which provide protection against mammals and botflies (*Philornis* spp.) that parasitize nestlings (Smith, 1968). At least one species, the russet-backed oropendola, feeds its young mostly on the same kinds of insects that caciques feed their young (Robinson, 1985b). A critical difference between oropendolas and caciques is that oropendolas are also gregarious away from the colony (Robinson, pers. obs.). It is not clear why oropendolas feed more socially than caciques. Oropendolas are considerably less agile than caciques and may therefore need more advanced warning to escape aerial predators such as forest falcons (*Micrastur* spp.) and hawks (*Accipiter* spp.). Females also spend more time searching in suspended dead foliage than caciques (Robinson, 1985b), which may increase their vulnerability to predators. Searching dead leaves required probing maneuvers in which the head is often buried in foliage, and searching dead palm fronds produces a great deal of noise, which could attract predators. Males feed higher in the foliage than females ((Robinson, 1985b), and often give loud alarm calls, which suggests that they may act as sentinels. Therefore, female oropendolas may benefit from selfish herd effects and increased vigilance by foraging in flocks, though this is highly speculative.

Because females nest and feed together, males can potentially guard all of the females from a colony continuously throughout the day. Copulations occur at the colony in *P. montezuma* (Skutch, 1954), *P. decumanus* (Tashian 1957; Drury, 1962), and *P. angustifrons* (Robinson, pers. obs.), though not in *P. wagleri* (Chapman, 1928; N. G. Smith, 1968). Unfortunately, there are no published studies of color-marked populations of oropendolas. One male *P. angustifrons* that I marked in 1981 copulated with at least three females during a four-day period and was dominant to at least two other males. All copulations occurred only when one male was present at the colony, a situation that virtually never occurs in yellow-rumped cacique colonies. The ability of lone males to exclude all other males from a colony tree greatly enhances the ability of males to capitalize on the environmental potential for polygamy since there are no other males to disrupt copulations. This marked male left the colony when the females left and returned to the colony when they did, suggesting that it may have also guarded females away from the colony. It also stayed in the colony during the entire incubation and nestling-feeding period when it chased away giant cowbirds, which are brood parasites, potential nest predators such as toucans, and, inexplicably, attacked and chased away any yellow-rumped caciques that approached an oropendola nest ((Robinson, 1985b). Tashian (1957) and Drury (1962) also describe dominance interactions among males and "associations" between particular males and groups of females in *P. decumanus*. In *P. wagleri*, however, males apparently do not interact aggressively or form dominance hierarchies, and copulations occur away from the colony. Chapman (1928) stated that males sequentially associate with females during the late nest-building period, in which case *P. wagleri* may have a mating system like that of the yellow-rumped cacique.

The extraordinary sexual dimorphism in weight found in oropendolas provides further, albeit indirect evidence that sexual selection is intense among male oropendolas and that they may show pronounced polygynous mating systems such as harem defense. Most males oropendolas weigh at least twice as much as females (ffrench, 1980, N. G. Smith, 1983, Robinson, 1985d). *P. decumanus*, in which males weigh nearly three times as much as females (ffrench, 1980), is one of the most sexually dimorphic in weight of all birds. Such extreme sexual dimorphism is associated with harem defense in a wide array of taxa (Leutenegger and Kelly, 1977; Alexander et al., 1979). In sequentially polygynous species such as red-winged and yellow-headed blackbirds and the yellow-rumped cacique, males weigh only 50-90 percent more than females (Willson, 1966; Searcy, 1979; Robinson, 1985c). Based on these considerations, I predict that most oropendolas will prove to have female defense-based mating systems. Studies of color-marked populations are clearly needed, as are data on the spatial distribution of females during the egg-laying period.

The mating system of the giant cowbird (*Scaphidura oryzivora*) may be somewhat similar to that of oropendolas. Giant cowbirds only parasitize the nests of caciques and oropendolas, both of which are colonial (N. G. Smith, 1968). Females feed together and raid colonies in small flocks that are accompanied by one to three males (Robinson, pers. obs.). Because females are social at all times, it may be possible for males or coalitions of males to defend them against all other males. Two color-marked males that I observed in southeastern Peru appeared to be residents in an area in which there were several cacique and oropendola colonies. Females, on the other hand, only appeared when there were oropendolas building nests. One of these males accompanied a small flock of females during the entire time they were present, but at least two other unmarked males also accompanied these females. I saw no dominance interactions among these males, so I do not know if these males were guarding the females or simply following them. Similarly, males are only 15-25 percent heavier than females (Robinson, pers. obs.), which indicates reduced intrasexual competition among males.

3. Resource Defense Polygyny

All the systems discussed so far involve males guarding females, either singly or in groups. The Icterinae discussed below show very different behavior: they defend the areas in which females nest rather than the females themselves. This mating system is especially prevalent in marsh- and field-nesting species such as the red-winged, yellow-headed, and yellow-hooded blackbirds, and the bobolink (Table 9.1). Various authors have hypothesized that territory quality correlates with the number of nesting females in red-winged blackbirds (Holm, 1973; Searcy, 1979; Lenington, 1980; Orians, 1980; Picman, 1981; Yasukawa, 1981), yellow-headed blackbirds (Willson, 1966), and bobolinks (Wittenberger, 1980) but there are few data on what determines territory quality. Territory quality may be determined by the availability of safe nest sites (e.g., in red-winged blackbird: Holm, 1973; Searcy, 1979; Lenington, 1980; Picman, 1981), or the quality of food resources (e.g., in yellow-headed blackbirds: Willson, 1966; bobolinks: Wittenberger, 1980; and, occasionally, red-winged blackbirds: Ewald and Rowher, 1982). The best evidence for the importance of food was gathered by Ewald and Rohwer, who increased the number of females nesting in a territory by artificially supplementing the food available. Alternatively, females may choose males on the basis of some aspect of male quality such as plumage, song type, or the expected quality of parental care provided, as Weatherhead and Robertson (1979) suggested.

Why males do not guard females during the egg-laying period remains to be resolved in these species. Females of many species with resource defense-based mating systems feed extensively outside the territory, and therefore have ample opportunities to mate with other males. For example, in red-winged blackbirds, there are many young and subordinate males that defend territories around the periphery of marshes where females seldom nest, but often forage (Roberts and Kennelly, 1977; Orians, 1980; Searcy and Yasukawa, 1983). Roberts and Kennelly (1977) found that females often feed outside male territories during the egg-laying period. Indeed, as Bray et al. (1975) have shown in a study in which territorial males were vasectomized, many females do not restrict their matings to territorial males.

In order for it to be worthwhile for males to defend territories, there must be some way that males can be assured of paternity. One possibility is that males may force females to copulate with them, providing parental care to only these females. There is some evidence that female red-winged blackbirds may select territories of older males, which generally provide better parental care (Yasukawa, 1981). Since territorial males exclude all other males from their territories, no other males could provide parental care. A somewhat analogous situation exists in the orange-rumped honey-guide (Indicatoridae) in which males defend bee nests and allow only their mates to forage in their territories (Cronin and Sherman, 1976). Females usually mate only with a resource-holding male in the male's territory. In Anna's hummingbirds the males allow females to feed in their territories only if they copulate with them (Wolf, 1975). There are, however, no data to suggest that these behaviors occur in the Icterinae. In this context, it would be interesting to know where females forage during the egg-laying period, and where, when, and with whom copulations occur in red-winged blackbirds.

In boat-tailed and great-tailed grackles, however, males provide no parental care (Selander and Giller, 1961; Kok, 1972). In these species, males defend portions of colony sites. Females nest in the territories of the males with which they copulate (Selander, 1972). Females could choose males based on some aspect of their plumage or display. The long, v-shaped tails of males, for example, may be a trait that is under sexual selection by female choice. Alternatively, females could choose males based on territory quality, especially if some trees or parts of trees provide better protection against predators than others. It is, however, not clear what determines territory quality or female choice in either species (Kok, 1972).

Selander (1972) argued that resource defense polygyny may be advantageous when nesting is synchronous and males could therefore consort only a few females. By remaining at a colony and defending a territory, males may maximize the number of females with which they mate, especially since copulations occur at the colony. Nevertheless, territorial males cannot be assured of paternity since females feed away from the colony. However, if there is last-sperm precedence, then perhaps the best strategy for males would be to copulate with females after each prolonged absence, as Power (1980) and Power et al. (1981) have observed in mountain bluebirds and starlings.

4. Resource Defense Monogamy

Marsh-nesting blackbirds in relatively aseasonal areas of the tropics tend to be monogamous, or even cooperative breeders. In aseasonal parts of Venezuela, male yellow-hooded blackbirds have a succession of mates in their territories, but rarely have more than one female nesting at any one time (Wiley and Wiley, 1980). Pale-eyed and scarlet-headed blackbirds also defend territories and are monogamous (Orians, 1980; Robinson, pers. obs.). Orians (1980) showed that the large seasonal resource pulse characteristic of temperate marshes does not occur in most tropical marshes. Most species feed on large Orthopterans, which are difficult to capture and widely scattered. Orians argued that these resources are defendable and both sexes are needed to provision nestlings. The low breeding density of females means that there is a low environmental potential for polygyny. Indeed, the carrying capacity of tropical marshes may be low enough that safe nest-sites are not in short supply.

An exception to the above patterns is the brown-and-yellow marshbird, in which nests are sometimes clustered together in patches of dense vegetation (Orians et al., 1977; Orians, 1980). Marshbirds are cooperative breeders in which presumably family groups establish nest-sites. In areas

NEST (•) DISTRIBUTION

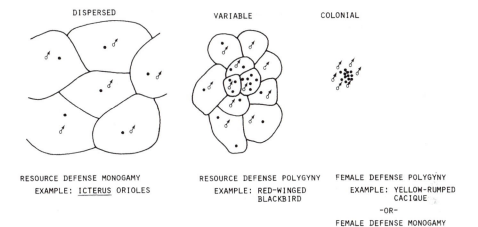

Fig. 9.7. Schematic summary of the relationship between nest
distribution and mating systems in blackbirds.

where both food and nest-sites are in limited supply, cooperative breeding
may be particularly advantageous. Groups may be better at competing for
nest-sites, chasing away predators, and provisioning nestlings. Cooperative
groups may therefore experience some of the benefits of coloniality without
the costs of reduced paternal care and increased competition for nestling
food near the colony. The data for this species, however, are too few to
evaluate this hypothesis.

Conclusions

The spatial distribution of females and the extent of parental care required
appear to be the major determinants of mating systems in the Icterinae (Fig.
9.7). Species in which females nest near each other tend to be polygynous,
while those in which females nest far apart tend to be monogamous. Highly
colonial species in which females feed away from the colony sites tend to
have female-defense-based mating systems. Nesting synchrony and the degree of sociality among foraging females determine the kind of female defense mating system that occurs (e.g., harem defense, dominance defense,
female defense monogamy). Species in which females nest at high densities
in particular habitats tend to have resource-defense-based mating systems.
The spatial distribution of females in turn is determined either by the availability of nest-sites that are safe from predators or the distribution of food.

For most colonial Icterinae, safe nesting habitat is extremely scarce relative to the populations that can be sustained by the food resources of a habitat. Lack (1968) also concluded that coloniality in grassland-nesting Ploceinae results from a scarcity of safe tree nest-sites relative to the populations of seed-eating species that can be sustained.

The distribution, abundance, and kinds of food available determine the extent of male parental care needed. Males play an important role in nest-guarding and nestling-feeding in species that commute long distances to feeding areas, and in those with open-cup nests. Similarly, males play an important role in parental care of young in tropical marsh-nesting black-birds, which live in much less productive habitats than temperate marsh-nesting species. The ready availability of fruit and nectar, however, may make it possible for female caciques and oropendolas to use energetically costly insect-searching tactics when feeding nestlings. Food resources therefore play a critical role in determining the extent to which males can capitalize on the environmental potential for polygamy.

FUTURE INVESTIGATIONS

There are a number of questions regarding blackbird mating systems that have not yet been addressed, some of which I discuss below.

1. To what extent do females defend their nests as a group and how does this affect the spatial distribution of nests? Studies of the yellow-rumped ca-cique and red-winged blackbird suggest that group defense may be an important advantage of clustering nests (Picman, 1981; Robinson, 1984). If group defense is important, then clustered nests should have a higher success rate than dispersed nests, in spite of reduced parental care from males and the increased food competition near the colony (Altmann et al., 1977).

2. How do males assure their paternity in species in which males defend territories rather than females (i.e., resource defense polygyny)? In most species with this mating system, females regularly feed outside of male territories and there is evidence to suggest that females do not always copulate with territorial males (Bray et al., 1975). Answering this question will require extensive observations of where and when females copulate and where they feed during the egg-laying period. It would be interesting to determine if males allow only females that copulate with them to nest in the territories they defend, as has been found in honey guides (Cronin and Sherman, 1976) and hummingbirds (Wolf, 1975).

3. What role do females play in guaranteeing the fidelity of males in monogamous, colonial species such as the common grackle and Brewer's blackbird? Males of both species consort single females (Horn, 1968; Wiley, 1976), though in some areas Brewer's blackbirds are polygynous (L. Williams, 1952). Horn's (1970) study of the timing of precopulatory

displays in females suggests that females may exert considerable control over the extent to which males can monopolize females by nesting synchronously (see also Lumpkin 1981).

4. What phenotypic traits (e.g., age, size) determine female reproductive success, and how do individual mating strageges affect the temporal and spatial distribution of females? There are few data on what factors determine access to the best nest-sites for females. Fighting among females at or near nest-sites has been reported in the yellow-rumped cacique (this study), russet-backed oropendola (Robinson, pers. obs.), chestnut-headed oropendola (Chapman, 1928), Brewer's blackbird (L. Williams, 1952), and the red-winged blackbird (LaPrade and Graves, 1982). In the yellow-rumped cacique, small young females are excluded from the best nest-sites and are forced to nest asynchronously or in poor colony sites. In red-winged and yellow-headed blackbirds, young females nest later than older females (Crawford, 1977). In red-winged blackbirds, long-term residents tend to nest together (Picman, 1981). These phenotype-limited strategies have a major effect on the temporal and spatial distribution of females, and therefore play a critical role in determining the "environmental potential for polygamy."

5. What phenotypic traits determine male reproductive success and how do phenotype-determined strategies affect the extent to which males can monopolize females? In most Icterinae, males compete among each other for access to females or territories used by females. Sexual dimorphism is strongly correlated with polygyny in the Icterinae (Selander, 1958), presumably because larger males win the fights that determine dominance and/ or access to the best territories. There are, however, few data on the effects of size on dominance and reproductive success in most Icterinae. Size has been shown to correlate with reproductive success in the yellow-rumped cacique (this study), brown-headed cowbird (Ankney and Scott, 1982) and, at least occasionally, in the red-winged blackbird (Yasukawa, 1979, 1981). In the yellow-rumped cacique, however, weight declines with age, which causes dominance to decline with age (Robinson, 1985c). Large brown-headed cowbirds also experience greater winter mortality (D. M. Johnson et al., 1980). These data suggest that dominance incurs costs that may limit the extent to which individual males can monopolize females over an entire lifetime. Therefore, studies of reproductive success based on a single season or observation may overestimate the extent to which males monopolize females.

SUMMARY

1. Coloniality in the yellow-rumped cacique and possibly in many other Icterinae results from a scarcity of safe nest-sites and the advantages of group defense against predators.

2. The mating system of the yellow-rumped cacique can be characterized as dominance defense polygyny in which dominant males sequentially consort and guard females at and away from the colony.

3. Phenotype-limited strategies of males and females have a major effect on the degree to which males monopolize females.

4. Most other colonial Icterinae are also polygynous, and most species with dispersed nests are monogamous.

5. Oropendola females nest and feed colonially, and may have a harem-defense-based mating system, though available data are insufficient to confirm this possibility.

6. Marsh-nesting species and two species of open-country inhabiting grackles have resource-defense-based mating systems in which males defend territories in which females nest.

7. A brood parasite, in which females tend to be social some times of the day and solitary at others, also has a dominance-defense-based mating system, but males defend only one female.

8. More data are needed on where and when copulations occur, especially in species with resource-defense-based mating systems, and on the phenotypic determinants of male and female reproductive success.

10. Ecological and Social Determinants of Cercopithecine Mating Patterns

●

SANDY J. ANDELMAN

TRADITIONALLY, primate social systems have been characterized according to patterns of male residence: that is, as monogamous, uni-male (harem), or multimale (e.g., Crook and Gartlan, 1966; Crook, 1972; Eisenberg et al., 1972; Goss-Custard et al., 1972; Clutton-Brock and Harvey, 1977). However, because of the diversity and complexity of primate social relationships, these classifications are of only limited value in predicting the type of mating system a particular population or species is likely to display.

The mating patterns within any breeding population are determined by the interactions between the reproductive behavior of females and males. Consequently, attempts to understand the causes of differential mating success in any group must focus on the effects of particular traits on the breeding success of both males and females (Clutton-Brock, 1983). Unfortunately, most attempts to explain the occurrence of observed mating patterns among primates have not used this approach (e.g., Eisenberg et al., 1972; Goss-Custard et al., 1972; but see J. G. Robinson, 1982). They generally have focused on either the importance of gross categories of social structure or the importance of male competition.

In this chapter, I consider social and ecological factors that influence mating patterns in several species of Old World monkeys (Cercopithecinae). Since the energetic costs of reproduction differ for the two sexes, female reproductive success is usually limited by ecological resources such as the availability and quality of foods, whereas male reproductive success is limited by the availability of suitable mating partners. Proceeding from these assumptions, I first explore the ecological factors influencing female group size in Cercopithecines. I then consider patterns of male interaction with female groups, and the resulting social systems. Finally, I discuss the overall mating patterns in these species in terms of the relationships between female reproductive decisions concerning the rate or timing of births and partner preferences, and male social and reproductive strategies.

Primates often have long life spans, which therefore necessitate long-term studies. As a result, data relating to female and male reproductive strategies frequently are limited, and, for many species, nonexistent. In the field, estimates of mating success depend on indirect measures, and different studies have used different indices. The frequency of copulation or the

duration of consortships for males do not necessarily provide good measures of reproductive success, since female sexual receptivity does not always relate to her physiological state (e.g., Andelman, 1985; see also Hrdy
and Whitten, in press). At best, it is difficult for human observers to detect
ovulation in wild female primates (Andelman et al., 1985); at worst, it is
impossible, even in species such as baboons, where sexual swellings are
often used by researchers as indicators of reproductive status (e.g., Wildt et
al., 1977). Methodological problems of particular studies are discussed below in the contexts in which they arise. Where appropriate, I have included
anecdotal data from provisioned populations or captive groups, but since
such conditions may affect many aspects of behavior and reproductive biology (e.g., Sugiyama and Ohsawa, 1982), unless stated otherwise, I rely
on data from unprovisioned wild populations.

FEMALE GROUP SIZE AND DISTRIBUTION

The subfamily Cercopithecinae consists of four main groups: *Macaca, Papio* (including *Theropithecus*), *Cercocebus*, and *Cercopithecus* (including
Erythrocebus, Miopithecus, and *Allenopithecus*). The basic female group
structure is very similar for all Cercopithecines that have been examined,
with the exception of *Papio hamadryas*. Following Wrangham (1980), Cercopithecines may be described as "female-bonded" in that females typically remain in their natal groups for life and female social interactions are
characterized by pronounced dominance hierarchies and well-developed
networks of grooming relationships. Males usually transfer out of their natal
groups upon reaching sexual maturity. Consequently, most Cercopithecine
social groups consist of a stable core of related females and their immature
offspring, while the membership of adult males in such groups is more transient (*P. hamadryas* is exceptional in that females do not breed in their natal
groups).

Female group size in Cercopithecines ranges from two in hamadryas reproductive units (although hamadryas sleeping groups may contain hundreds of individuals) to more than thirty in some macaque and baboon social
groups (see Fig. 10.1 for average female group sizes in unprovisioned Cercopithecine populations). In the following section, I briefly outline the factors influencing group size among Cercopithecines. Because pertinent ecological data are not yet available for many species, this discussion must
remain largely qualitative (see also Klein and Klein, 1975 and Terborgh,
1983 for detailed discussions of group size in New World monkeys).

Terborgh (1983) has recently argued that, although the upper limit on
group size is set by intragroup competition over scarce feeding sites (Krebs,
1974; Pulliam et al., 1974; Dittus, 1977; Wrangham, 1980; Caraco, 1979;
Struhsaker and Leland, 1979), mutual interest in avoiding predation is the

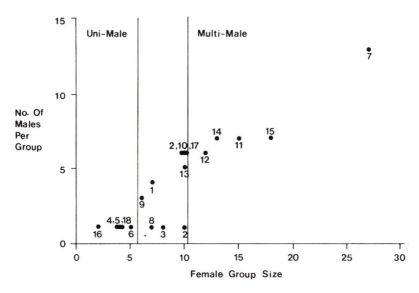

Fig. 10.1. Cercopithecine social systems as a function of female group size. Female group sizes of less than six lead to uni-male social groups; those greater than ten lead to multimale social groups. Female group sizes between six and ten may result in either uni-male or multi-male social systems, depending on ecological factors (see text). Data points represent average (modal) numbers of females and males for each species, based on published reports, and therefore may not adequately represent intraspecific variation in group size. Note: species (2) appears as two separate data points because the two available published reports of group composition differed greatly. Species are indicated sequentially: 1) *Cercopithecus aethiops* (Struhsaker, 1967; Andelman, 1985); 2) *C. ascanius* (Struhsaker and Leland, 1979; Cords, 1984); 3) *C. mitis* (Struhsaker and Leland, 1979); 4) *C. mona* (Struhsaker, 1969; Gautier, 1971); 5) *C. nictitans* (Struhsaker, 1969; Gautier, 1971); 6) *C. pogonias* (Struhsaker, 1969; Gautier, 1971); 7) *Miopithecus talapoin* (Gautier-Hion, 1970; Rowell, 1973); 8) *Erythrocebus patas* (Hall, 1965); 9) *Cercocebus albigena* (Waser and Floody, 1974; Waser, 1977); 10) *Macaca fascicularis* (van Schaik and van Noordwijk, in prep.); 11) *M. mulatta* (Southwick et al., 1965; Lindburg, 1977); 12) *M. nemestrina*; 13) *M. sinica* (Dittus, 1977); 14) *Papio anubis* (DeVore and Hall, 1965; Hall, 1965; Rowell, 1966; 1968; Altmann and Altmann, 1970; Smuts, 1985); 15) *P. cynocephalus* (Altmann and Altmann, 1970; Hausfater, 1975; K.R.L. Rasmussen, 1980); 16) *P. hamadryas* (Kummer, 1968; Sigg et al., 1982); 17) *P. ursinus* (Stoltz and Saayman, 1970; Seyfarth, 1978a, b); 18) *Theropithecus gelada* (Dunbar and Dunbar, 1975).

primary factor selecting for cohesive primate aggregations. This is an old argument (e.g., Crook and Gartlan, 1966), but as I indicate in the following section, predation cannot be considered the primary pressure for group-living among Cercopithecines.

Leighton and Leighton (1982) and Terborgh (1983) suggest that primate group size is determined by the size of the average resource patch on which the species characteristically feeds, or on the number of available feeding sites. The weakness of this proposition is that it fails to consider the causes of variance in group size. Even within a single habitat, intraspecific variation in group size is expected (e.g., see Wrangham, 1980, pp. 271-72). Several recent models address the question of optimal group size in vertebrates (e.g., Rodman, 1981; R. M. Sibley, 1983; Vehrencamp, 1983; Pulliam and Caraco, 1984). Despite differences in their assumptions, three of these models (Rodmañ, 1981; Sibley, 1983; and Pulliam and Caraco, 1984) conclude that groups that are optimal in terms of maximizing foraging efficiency or individual fitness will not be stable in nature (see also Packer, Chapter 19 this volume). In part, this instability is caused by conflicts of interest among conspecifics. Since both social and environmental factors affect the means by which animals compete (e.g., Schoener, 1971; Gill and Wolf, 1975; Maynard Smith, 1979; Rubenstein, 1981b), the actual costs and benefits of different group sizes or structures will vary for different individuals as a function of factors such as age, sex, relative dominance rank, proportion of kin within the group, and population density. Observed group sizes and structures thus will represent a compromise among a variety of individuals of differing reproductive perspectives, and as such may be subject to considerable fluctuation. However, despite the limitations of the resource patch size model, given the crudity of available data, it is useful as a starting point.

For arboreal frugivorous and folivorous primates, the number of available feeding sites is limited by the size of the feeding tree (Leighton and Leighton, 1982; Terborgh, 1983). As expected, the smallest female group sizes among Cercopithecines (< 5 adult females) are found in arboreal species, such as the mona monkey (*Cercopithecus mona*) or *C. nictitans*, which feed on fruits or shoots, generally occurring in small, high-quality patches. Both hamadryas and gelada (*T. gelada*) baboons, though primarily terrestrial, also have small female groups. These species exhibit several levels of group organization, but reproductive units usually contain two or five females, respectively. In hamadryas, a single unit of approximately two females is probably optimal for feeding on a single *Acacia* tree (Kummer, 1968). Geladas, however, appear to be an exception to the general pattern, since they feed on grasses (Dunbar, 1984a).

Large female group sizes (> 11 adult females) are limited to terrestrial or semi-terrestrial species which typically exploit foods that are continuously distributed over wide areas. For example, Lindburg (1977) reports that rhesus macaques (*Macaca mulatta*) in India are heavily dependent upon sal (*Shorea*), a tree species that he characterizes as having a virtually continuous distribution. He does not indicate, however, whether individuals are distributed over several trees when feeding. Grasses (blades, seeds, and rhizomes), which are also continuously distributed, comprise the subsistence diet (Wrangham, 1980) of olive (*Papio anubis*) and yellow (*P. cynocephalus*) baboons (DeVore and Hall, 1965; Altmann and Altmann, 1970; Lieberman et al., 1979; Strum and Western, 1982). Likewise, the gelada baboon, whose reproductive units consist of small female groups, is found in large bands when feeding on grass (Dunbar and Dunbar, 1975; Dunbar, 1977).

In summary, female group size—at least at the extreme ends of the continuum—does seem to depend roughly on the number and distribution of feeding sites (a similar argument for the relationship of overall group size to food distribution is presented in Terborgh, 1983). However, additional factors probably also influence female group sizes and social systems, particularly for intermediate-sized groups. These are discussed in the following section.

SOCIAL SYSTEMS

Which social system will be most prevalent for a given species or population relates ultimately to selection pressures on female cohesiveness (Wrangham, 1980), since the ability of an individual male to maintain exclusive access to a group of females depends in large part on the size and dispersion of female groups (Bradbury and Vehrencamp, 1977b; Emlen and Oring, 1977; Terborgh, 1983).

The relationship between female group size and social system in Cercopithecines is depicted in Figure 10.1. Female groups of less than six are associated with uni-male social groups; groups with more than ten females are generally multimale. Linear regression indicates that the number of males in multimale social groups is positively related to the size of the female group ($F_{7,3} = 33.456; p < .01$). Figure 10.1 also suggests that there is little interspecific variation within this subfamily in the relative proportion of adult males and adult females in multimale social groups (approximately 1:2.4). This lack of variability presents a problem for the hypothesis that multimale social groups are an adaptation to high predation pressure (e.g., Crook and Gartlan, 1966). Vervet monkeys are subject to predation by numerous vertebrates (e.g., leopards, baboons, eagles, and snakes; Cheney and Seyfarth, 1981). In Amboseli, 67 percent of all vervet mortality is due to predation—a rate unequaled by any other primate population studied to date (Cheney et al., 1986). Yet, despite their vulnerability to predation, vervet groups do not have more males than other Cercopithecines with similar-sized female groups (6-10 adult females; Fig. 10.1).

At intermediate group sizes (6–10 adult females), both uni-male and multimale social systems occur. In this range, social systems should vary—both between different populations of a single species and between different species—as a function of factors that influence the ability of an individual male

205

to monopolize a group of females (e.g., habitat continuity, predation, and population density). Data from different populations are insufficient to permit interspecific comparisons across this range of female group sizes. However, analysis of intraspecific variation in social structure in the redtail monkey (*C. ascanius*) provides a test case.

In the Kibale forest in Uganda, Struhsaker and Leland (1979) report that redtails occur in uni-male social groups, while Galat-Luong (1975) and Cords (1984), in the Central African Republic (C.A.R.) and Kakamega, Kenya, respectively, typically find the same species in multimale groups. Variation in habitat continuity between populations does not seem to be important here, since redtails feed preferentially on mobile arthropods, and group dispersion does not differ significantly between Kibale and Kakamega (no data on dispersion are available from the C.A.R.). Differences in predation pressure between these populations might potentially account for differences in social structure. With increasing levels of predation, lone males, or unassociated male cohorts, might be more vulnerable, and consequently such males might be under more pressure to associate with female groups (Terborgh, 1983). Unfortunately data on the relative intensities of predation in these populations are not yet available. However, since lone males are common in vervets (pers. obs.)—a species with high predation rates—the increased vulnerability hypothesis seems unlikely.

Another factor that might influence the ability of an individual male to monopolize a harem is the occurrence of cooperative male defense of females, or of male coalitions (Terborgh, 1983). In baboons, low-ranking males frequently form coalitions to increase their probabilities of gaining access to estrous females (Hall and DeVore, 1965; Packer, 1979a; Rasmussen, 1980). Similarly, the costs of maintaining exclusive possession of a female group might be greater for a harem male when faced with a coalition of males than during encounters against a single opponent (e.g., hanuman langurs; Hrdy, 1977). Thus, in redtails, multimale groups should be more common when extratroop males form bands or coalitions than when they occur singly. Currently available data are consistent with this prediction, since, in Kibale, no all-male redtail bands have been found and uni-male social groups are prevalent (Struhsaker and Leland, 1979). In Kakamega, there are both male coalitions and multimale redtail groups (Cords, 1984).

Variation in the tendency of extratroop males to form coalitions is unlikely to provide a unitary explanation for either intra- or interspecific variation in the social structures of primates with intermediate-sized female groups. For example, the modal female group size for vervets is smaller than that for redtails (seven and ten, respectively; see Fig. 10.1), yet vervets typically have multimale social groups and no all-male vervet bands have been reported. Differences in social structure between vervets and redtails may relate ultimately to differences in the dispersion rather than the size of female groups. The median group spread for redtails in both Kibale and Kakamega was 51-55 meters. Vervet females are less cohesive: in Amboseli the group spread often exceeded 300 meters (Struhsaker, 1967; pers. obs.). An individual male redtail should be able to monopolize ten females in a 51-55 meter area more easily than a single male vervet might monopolize seven females dispersed over 300 meters.

As this example illustrates, no single factor is likely to explain the adaptive significance of variation in social structure. More probably, as further data become available, we will find that several factors, varying in their importance, are responsible for differences in primate group structure. If we are to distinguish between alternative models, we need information on the factors influencing the proclivity of unassociated males either to form bands or to remain solitary, on the differences in the dispersion of feeding sites and female groups, and on differences in predation pressure.

SOCIAL STRUCTURE AND REPRODUCTIVE STRATEGIES

If Cercopithecine social systems result primarily from ecological pressures on female behavior and grouping patterns, what are the consequences of this for male and female reproductive strategies, and for the resultant mating systems? For primates, as well as for other mammals, it has generally been assumed that in uni-male social groups it is the harem male who is responsible for most conceptions within his group. In multimale groups, it is usually assumed that the highest-ranking male accounts for most conceptions, or that the probability of insemination is related linearly to male dominance rank.

Such generalizations are often inferred from indirect evidence rather than observed directly, and primate field studies often differ substantially in their measures of probable paternity. As a result, classifications of social structure do not necessarily reflect actual variation in individual mating success across years or among groups or species. Nor are they likely to reflect the actual variation in "genetic" mating patterns (Wickler and Seibt, 1983; Gowaty and Karlin, 1984). Realistic measures of these variables must await further studies and the use of more sophisticated techniques, such as electrophoresis, in the field. Nevertheless, by first examining components of female reproductive strategies that are likely to influence male behavior, and, subsequently, by examining the observed variation in male behavior and mating success in these two social systems, it may be possible to extract some general patterns.

Female Reproductive Biology

Both ecological and social factors influence female reproductive biology. In turn, two aspects of female reproductive biology, which are probably important determinants of male social and reproductive strategies, are fecundity, or birth rate, and the degree of synchrony or asynchrony in births.

Although, strictly speaking, demographers define fecundity as the average expected number of *female* offspring born to a female at different ages (e.g., Caughley, 1977), few primatologists distinguish the sex of the offspring in their measures of birth rate. Consequently, for the purposes of this discussion, fecundity will refer simply to the average number of offspring produced by females of different populations or species. Furthermore, since the litter size for Cercopithecine females is virtually always one, differences among females in fecundity are due to differences in interbirth intervals and *not* to differences in litter size.

Fecundity is influenced by a variety of factors, including phylogeny, ecology, individual age, and social relationships. Fluctuations in birth rate (i.e., the interval between successive births for an individual female) has a profound influence on demographic patterns such as operational sex ratio (Dunbar, 1979b). This, in turn, has important implications for individual mating strategies. In particular, for species with synchronous or tightly clumped births, female reproductive rates will set a lower limit to the length of male tenures, which might potentially lead to successful inseminations by incoming males.

Available information concerning fecundity in free-ranging, unprovisioned Cercopithecines is summarized in Table 10.1. These data are relatively crude, since it was not always clear how interbirth intervals were computed in different studies. For example, Strum and Western (1982, p. 62) incorrectly attribute to Caughley (1977) the definition of fecundity as "the number of offspring per unit time," rather than as the number of *female* offspring per unit time. Similar inconsistencies are prevalent in the literature, making it difficult to draw accurate comparisons between studies. On the other hand, Strum and Western *do* make the important distinction between what they term "survival" interbirth intervals—the interval between births of infants surviving to a specified age—and "absolute" interbirth intervals. Since other studies do not make this distinction (e.g., Dunbar, 1980b), comparisons between species or populations are necessarily imprecise.

Fecundity is positively correlated with the distribution and availability of food resources in two populations of Kenyan vervet monkeys (Whitten, 1983; Cheney et al., 1986), among toque macaques (*Macaca sinica*; Dittus, 1980), and olive baboons (*Papio anubis*, Packer, 1979a, b; Strum and Western, 1982). Data for other Cercopithecines are insufficient to determine whether there was a similar relationship between ecology and fecundity.

Fecundity is also related to female age in some species. Birth rates are highest among females of mid-reproductive age in hamadryas (Sigg et al., 1982) and olive baboons (Strum and Western, 1982), and in toque ma-

TABLE 10.1
Fecundity in female cercopithecines

Species	$\bar{X}$ no. infants/yr	Varies w/ecol.	Varies w/age	Varies w/ soc. rel.	References
C. aethiops (Amboseli B&C/A)	0.80/0.56	Y	N	N	Cheney et al., 1986
C. aethiops (Samburu RR/LM)	0.69/0.48	Y	?	Y/N	Whitten, 1983
C. ascanius	0.50	?	?	?	Struhsaker and Leland 1979
P. anubis (Gilgil)	0.66	Y	Y	?	Strum and Western, 1982
P. anubis (Gombe)	0.54	?	?	?	Packer, 1979a,b
P. cynocephalus	0.57	?	?	N	S. A. Altmann et al., 1977
P. hamadryas	0.50	?	Y	N	Sigg et al., 1982
T. gelada	0.42	?	?	Y	Dunbar, 1980b
M. sinica (Polonnaruwa/ Anuradhapura)	0.69/0.51	Y	?	Y	Dittus 1975

caques (Dittus, 1980). No relationship between age and reproductive rate is found among female vervets (Cheney et al., 1986).

Data concerning the relationship between dominance rank and fecundity are also equivocal. No relationship between these factors was found among vervet monkeys during a seven-year study in Amboseli (Cheney et al., 1986), although Whitten (1983) reported a positive relationship between dominance rank and reproductive rate in one group of vervets, but no relationship between these two variables in another group, during a twenty-six month study. In gelada baboons (*Theropithecus gelada*; Dunbar, 1977), and toque (Dittus, 1979) and rhesus (*M. mulatta*, Drickamer, 1974; Sade et al., 1976) macaques, fecundity is correlated with dominance rank, although these findings should be interpreted with caution since often it is not possible to separate the effects of age and dominance. In addition, extensive research on captive talapoin monkeys (*Miopithecus talapoin*) has also demonstrated a positive relationship between dominance rank and fecundity (Keverne et al., 1982), but, unfortuntely, comparable data are not available for free-ranging individuals.

In summary, birth rates among Cercopithecines range from a low of one birth every 28.57 months for gelada females to one birth every 15 months for some female vervets. Habitat quality seems to have the most consistent

TABLE 10.2
Timing of births among cercopithecines

No seasonality	Birth peak	Seasonal
Papio anubis	*P. anubis*	*Cercopithecus aethiops*
P. ursinus	*P. cynocephalus*	*C. ascanius*
Macaca silenus	*P. hamadryas*	*C. mitis*
	Theropithecus gelada	*C. neglectus*
	M. arctoides	*C. nictitans*
	M. assamensis	*Allenopithecus nigroviridis*
	M. fascicularis	*Miopithecus talapoin*
	M. nemestrina	*Erythrocebus patas*
	M. radiata	*M. assamensis*
		M. cyclopis
		M. fuscata
		M. mulatta
		M. sinica
		M. sylvanus

SOURCES: See Fig. 10.1.

effect on fecundity, while relationships with female age or social status are less clear. Further data concerning reproductive rates of different Cercopithecine populations, particularly for the forest-dwelling *Cercopithecus* species, are needed.

Female decisions concerning when to reproduce in relation to the parturition of other conspecifics may also be influenced by both ecological and social constraints. The selective pressures exerted by these two factors may be either complementary or conflicting. Data regarding the degree of synchrony or asynchrony in births in wild populations of Cercopithecines are summarized in Table 10.2.

Extreme synchrony is seen in vervets, where 87 percent of all births in Amboseli occurred within a six-week period, and total births were restricted to four months during the year (Cheney et al., 1986). In fact, despite considerable variation in the habitats in which they are found, all *Cercopithecus* species that have been studied are seasonal breeders (i.e., there is a distinct period during the year during which no births occur, and most births are clumped within two to five months; Table 10.2). By contrast, two species (olive baboons; Ransom and Rowell, 1972; and chacma baboons; DeVore and Hall, 1965) showed no clumping of births, with parturition occurring in all months of the year. For the purposes of this chapter, we need only consider the general patterns of clumped or continuous births, and can ignore the proximate mechanisms that regulate synchrony.

Uni-Male Social Groups

With the exception of vervets, talapoins, and *Allenopithecus*, most guenons (including patas) typically occur in uni-male groups (Struhsaker, 1969, 1975, 1977, 1980; Struhsaker and Leland, 1979), as do hamadryas and geladas. Uni-male social groups are generally equated with female defense polygynous mating systems (Emlen and Oring, 1977), in which reproductive variance among males is high (e.g., LeBoeuf, 1974) and intrasexual competition among males is usually intense. As a result of this competition, the tenures of individual males as harem leader may be relatively short, with mainly young males in their prime maintaining such positions. Consequently, the actual variation among males in lifetime reproductive success may be substantially less than is often implied by the apparent variance observed during a one- or two-year field study (Clutton-Brock, 1983).

Determining male reproductive success is very difficult, since relatively few copulations have been directly observed in most uni-male species. For example, during thousands of hours of observation on redtails and blue monkeys in Kibale, Struhsaker and his colleagues (Struhsaker and Leland, 1979) witnessed only six and twenty-one copulations in these two species, respectively.

Also, both theoretical and empirical evidence suggest that harem males are not likely to be the exclusive copulators in their groups. Births are highly clumped in all uni-male Cercopithecines for which data are available (Table 10.2). Female redtails, blues, mona monkeys, *C. nictitans*, and patas give birth seasonally, and synchronous birth peaks are also evident in geladas and hamadryas. Such synchronous conceptions should, theoretically, reduce the potential for a single male to sexually monopolize multiple females (Emlen and Oring, 1977). Limited empirical evidence suggests that this is in fact the case, at least in some uni-male Cercopithecines.

In a field study of seasonally breeding patas monkeys, Chism and Olson (1982) found that while only one male typically remains with a female group throughout the year, during the breeding season, large influxes of all-male bands may occur. Several males were responsible for matings in some groups during some breeding seasons. Chism and Olson suggest that the male who was resident in a particular group outside the breeding season was not necessarily likely to have fathered infants in the group (based on observed patterns of copulations).

Similarly, the resident redtail in Kibale was not responsible for all copulations. During a multimale influx, five extragroup males temporarily joined the group and were observed copulating with females (Struhsaker, 1977). In Kakamega, where uni-male social groups were not considered typical (Cords, 1984), during 9.5 months of uni-male domination, extragroup males copulated with females in one troop during temporary influxes.

In hamadryas and geladas, females apparently do not copulate regularly with extragroup males (Kummer, 1968; Dunbar, 1978a). In the former, however, females do solicit copulations with subadult and juvenile male troop members. Hamadryas harem leaders attack such copulating pairs, suggesting a possible threat to their status as sole inseminators. The relative capabilities of adult and subadult male hamadryas to produce sperm are not known.

In addition to the effects of birth synchrony and competition from extragroup or subadult males on variance in male reproductive success, the frequency with which females give birth may also be important. Unfortunately, as shown in Table 10.1, little information concerning fecundity of females in uni-male Cercopithecine groups is presently available. Female redtails and hamadryas give birth, on average, every 24 months, and female geladas about every 28.57 months. Since births in uni-male species are generally synchronous, this suggests that male tenures in such groups must be a minimum of about two years in order to pay off in terms of inseminating females. Again, few relevant data are available. In Kibale, the average tenure for male redtails is 22.5 months (Struhsaker and Leland, 1979) and the average tenure for blue monkey males in Kibale is 22.7 months (Rudran, 1976, 1978; Struhsaker and Leland, 1979). The probability of transfer between reproductive units or bands for gelada males was estimated as 0.037/male/year in one population (Dunbar, 1980b) and as 0.058/male/year in another (Ohsawa, 1979).

Based on this limited information, it seems that in redtails, and perhaps in blues, where male tenure is generally shorter than interbirth intervals, male tenures are often insufficient to ensure insemination of females. Not surprisingly, then, infanticide by incoming males, a reproductive strategy that usually shortens female interbirth intervals, occurs in these species. In some cases it may increase the reproductive success of resident males (see Hausfater and Hrdy, 1984 for references and detailed reviews). Given present data, it appears that harem male geladas, and possibly hamadryas, are more likely to have exclusive mating relationships with females than are redtail and possibly blue or patas males. In redtails, blues, and patas, being a male resident in uni-male groups versus being an extragroup male making temporary incursions into female groups might be viewed as two alternative reproductive tactics whose relative success in different circumstances remains to be assessed. Clearly, further research is needed to establish reasonable estimates of birth rates, male tenure in groups, copulation rates for resident and extragroup males, and the relationship between copulation rates and the probability of insemination in uni-male Cercopithecines as well as in other primates. Such estimates will permit more accurate measurements of the factors that influence lifetime reproductive variance for males.

Multimale Social Groups

Females in multimale Cercopithecine groups exhibit more diverse reproductive patterns, both with respect to reproductive rates and synchrony of births, than do females in uni-male groups (see Tables 10.1 and 10.2). Thus multimale social groups potentially offer a greater variety of demographic and social circumstances than do uni-male groups. In addition, multimale groups potentially provide a greater opportunity for the operation of epigamic selection or female choice. In this section I review evidence concerning the diversity of male reproductive strategies, and their relative success, as a function of different patterns of female behavior.

Dominance. Intrasexual competition among males over access to fertile females may assume a variety of forms. In primates, the effect of dominance rank on differential male access to females was one of the first variables to be examined (e.g., Carpenter, 1942; DeVore, 1965), and S. A. Altmann (1962) constructed a priority-of-access model of mating behavior, which has been tested in a number of species (e.g., rhesus, Suarez and Ackerman, 1971; yellow baboons, Hausfater, 1975). Although in most Cercopithecines males can be ranked according to a linear dominance hierarchy (or a reasonable approximation of a linear hierarchy), male dominance ranks are generally less stable than those of females. For example, during a three-year period in Amboseli, adult male vervets changed ranks at an average annual rate of 0.75, while the comparable figure for female rank changes was 0.11 (Cheney, 1983). For yellow baboons in Amboseli, during a fourteen-month study by Hausfater (1975), the average annual rate of rank change for males was substantially higher, 3.09. We might then expect male strategies to be rank-dependent, and the strategy of any given male would then vary during his lifetime.

High dominance ranks should be most advantageous in situations where females are relatively asynchronous in their sexual receptivity or fertility (Emlen and Oring, 1977). However, the effects of male dominance on mating success are not always clear-cut. In olive and yellow baboons, where females are relatively asynchronous in both their conception and receptivity patterns, a variety of studies have found that high-ranking males are able to maintain priority of access to females during the peak of estrus (e.g., Hall and DeVore, 1965; Hausfater, 1975; Packer, 1979a, b), maintain consortships with females over more cycle days than low-ranking males (e.g., Rasmussen, 1980), or may simply obtain the greatest proportion of copulations (e.g., baboons, Hall and DeVore, 1965; Hausfater, 1975; Popp, 1978; Packer, 1979a, b). In other studies, high-ranking males actually did worse than low-ranking males: DeVore (1965), Smuts (1982), and Strum (1982) found that male dominance rank was *inversely* correlated with consort success in some groups of olive baboons.

Where females are relatively synchronous in their conception patterns, male dominance rank should be less important in obtaining fertile copulations. This appears to be the case. In bonnet macaques, Simonds (1965) found no relation between dominance rank and mating success among males. In vervets, the frequency of male copulation *attempts* was correlated with dominance, and high-ranking males obtained a greater proportion of copulations than did middle- or low-ranking males in *some* groups in *some* years. However, in three groups, over six years, there was a nonsignificant inverse correlation between male dominance rank and success rate of copulation attempts (i.e., the probability that a male-initiated copulation attempt would not be rejected by a female). Nor did high-ranking male vervets obtain a greater proportion of copulations in the week when conception was most likely to occur (see Andelman, 1985). In one captive group of synchronously breeding macaques, however, male dominance rank was significantly correlated with mating success (Hanby et al., 1971).

Size. As with dominance rank, only in asynchronously breeding species would we expect size to contribute to a male's ability to monopolize fertile females. In some cases the two measures (i.e., rank and size) may be closely related. There are few field studies, however, for which weights of males have been obtained. Male rank was positively correlated with body weight in wild olive baboons (Packer, 1977) and captive rhesus (Maroney et al., 1959) and captive pigtail (Tokuda and Jensen, 1969) macaques; thus the interactions between these two variables in determining male reproductive success may be difficult to disentangle in these species. In only two studies, however, was male size positively correlated with consort success. Smuts (1982), who measured the amount of time males spent in consort, rather than the number of copulations obtained, found that among olive baboons at two sites in Kenya (Smuts, 1982 and Popp, 1978), male body weight was positively correlated with male consort success, but only among those males residing in a troop for longer than one year. This correlation may not tell us much about reproductive success among males, however, since in yellow baboons the duration of consortships was a poor predictor of copulatory rates (Hausfater, 1975). Since olive baboons are not seasonal or synchronous breeders, this seems counter to the predictions, and more data from other species would be useful.

Coalitions. K.L.R. Rasmussen (1980) has suggested that coalition formation might be more effective, and consequently more common, in open terrain where good visibility enhances coordination and communication between males. By contrast, she suggests that individual dominance in dyadic interactions would be more important in forests or densely vegetated areas where maneuverability may be reduced. In redtails and blue monkeys, both forest-dwelling species, coalitions were not uncommon (see p. 206). However, in savannah populations of both olive and yellow baboons (Hall and DeVore, 1965; Packer, 1979a, b; Rasmussen, 1980), where visibility is presumably good, low-ranking males formed coalitions to take over females at the peak of estrus. The success of such alternative strategies may depend not so much on suitable habitat as on demographic factors, which may influence the probability that males will aid one another in coalitions of this sort. Males who have interacted previously, and those who are of similar rank, are more likely to form coalitions than are unfamiliar males or males of highly dissimilar competitive ability (Noë, 1984). The value of gaining access to an estrous female would also influence the potential payoffs of such coalitions (Packer and Pusey, 1985).

Grooming. In Cercopithecine primates that form consortships, grooming is known to be an important part of the copulatory sequence (Hall, 1962; Kaufmann, 1965; Saayman, 1970). By contrast, in some uni-male and multimale species which do not form consortships, male grooming of females, and cyclical changes in grooming of females, is reduced or absent (e.g., patas, Hall, 1965; hamadryas, Kummer, 1968; geladas, Dunbar, 1978a). Saayman (1970) suggests that the primary function of sexual grooming is to maintain the spatial proximity of consort partners in serially mounting species. More generally, sexual grooming may represent a male strategy to increase the receptivity and cooperation of estrous female consorts (K.L.R. Rasmussen, 1980), and Kummer (1968) reported that male grooming of females is sometimes more effective in eliciting female cooperation than is herding.

In both olive baboons (Packer, 1979a, b) and hamadryas (Kummer, 1968), there was a positive correlation between male age and rates of grooming of females, suggesting, perhaps, males who are less successful in direct agonistic interactions over females may use alternative means such as grooming to increase female cooperation, and perhaps preference. In hamadryas baboons, such cooperation can apparently create an asymmetry whereby even when the challenging male is dominant to a male in consort with a female, he will "respect" the social bond between such male-female pairs, and takeover attempts do not occur (Kummer et al., 1974; Bachmann and Kummer, 1980). In chacma (Seyfarth, 1978a, b) and olive baboons (Smuts, 1985), females form close social bonds with particular males, and these persist over long periods of time. This strategy may be particularly successful with low-ranking or young females who, by associating with a male, may reduce the risk of encountering aggression from a higher-ranking female.

The available evidence provides some support for the predictions that direct methods of male–male competition (e.g., dyadic or tryadic agonistic interactions) are most common in asynchronously breeding species, and indirect competitive methods (e.g., soliciting female cooperation) are more prevalent in synchronously breeding species. More data are available for baboons than for any other multimale species, however, so it is difficult to generalize. Further research on other multimale species, particularly those that are seasonally breeding, is needed.

SUMMARY

In summary, among Cercopithecines, female group size is closely related to the distribution of feeding sites. Male membership in groups, and the resulting social system, are strongly dependent on the size of female groups. In general, small female groups lead to uni-male social systems. There is a significant linear relationship between the size of the female group and the number of male members in multimale groups. At intermediate female group sizes, both social systems occur, and a variety of factors, including dispersion of females and cooperative male behavior, appear to be important in determining which system prevails. The ratio of adult males to adult females in multimale groups (for which data are available) is virtually constant across species, suggesting that multimale social structures are not primarily an adaptation to high predation pressure.

Cercopithecine mating systems are related to social structure, but this relationship is less absolute than has been implied previously. For uni-male groups, in particular, resident males often are not the only copulators, and further data concerning alternative strategies, such as being an extragroup male and making incursions into female groups during the breeding season, are needed. The relative success of such strategies will probably depend on the length of female interbirth intervals (fecundity) and the frequency of infanticide.

In multimale social groups, a greater diversity of both female and male reproductive strategies is apparent. In the past, much data has been collected on the importance of male dominance in obtaining copulations. However, the relationship between male copulation success and breeding success (i.e., the probability of insemination) is often imprecise (e.g., Andelman, 1985). Although monopolizing estrous females is probably more important in asynchronously than synchronously breeding species, reality is more complex than this. Again, further data on alternative male strategies are needed. In particular, there is a large and growing data base on baboon social and reproductive strategies, but comparatively little is known about the breeding behavior of undisturbed populations of other species, such as many of the macaques, vervets, and talapoins. Further long-term studies of these species are needed before we can form conclusive generalizations.

11. Resource Distribution, Social Competition, and Mating Patterns in Human Societies

———————————————— • ————————————————

MARK V. FLINN AND BOBBI S. LOW

HUMAN SOCIAL systems have been studied extensively. We probably know more about the particular patterns of human pair-bond formation than about the patterns of any other species. Nevertheless, there have been few attempts to use this knowledge of human behavior to test hypotheses from biological theory.

Other chapters in this volume summarize current knowledge of male and female reproductive strategies in a number of species of birds and mammals. In this chapter we examine data from 849 human societies. Our aim is to test predictions about human mating and marriage systems derived from the same evolutionary principles used in the other chapters.

Anthropologists and other social scientists have expressed skepticism about the relevance of evolutionary theory for human behavior. For example, Sahlins (1976, p. 26) argued, "There is not a single system of marriage, post-marital residence, family organization, interpersonal kinship, or common descent in human societies which does not set up a different calculus of relatedness and social action than is indicated by the principles of kin selection."

We agree that the application of current evolutionary theory to human social behavior is not a simple or easy task. However, it seems to us a promising endeavor, if appropriate modifications are made for the unique aspects of human sociality. As Chagnon (1979a, p. 88) suggested: "The direct application of theory from evolutionary biology to human marriage behavior and mating strategies is . . . not possible until the theory is modified to take into consideration the interdependency of individuals . . . and how their interdependency—coalition alliances—structures human mating behavior."

Alexander (1974, 1977, 1979), Chagnon (1979a, b, 1982), and Irons (1979a, b, 1983) have examined human marriage systems using current evolutionary theory, integrating knowledge about resources and the importance of coalitions; here we hope to continue that process. Focusing on patterns of mate exchange and marriage, we analyze, first, the extent to which observed patterns are predicted from the theory; and, second, the points at which the uniqueness of humans makes the unmodified theory inapplicable.

Trends in Nonhuman Species

All organisms acquire resources from the environment in order to survive and reproduce. Not surprisingly, the temporal-spatial distribution of resources is crucial to mating systems theory (Borgia, 1979; Bradbury, 1981; Breder and Rosen, 1966; J. L. Brown, 1964; Bygott et al., 1979; Cade, 1979; Clutton-Brock and Harvey, 1977; Emlen and Oring, 1977; Jenni, 1974; Low, 1976, 1978; Orians, 1969; Rubenstein, 1980; Selander, 1972; Thornhill and Alcock, 1983; Trivers, 1972; Verner, 1964; Verner and Willson, 1966; Wade and Arnold, 1980; Wilbur et al., 1974; Wilson, 1975; Wittenberger, 1979, 1981; Wrangham, 1980).

Because mating systems are a summation of the behaviors of individuals, including conflicts of reproductive interest, most theorists begin with an analysis of individual reproductive strategies, emphasizing differences in male and female strategies. The amount of parental investment each sex provides has important consequences for reproductive strategies. In general, the sex investing less parentally (usually male) exerts the most effort in intrasexual competition for mates (Darwin, 1871; Bateman, 1948; Trivers, 1972), whereas the sex providing more parental investment (usually female) is more discriminating in mate choice (Darwin, 1871; Fisher, 1930). These generalizations about male and female reproductive strategies appear valid for humans (Symons, 1979; Davenport, 1976; Van den Berghe, 1979; Daly and Wilson, 1983; Irons, 1983), although males in most human societies provide considerable amounts of parental investment (e.g., Irons, 1979, 1983; Flinn, 1981, 1987), and females may exert significant effort in mate competition (e.g., Schlegel, 1972; Low, 1979; Hrdy, 1981; Irons, 1983).

Studies of nonhuman mating systems (e.g., Borgia, 1979; Emlen and Oring, 1977; Thornhill and Alcock, 1983) suggest that males follow a limited number of strategies in mating effort: 1) competing for control of resources such as nest-sites or sources of food in order to attract females (e.g., red-winged blackbirds, elephant seals, blowflies); 2) competing for control of mates, for example, by actively preventing competitors from having access to females (e.g., elk, hamadryas baboons); 3) competing for mates by advertising phenotypic indicators of possible heritable genetic qualities (e.g., lek-breeding species such as sage grouse, and scramble competition species); and 4) competing for matings by searching for receptive females (e.g., bumblebees). In the last two cases, competition for mates is essentially resource-free, and it is often presumed that females choose mates solely on phenotypic indicators, giving rise to the "lek paradox" (Taylor and Williams, 1982); that is, decreased heritability of traits. When intrasexual competition is intense, males may use alternate strategies. More than any other species, however, human males appear to use all four of the above

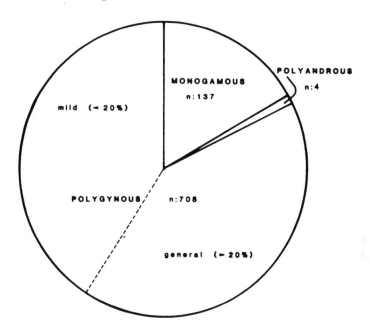

Fig. 11.1. Marriage patterns in 849 societies.
Data from Murdock, 1967.

strategies to varying degrees in different populations and sometimes within the same population.

General Trends in Human Societies

Human populations exhibit a variety of different marriage patterns. Many societies are polygynous, some highly so; some are monogamous; and a few may be slightly polyandrous[1] (Fig. 11.1). Humans have extensive social networks based on kinship and reciprocity. Male–female bonds include

---

[1] Unfortunately, social scientists often do not gather detailed information about reproductive strategies and environmental conditions. For example, even the usual measure of polygyny, common in studies of other species, requires the computation of the variance in reproductive success of males versus females (e.g., Clutton-Brock et al., 1982); in fact, many biologists define polygynous systems as those in which the variance in male reproductive success exceeds variance in female reproductive success (cf. Wickler and Siebt, 1983). Few studies of human societies provide the necessary information for such computations (but see Chagnon, 1979a, b; Flinn, 1983; Borgerhoff Mulder, 1987). Consequently, cross-cultural studies of human mating systems must rely upon ethnographic descriptions of marriage and pre- and extramarital mating (see Appendices A, B).

such important relationships as brother–sister, father–daughter, and husband–wife. Female–female bonds include sister–sister, mother–daughter, aunt–niece, grandmother–granddaughter, cousin–cousin, and co-wives (reviewed by Irons, 1983). Male–male bonds include brother–brother, father–son, uncle–nephew, grandfather–grandson, and cousin–cousin.

The relative importance of each of these bonds varies from society to society.[2] Several patterns emerge, and are commonly associated with residence practices (see Murdock, 1949). Societies with patrilocal residence ("married" couples take up residence near the husband's father) tend to have strong bonding among brother–brother, father–son, paternal uncle–nephew, and sometimes husband–wife. Societies with neolocal residence (a new location, not near either spouse's kin) emphasize husband–wife bonds. Societies with avunculocal residence (near the husband's maternal uncle— his mother's brother) emphasize brother–sister and maternal uncle–nephew bonds. Societies with matrilocal residence emphasize brother–sister, sister– sister, mother–daughter, maternal aunt–niece, and grandmother–granddaughter bonds.

Murdock (1949) attempted a broad study of human sociality by comparative analysis of a large number of societies. He proposed that patterns of residence and pair-bonding were related to subsistence activities; hunting and other male-dominated activities required males to remain in their familiar natal locality (patrilocal residence), whereas certain types of horticultural practices dominated by females favored matrilocality (Murdock, 1949). More recent studies suggest that in addition to subsistence, intergroup competition (e.g., warfare) and patterns of mating (e.g., stability of mating relationships, degree of polygyny) are important factors associated with residence (Ember and Ember, 1971; Otterbein and Otterbein, 1965; Aberle, 1961). This variability provides an excellent basis for comparative analysis of the factors associated with different marriage patterns.

Previous cross-cultural studies suggest there are no obvious general large-scale geographic or environmental correlates of human mating systems (Murdock, 1949). The predictability of effective rainfall does show a correlation with the degree of polygyny (Low, 1987), but polygyny does not appear to correlate with extremes of temperature or rainfall. Risk of serious parasites shows a positive correlation with degree of polygyny (Low, 1987). None of these factors correlates with the degree of outbreeding, or particular types of marriage preferences, which we analyze in this paper. Several cor-

[2] Anthropological studies have only recently begun to use systematic observation techniques to record quantitatively behavioral information (e.g., scan sampling methods—Hames, 1979 and Flinn, 1983). Consequently, the sorts of data we would like to have on the patterns of behavioral interaction are not available for a large cross-cultural sample. We rely, therefore, on ethnographic descriptions of the residence practices and general patterns of interaction between and within the sexes in different human societies.

relates of marriage systems involving the interaction of social and ecological variables (some of these have obvious nonhuman analogues), have been proposed in previous studies. These include: type of subsistence economy, level of political stratification, residence, exchange of considerations at marriage (e.g., bride price), and warfare. These conditions appear to be associated with differences in ability to control mates and resources.

Just as correlations between the level of resources and such phenomena as population density are known for nonhuman species, there are associations among resource base, population density, and degree of polygyny in human societies (Osmond, 1966; Alexander et al., 1979. Osmond's (1966) analysis of cross-cultural data indicates a "U curve" association between the incidence of monogamy and the level of social complexity. Societies with a low level of social complexity—hunting and gathering subsistence economy, a community size less than 100, no craft specializations, and no hereditary classes (e.g., the Copper Eskimo, Jenness, 1959; and the !Kung Bushmen, Lee, 1982)—tend to have monogamous or slightly polygynous marriage systems. Alexander et al. suggest that monogamy is ecologically, rather than socially, imposed in these societies, in a manner analogous to much monogamy in nonhuman species. Moderately complex societies, with simple agriculture and/or animal husbandry subsistence economy, community size 100–1,000, more than one craft specialization, and class stratification based on a hereditary aristocracy (e.g., the Ashanti, Rattray, 1927; and the Lau Fijians, L. Thompson, 1940), tend to be polygynous. Large, complex, "state" societies—community sizes greater than 50,000, intensive agriculture subsistence economy, three or more social classes or castes, and three or more craft or industrial specializations (e.g., the Irish, Kennedy, 1975)—tend to have monogamous mating systems.[3] Alexander et al. (1979) suggest that monogamy is socially, rather than ecologically, imposed in these societies.

In the following section we examine the variety of human mating systems in more detail. We identify several non-exclusive "types" of human mating systems. To an extent greater than most other organisms, humans exhibit considerable variability in reproductive strategies within as well as between populations. Nonetheless, generalization is sometimes useful, especially when data detailing differences in individual strategies are unavailable.

TYPES OF HUMAN MATING (MARRIAGE) SYSTEMS

Resource Limited ("Ecologically Imposed") Monogamy
For human populations living in harsh, unproductive regions of the world

---

[3] Some of the Chinese dynasties and some Islamic states such as the Ottoman Empire are notable exceptions.

(e.g., the arctic, the Kalahari Desert, and the Ituri rain forest), resources such as food are sparsely distributed and difficult to accumulate or defend. Consequently, social groups are very small (less than 100), and there is little variability among individuals in resource control or social power (see also Dyson-Hudson and Smith, 1978). The type of subsistence economy and the degree of sexual division of labor also appear to be important factors that influence whether males can control sufficient resources or social power to attract more than one wife. Among the Copper Eskimo, for example, it appears difficult for a man to provide enough food to sustain more than one wife plus offspring for long periods (Jenness, 1959; see also R. Bailey, 1985 and Peacock, 1985 for data on the Efe). Alexander et al. (1979), in fact, called these societies "ecologically monogamous," recognizing the role of resource scarcity. All monogamy known in nonhumans falls into this category.

Some men, through their skill in hunting and establishing reciprocity, are able to have multiple wives in these societies, but in general there appears to be relatively little difference in the reproductive variance of the sexes. For example, among the Kalahari Bushmen about 5 percent of the men have two wives, and very rarely do men have more (Lee, 1979).

The inability to accumulate excess resources seems to be a critical factor restricting polygyny in these ecologically monogamous systems. Most males appear unable to control sufficient resources or social power to surpass the "polygyny threshold" (Verner and Willson, 1966; Orians, 1969); that is, to create a situation in which it is reproductively worthwhile for a woman to become a second wife.

Mate Control Polygyny

In some less harsh, more resource-abundant areas of the world, polygyny is common. Among human societies in such regions with hunter-gatherer (e.g., Australian aborigines) and simple agriculture (e.g., Yąnomamö Indians) subsistence economies, material resources are usually rather evenly distributed among members of a residential group: individuals wielding power are not materially better off than others. Where population densities are low there is very little individual ownership of resources such as land (Murdock, 1949), although groups may claim and defend them; and usually only a limited amount of labor is required for subsistence purposes (Sahlins, 1972). Such conditions apparently mean that the costs of territorial defense or other sorts of resource control behaviors would outweigh the benefits (see Chagnon's 1983 discussion of Yąnomamö warfare). If human interactions were no more complicated than those of other species, such a situation might be analogous to the strategy of female-control polygyny (as in many nonterritorial ungulates, e.g., elk).

Among many of these societies, some "headmen" wield more influence or power than others. Headmen typically are polygynous and have high reproductive success (Chagnon, 1979a, b; Betzig, 1986); in societies such as the Yąnomamö, headmen tend to be men who have relatively large groups of male kin locally resident. Male–male coalitions are critically important for acquiring and keeping wives (Chagnon, 1982).

In some societies, gerontocracy is a common basis for polygyny. Older males frequently wield considerable power over the distribution of young, marriageable females. Among the Tiwi, men do not even acquire their first wives until they are forty or fifty; successful men may accumulate as many as ten wives by their seventieth birthday (C.W.M. Hart and Pilling, 1960).

Among these societies, polygyny apparently results from social competition for mates; control of "ordinary" resources does not seem to be significant, but control of human "resources" may be (see Ember and Ember, 1983). Males exert considerable effort not only to control their mates and their male competitors, but also their female relatives:

Put bluntly, in Tiwi culture daughters were an asset to their father, and he invested those assets in his own welfare. He therefore bestowed his newly born daughter on a friend or an ally, or on somebody he wanted as a friend or an ally . . . or, the father might bestow an infant daughter on a man—or some close relative of a man—who had already bestowed an infant daughter upon him, thus in effect swapping infant daughters. (Hart and Pilling, 1960, p. 15)

Several years ago in Village 18, the wife of one of the village headmen began having a sexual affair with another man. She came from the other lineage in the village, and her brother, also one of the village headmen, attempted to persuade her to stop the affair. The two headmen were brothers-in-law and had exchanged sisters in marriage. The woman in question refused to follow her brother's advice, so he killed her with an ax. (Chagnon, 1979b, p. 394)

Frequently there are organized patterns of mate exchange (Chagnon, 1979b; Irons, 1981; Lévi-Strauss, 1949; Tylor, 1889). These practices often are associated with rules that stipulate that marriages can occur only between certain types of cousins, termed "cross-cousins." Here kinship is a critical component in social competition for mates (Chagnon, 1982). Cousin marriage is an important aspect of human mating systems that we will address later in this chapter.

In some polygynous human societies (about 20 percent; Murdock, 1967) sisters are preferred co-wives, resulting in sororal polygyny. Joint residence (cohabitation) of co-wives is much more common with sororal polygyny (Murdock, 1949), perhaps because the reproductive interests of sisters are more similar than unrelated co-wives (Alexander, 1979; Daly and Wilson, 1983). Sororal polygyny facilitates cooperation among female kin, especially where marital residence is matrilocal (e.g., the Tiwi; Hart and Pilling, 1960).

Resource Control Polyandry

Polyandry is quite rare among humans. Only 4 of 849 societies in the *Ethnographic Atlas* (Murdock, 1967) are classified as polyandrous. Almost all polyandrous mating relationships are fraternal; that is, co-husbands are brothers (Hiatt, 1981). Among the Pahari, initially polyandrous unions later tend to become polygynandrous because groups of brothers usually try to acquire additional wives (Berremann, 1962).

Tambiah (1966), in his study of polyandry among the Kandyans of Sri Lanka, concluded that polyandry is "an arrangement by which two brothers, who singly cannot support individual families, may combine their land and maintain a common family, thereby bettering their living standards, and also ensuring that their land will go to a limited set of heirs" (p. 316). Parental pressure to cooperate is an additional factor; polyandrous marriages are usually arranged by parents, often while offspring are quite young (Goldstein, 1978).

The critical ecological factor associated with fraternal polyandry seems to be the tightly controlled distribution of an immobile and limited resource (land) which loses its value when divided. This favors male–male coalitions, usually cooperative groups of coresidential brothers. Most studies agree that nonpartable land inheritance and cooperation among brothers are important factors favoring polyandry (Alexander, 1974; Beall and Goldstein, 1981; Goldstein, 1978; Hiatt, 1981; Peter, 1963; Tambiah, 1966; Yalman, 1967; Durham, in press). There are, however, other more common mechanisms for preventing the division of resources (e.g., primogeniture), and there are no clear ecological differences.

Beall and Goldstein (1981) present data from Nepal which suggest that polyandrously mated males have lower reproductive success on the average than do monogamously mated males (although numbers of offspring are not the best measure of inclusive fitness). They conclude that individuals are maximizing economic rather than reproductive goals. Hiatt (1981) suggested that prevention of female initiative and cuckoldry are important factors favoring polyandrous matings. However, we have insufficient data as yet to reject the hypothesis that polyandry is an inclusive fitness maximizing strategy. Alexander (1974, p. 371) proposed that the effects on reproduction over several generations should be considered: "Polyandry . . . is related to the low and reliable productivity of farms, with the result that additional labor without additional children (thus, more than a single male per family) has come to be the best route to long-term maximization of reproduction because of the necessity of maintaining the minimal acceptable plot of land."

Goldstein (1978) pointed out some important differences between the polygynandrous marriage system of the Pahari and the fraternal polyandry of Tibet. Among the Pahari a group of poor and/or young brothers might be able to afford only one wife, and are making the best of a bad situation; these males might otherwise have no mating success at all. Wealthier groups of brothers, however, acquire many wives and are effectively polygynous.

Tibetan polyandry appears to work differently. Wealthy landowning families seem to prefer fraternal polyandry. Goldstein's analysis of the Tibetan marriage system suggests that parental and sibling manipulation, and long-term (multigenerational) reproductive maximization are important factors affecting reproductive strategies (Goldstein, 1978, pp. 320-36):

Because authority is customarily exercised by the eldest brother, younger male siblings have to subordinate themselves with little hope of changing their status . . . furthermore, in contrast to the expressed . . . ideal, the covert assumption and behavioral reality is that those members staying with and maintaining the initial family corporation get the major share of the estate. . . . under what conditions, then, might younger males perceive the opportunity costs of fission as not prohibitive? Traditionally, there are few indeed.

It seems likely that among fraternally polyandrous households, the first-born sons are usually the genetic offspring of the eldest co-husband brother. The eldest brother appears to have certain advantages over his younger siblings when their wife is first acquired; indeed, some of the "co-husbands" may not even be sexually mature yet (Goldstein, 1978). This situation appears to foster an intergenerational favoritism of first-born sons, and the manipulation of younger siblings—who are likely to be less closely related to the powerful eldest brother and eldest co-father—with limited options for inclusive fitness maximization other than cooperating within the fraternally polyandrous households. Such a situation may be analogous to the helping phase of the "helpers at the nest" phenomenon (Alexander, 1974) seen in species like acorn woodpeckers (Koenig and Pitelka, 1981) and scrub jays (Woolfenden, 1981), in which reproductive success is correlated with possession of difficult-to-obtain resources such as territories or "acorn trees" (Emlen, 1984).

Resource Control Polygyny

Polygynous mating systems frequently seem to be associated with ecological conditions (in combination with appropriate technologies) that allow for the development of surplus material resources controlled by males (Murdock, 1949). For example, in an extensive analysis of the mating patterns of native Americans in western North America, Jorgensen (1980, p. 167) found that "high incidences of polygyny occurred primarily in extractive societies in which large surpluses could be stored and in which men controlled the subsistence resources."

When resources can be stored and defended, they are seldom distributed equally among individuals. In societies in which some men control more resources such as land or animals than others, these prosperous males tend to have multiple wives (Irons, 1979b). Indeed, in many societies there are large economic prerequisites for marriage, such as accumulating sufficient resources for a "bride-price" (Goody and Tambiah, 1973). Among societies in which important mobile resources are controlled by individuals or kin groups, the resources often serve as the medium of exchange for marital transactions. Rather than exchanging women for women directly, material items such as pigs (e.g., the Tsembaga-Maring, Rappaport, 1968), sheep (e.g., the Yomut Turkmen, Irons, 1975), cattle (e.g., the Kanuri, R. Cohen, 1967), or cash (e.g., the Nyoro, Beattie, 1960) are used to acquire a bride. Data on Kipsigis bridewealth indicate that age, reproductive condition, and marital distance affect amount of bridewealth, with women of higher reproductive potential commanding higher bridewealths (Borgerhoff Mulder, 1987).

The association between resource control and polygyny among humans does not appear to be entirely analogous to the polygyny threshhold model (Verner and Wilson, 1966; Orians, 1969). That is, human females do not exercise free choice among males controlling resources. Parents or male relatives frequently make the exchanges, and in most such societies male–male coalitions are important, with an association between social power (e.g., political status) and resource control (Alexander, 1979; Chagnon, 1979a). Kings and chiefs control people as well as material resources, and one of the most common perquisites of high office is polygyny (Betzig, 1986). The degree of polygyny attained in despotic societies is often quite high (Dickemann, 1979):

Van Gulik's description of Chinese Imperial Harem procedures, involving copulation of concubines on a rotating basis at appropriate times in their menstrual cycles, all carefully regulated by female supervisors to prevent deception and error, shows what could be achieved with a well-organized bureaucracy. Given nine-month pregnancies and two- or three-year lactations, it is not inconceivable that a hardworking emperor might manage to service a thousand women.

Among humans there appears to be an association between socioeconomic status and degree of mate control (Irons, 1979b; Betzig, 1986; Dickemann, 1981; Flinn, 1981, 1986b; Hill, 1984). In general, high-status males guard their mates much more closely than do low-status males (Dickemann, 1981). Claustration, foot-binding, infibulation, virgin marriage, and chastity belts are some of the mate-guarding tactics used primarily by high-status males to reduce the risk of cuckoldry (cf. Low, 1979). Because high-status males in these societies usually have considerable heritable resources, they have much more to lose. Thus "resource control polygyny" usually involves social control as well (Dickemann, 1981).

Serial Monogamy

Males are not always able to control females, especially when their own sources of power are unpredictable and women have independent access to resources. Among many human societies, divorce is quire common (Stephens, 1963). For example, among the Cuna Indians adults typically have been divorced four or five times by the time they die (Nordenskiold, 1949). In some societies successful males will marry and divorce a series of young women, resulting in serial monogamy. Such "temporal polygyny" can generate high variance in male reproductive success.

The duration of the commitment of resources and other reciprocal transactions between mates or between their kin appears to be associated with the duration of mating relationships (Flinn, 1983). For example, consider Fortune's (1963, p. 61) assessment of mating relationships among the Dobu:

> She has no great dependence on her husband to care for her children, since a woman can nearly always get a new husband for future help, and her brother ultimately provides for them in any case. Consequently she behaves very much as she likes in secret.

In Dobuan society long-term reciprocity between mates, and nepotism from father and father's kin evidently are not very important or frequent. Patterns of nepotism are quite variable among human societies, and they appear to be associated with certain aspects of mating systems (Murdock, 1949; Alexander, 1979; Irons, 1979a, 1983; Flinn, 1981, 1983, 1986a; for nonhuman primates see Wrangham, 1980). Biases in nepotism toward uterine or maternal kin, for example, are associated with relatively low confidence of paternity (Alexander, 1974, 1977, 1979; Irons 1979a; Kurland, 1979; Flinn, 1981). The pattern of investment in sister's offspring and uterine kin arising from low confidence of paternity is seen not only in promiscuous systems, but in many serially monogamous systems.

A number of ethnographic studies in Carribean populations have found associations between frequency of divorce and economic status. In general, wealthier, landowning families tend to have more stable mating relationships, whereas poorer, landless families tend to have less stable mating relationships (R. T. Smith, 1956; E. Clarke, 1957; M. G. Smith, 1962). These studies suggest that resources have an important influence on the duration of mating relationships.

Many of these societies are characterized by variable and unpredictable resource losses, such as short-term employment or variable crop production. Males may have resources one month but be impoverished the next, resulting in opportunistic short-term mating relationships (e.g., E. Clarke, 1957; McCommon, 1983).

Promiscuity

Among the Nayar "it is not certain how many husbands a woman might have at one time; various writers of the fifteenth to eighteenth centuries mention between three and twelve" as well as receiving "occasional fleeting visits" (Gough, 1961a, p. 358); thus the adage "no Nayar knows his father" (Gough, 1961a, p. 364). Exceedingly fluid sexual associations also appear to have been the norm among the Kaingain (Henry, 1941), the Menangkebau (Loeb, 1934), and the Mehinacu (Gregor, 1974).

All of these societies are characterized by the relative importance of uterine (maternal) kin. Inheritances among the Nayar, for example, pass from a man to his sister's offspring rather than to his own putative offspring (Gough, 1961a; also see Alexander, 1977).

Several factors appear to favor promiscuous mating relationships. These include mate absenteeism due to trading or warfare, and females' reliance on their brothers, rather than their mates, for male parental effect (Alexander, 1977, 1979; Flinn 1981; Irons, 1983). Thus, as Alexander (1974) noted, when paternity certainty is low, males investing in their "own" children may lose reproductively compared to males investing in their sister's children, who will be at least one-eighth related.

This pattern is consistent with the evidence from other species, in which the expenditure by males of costly parental effort specific to the recipient offspring (Low, 1978) is accompanied by an elaboration of behaviors on the part of the male that serves to decrease the possibility of cuckoldry. Thus, males expend parental effort only on offspring virtually certain to be their own.

Socially Imposed Monogamy

As noted earlier, Osmond (1966) found monogamy to be correlated with intensive agriculture, complex stratification, a large-state level of political integration, community sizes of over 50,000, and craft or industrial specialization (see also Flannery, 1972; Alexander, 1979; Alexander et al., 1979). In brief, large complex state societies tend to be monogamous.

Monogamy in state societies does not result from a lack of variability in resource control or social power; most states have the equivalents of presidents and millionaires. Instead, monogamy in state societies appears to result from a socially imposed reproductive egalitarianism, for example, the modern Chinese system favoring one child per family (e.g., *Newsweek*, 26 April, 1984). Thus socially imposed monogamy has no homolog in nonhuman species, unlike ecologically imposed, or resource limited, monogamy (Alexander et al., 1979).

Visibly polygynous individuals or groups are persecuted by the larger society, for example, the plight of the Mormons during the 1800s in America. The first targets of democratic revolutions are despotic polygyny and nepotism (Betzig, 1983, in press). Intergroup competition appears to play an important role in the development and maintenance of large state societies (Alexander, 1971, 1979). Socially imposed monogamy appears to be an important unifying force in state-level societies that may increase the competitive ability of these societies (Alexander et al., 1979; Alexander, 1979, in prep.). We know of no analogy among nonhuman species to the relationships among social power, group competition, and the restriction of mating opportunities occurring in modern state societies.

Summary and Discussion of the Types of Human Mating Systems

Humans exist in such a variety of environments with different population densities, different economic technologies, and different historical backgrounds that assessing the reproductive values of resources and their influence on mating patterns becomes extremely complex. Land may be a critical, defendable resource in one society and a valueless, undefended resource in another, even though both societies are in similar ecological conditions (e.g., tropical rain forest). A combination of environmental, demographic, and historical conditions must be analyzed if we are to understand human mating systems better. As in many other species, residence patterns and resource control appear to be important factors affecting reproductive strategies (cf. Alexander, 1979, p. 186). Among humans, however, a unique reproductive strategy emerges: parental and kin control of mating arrangements.

KINSHIP AND THE CONTROL OF MATING

In nonhumans there is considerable evidence of a relationship between resource distribution and the ability of males to control resources or females (Emlen and Oring, 1977). Humans go further: males may control not only resources and mates, but also their sisters, daughters, nieces—their female kin. Female kin are an important resource that males may use to acquire mates for themselves and for their male kin. Lévi-Strauss (1949) posited that marriage is essentially a contract between men, a formalized exchange of women (cf. Paige and Paige, 1981). The ramifications of marital exchange for alliance considerations have been a persistent focus of anthropological theory (e.g., Tylor, 1889; Leach, 1961). More recent analyses have interpreted marital exchange in terms of reproductive competition for females among coalitions of male kin (Chagnon, 1974, 1979a, b, 1982; Irons, 1981; Flinn, 1981; cf. Paige and Paige, 1981). Human male coalitions, however, go much further than coalitions of males in other primates such as chimpanzees (deWaal, 1982), which simply compete directly, and do not manipulate and/or exchange females.

Females, of course, are not simple pawns of male reproductive politics.

Women may exercise considerable power of choice, although seldom overtly (Rosaldo and Lamphere, 1974). Physical size, exclusive male use of weaponry, and male control of legal and political mechanisms probably give males certain advantages in conflicts of interest (see also Alexander, 1979; Hartung, 1982).

The reasons why, in humans, individuals are able to exert control over the mating behavior of their relatives (for example, by arranging the marriages of adolescents, or even infants and unborn children!) probably involve several other important and extreme aspects of human sociality: 1) a long period of parental investment; 2) extensive, multigenerational networks of kinship and reciprocity; 3) group competition, including warfare; and 4) heritability of resources over many generations.

The importance of these aspects of human sociality is illuminated by a unique aspect of human mating systems: preferences for certain types of cousins as marriage partners. No other species shows such specificity regarding *categories* of mates. Anthropologists have considered cousin marriage preferences to be a purely cultural phenomenon. For example, Murdock (1949, p. 287) states:

The rules governing marriages with first cousins will serve as a test case. Since the daughters of the father's brother, father's sister, mother's brother, and mother's sister are consanguineally related to a male Ego in exactly the same degree, all intercultural differences in marriage regulations applying to the several types of cousins represent divergences from biological expectations.

Here, Murdock has a different concept of "biological expectations" than we do. We believe that the principles of modern evolutionary ecology suggest testable hypotheses about specific intra- and intercultural differences in cousin marriage patterns.

COUSIN MARRIAGES

All human societies have rules specifying which relatives may not marry, such as incest taboos. In addition, in some societies, certain relatives are stipulated as marriage partners. Typically, in small and intermediate-sized societies, these marriages are arranged by the older generation (e.g., Whyte, 1978). These preferences and prohibitions have been argued to have the following functions:

Mating prohibitions:
   a) inbreeding avoidance; putative parallel cousins may be half-siblings (Alexander, 1979, cf. Westermarck, 1891)
   b) reciprocal arrangements to exchange women (Lévi-Strauss, 1949; cf. Alexander, 1979; Chagnon, 1982; Irons, 1981)
   c) discouragement of inbred, powerful, nepotistic subgroups

Mating "preferences":[4]

a) creation and/or reinforcement of useful kin ties (Fortes, 1969; Barth, 1956)
b) optimal distribution of heritable resources to relatives (Barth, 1956; Aswad, 1971)

Some anthropologists (e.g., Needham, 1962) have argued that there is an important distinction between obligatory, "prescriptive" rules and nonobligatory "preferential" rules. This distinction has been questioned (Ackerman, 1964); here we are concerned with actual behavior—statistical tendencies in patterns of marriage.

There are four genealogical types of cousins that a man might marry (Fig. 11.2):

a) his father's brother's daughter (FBD)
b) his mother's brother's daughter (MBD)
c) his father's sister's daughter (FZD)
d) his mother's sister's daughter (MZD)

In our own North American and Western European societies (as in many other "monogamous" societies with neolocal residence) we do not distinguish among these four types; they are all just "cousins." However, in most human societies they are distinguished in various ways and called by different terms. Typically these distinctions involve marriageable and unmarriageable categories. The most common distinction made is between the offspring of same-sex siblings as compared to the offspring of different-sex siblings. Anthropologists use the following terms to distinguish them: "parallel" cousins are offsping of same-sex siblings (FBD and MZD), and "cross" cousins are offspring of different-sex siblings (FZD and MBD).

Among human societies there are two basic patterns of cousin marriage, "symmetrical cross-cousin marriage," and "asymmetrical cousin marriage." Symmetrical cross-cousin marriage allows for marriage with both types of cross-cousins (FZD and MBD). In many societies the *only* permissible mate is a cross-cousin (first, second, or more distant). Symmetrical cross-cousin marriage is commonly associated with mating systems in which females are exchanged between groups of male kin (Lévi-Strauss, 1949; Irons, 1981). The other basic pattern of cousin marriage, asymmetrical cousin marriage, involves preferences for a single type of cousin. This category of cousin marriage is usually found in societies where resource control and kin alliances are affected by marriages: in situations in which a

---

[4] "Marriage preference" is an accepted term in anthropology. We use it in quotes to reflect the fact that there is not likely to be unanimous agreement about appropriate marriage partners; in fact, we hypothesize that expressed "preferences," or stipulations, are the outcome of conflicts of interest among different relatives about particular categories of relatives as mates.

A. MOTHER'S BROTHER'S DAUGHTER MARRIAGE (MBD)

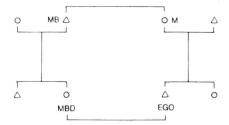

B. FATHER'S SISTER'S DAUGHTER MARRIAGE (FZD)

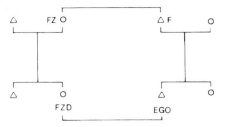

C. FATHER'S BROTHER'S DAUGHTER MARRIAGE (FBD)

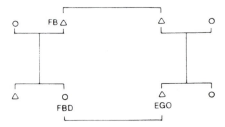

D. MOTHER'S SISTER'S DAUGHTER MARRIAGE (MZD)

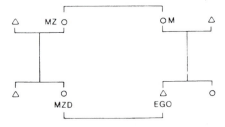

Fig. 11.2. (a–d). The four types of cousins. Tri-
angles represent males; circles represent females.

man's ability to control resources is affected by the alliances he forms through marriage.

If human cousin marriage preferences are manipulated in the reproductive interests of those who control them, that is, if they are not random cultural inventions, then we should find that they are associated with environmental differences in ways that are consistent with inclusive fitness maximization. Resource distribution, coalitions for mate competition, inbreeding avoidance, and nepotism are factors that appear particularly relevant to an evolutionary analysis of patterns of cousin marriage.

Unfortunately, reliable data on the frequencies of the different cousin marriages are not available for most human societies. There are, however, a number of notable exceptions (e.g., Aswad, 1971; Chagnon, 1974; Schapera, 1950). These studies indicate that the actual frequencies of the "preferred" cousin marriage type are quite high. Aswad (1971), for example, reported that during one generation, 41 percent of all marriages in a Turkish village were with the preferred type of cousin—father's brother's daughter. Goode (1970, pp. 94-95) reviewed a number of Muslim societies in which 10 to 20 percent of all marriages are father's brother's daughter marriages. For other societies, such as the Purum (cf. Ackerman, 1964; Needham, 1962), the frequencies of cousin marriages are nonsignificant even though there is an expressed preference for a type of cousin marriage. Rose (1960), Hajnal (1963), Charlesworth (1980), and others have noted that a bias toward certain types of cousin marriage (e.g., mother's brother's daughter) is likely due to the differences in age of marriage of the two sexes. This demographic bias, however, does not appear to have a large enough effect (about 4 percent in favor of MBD marriage) to account for the extent and variety of cultural preferences for cousin marriages, although additional data are required to establish this. An additional problem with interpreting ethnographic data on cousin marriage results from the failure by many ethnographers to distinguish genetic relatives from terminological relatives. In spite of these numerous methodological problems, we believe that analysis of the available ethnographic data on patterns of cousin marriage is useful to an understanding of human mating and marriage systems, even if the results are necessarily tentative.

We shall argue that resources and kin coalitions influence the choice of different types of cousins as mates in the following ways:

1) *Symmetrical cross-cousin marriages:* In the absence of significant material resources exchanged for women, we predict that men will exchange women for women (rather than resources for women) between male kin groups. This results in symmetrical cross-cousin exchange marriages; a prohibition of parallel cousin marriages may follow (Irons, 1981).

2) *Asymmetrical cousin marriages*: In the presence of important material resources, we predict the following associations between cousin marriage, kin coalitions, and residence:

a) FBD marriage, because it strengthens reciprocity and nepotism among male paternal kin, should be associated with patrilocal residence (Aswad, 1971; Barth, 1956; cf. Alexander, 1979).

b) MBD marriage, because it enhances nepotism and reciprocity among male maternal kin, should be associated with matrilocal and avunculocal residence (Alexander, 1979; cf. Flinn, 1981).

c) FZD marriage allows a male, if he is sufficiently powerful, to keep his son with him in a matrilocal society (in which husbands ordinarily go to live with their wife's kin). It should be practiced by high-status males in societies with matrilocal residence (Fathauer, 1961; Flinn, 1981).

d) MZD marriage does not enhance nepotism and reciprocity among male, but rather among female, kin. Because women seldom control important heritable resources, MZD marriages should be quite rare and never "preferred" on a societywide basis (see Alexander, 1977, 1979).

TESTS OF THE HYPOTHESES

1) Symmetrical Cross-Cousin Marriages

Rules of exchange prohibiting marriages within kin-group coalitions, and thus prescribing an exchange of women between kin groups (i.e., symmetrical cross-cousin marriages) are quite common (Lévi-Strauss, 1949; Murdock, 1949). These rules frequently take the form of parallel cousin incest taboos; that is, cross-cousins are the prescribed marriage partners.

Here we consider two hypotheses, not strict alternatives, about symmetrical cross-cousin marriages: the mate-exchange hypothesis (#1) outlined above, and Alexander's (1974, 1977, 1979) incest-avoidance hypothesis (Fig. 11.3). Alexander suggested that parallel cousin incest taboos function to prevent inbreeding in societies that allow or promote practices that result in putative parallel cousins sometimes being paternal half-siblings; he further suggested (1979, pp. 188-89) that symmetrical cross-cousin marriage patterns represent a mediation of conflicts of interests between the two spouses. Testing the inbreeding-avoidance hypothesis is limited by the lack of cross-cultural data on practices other than sororal polygyny that might lead to parallel cousins being half-siblings (such as more subtle wife- or husband-sharing among siblings). Our analysis is restricted to Alexander's (1979) use of cross-cultural data on sororal polygyny and the differential treatment of cross-cousins and parallel cousins.

INBREEDING AVOIDANCE
HYPOTHESIS: PARALLEL
COUSINS ARE TABOO
BECAUSE THEY MIGHT
BE HALF-SIBLINGS

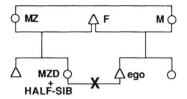

PATRILINEAL KIN COALITIONS

EXCHANGE HYPOTHESIS:
MATES ARE REQUIRED
TO BE CROSS COUSINS
BECAUSE FEMALES ARE
EXCHANGED BETWEEN
PATRILINES

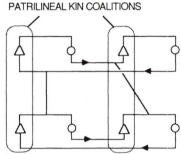

Fig. 11.3. Two hypotheses about symmetrical
cross-cousin marriage rules.

The mate-exchange hypothesis predicts a preference for cross-cousin marriage; the inbreeding-avoidance hypothesis predicts prohibition of parallel cousin marriage. The two hypotheses generate testable predictions. In some cases we cannot distinguish between the hypotheses, but in other cases the predictions derived from the two hypotheses differ, and we can test between them.

Prediction #1a. *Sororal Polygyny*: Both the ''mate exchange'' hypothesis and the ''inbreeding avoidance'' hypothesis predict an association between cross-cousin marriage preferences and sororal polygyny.

The inbreeding-avoidance hypothesis further predicts that parallel cousin marriage prohibitions should be associated with the levirate (rules prescribing that a widow marry her deceased husband's brother), the sororate (rules prescribing that a widower marry the sister of his deceased wife), and wife and husband-sharing between siblings. These are all conditions that might cause putative cousins to be half-siblings. As Alexander (1979) noted, cross-cultural data to test for the effects of these practices are lacking. The mate-exchange hypothesis makes no prediction regarding cross-cousin marriage and the levirate and sororate; although Lévi-Strauss (1949) and Radcliffe-Brown (1952) suggest that the levirate and sororate are likely to be adopted in societies with mate exchange.

TABLE 11.1
Cross-cousin marriage and the exchange of resources
(Preference for symmetrical cross-cousin marriage is
associated with the absence of exchange of material
considerations at marriage.)

	Bride price	Dowry/gifts	No exchange
Cross-cousin	39	6	67
Expected	48	11	52
Other	271	67	267
Expected	262	62	282

NOTE: $n = 717$, $\phi = 0.1193$, $\chi^2 = 10.201$, $p = 0.0061$.

Prediction #1b. *Resources:* The mate-exchange hypothesis predicts that the absence of important resources exchanged for mates should be associated with cross-cousin marriage preferences.

Prediction #1c. *Kin Objections:* The mate-exchange hypothesis predicts that the more distantly related males in the kin-group coalition will object most strongly to an "incestuous" marriage, whereas the close kin of the marriage partners will be the most supportive. The inbreeding-avoidance hypothesis predicts that "incestuous" marriages should be most strongly objected to by close kin.

Prediction #1d. *Extent of Prohibitions:* The mate-exchange hypothesis predicts that taboos against parallel cousin marriage will be extended to second cousins and third or more distant cousins, depending on the size of the initial exchange coalitions. The inbreeding-avoidance hypothesis predicts that only first parallel cousins should be prohibited as marriage partners, because only first parallel cousins could be half-sibs.

Prediction #1a, generated by both hypotheses, suggests that sororal polygyny should be associated with cross-cousin marriages. Alexander (1977, p. 327) found that "75 of 79 societies (95 percent) favoring or prescribing sororal polygyny treat parallel and cross-cousins differently, but only 35 of 101 monogamous societies (35 percent) do so ($p < .0001$)." In these societies, parallel cousins are prohibited as mates, and are often called by the labels used for siblings.

Prediction #1b suggests that the presence or absence of resources distinguishes the two hypotheses. Table 11.1 suggests that symmetrical cross-cousin marriages are more common in the absence of exchanges of material resources at marriage ($\phi = .119$, $\chi^2 = 10.20$, $df = 2$, $p < .006$), consistent with the mate-exchange hypothesis. A problem remains, however. The fact that resources are not exchanged at marriage may mean either that no exchangeable resources exist, or that exchangeable resources exist, but for some reason are not used in this context. To clarify this relationship, we examined exchange patterns in societies with and without domestic animals, with and without intensive agriculture (usually associated with land ownership), with and without stratification of wealth and hereditary class, and with and without rules concerning inheritance. Our reasoning was that the absence of animals, land ownership, wealth distinctions, and inheritance rules indicated a relative absence of physical resources exchangeable for women. We found that the exchange of women, or only a token exchange of goods, was significantly more likely in societies without rules of inheritance ($n = 571$, $\chi^2 = 58.348$, $p = .00001$, $\phi = .3197$). One hundred eighteen out of 181 societies without inheritance rules exchanged no goods at marriage, while 122 of 390 societies with some rules of inheritance exchanged no goods. Exchange of women rather than goods was more likely than expected in societies lacking any statifications of wealth or hereditary class, although the association was not as strong ($n = 763$, $\chi^2 = 18.724$, $p = .00001$, $\phi = .1567$). These results are consistent with Hartung's analysis of polygyny and inheritance, and suggest that when men can accumulate resources, they use those resources as mating effort (cf. Dickemann, 1982). Exchange of women versus goods shows no pattern with 1) the presence or absence of agriculture ($n = 802$, $\chi^2 = .0040$, $p = .9495$, $\phi = .0022$), or 2) the presence or absence of domestic animals ($n = 805$, $\chi^2 = .7332$, $p = .3918$, $\phi = .0302$). Thus, while the specific type of goods is not obvious, it does appear that in societies with inheritance rules and/or class distinction based on wealth or resource control, resources are far more likely to be used to acquire mates, whereas the direct exchange of women for women is less likely. This does not preclude the exchange of women in societies with exchangeable resources, which may occur for other reasons.

Thus far, predictions from the mate-exchange hypothesis have concerned *preferences*. What about *prohibitions*? Prediction #1c suggests that it is the closeness of relationship of those who object to the prohibited marriage type that distinguishes the two hypotheses. Unfortunately no cross-cultural data are available at this time. Chagnon (1982), however, indicates that among the Yąnomamö, close kin are most likely to be supportive of "incestuous" marriages, while more distant relatives (that are competing for the same women as mates) are more likely to object—sometimes to the extent of mortal combat. Such behavior really reflects the support of relatives in matters of mate exchange, and probably the manipulation of kinship terminology for the benefit of self and relatives (e.g., the same relationship will be called "incestuous" by competitors, and acceptable by relatives). This pattern is consistent with the mate-exchange hypothesis.

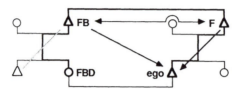

PREDICTION: **FBD** MARRIAGE IS ASSOCIATED
WITH PATRILOCAL RESIDENCE

Fig. 11.4. Father's brother's daughter (FBD) marriage and the flow of resources. Diagonal dashed
lines indicate the pattern of resource flow along paternal kin lines from father to son. Arrows indicate
nepotistic ties that are enhanced by FBD marriage.

Prediction #1d suggests that the extent of the prohibition on mating between parallel cousins distinguishes the two hypotheses. Many societies favoring cross-cousin marriages appear to extend the prohibition of marriage
between parallel cousins beyond first cousins (Murdock, 1967), supporting
the mate-exchange hypothesis; however, there are not well-documented detailed data cross-culturally on actual frequencies of these marriages (cf.
Hajnal, 1963).

To summarize our analysis of symmetrical cross-cousin marriage, it appears to be related both to the availability of resources and the avoidance of
close inbreeding. Additional data are needed, however, before we can determine the relative influence of each. Detailed, long-term demographic and
ethnographic studies such as Chagnon's work on the Yąnomamö are particularly useful.

In the following section we analyze the other major category of cousin
marriage, asymmetrical cousin marriage. The reasons why this kind of
cousin marriage is preferred appear to be different from the reasons for symmetrical cousin marriage.

2) Asymmetric Cousin Marriages

FBD (Prediction #2a) marriage appears to increase the nepotistic benefits
obtained from father's brother and other paternal kin (Fig. 11.4) (Barth,
1966; Aswad, 1971; Paige and Paige, 1981; Flinn, 1981). It may also be a
strategy in intrafamilial competition (Manson, 1983). The Tswana are a
good example (Schapera, 1950, p. 163):

In explaining the relative popularity of marriage with a father's brother's daughter,
the Tswana often add that it keeps the bogadi cattle in the same family circle and

TABLE 11.2
Residence and cousin marriage (Patterns of asymmetrical
cousin marriage are associated with residence.)

	Matri.	Avunc.	Neo/sym.	Patri.
FZD	3	9	0	1
Expected	2	1	2	8
MBD	9	10	3	35
Expected	8	6	7	36
Cross-cousin	24	18	8	62
Expected	17	11	14	71
Symmetrical	70	33	76	331
Expected	76	50	63	322
FBD	0	0	1	23
Expected	4	2	3	15

NOTE: $n = 716$, $\phi = 0.2083$, $\chi^2 = 93.188$, $p < 0.00001$.

prevents them from passing into the hands of outsiders. . . . Among nobles, where marriage with the father's brother is actually the most common form of cousin marriage, the dominant factor is certainly not bogadi but status. It is considered highly desirable that the chief's heir and other senior children should marry persons of rank. The motive is not so much "keep the blood pure" as to secure the political advantages attached to union with powerful or influential families.

We predicted FBD marriage would occur more frequently among societies with patrilocal residence, and this is in fact the case (Table 11.2). Alexander (1979) made the same prediction, for a very similar reason—because male alliances are stronger in patrilocal residence.

MBD (Prediction #2b) marriage appears to increase the nepotistic benefits obtained from mother's brother and other uterine kin (Fig. 11.5), as described by Alexander (1979, p. 175). It may also resolve certain conflicts about choice of marital residence (Gough, 1961b; Richards, 1950; Murdock, 1949; Flinn, 1981). As Fortes (1950, pp. 281-82) noted:

As has been mentioned, a cross-cousin [MBD] is regarded as the most satisfactory spouse. At least this is still the view of the older people. Many young men take a different view. They say that a cross-cousin [MBD] is "too-near," almost a sister, and so she is never as attractive as an unrelated girl. . . . But in rural areas there are still many young men who approve of marriage with a mother's brother's daughter (wofa ba) on the grounds that it creates an additional bond with their maternal uncles and that it is a more secure marriage than a match with an unrelated woman. Women often argue in favor of the custom. They say it strengthens their claims on their husbands and their children's claims on both paternal and maternal kin. The older people—parents, mother's brothers, and father's sisters—in whose interests and at whose insistence cross-cousin [MBD] marriages are arranged, defend the custom on

239

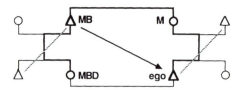

PREDICTION: **MBD** MARRIAGE IS ASSOCIATED
WITH AVUNCULOCAL RESIDENCE

Fig. 11.5. Mother's brother's daughter (MBD) marriage and the flow of resources. Diagonal dashed
lines indicate the pattern of resource flow along paternal kin lines from father to son. Arrows indicate
nepotistic ties that are enhanced by MBD marriage.

various grounds. The commonest argument is on the grounds of property and
wealth. Cross-cousin [MBD] marriage most often occurs between the children of full
siblings or of first uterine cousins. Such a marriage, it is contended, ensures that a
man's daughter and her children derive some benefit from the property he is obliged
to leave to his sister's son . . .

MBD marriage is predicted to occur more frequently in societies with avunculocal residence patterns; as Table 11.2 shows, this prediction is supported.

FZD (Prediction #2c) marriage allows a high-status male to keep his son
co-resident with him in societies where young men would normally reside
with their mother's brothers (matrilocal and avunculocal). With FZD marriage, father-to-son nepotism, which uterine kin are likely to oppose, also
entails mother's brother (husband's father) to sister's daughter (son's wife)
nepotism (Fig. 11.6), which uterine kin prefer. Among the Trobrianders
(Fathauer, 1961, pp. 242-45):

A man will ordinarily go to the village of his sub-clan after marriage, where he has
a claim to land. His mother's brother will bring pressure to bear upon him if he delays too long in coming, because the matrilineally related group of males is strengthened by each addition, which may be useful in feuds or inter-village disputes. However, because of the high rank of a chief, and particularly the chief of the Tabalu, the
uncle will not object if the chief wishes to keep his son in the village with him. The
chief grants his son the right to reside in the community, which is usually terminated
upon the death of the father. His residence right is strengthened if he was married in
infancy to his father's sister's daughter. This makes his right to remain in the village
even after his father's death almost, but not quite, inalienable. . . . Since the marriage is made in infancy, the members of the father's sub-clan are not worried about

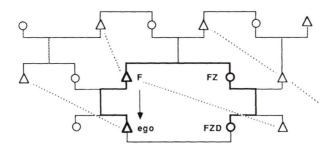

PREDICTION: **FZD** MARRIAGE IS ASSOCIATED WITH
MATRILOCAL AND AVUNCULOCAL RESIDENCE

Fig. 11.6. Father's sister's daughter (FZD) marriage and the flow of resources. Diagonal dashed lines indicate the pattern of resource flow along maternal kin lines from mother's brother to sister's son. Arrows indicate nepotistic ties that are enhanced by FZD marriage.

the advantages he gives his son since they know that they will get back these privileges in the next generation.

Table 11.2 indicates that residence and type of cousin marriage are significantly associated ($\phi = .208$, $\chi^2 = 93.19$, $df = 12$, $p < .00001$). While these cross-cultural data are at least suggestive, ethnographic research documenting associations at the individual behavioral level would be much more convincing. Unfortunately, such data have yet to be published.

The fourth type of assymmetrical cousin marriage, mother's sister's daughter (MZD), is virtually unknown, supporting Prediction #2d.

CONCLUDING REMARKS

Analysis of human mating systems from the perspective of modern evolutionary and ecological theory is still in a preliminary stage. In this chapter we present no new data; anthropologists have known for some time that there are nonrandom associations between ecological conditions and mating patterns in human societies. What, then, does evolutionary theory bring to the study of humans? We believe that it has two main uses.

First, most anthropological explanations are based on "proximate" factors, such as economic and/or status competition (e.g., Paige and Paige, 1981). Proximate explanations are incomplete. We still have no rationale for *why* males should choose to compete for status; is it "purely cultural" or random? Examination of comparative data from other species suggests that it is important to explore the relationship between proximate causes and long-term reproductive trends (e.g., Alexander, 1979; Daly and Wilson, 1983; Irons, 1979b). In the above case of status competition, testing for correlations between status and reproductive success (where comparison of different status groups is appropriate, i.e., in the absence of trends that differentially affect status groups, as for example with the "demographic transition" in industrial societies—see Turke, 1985) has proved useful (Chagnon et al., 1979; Chagnon, 1979a; Essock-Vitale, 1984; Flinn 1983, 1986b; Hill, 1984; Irons, 1979b; Turke and Betzig, in press).

Second, an evolutionary approach sometimes suggests new hypotheses to old problems (see Chagnon and Irons, 1979 and Alexander, 1979, for examples). Evolutionary hypotheses can make rather specific predictions that are quite testable, although the data required often have yet to be gathered.

In this study, both these uses of evolutionary theory were important. Our analysis indicates that resource distribution and social competition significantly influence human mating in ways that are consistent with evolutionary models. In some cases the models developed from the study of nonhuman species must be extended and modified to include the unique aspects of human sociality, notably the control of the mating of relatives such as daughters. This attempt at integration of anthropology and evolutionary biology seems to us a particularly fruitful endeavor. The lack of certain kinds of field data, however, has impeded this process.

A number of recent ethnographic field studies (e.g., R. Bailey 1985; Berte, 1983; Borgerhoff Mulder, 1987; Chagnon, 1982; Flinn, 1983; Hames, 1979; Hawkes et al., 1982; Hurd, 1981; Hurtado et al., 1984; Peacock, 1985; Turke and Betzig, in press; Wrangham and Ross, in prep.) are now beginning to provide new data useful for testing evolutionary hypotheses. Clearly, however, more field research examining individual behavioral strategies is critical to an understanding of human mating systems and sociality in general. We hope this chapter provokes such research.

APPENDIX A

Materials and Methods

Data were coded for 849 societies in Murdock's *Ethnographic Atlas* (Murdock, 1967) with regard to: degree of outbreeding, direction of marriage preference, exchange of goods at marriage, principal marital residence preference, class stratification, direction of inheritance of real property, type and intensity of agriculture, and type of animal husbandry. The coding rules are given in Appendix B. Nonparametric statistical analyses were used, principally Kruskal-Wallis and χ^2. χ^2 results were analyzed for contribution of each cell (robustness), and any nonrobust results are noted.

APPENDIX B

Coding Rules

All data are from the *Ethnographic Atlas* (Murdock, 1967)

1. Outbreeding (col 25)
    1. Qp, T—any first cousin allowed as marriage partner
    2. C, E, D, F, G, M, P—some first cousins
    3. R, S, O—no first cousins
    4. N—no first *or* second cousins
2. Marriage preference (col 25)
    1. Tp, Qp, P, Dp, Cp, Bd—FZD preferred (uterine preference)
    2. CM, Fm, M, Em, Tm, Qm—MBD (weaker uterine preference)
    3. C, Qc, Cc, Tc—cross-cousins (parallel cousin marriage prohibited)
    4. N, O, Q, T, R, S—(prohibition, equivalent treatment)
    5. Qa, Da, Fa—FBD (agnatic preference)
3. Marital goods exchange (col 12)
    1. B—bride-price (substantial transfer of goods from groom to bride's family
    2. D, G—dowry, or reciprocal gift exchange
    3. O, S, T—absence of payment, service, or token payment only
    4. X—exchange of women rather than goods
4. Principal marital residence (col 16)
    1. M, O, U—matrilocal, no common residence, uxorilocal
    2. A, C, D—avunculocal; optionally uxorilocal, virilocal, or patrilocal
    3. N, B—neolocal, ambilocal
    4. P, V—patrilocal, virilocal
5. Type and intensity of agriculture (col 28)
    1. O—complete absence of agriculture
    2. C, E, H—casual, extensive, shifting, horticulture
    3. I, J—intensive agriculture, with or without irrigation
6. Type of animal husbandry (col 39)
    1. O—absence or near absence of domestic animals
    2. P—pigs are the only major domestic animal
    3. B, C, D, E—bovines, camels, equines, deer
7. Class stratification (col 67)
    1. O—absence of significant stratification among freemen
    2. W—wealth distinctions
    3. D, E—dual (hereditary aristocracy and commoners), elite: in both of these one class has significantly more control over scarce resources
8. Direction of real property inheritance (col 74)
    1. P, Q—patrilineal: sons (P) or other patrilineal heirs
    2. C, D—both sexes inherit, either equally (C) or sons more (D)
    3. O—absence of property rights
    4. M, N—matrilineal: sister's sons (M) or other matrilineal heirs

12. The Evolution of Mating Strategies in Male Antelopes

———————————— • ————————————

L. M. GOSLING

THE THEME of this chapter is that variation in the mating strategies of male antelopes is a consequence of differing adaptations to maximize encounters with sexually receptive females. The argument thus depends to a large extent on female behavior and, while this has not been studied in detail, it is possible to make generalizations which are sufficient for the present aims. The first is that almost all parental care is by females; males that accompany females might contribute to the survival of their own offspring by detecting or repelling predators but, in general, male parental investment is limited to the production of gametes. In contrast, females suckle their young, groom them, consume their excreta to remove odors that might attract predators, and protect them from attacks by predators and conspecifcs (Gosling, 1969; Lent, 1974; Estes and Estes, 1979). Offspring stay with their mothers for at least months and at most for many years. The mother-offspring bond is the most enduring relationship for females and, where long-term associations have been reported between adults, they are probably between mothers and daughters. Mothers may continue to invest in their adult female offspring (Clutton-Brock et al., 1982), although the relative costs, such as those from feeding interference, and the benefits, such as those from predator detection, have not been quantified.

Given 1) that selection should maximize the production of offspring that survive to independence, 2) the large nutrient requirements of pregnancy, lactation (e.g., Sadlier, 1969), other kinds of offspring care, and 3) the observation that female ungulates need to spend the great majority of their time feeding or ruminating, it can be argued that females should try to maximize their nutrient intake for reproduction and that this will involve a large part of their active behavior. They should also minimize risks to themselves and their offspring, particularly those from predation, and this may be the main reason that they join groups. In this chapter I explore the influence of female behavior, particularly their patterns of foraging and their tendency to form groups, on male mating strategies.

The chapter mainly (but not exclusively) considers the antelopes of savannah habitats in east and southern Africa. The social organization of these animals has been reviewed by Hendrichs (1972), Estes (1974), Jarman

(1974), and Leuthold (1977), among others. Jarman established main categories of social organization and related these to the feeding ecology of the species included in each. Here I extend this treatment by considering the strategic decisions that face individual males.

Male reproductive success (RS) depends initially on intrasexual competition for access to females, and secondly on how they maximize the number of matings they achieve. These two problems, of sexual and environmental selection respectively, are introduced in the two sections that follow.

Competition between Males

The outcome of contests for a resource depend on the relative competitive ability of the contestants and the value of the resource to each contestant; when a resource is more valuable to one animal than another it will fight harder. The other contestant may avoid wasteful combat and the risk of injury by giving up early (Maynard Smith, 1974; Maynard Smith and Parker, 1976). A third possible factor, uncorrelated asymmetries between the contestants, will not be considered here because such asymmetries always depend on the value of the resource—for example, the investment in obtaining the resource, to the extent that this affects future investment, and the owner's superior knowledge of the resource (Parker and Rubenstein, 1981).

Competitive ability is often correlated with morphological and physiological attributes. Body size affects the outcome of contests in various invertebrates (e.g., spiders, Riechert, 1978) and in the antelopes, whose fighting techniques involve violent head clashes, encounters are probably influenced by age, body size, horn size and shape, strength, bodily condition (nutrients available for active behavior), and so on. Of these factors only the effect of the relative age of the contestants has been documented (e.g., greater kudu, Owen-Smith, 1984) but it seems intuitively likely that the factors listed above are also involved.

If a male is regularly defeated by another in a series of contests it would be advantageous for it to avoid, or reduce, encounters with that individual until its competitive ability has improved. The existence of consistent dominance relationships between males, with subordinates often withdrawing before contact is made (e.g., hartebeest, Gosling, 1975) suggests that animals do behave in this way and also that they know each other individually and remember the outcome of previous encounters. Where males live and mate in groups, the dominance status of individuals determines their access to mates (e.g., buffalo, Sinclair, 1977). Because this form of dominance depends on potential competitors' remembering past encounters with particular individuals, it is called an "individual reference for dominance." In a sense this reference is a simple extension of the individual's competitive ability, but alpha males may sometimes maintain their status by threat when their competitive ability has declined below that of other group members.

Mating strategies often depend on the advantage that the owner of a territory or a group of females has over competitors. Thus, rather than establishing an individual reference for dominance a male must establish itself as an owner. In practice these references for dominance are linked because only males that have achieved high status leave male groups and try to take over territories (impala, Jarman, 1979; hartebeest, Gosling, 1975). Once established in a territory, males win all agonistic encounters—up to the time when they are defeated by a high-status intruder and thus lose the territory. Why do owners always win? The reason may be that they have more to lose than intruders have to gain and they may thus be prepared to escalate a contest further. Owners probably have a better knowledge of the territory's resources, including the distribution of food, the places where predators occur, and the timing and direction of visits by females. Most intruders will not know the territory so well and, at the time of the contest, would have less to gain.

Another reason that owners may fight harder is that their chance of survival and future RS may be affected more seriously if they lose than that of the intruder. Firstly they may be at risk from predation when expelled into the unfamiliar world outside their territories (Gosling, 1975), whereas the intruder would not have this disadvantage. Secondly, and particularly in owners that are declining in competitive ability, it may be very difficult for a male to regain its position once it has been displaced. Owners may thus be prepared to escalate to risky levels while younger challengers, with opportunities for future less damaging contests, withdraw early. The second argument may also apply to males with individual references for dominance. An alpha male may have a heavy investment in the relationships that its status depends on, and, to the extent that this investment affects its future RS, it has more to lose by giving up status than a competitor has in not gaining it. In an extreme case where it would be impossible for a male to regain status it would have nothing to lose by escalating to very dangerous levels—a form of terminal investment.

Where a male has an owner's advantage it will pay an intruder to avoid the costs of an encounter by identifying it and withdrawing early. The link between the owner and the resource distinguishes it, and the owner can thus be said to have a "resource reference for dominance." The identification of owners (a special case of competitor assessment) is thus critical and selection would be expected to favor the ability of owners to advertise their status. Most, perhaps all, territorial male antelopes comprehensively scent mark their territories and intruders may identify the owner by matching the scent of the males that they meet with that around them. Owners may thus advertise their identity to their own advantage since intruders will usually avoid an escalated contest with a territorial male (Gosling, 1982). Other cues are also used, particularly where there is a wide difference in competitive ability between intruder and owner: low-status intruders often turn and flee when they see a large male approaching directly in a confident posture (Gosling, 1975).

It is not known how important "resource references for dominance" are for males that accompany and defend groups of females. Such males certainly establish their individual dominance status before joining female groups (e.g., greater kudu, Owen-Smith, 1984) and the extent to which this is augmented by an owner advantage is unknown. Males that accompany female groups could have information analogous to the territorial male's knowledge of its territory, that is, of the group's feeding routes, watering places, etc., and, in addition, it may know the reproductive state of the females and perhaps the chance that they will become receptive. More generally, multiple references for dominance are probably the norm: a territorial male may depend mainly on advertising its identity as an owner to win contests but it may also be individually known to many intruders who remember the outcome of past encounters in male groups.

Strategies to Maximize Matings

If females' movements are mainly designed to optimize their nutrient intake, males have only two main options if they are trying to maximize the number of times they mate. They can either follow one or a group of females over a part, or all, of their foraging movements, and maintain dominance over other males that are also following, or, they can wait in a part of the range that females visit and defend this area against other males. The latter, a sit-and-wait strategy, is usually called resource defense territoriality because it involves defending resources that females need. Its success depends on the ability of a territorial male to defend a territory whose resources will attract either more females, or a few females more often, than those of competitors.

Another, highly specialized, sit-and-wait strategy is to acquire a lek territory. Leks are areas with clumps of males which attract females when they become receptive but which have insignificant food resources (e.g., topi and Uganda kob). The reason that females visit leks for mating is a matter of debate (Bradbury, 1981; Bradbury and Gibson, 1983; and others): one possibility is that, on leks, females can choose between males and select those with heritable traits that will increase the likely RS of their offspring. However, even in this case female food requirements may play a role since leks near a good food supply should be most successful in attracting females.

The chapter explores the circumstances that lead to the evolution of these strategies. Some predictions about the effects of different patterns of resource distribution can be checked from the literature but the biggest problem in designing real tests of theory is the scarcity of information about intraspecific variation in mating strategies: interspecific variation can suggest the main factors that affect the evolution of social behavior but specific information at the level of the individual social animal, that is, at the level at which sexual and natural selection operates, will be needed before the evolution of particular strategies can be unraveled in detail.

FOLLOWING STRATEGIES

Under What Conditions Would Selection Favor Following?

Following is more likely to be an effective strategy than a static strategy when the chance of a receptive female's occurring in any one location is very low. The conditions that would contribute to this are: 1) a low density and unpredictable food resource, so that the time spent by females in any one area and the tendency to form static feeding aggregations is unpredictable and low, and 2) a low degree of breeding synchronicity, with females becoming receptive during a large part of the year perhaps in response to unpredictable environmental change. Unsynchronized breeding on its own would not be expected to select for following but it should have an additional effect when combined with the influence of a very low density food supply.

The benefits of following would increase if: 3) there was a tendency for females to form groups so that males could monopolize a number of females by following, and if 4) there was an advantage for males in being in a group (in addition to any mating advantage) and a lower chance of isolation for males that adopt following strategies than those that adopt other strategies.

Following should also occur when: 5) potential mates and competitors are too concentrated in space and time for any sit-and-wait strategy to be economic. This might occur when populations are divided into large, mobile, mixed-sex groups: males defending territories in part of a group's range would spend long periods alone and short periods swamped by large numbers of conspecifics.

Feeding Ranges and Male Behavior

A number of studies have shown that females, unrestrained by male herding behavior, move in relation to local variation in the food supply. Perhaps the best-known species where males normally adopt a following strategy is the buffalo, whose movements between wet and dry season food reserves in the Serengeti National Park (NP) have been documented by Bell (1970) and Sinclair (1977). Nobody has compared the nutrient intake of an animal on a particular foraging route with that which it would have obtained on another route, although a number have shown that animals select the most nutritious plant species (D.R.M. Stewart, 1967; Field, 1968, 1972) or plant parts (Gwynne and Bell, 1968; Bell, 1970) out of swards that also contain less nutritious elements; both aspects are reviewed by Jarman and Sinclair (1979). The consequences of alternative foraging routes would be interesting because although many aspects of broad-scale movements clearly do occur in relation to local variation in food quality, some aspects, notably the avoidance of habitats that might conceal predators, may interfere with the optimal foraging requirements of females.

The species in which some, perhaps most, males adopt following strategies include roan antelope and oryx. Both live in arid habitats and are adapted to low-quality diets; oryx range over very large areas and are the only East African ruminants which are independent of free water and feed mainly on grass (Hofmann, 1973). Their food supply has a low-standing crop and a high concentration of indigestible structural elements, and its nutrient concentration is probably among the lowest of the antelopes. In both cases large adult males accompany groups of females and their offspring. These males influence the degree to which the group is clumped by herding behavior but they do not usually dictate the route taken. In a group of oryx that contained eleven females, three acted as leaders during marches in single file and the alpha male frequently walked at the back (Walther, 1978c). In the Kruger NP roan live in ranges which, over the whole year, are 64-104 square kilometers in area. The same groups seem to occupy the same ranges for very long periods, sometimes for years, and, interestingly, because the whole range is far too large to defend at any one time, there is little overlap between ranges. At any one time the group lives in an area of about two-four square kilometers which probably contains the best food (and water) supply for the season in question. Alpha males do not defend this entire seasonal range but they defend the area around the female group, wherever it happens to be, against other males that approach to within 300 to 500 meters. Joubert (1974) speculates that the year's range is kept free of other groups by scent marking (particularly from the interdigital gland) but, in view of the frequent intrusion by other antelopes into more intensely defended and scent marked areas (Gosling, 1982), it is more likely that groups occupy the seasonally optimum part of their range at the same time (for example, near water in the dry season) and the limited defense mentioned maintains spatial exclusiveness at each successive season. Oryx groups are adapted to a more arid environment than roan and cover much larger areas.

Following strategies are also characteristic of two large browsing species, eland and greater kudu. Eland live at low densities in seasonally arid areas and are selective, concentrate feeders adapted to a high-protein, low fiber diet (Hofmann, 1973). Females live in groups that average eleven (Hillman, 1976) and, given the likelihood of feeding interference in such selective feeders and their great size (females average 450 kg) it is not surprising that their foraging range is large: females often have ranges of over a thousand square kilometers and are highly mobile in response to seasonal variation in food quality (Hillman, 1982). Males establish dominance relationships through agonistic behavior and these presumably dictate access to receptive females. The ranges of adult males, which live alone or in small groups, are much smaller than those of females, often about thirty square kilometers, and males follow groups of females when they are within these areas.

Greater kudu live at low densities in scrubland habitat and feed on the leaves of woody plants and creepers (Owen-Smith and Novellie, 1982). In a population in the Kruger NP studied by Owen-Smith (1984) females live in groups which average 9.7 members (including young) and have overlapping ranges of four to twenty-one square kilometers. In the nonbreeding part of the year males join male groups where they may establish individual references for dominance. In the breeding season, single adult males join and, for periods of weeks or days, follow groups of females. One or more younger males (up to five years old and almost fully grown) sometimes accompany groups that contained an alpha male. The younger males are usually ignored although they were subordinate in occasional low-intensity interactions. Fights between males are rare and usually occur when a receptive female is nearby. Alpha males sometimes switch between female groups but Owen-Smith saw few interactions between males, perhaps because of the low adult sex ratio in the population (twelve females per male over six years old).

Neither eland nor greater kudu males could economically defend territories, given the low densities of females and the large ranges. But in neither species do males spend all their time following. Both show behavior with elements of a sit-and-wait strategy: male eland live in relatively small ranges and adopt a following strategy only when they are visited by females. Some male kudu switch between female groups, which suggests an element of optimizing contact with females within the male's range rather than following one group. Why do males not simply follow one group of females? One possibility is that by limiting their range, males reduce the costs of competition by limiting male competitors to those known individuals which also use the area.

Following strategies sometimes occur when females occur singly or in small groups. Here males may attempt to maximize their RS by switching between female groups. Examples include bushbuck, which live in dense cover and are usually solitary (Waser, 1974) and lesser kudu (Leuthold, 1974). Lesser kudu occur at low density in arid scrubland and feed mainly on the leaves of woody plants. Males and females have ranges of a similar size in Tsavo NP (0.8–4.3 and 0.4–3.5 km^2 respectively, Leuthold, 1974).

Female ranges show almost complete overlap and they occur most commonly singly or in groups of two or three (average 3.0). Adult males also show almost complete range overlap and they occur singly or in groups of two or three. Leuthold did not observe agonistic behavior but males probably establish individual references for dominance where they overlap. It is not known how much time alpha males spend with females, but in twenty-seven observations of one known female, at least three different adult males accompanied it on thirteen occasions. The males did not spend much time with the female and presumably divided their time among the females within their range. Probably all male ungulates get information about female receptivity from the odor of their urine and male lesser kudu might thus monitor the reproductive status of the females in their range without making direct contact. A similar pattern of social behavior occurs in bushbuck (Waser, 1974). No data are available to test the hypothesis that males join females in relation to the chance that they will be receptive.

In the cases of buffalo and one topi population, following seems linked to a situation where the population is divided into a few very large and mobile groups. As discussed later, joining groups is probably an antipredator adaptation and, in these cases, large groups are possible because of an abundant food and a mobile feeding strategy.

The most consistent units in buffalo herds are groups of females, which probably consist of genetically related individuals. These units, of 30–60 animals often collect into larger herds of up to 2,000, particularly in the wet season when food is abundant (Sinclair, 1977; Hofmann, 1973). Buffalo feed preferentially on tall sump grasses and have ranges of 126–1,075 square kilometers (Mloszewski, 1983). Males establish dominance relationships and some follow female kin groups. However, males sometimes form separate groups, possibly because of the differing food requirements of males, which are 42 percent heavier than females. Individual males closely follow receptive females and lower-ranking males sometimes copulate with females early in estrus. Only at peak estrus do alpha males displace other males and mate with the female themselves (Grimsdell, 1969; Sinclair, 1977; Mloszewski, 1983).

A population of topi at Inshasha in Uganda show similar behavior (Jewell, 1972). Large groups of adults and juveniles graze while moving rapidly over wide expanses of homogeneous grassland. Although some males defend static resource territories, others defend sectors of the mobile groups and subordinate males are driven to the periphery. In buffalo and these topi the advantages of following as opposed to resource defense territoriality might depend on the economics of defending a static territory and the chance of mating with receptive females when very large numbers of potential mates and competitors move quickly through the area. Selection might favor following in these circumstances and, in the case of buffalo, all males appear to follow although no good studies have been carried out in low-density populations. Some topi at Inshasha persist in defending resource territories and Duncan (1975) found no evidence of following strategies in the western part of the Serengeti NP where females live in large mixed herds. Perhaps large mobile herds only disrupt resource defense partially, but to an extent where following becomes profitable as an alternative strategy. If so, we need to know why this threshold is passed at Inshasha and not in the western Serengeti.

Although the question has not been addressed specifically, there is no evidence for exclusiveness in group ranges by oryx or eland. Sinclair (1977) showed that buffalo groups are more spaced out than would be expected by chance but the mechanism is unknown; it could consist of avoiding areas grazed by other animals rather than area defense. Owen-Smith (1984) attributes the spacing of alpha male greater kudu to the spacing of the female groups that they follow, although males do switch between groups. Assuming that male spacing depends on female spacing, then its ultimate cause is probably linked to defending, or avoiding the joint use of, the food supply; however, the proximate mechanism of this separation is unknown.

Breeding Synchronicity

The timing of breeding in antelopes is strongly influenced by the seasonal patterns of rainfall, which dictate the availability and quality of food plants. While there is no consensus in the literature, it seems likely that the timing of conception has been selected to maximize the chance of good-quality herbage for offspring during weaning. Whether or not females breed at the appropriate time may also depend on their condition when they need to conceive. Most African antelopes breed through the year with peaks that reflect single or bimodal peaks of rainfall. These peaks are sometimes sharper than expected, probably because of the selective effects of predation against young born alone (Estes and Estes, 1979; Gosling, 1969), but this is not the case in species with following strategies. In these, offsping either hide alone in long grass when very young or are protected by the group defense of their mothers.

The chance of females becoming receptive in a group accompanied by a male thus reflects the incidence of rainfall. As expected, the species living in arid habitats, such as oryx, do breed throughout the year and probably in a way that is unpredictable. In combination with a large feeding range these circumstances should favor following strategies since a sedentary male could easily miss receptive females. This logic does not apply to buffalo, which have a restricted breeding season with most conceptions in a three-month season (Sinclair, 1977) and where males accompany female groups irregularly throughout the year. Males may need to maintain dominance over potential competitors throughout the year rather than just before the breeding season, but this does not explain why they follow females. The alternative explanations include being available for female choice and the antipredator advantages of group membership. In contrast, adult male greater kudu follow female groups during a two-month breeding period but spend the nonbreeding part of the year in male groups (Owen-Smith, 1984). In this case the timing factor may influence why males are more consistently with females than in the case of lesser kudu (Leuthold, 1974) where there is year-round breeding and monitoring the reproductive state of individual females may be a relatively important element of the following strategy. It could pay a greater kudu male to stay with a female group when its members are more likely to become receptive in the near future. However, groups of female greater kudu are also larger. It is difficult to distinguish the importance of these two factors. The dividing line between the two species of kudu may prove artificial when information on intraspecific variation becomes available. However, while the timing of female receptivity could effect the adoption of a following strategy by greater kudu, the spatial distribution of the food supply and the tendency for females to form social groups are generally more important.

The Tendency to Form Groups

The tendency for females to form groups and the advantages for a male in group membership are special cases of the general problem of why animals join groups (see Bertram, 1978; Rubenstein, 1978; Gosling and Petrie, 1981). Among plains antelopes the suggested ideas are that by joining groups individuals could benefit a) from enhanced sward quality due to the facilitatory effects of prior grazing or by imitating food selection by others, or b) by reducing the chance that they will be killed by predators. The existence of multispecies groups with differing feeding requirements is strong evidence for the antipredator hypothesis (Gosling, 1980) and similar arguments can be used here (greater tendency to group during resting than feeding, reduced individual distance during resting). There is also evidence that antelopes use information obtained by group members about predators. I have seen one member of a group of male hartebeest detect a stalking lion, perform the characteristic antipredator snort and look intently toward it: other animals responded immediately in a similar way and turned in the direction that the first male was facing (Gosling, 1975). Mloszewski (1983) observed similar behavior in buffalo. In addition, the logic of Hamilton's predation dilution argument (W. D. Hamilton, 1971) seems inescapable. It is the most likely explanation of why juvenile eland, and other young antelopes, are so strikingly gregarious (Hillman, 1982). Another advantage for group members is that they can potentially cooperate in antipredator behavior: female eland cooperate in this way (Kruuk, 1972) as do buffalo (Sinclair, 1977; Mloszewski, 1983).

There is some evidence that isolated animals are more vulnerable to predators than animals in groups. Lions are more successful when hunting single wildebeest and Thomson's gazelle although the success rate also increases when these prey are in very large groups (Schaller, 1972). Individuals of nonterritorial species appear to be more vulnerable when alone: Mloszewski (1983) saw lions kill a male buffalo that strayed outside a previously compact group. In species where males become territorial it is often assumed that territorial males may be at greatest risk. In fact they may be less vulnerable than isolated nonterritorial males. In a study of the flight distances of Thomson's gazelle (from a car) Walther (1969) found that solitary wandering males fled at the greatest distances while animals in groups and territorial males could be approached more closely. Fourteen hartebeest males of known social status were killed by predators during my study in Nairobi NP; of these only five were in permanent territories in spite of the fact that the sixty-two to eighty-four territories in the study area, and their owners, were closely monitored for three years; one was in a peripheral scrubland territory and eight were isolated high-status males which were unable to regain territories (Gosling, 1975). These males are comparatively rare but they may be vulnerable because they are often injured in fighting for territories and are in unfamiliar habitat. Although relatively secure, territorial males are probably less good at detecting predators than groups. One territorial male hartebeest was killed by two lions, which attacked from the scant cover of a low termitarium. They would probably have been detected by a group. A territorial male wildebeest was stalked and killed by a lioness as it rolled on its back during scent-marking behavior (B. Foster, pers. comm.). Whether males are killed in a territory or as a result of competition for territories, these observations suggest a cost that could be reduced by a following strategy.

Unfortunately, it is not known whether males that adopt following strategies spend more time alone than territorial males or to what extent they isolate before and after a period as an alpha male in a group. However, such periods are probably shorter where subordinate males remain with the group, as is sometimes the case in oryx, eland, greater kudu, and buffalo. An antipredator advantage seems the most likely explanation of the remarkable fact that alpha male oryx herd adult subordinate males as well as females (Walther, 1978c). Oryx also occur in groups containing single males and it would be interesting to explore the tradeoff between the costs of allowing subordinate males into groups (feeding interference and competition) and their antipredator benefits. Males also compete for dominance in male groups and the opportunity for establishing an individual reference for dominance could also explain why males join groups. The formation of groups where intrasexual competition is less common (e.g., in groups of females) and available information on the antipredator value of group membership suggest that this function is secondary.

In all of the species mentioned, males are larger than females and they might thus have different foraging requirements. Larger animals have a lower metabolic rate but they always require absolutely more food (Jarman, 1974). There may thus be a conflict for males between staying with a female group and their own foraging requirements. If this were so, then one aspect of sexual selection (time spent with females) might operate in the same direction as natural selection against large body size in males. This tendency would be opposed by selection for large body size where this affected success in intrasexual competition but, even so, it would be expected that male body size would be more similar to that of females in species that adopt following strategies than in those where the mating strategy allows separate foraging (this might be the case in species with restricted breeding seasons and perhaps in those with year-round resource territoriality since here females may normally be absent from the territories for the majority of the time). Some information is available about the separation of males and females in species where males adopt following strategies. Groups of adult males occur in most cases. This is predictable where groups of females are accompanied by single males but not where a number of males normally live with females and where subordinates are not excluded (e.g., buffalo, Mloszewski, 1983). In the latter cases feeding needs may be responsible for the observed separation and may limit the time that alpha males stay with female groups.

DEFENDING A RESOURCE THAT FEMALES NEED

Under What Circumstances Would Selection Favor Defending a Resource That Females Need Rather Than Following?

If male antelopes generally attempt to maximize the number of times they mate (and thus their lifetime RS) they should start to defend resources that females need when the attraction of females to an economically defendable resource allows a male to mate more often than if it adopted other strategies. Males can potentially achieve a high number of matings either by mating once, or a few times, with a large number of females or by mating many times with one or a few females. The number of matings and consequent RS of a male will presumably pass some threshold level above which the benefits of the defended area, or territory, at a given size, outweigh the costs of establishing and defending it. "Resource defense polygyny" (Emlen and Oring, 1977) is the most common mating system among antelopes and the conditions responsible for its appearance must be widespread. However, not all males that defend mating territories are polygynous and a general account must also explain the evolution of resource defense monogamy (e.g., dik-dik and klipspringer).

Whether or not males defend territories will depend on the costs of area defense, as well as the benefits (Davies, 1978). In addition to energetic costs, these may include the risk of injury and death in fighting (Gosling and Petrie, 1981). Establishing a spatial reference for dominance may avoid the high costs of agonistic behavior when males try to establish an individual reference for dominance when there are large numbers of competitors.

On a priori grounds the conditions that might lead to resource defense strategies are: 1) a high-quality and clumped food resource so that the numbers of females and their tendency to form feeding aggregations are high *or* a heterogeneous food resource consisting of more than one season's food supply so that time spent by one or a few females in a reasonably small area is high, and 2) a high degree of breeding synchronicity, with many females becoming receptive during a relatively small part of the year, perhaps in response to predictable environmental change. Emlen and Oring (1977) argue that males could not monopolize mates if all became receptive in unison: this is true to some extent but the argument only applies to simultaneous receptivity, a rare condition among mammals. My argument here is based on the assumption that if females become receptive during a limited season, then males can potentially achieve many matings by defending a resource that females need at this time; 3) males in familiar areas are less likely to be killed by predators than those in areas which are relatively unknown because in familiar areas they would have knowledge of previous predation attempts, including likely ambush places, as well as escape routes. In general, animals that live in small areas should be more familiar with such details than those in larger areas; 4) the costs of resource defense territoriality are lower than those involving individual references for dominance, particularly when the number of competitors is high, and/or males with spatial references for dominance have an owner advantage over males that adopt following strategies.

Defending Resources That Attract Large Numbers of Females

Most males which defend territories that are only a small part of the range of visiting females show herding behavior (e.g. hartebeest, Gosling, 1974; gazelles, Walther et al., 1983). Herding males try to stop females that move toward the territory boundary by positioning themselves between the females and the boundary or by using threats to drive them back into the territory. This is sometimes effective for short periods and particularly when a female is in estrus, because then the male tries very hard to stop it. Otherwise females eventually get past the male, so that although the male may delay movements between territories it does not prevent them. Most territorial males have a minimal influence on the foraging movements of females.

The foraging movements of females in species where males show resource defense territoriality have been studied mainly at the level of population movements in relation to major habitat divisions. Examples are the feeding migrations of wildebeest in the Serengeti (Pennycuick, 1975; Inglis, 1976) and in the Amboseli basin (Andere, 1981), the movements of various members of the ungulate community in relation to sward characteristics at different levels of the grassland catena (Vesey-Fitzgerald, 1960; Bell, 1970), and the related movements between wet season dispersal areas and the dry season food reserves of the swamp area in the Amboseli basin (Western, 1973). These studies of population and community ecology show that females are typically clumped in space and suggest a potential for males to monopolize part of the resource that females are utilizing when they become sexually receptive.

There is little direct evidence needed to show that males compete to monopolize areas of habitat in relation to the chance that they will be visited by receptive females or even simply in relation to the frequency of visits by all females. However, males observably establish territories in areas that females visit and not in areas that they avoid. This is most clear in species with very specific habitat requirements. Waterbuck are confined to riverine vegetation and males compete to establish territories along strips of this habitat (Kiley-Worthington, 1965; Spinage, 1969); there are no waterbuck territories in open grassland areas nearby. Similarly, there are no hartebeest territories on the steep valley slopes in Nairobi NP whose long coarse grasses are eaten only in the late dry season when few females are receptive. Instead territories are established in areas of more nutritious grassland which females visit in large numbers throughout the year (Gosling, 1974). It could be argued that males establish territories in relation to their own needs, which happen to be similar to those of females. This will be discussed later.

A more important source of information than the presence or absence of territories is whether males compete more intensely for territories in areas that females visit more frequently (and/or in larger numbers) than in those they visit less often (and/or in smaller numbers). If males did behave in this way a number of predictions could be made. Because of greater competition in the habitats that females preferred, territories in these areas should be smaller than in the less-preferred areas (from Huxley's "rubber disk" theory of territoriality) and should have a higher rate of ownership changeover. The first result would also be predicted if food quality influenced territory size, but the second result could be independent of this. With increasing numbers of competitors preferred habitats should also become saturated with territories more quickly than less-preferred habitats. And, in spite of the costs of reduced territory size and ownership tenure, males in preferred

TABLE 12.1

Competition among male hartebeest for territories that attract
different numbers of females

	Ecotone territories ($n = 31$)	Scrubland territories ($n = 13$)	
Female presence (°₀)	32.7	18.9	$\chi^2 = 54, p < 0.001$
Mean female group size ($\pm$ SD)	11.4 ± 3.7	8.2 ± 3.7	$t = 2.61, p = 0.012$
Owner presence (°₀)	86.3	50.5	$\chi^2 = 413, p < 0.001$
Mean number of owners ($\pm$ SD)	3.1 ± 1.2	2.2 ± 0.8	$t = 2.51, p = 0.016$
Number of owner changeovers ($\pm$ SD)	$2.7 + 1.9$	1.6 ± 1.9	$t = 1.82, p = 0.076$
Territory size (km²)	0.27 ± 0.07	0.51 ± 0.27	$t = 4.74, p < 0.001$
Rate of increase ($\pm$ SE; owners per adult male)	0.060 ± 0.009	0.096 ± 0.015	$F = 4.30, p < 0.05$

The data show the use by females of ecotone and scrubland territories, the proportion of time that the territory owners were present, and various measures of male competition including the number of males that successively occupied the territories and their changeover rate. Territory size is smaller in ecotone territories, and although this may be partly because of a better food supply, the faster increase in numbers of this type of territory as male density increased suggests that they are also more compressed through intrasexual competition. The territories considered (except in the bottom row) are those from a larger sample which were a) in the two defined habitat types and b) retained similar boundaries through the three-year study (Gosling, 1974, 1975). Temporary territories are excluded. The analysis in the bottom row considers the rate of increase of all permanent territories. including those that became permanent after being newly established, in relation to the increasing number of adult males in the study population. Values summarize data collected in sixty-four fortnightly censuses of all territories. All territory owners were individually known. The number of consecutive owners and of changeovers are for the three years of observation. From Gosling, 1975.

areas should still monoplize more females than elsewhere. Data on harte-beest are consistent with these expectations (Table 12.1).

The shape and orientation of hartebeest territories suggest that males try to defend a long-term food supply by maximizing the number of grass communities inside their boundaries. Many are elongated in shape with their longitudinal axis at right angles to the grassland ecotone between scrubland and short grassland and the ecotone between short grassland and valley habitat. Some crossed both of these major ecotones which are rich in medium-length, nutritious grasses, particularly *Themeda triandra*. Males utilize these diverse communities in a seasonal pattern and each day move between the medium or long grass areas for feeding and short grass areas for resting and ruminating. When, in an increasing population, territories are divided between two males, both usually manage to keep parts of all the grass communities of the original territory (Fig. 12.1). M. V. Jarman (1979) recorded related patterns of foraging by territoral male impala in Serengeti NP. These moved in relation to a grassland catena, feeding on the catena apex in the

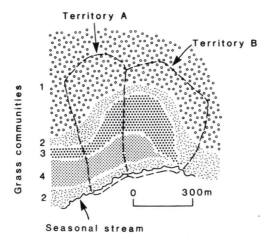

Fig. 12.1. The way in which hartebeest territories split when they are divided between an invading male and the previous resident shows that males try to maximize vegetation diversity within their territory. In this example the invading male has taken over territory B and the male that previously occupied A + B is now confined to A. Typically the split is at right angles to the series of parallel grass communities so that the males respectively acquire and retain a part of each. The numbered communities were dominated by 1) *Setaria/Ischaemum*, 2) *Themeda/ Pennisetum*, 3) *Themeda triandra*, and 4) a short grass assemblage including *Cynodon dactylon* (Gosling, 1975; redrawn from Gosling and Petrie, 1981).

wet season and moving down to longer grasses in the dry season.

The scale of grassland heterogeneity also influences the type of resource defense territoriality shown by topi in different parts of the Serengeti (Duncan, 1975). There is a single wet and dry season each year and topi attempt to maximize food quantity and protein content by feeding from particular grass communities in each season (Duncan, 1975; Jarman and Sinclair, 1979). Where the communities needed in both wet and dry seasons occurred in a finely divided heterogeneous pattern they were used by females throughout the year and males defended (and fed from) areas that included both. Where the wet and dry season food reserves occurred in large homogeneous areas that were more widely separated, males were territorial in the area occupied by the females in the wet season (when they became receptive) but joined mixed groups to utilize the dry season food reserve. These observations, which are confirmed in another area by Montfort-Braham (1975), throw into focus the relationship between territoriality and the timing of mating. Why do topi in the western Serengeti abandon their territories in the dry season (except for brief visits to scent mark: Duncan, 1975) while territorial male hartebeest in Nairobi NP and wildebeest in Ngorongoro Crater attempt to keep them through the year? Apart from the benefits of owner advantage at the start of the breeding season, discussed elsewhere, this may be a consequence of the availability of food and water within economic range of the territory. In Nairobi NP dry season, food and water reserves are within range of all hartebeest territorial males even though there is a cost in traveling (energy expended, time lost in feeding, risk of predation while traveling through relatively unfamiliar areas, risk that the territory will be occupied by another male). The same argument applies to territorial male wildebeest in Ngorongoro but not to migratory wildebeest in the Serengeti. In the Serengeti, wet and dry season food reserves are widely separated and males have no alternatives but to follow the seasonal movements of the population.

It can be argued that territorial males simply defend a food resource for themselves rather than one that females need. This seems unlikely because territorial male hartebeest and impala become thinner than nonterritorial males during dry periods (Stanley Price, 1974; Jarman and Jarman, 1979). This is to be expected, because although territorial males attempt to maximize grassland diversity within their territories, they could not defend a food supply of the same diversity as that within the ranges of nonterritorial males which, in hartebeest, are over thirty times larger on average, and they also incur the costs of walking to water and other resources that the territory lacks; nonterritorial male hartebeest rest in groups near water holes and thus minimize the cost of traveling to this limiting resource (Gosling, 1975). However, modifying the original argument, do territorial males defend a long-term food resource for themselves so that they can maximize mating with females that visit the area? This formulation is consistent with the known mating function of territorial behavior in male antelopes: territories appear to be spatial references for dominance during reproductive behavior and in all territorial antelopes, except waterbuck (Wirtz, 1981, 1982), only territorial males have been seen mating (e.g., gazelles, Walther et al., 1983; wildebeest, Estes, 1969). The question raises a logical problem which is also partly semantic. Hartebeest territories (in my Nairobi NP study area) cover all preferred habitat, so that females spend nearly all their time in male territories (there are usually twenty to thirty territories in each female's range) and observably feed there. Thus females obviously need the food resources that males defend and males clearly defend much larger amounts of food than they need themselves. But do the males attempt to defend more than they need to attract females or do they defend more to compensate for the food lost to females (and other grazing ungulates) that visit the territory? This question cannot be answered with the information currently available. Selection would probably favor both the defense of a personal food supply and one that females need and the relative importance of these factors is a problem for the future. It would be best approached in a species where males and females have different food requirements (for example, when the sexes differ in body size).

Defending Resources That Are Used by Resident Groups of Females

In the previous section it was argued that males compete to establish territories in high-quality food patches where females aggregate. The tendency to form social groups is less important because females of the species discussed move freely through the male territories within their home range and because mating interactions are typically prolonged so that males could not cope with more than one receptive female at a time (Murray, 1982b, records a case of a territorial male impala mating with eight females in one day but this is exceptional). However, grouping behavior could be an important determinant of resource defense territoriality where females feed on a food supply that is heterogeneous at a fine scale over areas that are large in relation to individual feeding ranges. An example is the gerenuk, which inhabits arid scrubland and feeds on the leaves of a large number of species in a complex community of shrubs and trees (Leuthold, 1970, 1978a). Females occur at low densities and the potential for polygyny is low. However, males defend and scent mark territories (these may be 2-3 km^2 in size although the area defended at any one season may be smaller: Leuthold, 1978b; Gosling, 1981). Similar cases include the observation by Backhaus (1959) of a male Lelwel hartebeest which defended an area of at least three square kilometers (probably imperfectly) and accompanied a group of five females and young which remained within the territory. Duncan (1975) and Montfort-Braham (1975) also observed male topi that defended either the entire home range of a female group, or large parts of it, for long periods. Long-term residence was possible because the grassland was finely heterogeneous and female groups could find a year-round food supply in a limited area.

This sort of area defense may be economically possible because of the tendency for females to form small groups. A male that defended either a large part, or all, of the range of such a group would have a good chance of mating with most of its females when they became receptive. Such males could be adopting a strategy of monopolizing a small number of females with a high chance of mating with each. This tendency reaches an extreme in the case of the dik-dik, klipspringer, and blue duiker where pairs or very small family groups occupy territories (Hendrichs, 1975; Dunbar and Dunbar, 1974b; Dubost, 1980). Crypsis is important in the antipredator behavior of such small antelopes and this may be why females do not form groups. However, individuals can live in small areas, apparently permanently (pair territories of dik-dik are 5-20 ha in the Serengeti NP: Hendrichs, 1975), and males attempt to monoplize all matings with one female by defending its entire range. Under these circumstances a male's lifetime RS becomes heavily dependent on the length of time that it survives to breed and this factor may be decisive in preventing polygyny: males would be more conspicuous to predators if they ranged more widely to search for additional mates or if they tried to retain more than one female. The reason that these small antelopes live in pairs, as opposed to living alone with overlapping ranges, may be because of the large increase in the chance of detecting predators when a group increases from one to two (Kenward, 1978; Bertram, 1978).

The distinction between the ecological conditions that select for strategies where males defend all or part of the range of one or a small group of females and those that select for following is a fine one. The most important factor is probably the density of available nutrients and its effect on female range size and numbers in relation to the area that males could economically defend. Using Hofmann's (1973) terminology to separate major categories of herbivore diet, "concentrate feeders" such as gerenuk might tend to adopt resource defense territoriality and "roughage feeders," such as oryx, to adopt following strategies. However, body size will also influence this outcome so that eland are more likely to be followers because they are big and have to forage over large areas to satisfy their requirements. The economics of territory defense are affected by the area to be defended and the amount of competition. The latter may be mainly a consequence of the number of males in the area (more generally, by the operational sex ratio: Emlen and Oring, 1977) but it may also depend on scent marking, which allows territory owners to advertise their status and thus to reduce the costs of defense (Gosling, 1982).

The fine distinction between males that defend territories of this kind and males that follow females is highlighted by the contrast between gerenuk and the lesser kudu, a species where males monitor and follow small groups of females. Both species have been studied in the same area of arid scrubland in Tsavo NP (Leuthold, 1974, 1978b; Gosling, 1981) and in both cases the preconditions for a male strategy appear to be similar: food at a moderate density and finely heterogeneous; females at low density and in small, moderately sedentary groups; year-round breeding; and a low male density. Two explanations are possible. One is that the similarity between lesser kudu and gerenuk ecology is superficial and that detailed study of factors such as the amount of time males could spend with individual females under different ecological circumstances and at different male densities will reveal critical differences. Because of the various problems of interspecific comparison (Jarman, 1982) such studies should be intraspecific: they may give the incidental result that the mating systems currently assigned to either gerenuk or kudu are as flexible within as between species. The second possibility is that the two species represent arbitrary and equally effective evolutionary solutions but that certain anatomical specializations preclude intraspecific flexibility. These specializations could be a trend toward generalized status advertisement in the gazellinae with well-developed scent marking in territories which allows owners to be identified, that is, assessed as potential adversaries (Gosling, 1982), and a trend toward individual references for dominance in the tragelaphine with conspicuous visual patterns and close distance olfactory signals (such as those provided by inguinal glands).

Breeding Synchronicity

Although breeding synchrony may affect the degree of polygyny in antelope populations (e.g., impala, Murray, 1982a, b) there is no evidence that it determines whether or not resource defense territoriality occurs. Hartebeest, topi, and wildebeest all show classical resource defense territoriality of the type where territories are visited by females as part of a larger home range. However, hartebeest breed throughout the year with two peaks corresponding to a bimodal rainfall pattern (Gosling, 1969), topi usually have a restricted mating season of about four months (Duncan, 1975; Montfort-Braham, 1975), while wildebeest have highly synchronous mating with over 80 percent of conception within a three-week period (Estes and Estes, 1979; Watson, 1969). In all these species some males spend a large part of the year without visits by receptive females. In spite of this, some males try to defend their territories throughout the year even though their condition deteriorates faster than that of nonterritorial males during times of environmental stress. This costly behavior is probably selected for by the advantages of being in residence at the start of the mating season.

The importance of prior occupancy suggests that breeding synchronicity may generally have less importance than expected on a priori grounds. However, there is one case where breeding synchronicity may determine the occurrence of resource defense territoriality. Migratory wildebeest in the Serengeti have a highly synchronized rut at a time when they are in highly clumped aggregations. But, in contrast to most populations where males defend resources territories, these huge groups are continually on the move as the individuals attempt to maximize their intake of short grasses by moving to unused areas. The result is the spectacular annual migration of Serengeti wildebeest. The rut occurs during a period of rapid movement and in it, some males establish small static territories which they vigorously defend against other males and in which they attempt to retain and mate receptive females (Estes, 1966, 1969; also see discussion in preceding section). Such territories are established for a period from a few hours to a few days (Walther, 1972a) and are abandoned as the migration moves on. How does this sort of territoriality fit into the theoretical framework outlined above? The aggregation preconditions are satisfied but, with such rapid movement by females, why has territoriality evolved rather than following? Possibly the high costs of territoriality are outweighed by the high potential benefits when a large proportion of the females entering a territory, and being continually renewed, are in estrus.

Resource Defense and Intrasexual Competition

The density of hartebeest was very high during my study in Nairobi NP and the small territories were probably a consequence of intense male competition. Territories would almost certainly enlarge in circumstances where male density was reduced, but would males ever switch to a following strategy? A relevant observation was made in Amboseli NP in an area which contained only one male (D. Western, pers. comm.). Over eight months this animal remained with a group of about six females and juveniles during their movements in a range of eleven square kilometers. This male appeared to adopt a following strategy in the absence of male competitors.

The level of competition also appears to affect the intensity of area defense by territorial males. When large male groups invaded hartebeest territories, as often happened in territories next to pools of water in the dry season, the resident male usually gave up area defense, at least during the hot time of the day, and simply rested with the other males. In the evening when the group dispersed to graze the male often chased a few stragglers and thus speeded up their departure. Presumably the cost of attempting to expel the entire group was too high. The same principle may explain the occurrence of two types of resource defense territoriality shown by Grant's gazelle in the Serengeti (Walther, 1972a). In medium-size grassland clearings where groups of females and nonterritorial males are small, territorial males usually attempt to expel all intruding males. In the open plains where very large mixed groups sometimes pass through a territory, owners do not try to expel other males but assert their dominance over any that approach while they are mating. Territorial male impala also tolerate groups of subordinate males in their territories although they expel high-status males (which might compete for territory ownership) and neighboring territorial males that intrude (M.V. Jarman, 1979).

The costs of competition could also influence the unexpected occurrence of territoriality by migratory wildebeest. As discussed above, wildebeest occur in very large mobile aggregations during the rut and it seems unlikely that there are persistent social affiliations, except perhaps between mothers and their previous offspring (Watson, 1969). An individual reference for dominance might be favored by selection where males could establish dominance relationships at a reasonably low cost, presumably over a long period, and where these relationships were maintained when females become receptive. When, during competition for a receptive female, there was a high chance of meeting males for the first time (as in very large and highly mobile groups) then there would be selection for a strategy in which the reference for dominance did not involve individual recognition. If this is why migratory male wildebeest defend territories, why does this strategy not occur in buffalo or in Inshasha topi? Perhaps there is a lower degree of mixing in these cases and a greater opportunity to establish an individual reference for dominance. As discussed before, this may be the case in buffalo groups (Grimsdell, 1969; Sinclair, 1977; Mloszewski, 1983) but, on balance, it seems more likely that male wildebeest defend territories because of the enormous potential benefits when many females are in estrus and that in the case of Inshasha topi and buffalo, breeding is not sufficiently synchronous for this benefit to outweigh the costs of territoriality when large numbers of competitors are present.

Resource Defense and the Disappearance of Following

Even where ecological conditions are such that resource defense territoriality could evolve, these conditions do not in themselves account for the disappearance of following. Where resources are dense and female population high and predictable in its movements, it would still seem to be a reasonable strategic alternative for males to follow and, perhaps, defend a group of females. Most females show a sufficient tendency to form groups to satisfy this prerequisite. Why does following not exist as a strategy in an area where territorialty occurs? Perhaps territorial males would be at a competitive advantage for reasons that are analogous to the owner advantage when territory residents encounter intruders. For reasons discussed earlier owners may have more to lose by losing their territories than intruders have in taking them over. Thus, where both strategies occurred, owners would be more prepared to escalate in contests with followers and more likely to win.

If this scenario is correct then following strategies would disappear in areas where territoriality arose. This would not be because they were nonadaptive in relation to a particular distribution of resources and females, but because they would be inferior to a competing strategy. This may generally be the case but there are a number of circumstances where following strategies could survive. They could persist in a modified form if some males mated covertly when territorial males were absent or distracted. They could exist as an equally successful alternative strategy (see later section) when territorial males were regularly, but only partially, swamped by potential mates and competitors (possibly, topi at Inshasha). Lastly, following might become an obligatory strategy where populations are grouped into very large aggregations which would completely swamp males that defended territories (all buffalo studied to date).

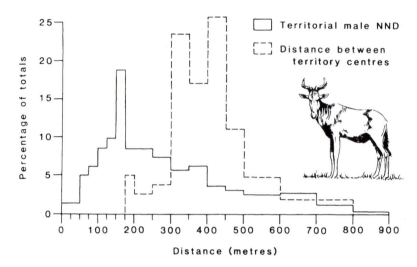

Fig. 12.2. A comparison of the nearest-neighbor distances of a territorial male hartebeest with the distances between the geometric centers of their territories. The distinct tendency for the males to be more clumped than expected is a result of (1) males resting on short grass areas in their territories and the proximity of these sites in small grass plains, (2) a relatively large number of territories being compressed into and around these plains, and (3) a gregarious tendency among territorial males: neighbors often rested 50-150 m apart when factors (1) and (2) were not a sufficient explanation. (1) and (3) are probably antipredator behaviors and (2) a result of preference for territories that attract large numbers of females.

The data on male nearest neighbor distances were averaged from four censuses of a 30 km² study area in Nairobi NP. All territorial males were individually known and the territories had been monitored for over two years. The geometric centers of the territories were measured from a map using the distance between the point in each territory where a line along the longitudinal axis was bisected, and its nearest neighbor. There were eighty-four territories on the map and an average of sixty-one territorial males present in each census; most were resting when inspected. From Gosling, 1975.

LEKKING

Leks occur in at least three species of antelopes, Uganda kob (Buechner, 1961; Leuthold, 1966), topi (Duncan, 1975; Montfort-Braham, 1975), and lechwe (Schuster, 1976). They consist of tight clusters of territories, which, at the center, have diameters of fifteen to thirty-five meters in kob (Leuthold, 1966), twenty-five to forty meters in topi (Montfort-Braham, 1975), and down to fifteen meters in lechwe (Schuster, 1976). Topi and lechwe leks contain up to about one hundred males. In the Toro game reserve in Uganda, with a population of 18,000 kob in an area of 400 square kilometers, there are eighteen leks each with an average of thirty to forty males (Buechner and Roth, 1974). Territories are larger around the lek perimeter and in all species some males defend resource territories.

What Circumstances Favor the Evolution of Lek Territoriality?

A number of authors (e.g., Wiley, 1974, and Wittenberger, 1978, discussing grouse leks) note that leks occur in patches of open habitat which seem ideal for antipredator scanning. They suggest that this factor may be involved in the evolution of lekking behavior. Bradbury (1981) has criticized this idea because it does not explain the occurrence of leks in particular species of birds in forests. However, it may be that we should not look for universal selection pressures to account for the evolution of lekking behavior but rather attempt to document the various pressures which, in different species, are responsible for the universal occurrence of some lek precursor, presumably some sort of male aggregation phenomenon.

Previous work has concentrated on the effect of territoriality in spacing out males; there has been no analysis of the tendency for territoral animals to form clumps. In hartebeest this tendency takes a number of striking forms, which will be described below. Once territorial males form clumps, then more intense intrasexual competition for territories in clumps and female choice of males that succeed in getting such territories may lead to a runaway process involving choice and further territory compression. At some point the resources of the territories may become less important in attracting females than the quality and local abundance of the males. Some aspects of antelope behavior may contribute to the debate about this hypothetical chain of events, because the behavior of individuals, including their antipredator behavior, is easy to observe in detail. Some aspects of hartebeest behavior suggest ways in which habitat selection and cohesive behavior may prove to be precursors of lekking. These are discussed before considering why hartebeest do not lek and why kob, topi, and lechwe do.

Aggregations Caused by Habitat Selection to Avoid Predation during Resting Periods

Nearly all hartebeest select open, raised areas when resting and ruminating (i.e., when they do not have to enter closed habitats to feed) and this is most simply explained as antipredator behavior. It is not clear whether this factor explains why males compete for territories in particular locations, although territories are often clumped around small open grassland plains. This may be partly because males attempt to secure sections of the rich ecotone between grass and scrubland. In contrast, the selection of open patches within territories for resting seems less ambiguous. Territorial males and females usually feed from medium and long grasses, spending most time in the grass/scrub ecotone. At the end of each grazing period they walk back to open, short grass areas, where there is little cover for stalking predators, to rest and ruminate. Hartebeest routinely detect potential predators (lion, cheetah, and blackbacked jackal) from such locations at distances which would be impossible in scrubland. Most adult hartebeest are probably killed by lion and the distribution of their remains suggests that at least twice as many are killed per unit area in scrubland than in grassland (Gosling, 1975).

This sort of habitat selection is a common feature of the behavior of most savannah antelopes. In Nairobi NP species including wildebeest, Thomson's and Grant's gazelle, and impala behave in a similar way during the resting times of the day. In the Serengeti, Duncan (1975) found that groups of females with a territorial male topi habitually returned to patches of open habitat and he also interpreted this as antipredator behavior.

Where patches of open habitat are large enough to accommodate the resting locations of a number of territorial males but not so large (or dissected, for example, by drainage channels) that these locations are widely dispersed, the males tend to aggregate while resting. This tendency is conspicuous among territorial male hartebeest in Nairobi NP in small grass plains.

Social Clumping by Territorial Males

Like most plains ungulates hartebeest are fundamentally gregarious, but, in spite of the antipredator advantages of group membership, most territorial males spend a large part of their time alone (62 percent on average for hartebeest: Gosling, 1975). Presumably they trade off the risks of isolation against the reproductive benefits of territoriality. However, there were a number of indications of a gregarious tendency in the behavior of such isolated males. One was the tendency to form compact groups with members of other species; a particularly conspicuous case is the association between isolated territorial males of up to three different species (Gosling, 1980). These usually occur during resting periods on patches of open habitat that are probably selected for antipredator reasons, although aggregation on these patches (which were often large) was not a sufficient explanation for the closeness of the association. A second indication of gregariousness is the tendency for territorial males to associate:

1) In the late dry season, groups of female (and nonterritorial male) hartebeest concentrated in areas with permanent water supplies and abandoned those where local water sources had dried up. Territorial males remained in the abandoned areas although they eventually had to make visits of variable length to water holes. Those males that remained in their territories grazed in their territories during feeding periods but, during the daytime resting period most, and sometimes all, of the males joined together in a small group at a position where a number of the territories adjoined and rested together. At the start of the afternoon grazing period the males (which were all individually known) moved back into their territories (Gosling, 1975).

2) When the short grass resting areas of neighboring males were adjacent to, or contiguous with, each other there was a distinct tendency for the resident males to rest far closer to each other than could be explained by the aggregation in open habitat. Most instances of this behavior involved indi-

vidual distances of 50–150 meters in territories that were up to 800 meters long. Sometimes, when the arrangement of the territories permitted it, three males rested together in this fashion. I often saw territorial males respond to the antipredator snorts of neighbors (although this also occured with more widely dispersed males) and once some males were orientated toward a lion that was out of their own line of vision, apparently by inferring its location from the orientation of neighbors that could see it (Gosling, 1975).

Temporary Territories

Hartebeest males gain territories either by finding a vacant territory, by fighting for ownership with the previous owner, or by establishing small territories between the disputed territory and its neighbor and gradually wearing down the resistance of the resident male by a protracted series of intrusions and agonistic interactions (Gosling, 1974, 1975). These small territories are established next to the short grass, open, part of the territory where most reproductive and agonistic behavior occurs and where, as described, any animals in the territory rest and ruminate. The function of this behavior is most likely to attempt to take over a permanent territory. They are probably established next to the "activity center" of the territory to maximize encounters with the male in the disputed territory. Because of the continual agonistic interactions and the lack of medium-length grassland, these territories are occupied for only short periods. Sometimes the males abandon the temporary territories, sometimes they displace the male in the permanent territory, and sometimes the permanent territory becomes split between the new male and the previous occupant. The split is usually along the axis running from the open (resting) area part of the territory to the long grass part, giving a characteristic elongated shape (Fig. 12.1). The elongated, radial arrangement of territories in the center of Uganda kob leks (Floody and Arnold, 1975) suggests that a similar process has occurred and that this is the mechanism of territory size reduction that accompanies the formation of leks.

Female Behavior and Mating

Female hartebeest usually rest in the same open parts of territories as their resident males, although a male with a group of females is less likely to rest near to a territorial neighbor. Although some reproductive behavior occurs in all active periods there is a well-defined peak of mating behavior around midday at the resting locations (Gosling, 1975). This may be because the risk of predation is lower than in the evenings, when more predators are active and when females move to feed in long grass areas, and also because the resident male may take advantage of the presence of females, which often graze into other territories in the evening.

Why Don't Hartebeest Lek?

The above information suggests how clumps of males could appear (see Fig. 12.2) and it is possible to see how, given an advantage for females in choosing between males, the runaway process outlined in the initial model could lead to the evolution of lekking. Given these precursors, why has lek territoriality not evolved in hartebeest? The answer may be that females are never sufficiently aggregated because hartebeest are adapted to feeding in a dispersed fashion on a low-quality, high-fiber diet (Stanley Price, 1977, 1978). Females are thus not usually available in sufficient numbers near any concentration of males for the runaway process of female choice to be initiated. Again related to their feeding strategy, hartebeest also mate throughout the year so that the value of any concentration of females to males will generally be reduced by the low proportion in a receptive state. The population density of hartebeest in Nairobi NP during the late 1960s was very high for the species because of heavy rainfall and low numbers of the competing wildebeest. If lekking is ever to occur in this species, then it would probably have done so under these conditions. However, at this density, fighting injuries were common among males and a few were killed, probably during fights for territory ownership. Average territory size reduced as density increased but it appeared to reach maximum compression in the highest-quality habitat (Gosling, 1974). Leks did not form.

Why Do Topi, Lechwe, and Uganda Kob Lek?

In contrast to hartebeest, topi select a high-quality diet: the protein content of the grass selected in the Serengeti at the driest time of year is higher than that selected by hartebeest in the wettest time on the Athi-Kapiti plains to the south of Nairobi NP (Duncan, 1975). Topi feed in medium-length swards and select mainly leaf blade as opposed to stem and sheath. In high-quality swards, such as those that grow following rainfall on burnt areas, topi reach very high densities. They also have a restricted breeding season: in the Serengeti. Duncan (1975) concluded that most mating occurs in December and January and Montfort-Braham (1975) gives November to March as the duration of the rut in Akagera NP. Lechwe are also concentrated at the peak of the rut in December/January because available pasture is reduced as floods force them out of lowland areas.

The food supply and the timing of breeding thus lead to circumstances where a high proportion of females become receptive in a limited area and time period. These characteristics may be linked aspects of the female reproductive strategy, which probably consists primarily of an adaptation to maximize nutrients for offspring production. The resulting concentrations of receptive females create a potential for extreme polygyny, but are not sufficient to account for the appearance of leks. This requires some spatial heterogeneity, which attracts males in an unusually high concentration.

This is unlikely to be simply a patch of high-quality food, because as males concentrate, their own feeding will tend to reduce sward quality. Instead, leks may originate where territorial males cluster as an antipredator adaptation. This data accords with the observation that Uganda kob and lechwe leks occur on raised open sites that appear to be more suitable for antipredator scanning than for food (Buechner, 1961; Schuster, 1976). Leuthold (1966) observed clusters of resource territories at such sites and thought that they might represent the initial stage of lek formation. The antipredator value of a lek site is probably not the only factor responsible for the appearance of a lek in a particular location: males would boost their fitness by selecting sites that were also close to a good food supply for females or to a route that females followed. Montfort-Braham (1975) found that topi leks were often near favored grassland ecotones and on migration routes while lechwe leks form when the population is compressed into a narrow zone by rising flood waters (Schuster, 1976).

This hypothesis takes account of the fact that females generally concentrate in optimum feeding areas around a lek and that they leave this food in order to visit the lek for mating. Why is there not selection for males to establish lek territories in the areas where females feed? Good visibility for mate choice can be discounted because visibility is good in feeding areas. Any disruptive male combat during mating could be avoided by visiting males in resource defense territories, which are available in all three species. Hartebeest territorial males clumping in the short-grass resting areas of their territories and the movements of female groups from feeding locations to these same areas provide a precise analogy for the behaviors that could precede lekking. Selection would then favor females that took advantage of the choice available by visiting the clumped males when they became receptive, and males that competed for territories with access to the clump.

While interspecific comparisons have only limited value in exploring the evolution of mating systems (Krebs and Davies, 1978; Jarman, 1982), it may sometimes be instructive to ask why systems that exist in one species have not evolved in another when similar selection pressures appear to be operating. The answer may be that there is no genetic variation in the population for selection to act on, but the question may also bring into focus selective pressures that otherwise seem trivial. For example, why do Thomson's gazelle not lek? The species occurs at high density and males form resource defense territories which tend to become clumped on productive short-grass areas. How does this species differ from the lekking species in its tendency for females to aggregate? The answer may be linked to the sward characteristics from which these small gazelle are adapted to feed and the topographic features that support these swards. Thomson's gazelle typically crop short grass swards that occur over large, well-drained and homogeneous areas; the preferred habitat is the catena apex (Bell, 1970) but the characteristics of this zone are very widespread in areas such as the Serengeti plains, where these gazelle are abundant. Water is not universally available and Thomson's gazelle have a moderate degree of water independence. Thus, although abundant, this species is relatively uniformly dispersed over large areas. In contrast, both kob, topi, and lechwe show a higher degree of local concentration. In the case of kob this is at first surprising, because like gazelle they prefer to graze from short-grass catena apices. However, unlike gazelle they are strongly water-dependent and are thus spatially tied to permanent water supplies. In the dry season large groups of kob drink at particular locations, then disperse along traditionally used routes to grazing locations (Bindernagel, 1968; Kingdon, 1982). These are on catena apices which tend to be small because drainage channels dissect most landscapes in the vicinity of watercourses large enough to provide permanent water. Leks occur on these dispersal routes. Kingdon (1982) also suggests that leks occur at positions where females tend to hesitate before crossing an obstacle in these routes, such as a valley with cover that would conceal predators. Topi feed from medium-length grasses and especially from those in sumps or flood plains. Such areas occur between areas of surrounding higher ground, which channel the topi into high concentrations. Leks form where this effect is maximal.

Breeding synchrony may be involved in the evolution of lekking in topi and lechwe, both of which have restricted breeding seasons but, in contrast, kob breed throughout the year (Buechner, 1974). This inconsistency suggests that lekking may be primarily a consequence of spatial factors and their aggregating effect, although further information on the reproductive ecology of kob and its effect on the economics of lekking could change this assessment.

To summarize, a revised model of the evolution of lekking might involve the following stages: 1) There is a tendency for resource defense territories to become compressed in patches of high-quality food because females visit these in large numbers; 2) both territorial males and females rest in patches of open habitat where the chance of predation is low; 3) where open habitat crosses a territory boundary, territorial males tend to clump. This tendency is increased for social (probably antipredator) reasons and because males attempting to take over territories establish small temporary territories next to patches of open habitat in existing territories because these are activity centers where most mating and defense occurs; 4) Given only these precursors, leks do not appear, probably because a) the fitness costs of competition between males are higher than the benefits from increased mating, or b) because the benefits to females from mate choice are lower than the costs of aggregation; 5) the runaway process involving female choice, further territory compression, and eventually lekking, is initiated when female and/or male numbers are boosted by additional factors such as a funneled migration route.

Animals are believed to adopt varying mating strategies within a population for at least three main reasons (Rubenstein 1980; Dunbar, 1982b). The first is when different strategies are more successful than their competitors under certain environmental conditions. In such cases it would be expected that each strategy would be associated with a characteristic set of environmental conditions. Secondly, the best strategy for an individual to adopt may depend on the strategies adopted by other individuals. Under these conditions a mixture of strategies may evolve by frequency-dependent selection. These are known as mixed evolutionary stable strategies (ESSs). Alternative strategies of this kind should yield the same fitness payoff but, in practice, this is difficult to measure. Both sorts of alternatives can consist of simple options such as "follow" or "defend a resource that females need" or they can show continuous variation, such as defending a territory of one size at one intensity of competition and chance of mating and of another size as these factors vary. The third main reason for adopting a different strategy is that individuals may simply be doing the best that they can under adverse circumstances. For example, they may be poor competitors because their growth was retarded or they may be too young or too old.

The mating strategies of following, defending a resource that females need, and lekking do not occur exclusively in single species or in single populations. In a number of species two of these strategies exist and all three may exist in the case of topi. There is also variation in resource defense territoriality within populations and one case of a satellite strategy in waterbuck (Wirtz, 1981), which otherwise show classical resource defense territoriality (Kiley-Worthington, 1965; Spinage, 1969).

There are two species, topi and oryx, in which both following and resource defense strategies occur, although other species will undoubtedly be added in the future. In topi, following is the less common strategy and the switch from resource defense occurs when very large, mobile, mixed herds make territoriality uneconomic. Following is the usual strategy for male oryx (e.g., Walther, 1978a) but a recent study at Galana in Kenya has shown that where resources are relatively predictable and abundant, males defend and scent mark large territories of four to five square kilometers (T. Wacher, pers. comm.). Females and nonterritorial males live in mixed groups with ranges of two square kilometers up to three hundred. Territorial males sometimes threaten the males in these groups as they enter a territory, then check the females for estrus; in other cases intruding males are ignored. As in those Grant's gazelle with large territories in the Serengeti plains (Walther, 1972b), subordinate males are not driven out and the territory is a spatial reference for dominance rather than an area that is kept free of competitors; presumably the exclusive defense of such large areas would be hopelessly uneconomic. In a study period of above average rainfall, Wacher found that most conceptions occurred at a postpartum estrus when females were isolated with their young calves. Following a group of females would clearly be ineffective under these circumstances. There are some indications that the defended resource may include the dense ground cover needed to conceal calves, which lie out as an antipredator adaptation during the first weeks of life. Only territorial males were seen mating but, as in the Grant's gazelle mentioned above, the males that follow could have some chance of mating: if so, they might be making the best of a bad job and following females in the hope of one becoming receptive away from a territorial male. Against this, Wacher saw females reject the precopulatory advances of non-territorial males. Males may simply join groups for the advantages of group membership when they are unable to get territories.

Hartebeest territories in Nairobi NP differ in size in relation to habitat variation. As well as the ecotone territories described earlier, there are larger territories in *Acacia drepanolobium* scrubland. The grassland in these territories contains taller and less nutritious species than ecotone territories. Females occur in smaller groups and less frequently in scrubland and the resident males leave their territories more often, usually to drink (Table 12.1). Presumably males in these territories achieve a lower rate of mating than males in ecotone territories. Against this males in scrubland territories have longer periods of residence almost certainly due to less intense competition. Perhaps the defense of scrubland territories is an alternative strategy in which males reproduce at a low rate over a long period, with low costs due to area defense and herding females, and with ample time for feeding. In ecotone territories males invest heavily in the expectation of a high level of reproductive success over a shorter time. During a three-year period the density of male hartebeest in Nairobi NP increased by 34 percent. As predicted from Huxley's rubber disk theory of territoriality, territory size was reduced as more territories were established. Also, some territories were established in areas where they had not previously occurred. Both of these processes were more common in scrubland than ecotone territories, which was consistent with other evidence that the latter were preferred. Males in scrubland territories may thus have been simply making the best of a bad job when the ecotone territories were saturated. Unfortunately the evidence is equivocal because it also supports the hypothesis that males only establish territories in scrubland when increasing numbers of females raise the benefits above a level where the low cost/low benefit strategy is viable.

In other cases old male hartebeest established territories in dense *A. dre-*

panolobium scrub or in valley scrubland after defeat in a fight for ownership of an ecotone territory (Gosling, 1974, 1975). Such areas attracted few females and their resident males had a high risk of being killed by lions. Males in these territories were clearly making the best of a bad job since they would be unlikely to regain a high-quality territory. On the other hand their chance of mating would probably be higher than if they joined a male group.

When females are attracted to high-quality males or resources defended by territorial males, other males may get some matings by adopting a satellite strategy. These strategies consist of waiting nearby and attempting to intercept females that are attracted to the resource or high-quality male. They occur in several animal groups, although there are few cases where the type of strategy (habitat dependent, mixed ESS, etc.) has been identified with certainty. Only one case has been reported among antelopes. Most breeding male waterbuck defend resource territories (Kiley-Worthington, 1965; Spinage, 1969) but some males become satellites within territories. They behave submissively and are probably tolerated by the territory owner because they share in territory defense. Satellites have a small chance of mating but a good chance of eventually taking over the territory (Wirtz, 1981, 1982).

Male kob that defend large territories away from leks are often regarded as animals that failed to get central lek territories. If so, they are simply making the best of a bad job. Such behavior would be selected for if the reasons that males fail to get lek territories are not under genetic control, for example, if they are injured or old. However, there is no evidence that this is the case. Kob on large territories look similar to those on leks and while there is a slight tendency for large territories to contain older males (as would be expected) the age distribution of males in the two types of territories is generally similar (Leuthold, 1966). Instead, the males in large territories may be defending a food resource that females need, as an alternative strategy to lekking (Gosling and Petrie, 1981; Dunbar, 1982b, 1983a).

Males that obtain central lek territories achieve a high level of mating but this must be weighed against the costs involved. Competition for these territories is intense, and, apart from the energy spent, males risk injury and death; Buechner (1961) records two deaths in fighting for lek territories. The territories are held for only a few hours or days (Leuthold, 1966) and, if a male fails to win a central territory where most matings occur (Buechner and Schloeth, 1965; Floody and Arnold, 1975) it may incur high costs without commensurate benefits. Alternatively males can establish resource defense territories in areas with fewer females but where the costs of competition are lower and tenure may be a year or more (Leuthold, 1966). Males in resource territories spend 1 percent of their time on territory defense compared with 8 percent for males on leks (Leuthold, 1966). It could also be an ESS for individuals to compete for lek and resource territories at different ages but this and other possibilities can only be resolved by long-term observations of individuals and with estimates of lifetime RS for males that adopt different strategies.

Similar considerations apply to the various topi territories described by Duncan (1975) and Montfort-Braham (1975) and I will concentrate on aspects that differ. The first concerns the relative survival of the offspring of males that adopt different mating strategies. In most antelopes, offspring survival is not directly affected by male behavior but, where juvenile survival is linked to the factors that influence the occurrence of a particular mating strategy, then the fitness of the males that adopt the strategy will be affected. In the case of topi fewer offspring survive in high-density areas where leks occur (Montfort-Braham, 1975). This is probably because there is a higher risk that mothers and their calves will become separated in large concentrations of animals. As in wildebeest concentrations (Estes and Estes, 1979) this is a major cause of calf mortality (Montfort-Braham, 1975) and, if the females that visit the leks continue to use the same areas later in the year, then the increased frequency of mating that a male achieves in a lek may be devalued by the poorer chances of survival of its offspring.

Topi also provide a case where female agonistic behavior may directly influence the payoff of resource defense territoriality. Where grassland is very heterogeneous with respect to seasonal food requirements and female ranges are thus small, Duncan (1975) sometimes found that males defended the entire range of a female group. This situation appeared to be continuous with cases where female ranges were larger and males defended only part of the range. However, it assumes particular significance in the first case since females also threatened and chased away strange females that tried to join their group. As a result group membership was very constant over long periods. This situation bears some similarity to that with a following strategy, particularly in the sense that the males' reproductive success is both limited and guaranteed by the number of females in the group. It also raises the problem of the extent to which female agonistic behavior excludes other females since this could conflict with a male strategy that attempted to maximize matings.

CONCLUSIONS

Three themes, reflecting different selection pressures on male reproductive behavior, have been addressed. The first is the ecological problem of where and when a male should be in relation to the feeding movements of females and the timing of their sexual receptivity. The second is the response of males to various aggregations and social groups of females. To some extent these behavioral tendencies dictate the potential for polygyny in a population (Emlen and Oring, 1977), although males may derive benefits, other than those of mating, and costs, from group membership. The third theme is that of intrasexual male competition. Males can achieve fitness benefits by reducing the costs of competition as well as by increasing the number of times they mate, and so this issue is not simply linked to the benefits of increasing levels of polygyny.

The problem of where a male should be in order to mate with a particular female is most easily envisaged in the following strategy. The simplest form of following is for a male to follow one female and thus ensure that it mates with it. In practice, harem defense monogamy does not exist in following species, and males generally have to make decisions about *when* to join a number of different females. This may involve monitoring each female's reproductive state and sometimes making decisions about when to leave or join a group, even when no positive information is available. When females show signs of becoming receptive males form a "tending bond" and try to defend the female against any other males. But how do males decide whether or not to join females when this sort of information is not available? Unfortunately, no attempt has been made to study the problem even though some aspects could be approached quite simply.

Empirical information confirms the theoretical expectation that following strategies occur where females are forced to forage over very large areas. Oryx, eland, and, to a lesser extent, greater kudu, roan, and buffalo all utilize plants whose nutrient yield is dispersed at a low density, either because of low production in arid areas or because of a high proportion of structural elements in sump grasses. As a result, females, which need to maximize nutrients for reproduction, are forced to forage over large areas and they spend little time in any one place. It seems unlikely that any inidividual male could defend a significant part of their food supply without the costs of defense easily exceeding the low level of benefits. For an alpha male attempting to maximize time with receptive females, this problem is compounded by the extended breeding seasons of oryx and roan. Even if males could survive on a defended foraging area the chances that the area would be visited by receptive females would be low. Buffalo, eland, and greater kudu have moderately synchronized breeding seasons, which suggests that this factor may be less important in selection for following than the spatial distribution of the food supply.

Following strategies would be expected to be more profitable for males when females form social groups. While this proposition has not been tested (for example, by intraspecific comparisons of following behavior in areas where group size varies) existing observations do not allow it to be refuted. The tendency to form groups is probably a universal antipredator behavior among savannah ungulates and its occurrence in females probably makes a following strategy more profitable for males. The number of animals available to form groups is probably determined by the nutrient density, or "carrying capacity,'' of the catchment area for each social group (this will be determined by the distance at which animals can perceive and respond to each other, in the widest sense), and this factor will determine the size of the group that forms. This interpretation differs from that suggested by Jarman (1974), where nutrient density is regarded as having a direct limiting effect on the number of animals in a group. Alpha males also benefit from the antipredator consequences of group membership and this would explain why alpha male oryx diligently prevent subordinate males from leaving their groups (Walther, 1978a). However, males may incur a feeding cost when, as a result of sexual selection, they are larger than females and could thus have different niche requirements.

Following strategies are rare among male antelopes and most solve the problem of making contact with females by some sort of sit-and-wait strategy. Resource defense territoriality is very common and it generally appears to be an adaptation to intercept female movements by defending a food resource that they need. However, only in hartebeest and topi have specific relationships been established between food requirements and details of the food supply within defended areas (Gosling, 1974; Gosling and Petrie, 1981; Duncan, 1975). In the first case males try to maximize grassland diversity within their territories and thus to monopolize areas that have food for females at a number of seasons.

There are two types of resource defense territoriality although these should be regarded as extremes in a continuum. In the first case males attempt to monopolize sectors of vegetation patches which attract large numbers of females. As a corollary of large female numbers, there are generally large numbers of competitors. Territories are thus small in relation to the range of the females that visit. In the second type males try to defend areas of low density but heterogeneous vegetation which is a long-term food supply for one or a small group of females. In these two situations males may attempt to maximize matings by respectively monopolizing large numbers of females, although with a small chance of mating with each (e.g., hartebeest, impala), and monopolizing one or a few females with a high chance of mating with each (e.g., dik-dik, klipspringer).

Sit-and-wait strategies depend on selecting a location that will be visited by receptive females in the future and this must depend on predicting female foraging behavior. This could be done simply, for example, on the basis that females will continue to occur where they were first found. However, this is not the whole story because male hartebeest often establish territories in areas which, at the time, are completely empty of females. Again, this problem has not been investigated in detail and, when it is, there will be at least two major obstacles. One is that the food requirements of most males are sufficiently close to those of females that it is hard to distinguish between the male's selection of its own food and those of the female: perhaps this can be resolved in the few African species where males reduce their food requirements during restricted breeding seasons (e.g., some impala, Murray, 1982a, b). A second problem is that most males slot into existing territory systems and adopt the boundaries imposed by their neighbors and their predecessor (e.g., hartebeest, Gosling 1974, 1975). They are thus strongly influenced by intrasexual competition. Male migratory wildebeest, which establish territories in a short breeding season and, presumably, in relation to the variable movements of females, might prove a good subject to unravel this problem.

The second theme of the chapter is the response of males to various aggregations and social groups of females. Emlen and Oring (1977) regarded these behavioral tendencies as important determinants of polygyny and this generalization appears to hold for the studies of antelopes that exist today. When females form groups it becomes possible for males to monopolize them. However, it may be confusing to link the factors responsible for this trend toward polygyny with the environmental factors identified above. Polygyny is universal among males that adopt the strategy of defending sectors of high-quality food patches but, among those that defend long-term, low-density food supplies, the degree of polygyny is generally low and monogamy is common. While monogamy could occur because the food supply places upper limits on group size, the mechanism of this limitation is not clear. It is more likely that where females (and males) depend to a large extent on crypsis to avoid detection by predators there will be selection against group formation and males will be forced to adopt either a monogamous strategy (e.g., dik-dik, blue duiker) or a strategy of monitoring the reproductive state of a number of dispersed females and following them when they become receptive promiscuously (e.g., bushbuck). Pair living may be selected for because two animals have a much better chance of detecting potential predators than one (Kenward, 1978; Bertram, 1978). When, as in gerenuk, large body size reduced the effectiveness of crypsis there will be selection for selfish herd behaviors, although nutrient densities may place upper limits on the numbers available to form groups (but not on the size of the group itself). If resource defense monogamy is to be analyzed in the same terms as resource defense polygyny, it may be better to regard polygyny and resource defense as separate evolutionary trends, although ones that clearly coevolve under some conditions. In the antelopes these conditions appear to be the formation of feeding aggregations by females on moderate to high-quality food patches and the formation of social groups as an antipredator behavior. The separation of these two aspects simplifies the analysis of breeding synchrony, which probably has its main effect on the potential for polygyny (e.g., impala, Murray, 1982b) rather than on whether or not resource defense occurs (except possibly in the case of migratory wildebeest).

The third theme, already mentioned above, is intrasexual male competition. Males can achieve fitness benefits by reducing the costs of competition as well as by increasing the number of times they mate. Owen-Smith (1977) has characterized ungulate territoriality as low cost/low benefit mating strategy compared with the high cost/high benefit strategy of dominance systems. While this view may be difficult to sustain across the entire range of resource defense strategies, there are some indications that males may defend territories to reduce the costs of establishing and maintaining an individual reference for dominance when the number of competitors is high (e.g., migratory wildebeest). The costs of competition include the energy costs of long damaging fights, injury, which may be mildly debilitating or crippling, and the risk of death either directly from fighting (e.g., hartebeest, Uganda kob) or from predation since males are easy to approach closely when fighting. These considerations may be important in the transition from an individual to an area reference for dominance. The former, characteristic of but not restricted to males involved in following strategies, is dependent on small numbers of males within the area where an individual male competes for mates. When this number passes a level where the fitness costs of agonistic behavior outweigh the mating benefits, males should switch to a generalized reference for dominance. The two possible references are one or a group of females and a territory. The reason that the latter is the most common in savannah antelopes may be linked to scent marking which occurs in most, perhaps all, cases of antelope territoriality. Scent marks in territories provide a means for owners to advertise their status to intruders (Gosling, 1982) and thus reduce the costs of agonistic behavior, possibly to the level characteristic of males which know each competitor individually.

APPENDIX. SCIENTIFIC NAMES OF THE SPECIES
MENTIONED IN THE TEXT

UNGULATES
Buffalo: *Syncerus caffer*
Bushbuck: *Tragelaphus scriptus*
Blue duiker: *Cephalophus monticola*
Dik-dik: *Rhynchotragus kirki*
Eland: *Taurotragus oryx*
Gerenuk: *Litocranius walleri*
Grant's gazelle: *Gazella granti*
Greater kudu: *Tragelaphus strepsiceros*
Hartebeest: usually Coke's, *Alcelaphus buselaphus cokei*
Klipspringer: *Oreotragus oreotragus*
Lechwe: *Kobus leche*

Lelwel hartebeest: *A.b.lelwel*
Lesser kudu: *Tragelaphus imberbis*
Oryx: usually fringe-eared, *Oryx beisa*
Impala: *Aepyceros melampus*
Roan antelope: *Hippotragus equinus*
Thomson's gazelle: *Gazella thomsoni*
Topi: *Damiliscus korrigum*
Uganda kob: *Kobus kob*
Waterbuck: *Kobus defassa*
Wildebeest: *Connochaetes taurinus*

CARNIVORES
Black-backed jackal: *Canis mesomelas*
Cheetah: *Acinonyx jubatus*
Lion: *Panthera leo*

13. Ecology and Sociality in Horses and Zebras

———————————————— • ————————————————

DANIEL I. RUBENSTEIN

FOR A FAMILY containing only seven species, the Equidae show a remarkable diversity of social systems. Horses (*Equus przewalskii* and *E. caballus*), plains zebra (*E. burchelli*), and mountain zebra (*E. zebra*) typically live in closed membership harem groups consisting of adult females, a single adult male, and their young (Klingel, 1974). In contrast, Grevy's zebra (*E. grevyi*) and the asses (*E. africanus* and *E. hemionus*) typically exhibit social systems in which female bonds are more ephemeral. Temporary aggregations of one or both sexes are common, but most adult males live alone in large territories (Klingel, 1974). Even though in both systems the young of both sexes reside with mothers until reaching sexual maturity and excess nonbreeding males live in all-male "bachelor" groups, in each details of these adult-juvenile and male-male relationships differ dramatically. The aim of this chapter is to show why these differences in social relationships have evolved in horses and zebras. Explanations relying on differences in niche or phylogenetic heritage are not likely to be important since horses and zebras are members of the same genus and make a living in more or less the same way, by grazing.

The social system of any species is shaped by the environment. As environmental forces change, group sizes, spatial dispersion, and mating systems also change. For example, Macdonald (1983) has shown that the size of fox groups increases as availability of prey increases; Gill and Wolf (1975) have shown that golden-winged sunbirds defend territories when flowers contain certain nectar levels but abandon the defense of resources after nectar levels reach a higher critical concentration; and Orians (1969) has shown that birds switch between monogamy and polygyny depending on the richness and patchiness of the resource. How this shaping process operates will depend on how environmental forces affect the behavioral "decisions" of individuals attempting to increase their reproductive success. Understanding how the process operates is likely to be complicated. There are many behavioral trade-offs, many environmental pressures pushing in different directions, and many differences among individuals constraining the behavioral options of some while altering the ecological perceptions of others. Moreover, every individual can potentially interact with every other member of the population and in ways that can range from extremely competitive to extremely cooperative. In order to understand how ecological circumstances shape social organization, a limited number of key paths through this maze must be found. By examining only a subset of social relationships—female–female, male–male, male–female, and breeder–nonbreeder—the major interactions can be described, and then examined to see how they are shaped by social and ecological pressures. In the next section a general model delineating how environmental forces structure these relationships is presented. It is then used to examine how particular environmental pressures shape horse and zebra sociality.

A GENERAL MODEL

Alexander (1974) proposed that living in groups was inherently disadvantageous. Costs associated with disease transmission and intensified competition automatically occurred and tended to increase as group size increased. To overcome these disadvantages at least one of three conditions had to be met. Group living either reduced an individual's chance of being preyed upon, enchanced feeding success, or was an unavoidable consequence of extremely localized resources. For ancestral horses and zebras the need to avoid being eaten was probably the major force leading to sociality. Even under contemporary conditions Schaller (1972) has shown that of all predatory attacks on zebra by lions, 35 percent were successful when zebras were solitary, whereas only 22 percent were successful when zebras lived in moderate-sized groups. But the need to reduce predation cannot be the sole ecological force shaping the social systems of horses and zebras today: no horses suffer predation yet all live in groups. Fulfilling other needs, such as acquiring food, water, and mates may also influence social relationships. But as Gakathu (1980) has shown, the interaction among ecological pressures may be strong. In Ambosoli Park, as the dry season commences, the plains zebra leave the bushy grasslands to graze on the more open plains. Given that vegetation quality on the plains is inferior to that of woodlands, and that quantity is lower than that of the swamp, their grazing movements seem surprising. But only if one ignores the fact that the number of predators is lowest on the plains and highest in the woodlands.

As Figure 13.1 shows, three ecological pressures, the need to avoid predators, and the needs to acquire food and water, can potentially affect female–female, male–female, and male–male social relationships. How they do so depends on: 1) major differences in the reproductive interests of males and females, and 2) a number of environmental and phenotypic constraints. The former influences the overall structure of the relationships, whereas the latter, by adjusting individual rankings of the ecological pressures, influences the magnitude and form of the relationships.

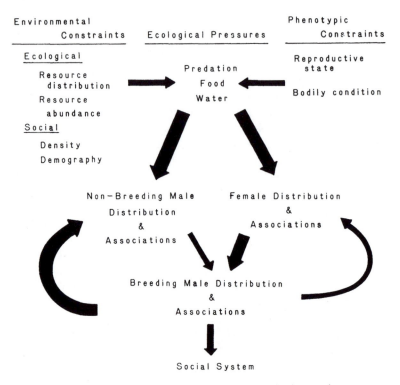

Fig. 13.1. Model of ecological pressures influencing intra- and intersexual relationship in mammals. Arrows depict social relationships and the width is proportional to relative importance.

Sex and Reproductive Strategies

According to Darwin (1871) investment in different reproductive activities will often have different consequences for each sex, especially in mammals. For males, reproductive success is usually limited by the number of progeny sired (Trivers, 1972). Reproductive effort put into acquiring mates will often have more effect on fitness than effort put into seeking food, water, or safety from predators. While it is true that males must eat and drink in order to maintain bodily condition, once a sufficient level has been reached, natural selection will favor subsequent investment into mating activities (Trivers, 1972). For females, the usual situation is different. Reproductive success is rarely limited by number of matings (Bateman, 1948; Trivers, 1972). Rather it is constrained mostly by the ability to acquire food, water, and find safe areas for offspring. Given these sexually differentiated reproductive interests, the ecological pressures exert a more direct effect on shaping female, rather than male, interactions. Consequently, as Figure

13.1 shows, female associations and distributions should be most strongly tied to resource distributions and abundances, whereas those of adult reproducing males are more tied to the distributions and abundances of females. This sexual dichotomy does not imply that females cannot augment fitness by investing in mating activities. Effort in discriminating among males can always enhance a female's reproductive success as long as males differ in attributes that enhance the condition or survival of the female or her offspring, or provide the offspring with superior genes (Bateson, 1983). Thus, as the loop in Figure 13.1 shows, male distributions and associations can potentially shape female-female interactions.

Environmental Constraints

The nature of the physical environment affects the operation of each ecological pressure listed in Figure 13.1. For example, when predators are concentrated, the need to reduce, or dilute, one's risk of being eaten favors joining groups (Pulliam, 1973; Triesman, 1975; Bertram, 1978; Rubenstein, 1978). Similarly, when food or water is concentrated in large patches, grouping is facilitated because competition is reduced (Jarman, 1974; Bertram, 1978; Rubenstein, 1978). But deciding how to behave when both pressures operate can be complicated, especially when the effects of food distribution oppose those of predation. Pulliam (1973) has shown that the conflict is reduced somewhat since scanning by others can lower vulnerability while allowing more time to be devoted to feeding. Nevertheless, the optimum level of each activity depends on the intensity of competition in relation to the probability of being eaten. Actual resource distribution affects these levels, but so does the social environment (Fig. 13.1). The size of a population, and its demography (e.g., operational sex ratio), determine the relative shortages of ecological resources and mates (Emlen and Oring, 1977). Since scarce resources are rarely apportioned equally among members of a population (Wilbur 1977; Rubenstein 1978, 1981a, b; Begon, 1984), phenotypic differences also alter the balance among an individual's behavioral options.

Phenotypic Constraints

More than anything else, phenotypic differences affect how individuals rank the environmental pressures. Among females, differences in reproductive condition alter dietary needs and susceptibility to predation. Maintaining bodily condition is important for survival, embryo development, and the raising of young to independence. Often, however, both quantitative and qualitative dietary requirements change throughout the reproductive cycle. While pregnant and nonpregnant, nonlactating females may place a premium on acquiring large quantities of high-quality vegetation, lactating females may place a premium on acquiring both food and access to predator-free sources of water. If the distribution of high-quality food and safe water do not coincide, then different reproductive classes of females may be distributed in different areas. Depending on the fertility of females found in the different locations, the effects of female reproductive condition on male associations and distribution might be dramatic.

Social relationships are also affected by differences in body size and condition, especially among males. Both vulnerability to predation and dietery requirements scale with body size. All else being equal, smaller species and smaller individuals are exposed to a wider range, and higher intensity, of predation. They also require smaller quantities, but higher qualities of food than larger species (Jarman, 1974). Thus size differences should lead to differences in feeding and antipredator behavior.

Among males the social consequences of body size are even more profound. Given the high variance in male reproductive success, sexual selection favors superior competitors, which usually are the largest or strongest males (Geist, 1974; Clutton Brock and Albon, 1979). For smaller, or poorer conditioned males the chances of obtaining matings by employing aggressive tactics is so low that selection favors delaying reproduction and the channeling of energy into rapid growth, or adopting alternative mating strategies (Gadgil, 1972; Rubenstein, 1980; Dunbar, 1982b). For these males acquiring forage and water might take precedence over acquiring mates. Depending on population size and demography, a variable number of youngish males will be induced to oust reproducing males. Thus associations and distributions among the nonbreeders might affect relationships among reproducing males (Fig. 13.1). Too often this effect is ignored, and exclusive emphasis is placed on the effects that reproductive males have on the "surplus" or "bachelor" males (e.g., the male-male loop in Fig. 13.1).

In summary, the model in Figure 13.1 provides a framework that delineates how the major types of social relationships are connected both to each other and to environmental pressures. How these pressures determine the details of each social relationship depends on constraints imposed by the nature of the environment and the distribution of phenotypes. In the next few sections the social relationships of horses and zebras are examined in relation to particular interactions among these constraints.

STUDY SITES AND METHODS

Behavioral and ecological observation on free-ranging feral horses have been carried out on Shackleford Banks, a barrier island off the eastern coast of North America, since 1973. Those on both common and Grevy's zebras were made in 1980 in the Samburu-Buffalo Springs Game Reserves of northern Kenya.

Whenever horses or zebras were encountered they were identified individually by morphological features such as coat color, position of the mane, presence or absence of distinctive facial markings, or stripes. Each individual's activity patterns, associations, and location were noted. Bodily condition was estimated by examining the loin area for amount of fat (sensu Mulvany, 1977; Pollock 1980). In well-conditioned horses (score = 5) the line of muscle on either side of the lumbar vertebrate appears flat when viewed from behind. As condition deteriorates the muscle takes on a triangular shape and eventually becomes concave (condition = 1). Detailed observations of interactions were monitored by focal animal sampling (Altmann, 1974). Vegetation was characterized into zones and the biomass and protein content of individual species was measured periodically. Data were analyzed using standard statistical tests (Sokal and Rolf, 1969).

THE ROLE OF ENVIRONMENT: A STUDY ON HORSES

The way environmental forces affect equid social relationships is most easily demonstrated by a population of horses inhabiting Shackleford Banks, a barrier island off the eastern coast of the United States. Of the three potential ecological pressures, only one, the need to acquire vegetation, is of any importance since there are no predators on the island and during most seasons fresh water is abundantly and evenly distributed along the northern edge of the island.

The island is approximately fifteen kilometers long and ranges in width from one to one and a half kilometers. Two features make the island unique. One is that its horses exhibit a variety of different types of adult association patterns. Some females form long-lasting bonds with other females and live in closed membership groups, whereas others form more ephemeral bonds and live in temporary membership groups. As for males, some wander widely, never attending a particular female for more than a few weeks, whereas others tend females continuously, thus forming harems. Some of these harem-forming males also establish territories, which they demarcate with dung and defend vigorously (Rubenstein, 1981c). The second is that each of the adult association patterns is limited to a region of the island that is ecologically unique in terms of habitat structure as well as vegetation abundance and distribution.

Female–Female Relationships

On the eastern end of the island there are three major vegetation zones: low-lying dunes border the ocean, a narrow salt marsh borders the back sound, and a swale or grassland lies between the other two. Each is fairly continuous and runs the length of the island (Fig. 13.2). At this end of the

SHACKLEFORD BANKS

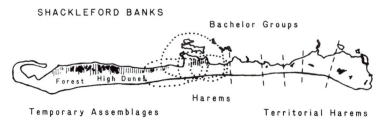

Fig. 13.2. Map of Shackleford Banks. Black areas represent fresh water and hatched areas represent forest. Dashed lines are territorial boundaries and dotted lines are home range boundaries. The western end of the island is on the left of the map; the east on the right.

island females live in permanent associations. Group movements are well coordinated and affiliative bonds among females, as measured by preferred grooming partners, are strong. Female dominance relationships exist, but are rarely expressed.

On the western end of the island the ecological situation is different. Tall dunes cover virtually the entire area except where they merge with a dense maritime forest. Swale occurs only in patches among the dunes, and some are no more than ten meters in diameter. Similarly, the marsh is restricted to small pockets that punctuate stands of forest. At this end of the island there are no permanent associations among females or males. Females leave or enter aggregations depending on patch size. When the patches are relatively large, groups grow as smaller groups fuse. But as soon as aggregations move to smaller patches, fissioning occurs. Such fission–fusion dynamics prevent permanent long-term associations from forming among females. As a result no strong grooming network or dominance hierarchy has appeared.

The existence of these two different patterns of female–female relationships is not easily explained. Grass is a resource that does not generally foster competition. Where it occurs it is relatively densely distributed, and one blade is not that much different from another, even if its neighbor is of a different species. Females should be able to associate for long periods, unless the overall structure of the habitat intensifies competition by reducing overall vegetation abundance or by partitioning the grassland into fragments too small to support average-size groups. On the western end of Shackleford patches of the swale and marsh grasses are few and variable in size and this explains why long-term associations are prevented. On the eastern end, however, the habitat is more continuous and could facilitate long-term associations. But showing that there is an environmental potential for closed membership groups is not the same as explaining why females live in them, especially since low levels of competition still occur, and there are also

288

costs of intensified endoparasitism (Hohman and Rubenstein, in prep.). Phylogenetic inertia could serve as one explanation. Although not adaptive under current conditions horses could still live in closed membership groups because in the past, when predators were abundant, the only way of ensuring that one's susceptibility was lowered was continuously to associate with others. By moving among open membership groups the possibility of being alone always remains. Given that some horses on Shackleford have already abandoned living in closed membership groups, the strength of this force seems limited. A more likely explanation is that females in closed membership groups derive some benefit that females in open membership groups do not. Such an advantage appears to exist, but is dependent on the nature of male–female relationships.

Male–Female Relationships

Given two distinct female distribution and association patterns, it is not surprising that there should be two different male–female relationships. What is interesting is that there are actually three: 1) single males tend single females, 2) single males tend groups of females, and 3) single males defend territories and tend groups of females. By comparing the benefits that females obtain from each of these sexual relationships it is possible to show that females living in closed membership groups derive an advantage that those living in open membership groups do not.

In theory males only have four behavioral options. They can either: 1) wander and tend individual females when they are in estrus; 2) tend groups of females, some of which are in estrus whereas others are not; 3) defend the resouces that females require; or 4) defend places along routes that females must travel in order to obtain critical resources. According to Emlen and Oring (1977), the behavior a male exhibits depends on the spatial and temporal distribution of females and resources. On the western end of Shackleford, where neither resources, females, nor the routes over which they travel are predictable, males have no alternative but to wander in search of estrus females and tend them when they can. As Figure 13.3 shows, the cost of this strategy, as measured in rates of aggression, is prohibitive. This option is almost always higher than the other strategies and shows no tendency to decline with age.

On the eastern end, tending groups of females is an alternative that all fully mature, reproducing males adopt. As Figure 13.3 demonstrates, the rate of aggression of this strategy is significantly lower than that of tending single females. But some males have modified this strategy. Not only do they defend females but they also defend an exclusive zone around their group of females. As Figure 13.2 shows, two-thirds (four of six) of the harem males defend the boundaries of the group's home range. Surprisingly, the costs of keeping other males far away from all his females are never as

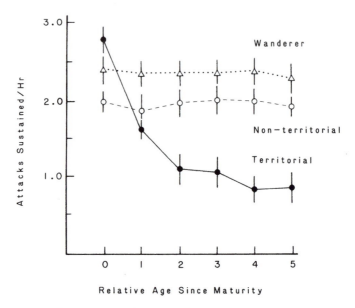

Fig. 13.3. Costs of aggression. Males tending individual females (Δ): males tending female groups (○): males tending female groups and defending territories (●). Two territorial. seven nonterritorial, and three wander males were observed each for at least one hundred hours over five years. Mean ± standard error.

high as those incurred by males tending individual females, and with age drop below those of males defending harems with overlapping home ranges. Only the most dominant stallions become territorial.

Ecological and demographic features of Shackleford account for these surprising cost relationships. First, the absolute number of potential intruding males is low because there are twice as many adult females on the island as there are adult males (Rubenstein, 1981c). High rates of juvenile male mortality and high rates of reproductive female mortality (Rubenstein, in prep.) account for this female-biased sex ratio. Second, approximately half of each territory is bounded by water. This severely limits access by intruding males (Rubenstein, 1981c). Third, the extreme openness of the habitat facilitates detecting intruders at a distance. And fourth, the ranges of bachelor males overlap only with those of nonterritorial males. All these factors contribute to lowering the long-term costs of establishing and maintaining a territory.

The advantages of territorial control do not stop at lowering a male's costs. In fact, as Figure 13.4a shows, the ultimate evolutionary measure of

(a)

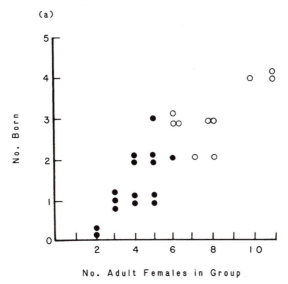

No. Adult Females in Group

(b)

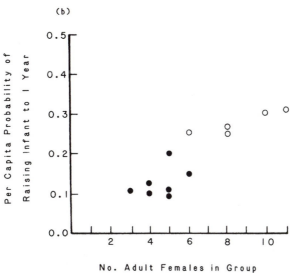

No. Adult Females in Group

Fig. 13.4. Male and female reproductive success, (a) males, (b) females.

(a) male success—measured by number of young born per group where groups differ in number of adult females residing there. No. born = 0.40 (no. adult females per group) − 0.22; $F_{1,23} = 71.9; p < .001$.

(b) female success—measured by the probability of each female's likelihood of bearing a young and rearing it to independence (one year old). Per capita probability = 0.03 (no. adult females per group); $F_{1,10} = 43.0; p < .001$.

● nonterritorial harems vs. ○ territorial harems.

291

success, reproductive success, increases when males defend territories. On average, territorial males sire more offspring than nonterritorial males largely because they have more adult females in their groups. Since both sexes disperse from the natal group in the Equidae, these increases cannot be the result of success in previous breeding seasons. For equid groups to grow, females must move to the territorial areas and then restrict their movements in order to remain within the territories. As the loop in Figure 13.1 suggests, females are responding directly to male distributions and associations.

The benefits for females of monitoring differences among males and adjusting their behavior accordingly are many. First, territorial females derive a feeding benefit. Females consistently require large quantities of high-quality vegetation if they are to survive, reproduce, and raise a foal to the age of independence. On average, horses on Shackleford feed for 70 percent of every hour during the summer and up to 80 percent of every hour during the winter (Rubenstein, 1981c). Despite this extensive commitment to foraging, the most frequent cause of death can be ultimately linked to malnourishment. All individuals are scored for bodily condition by examining the loin area for amount of fat (sensu Mulvany, 1977; Pollock, 1980). Among all Shackleford adults the average condition score is 2.8, whereas the average of the last score recorded before death for dead animals is 1.4. Although the difference was significant ($t_{17} = 2.73; p < .01$), it is possible that death was caused by some other factor that also caused condition to deteriorate. To account for this possiblity, the condition scores from one year prior to death were examined, and, again, the average value of 1.7 is significantly lower ($t_{16} = 2.02; p < .05$) than the average of overall population's score. In addition, nursing mothers show significantly lower condition scores ($\bar{x} = 2.0$) than do other females ($\bar{x} = 3.1; t_{23} = 2.96; p < .005$), and for mothers losing foals the score is significantly lower still ($\bar{x} = 1.7; t_7 = 2.91; p < .05$). Thus food limitation is a significant problem that has consequences for female longevity and reproductive success.

To measure whether females residing in territories derive a feeding benefit because of exclusive control of a renewable resource (cf. Davies and Houston, 1981), vegetation regeneration rates were measured. Sample plots (1 m²) in the open swale in both the ranges of territorial and nonterritorial horses were marked with hollow aluminum pegs. A square grid of wires spaced at five-centimeter intervals along both the x and y axes was placed on each plot. The height of each grass blade lying directly beneath the intersection of every pair of wires was recorded at two-week intervals. Figure 13.5 shows that during these intervals about 50 percent of the vegetation grew on average between one and three centimeters within territories, whereas within ranges that overlap with those of others, about 50 percent of the grass grew less than one centimeter during the same two-week period.

Vegetation Regeneration

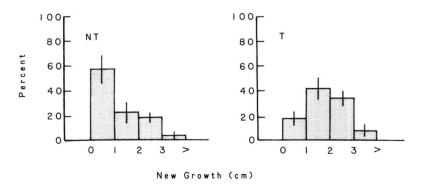

New Growth (cm)

Fig. 13.5. Vegetation regeneration. Regrowth of *Spartina patens* at two-week intervals for ten weeks during spring and summer 1976. $N =$ 1,000 for both territorial (T) and nonterritorial (NT) plots. Mean $\pm$ standard error.

At least with respect to swale vegetation, territorial females can expect to acquire more vegetation per mouthful than their counterparts that live in nonterritorial harems, or even open membership groups where home ranges also overlap.

A second advantage of living with territorial males is that harassment by *all* males is reduced. When males are grazing, or are on the lookout for intruding males, they monitor the estrus condition of their females. They also attempt to control the position and movements of their females, often with limited success, and they invade foreign groups in search of estrus females. These male activities interrupt the behavior of females, those both in and out of estrus. Since males are not discriminating in their sexual investigations, escape attempts by harassed females usually disrupt the behavior of nearby females. Table 13.1 shows the magnitude of the direct and indirect effect of male harassment; females residing with territorial males suffer significantly fewer disruptions than females residing with a nonterritorial male. An analysis of time budgets (Fig. 13.6) shows that the consequences of harassment are strikingly different for territorial and nonterritorial females. Territorial females graze about 9 percent more per hour than nonterritorial females, and this increase occurs despite the fact that territorial groups of females are larger and should suffer reductions in grazing time because of intensified female–female competition. Not only is the difference statistically significant ($p < .001$), it is biologically significant as well. A gain of about 5½ minutes per hour, when summed over the twenty to twenty-two hours a day

TABLE 13.1

Average hourly rate of interruptions sustained
by an individual female during the breeding season,
and caused directly by males, or indirectly by females
harassed by males

	Causes of interruptions	
Harem type	Male	Harassed female
Territorial	0.6	1.1
Nonterritorial	1.6	3.7

NOTE: Data derived from over 250 hours of observation on five territorial and seven nonterritorial harems. Male interruptions $t_{42} = 2.55$; $p < 0.01$; harassed female interruptions $t_{67} = 2.41$; $p < 0.01$.

that horses graze, is amplified to a feeding gain of about two hours per day. Thus by associating with males that are able to defend territories, and hence reduce intrusions (see Fig. 13.3 for rate of aggression), females derive another significant feeding advantage.

If foraging success is coupled to reproductive success, then territorial females should have a higher fitness than nonterritorial females. As Figure 13.4b reveals, this appears to be the case. Females residing with territorial males have higher per capita probabilities of raising foals to one year of age, the age of independence.

Thus on Shackleford Banks the maintenance of a territorial system benefits both males and females because both benefit from mate guarding. From the general model diagrammed in Figure 13.1, when food is the only critical resource, its distribution in a fairly continuous fashion is a prerequisite that permits females to live in closed membership groups. Whether or not they do appears to depend on the harshness of the environment, and on the negative consequences of male sexual activity. On Shackleford Banks, acquiring sufficient resources for survival and reproduction is difficult enough without the added stress imposed by males. As a result, selection seems to favor females that channel some of their reproductive effort into competing for access to high-quality dominant males. Such males not only effectively defend their own reproductive interests, but in doing so also augment a female's foraging and reproductive success.

PHENOTYPIC CONSTRAINTS: COMPARISIONS AMONG
HORSES AND PLAINS ZEBRA

Male–Male Relationships

The horses of Shackleford Banks show how female associations affect

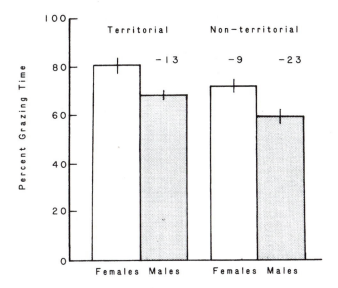

Fig. 13.6. Percent time spent grazing by males and female in territorial and nonterritorial harems. Over 1,000 hours observing five territorial and two nonterritorial harems. Mean ± standard error.

male associations and how male associations affect those of females. According to the model depicted in Figure 13.1, the associations and distributions of "surplus" or "nonreproductive" bachelor males should also affect the pattern of sociality by affecting the behavior of the reproductive males. In equids do such effects exist? They do, and in a significant way. But the impact can only be seen in comparision with situations where the demography or social behavior among bachelor males are markedly different. Two such comparative situations exist. One involves major demographic changes that occurred on Shackleford Banks in 1980. In general, male infants on the island die at a higher rate than female infants (Rubenstein, in prep.). As a result Shackleford Banks rarely supports a large bachelor male population. From 1973 to 1978 the bachelor male population varied from two to ten individuals. After a series of mild winters in 1976 and 1977, this number swelled to seventeen in 1980.

With the aging of the harem males, a demographic shift occurred in the population. By 1980 harems were headed by weaker individuals and, with the unprecedented number of young, agile males, the pressure imposed on the harem stallions by intruders increased dramatically. Virtually all the harem males were overthrown and the females were divided into fifteen new

TABLE 13.2

Patterns of association among six common zebra harems during
spring 1980

	Torch	Y^2	Zeus	Achilles	Saddle	Hermes
Torch		.63	.11	.00	.11	.00
Y^2			.06	.06	.17	.00
Zeus				.55	.06	.00
Achilles					.00	.08
Saddle						.46
Hermes						

NOTE: Index of association computed by the formula $IA = 2C/(A + B)$ where C is the number of times both harems are seen together, and A and B are the total number of times each harem is seen.

harems. Group sizes dropped from a mode of seven to a mode of two, and all the territories disappeared. Perhaps what is most striking is that the population birth rate was reduced from a mode of fourteen to a mode of eleven. Thus the dramatic increase in bachelor males and their intrusion rates led to a social revolution.

The second comparison that shows how bachelor male-breeding male relationships influence social systems involves a comparison among horses and plains zebras. Ordinarily, horses and plains zebras are classified as having identical social systems (Klingel, 1974). One striking difference exists, however, and that concerns the nature of the relationships that form among bachelor males. In horses the relationships are weak and ephemeral as males live in temporary associations of two to three individuals. At least in northern Kenya in the Sambura-Buffalo Springs reserve, plains zebra bachelor males have strong relationships, are organized by linear hierarchy, and live in closed membership groups averaging nine individuals (Rubenstein, in prep.). According to the general model this difference should result in differences in the distribution and relationships of the harem males. And indeed it does. By using an index of association $IA = 2c/(A + B)$ where A is the number of occurrences of group A alone, B is the number of occurrences of group B alone, and C is the number of occurrences where groups A and B were seen within twenty-five meters of each other, it is possible to measure how likely different harems are to associate together. As Table 13.2 demonstrates, there are strong associations among particular pairs of plains zebra harems.

To understand the possible significance of this result we have to appreciate that both horse and zebra harem males almost always charge away from their females when they detect an intruder male. The agonistic interaction that ensues is usually of low intensity with dominance determined quickly. Nevertheless, the two males often graze together until one, usually

TABLE 13.3
Numbers of successful and unsuccessful attempts by single
and paired common zebra harem males to prevent
bachelor males from entering a harem

Type of defense	Defensive outcome	
	Successful	Unsuccessful
Single male	0	8
Paired males	11	1

NOTE: Fisher exact proobability, $p = 0.00014$.

the subordinate, leaves, walking back slowly to his females. On Shackleford Banks if a single harem male meets a bachelor group he has little trouble dominating the two to three horses comprising it. He then remains and grazes with them, thus keeping them away from his females. In the plains zebra, a single harem male was never able to dominate all nine males and was never able to keep them from mixing with his females (Table 13.3). However, if a pair of harem males were already together because their two groups of females had come together, then they could almost always dominate the nine bachelor males and keep them away from their females (Table 13.3). So it appears that the nature of relationships among the younger subordinate males has a major affect on the nature of equid social organization. When the threat to male and female reproductive success is high a new level of sociality emerges. It involves strong intergroup male relationships and effectively neutralizes the reproductive challenge (Table 13.3). Why horse bachelor males live in open membership groups whereas plains zebra bachelors live in closed membership groups remains an exciting question. Perhaps, as argued earlier, the higher risk of predation faced by zebras necessitates that males also live in closed membership groups, so that the antipredator benefits due to dilution are assured.

REPRODUCTIVE CONDITION: A COMPARISON OF HORSES AND GREVY'S ZEBRA

The general model suggests that phenotypic attributes such as differences in reproductive condition should affect female social relationships. They do weakly in horses, but strongly in Grevy's zebras.

Female–Female Relationships

In horses, group movement is usually determined by females, and in particular by dominant females. From data gathered over a decade (1973–1984), 68 percent (± 12 percent) of movements from one vegetation zone

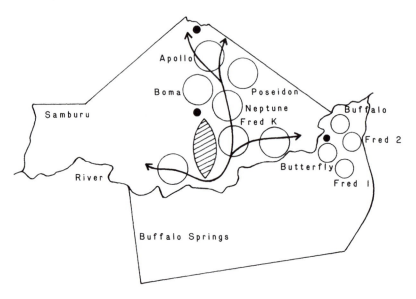

Fig. 13.7. Location of male Grevy zebra territories in relation to female ranging routes and predator-free standing water (●).

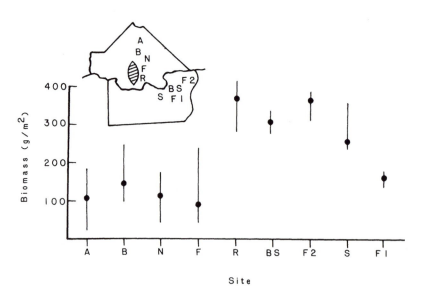

Fig. 13.8. Biomass within Grevy zebra territories. N = ten samples for each location. Mean ± range.

to another were initiated by dominant females (top half of hierarchy) in the season before the birthing begins. After the birth season commences, however, in those situations where the dominant female is without a new foal, her initiation of vegetation zone changes drops to 36 percent (± 14 percent). During this period, which lasts about six months, lactating females assume a more prominent role in controlling the foraging movements of the group. Thus reproductively induced changes in dietary requirements appear to have a profound effect on female distributions and associations among horses.

Female–Male Relationships

In Grevy's zebra, the effects of reproductive conditions are more pronounced since the bonds among adult females are less strong and individualistic tendencies are more readily expressed. Normally Grevy's females range between ten to fifteen kilometers per day (Rubenstein, unpubl.). In the Samburu Game Reserve there are two major movement routes (Fig. 13.7), and males establish territories along them. Some territories are nearer to peripheral water holes that afford protection from predators while drinking (Fig. 13.7), whereas others are nearer areas of highest biomass (Fig. 13.8). Although all types of territories are visited by the females that wander daily, only those having just borne a foal stop ranging long distances and take up long-term (two to three months) residence with males near safe standing water (Fig. 13.9). Again, reproductive condition appears to affect female associations and distribution. In the Grevy zebra old bonds are broken and new ones form, especially those with males.

Normally territorial Grevy's zebra males have a low confidence of paternity when copulating with wandering females; rarely does a wandering estrus female remain within male's territory for the duration of her heat. But since all female equids come into estrus shortly after giving birth, males associating, and copulating, with a newly arrived sedentary mother will increase their paternity confidence and gain a reproductive advantage (Rubenstein and Ginsberg, in prep.). Although the gain associated with exclusive mating access to these females is reduced somewhat because he protects another male's offspring, infanticide is not favored. Without the foal, and the mother's need for forage and access to safe watering sites, the female abandons the territory within one day (Rubenstein, pers. obs.). Thus the male is in a cruel bind: he has to assist another male's offspring to augment his own reproduction. The actual magnitude of this advantage will depend on whether or not territories near watering sites have enough good vegetation to attract as many estrus females that remain wanderers as do territories in other sites. From the biomass data shown in Figure 13.9, it appears that both types of territories on average offer females about the same amounts of vegetation. The territories in the proximity of water, however, are somewhat less variable, and appear to lack those few superrich patches.

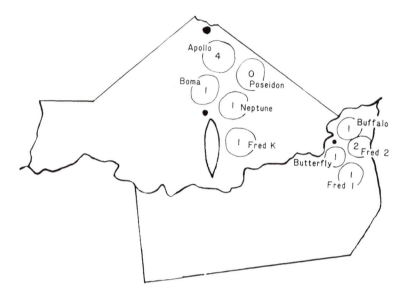

Fig. 13.9. Location of male Grevy zebra territories in relation to predator-free standing water (●) and the number (within circle) of resident
females with newborn foals.

Whether or not these differences in variance are important (sensu Rubenstein, 1982), is yet to be determined. In any event, it appears that in the
Grevy's zebra, female reproductive condition affects female relationships
and may have a significant effect on distributing males of different phenotypes in different areas, and influencing their reproductive success as well.

CONCLUSIONS: COMPARISONS WITH OTHER STUDIES

Horses and zebras exhibit few niche, or phenotypic differences, yet they
display a wide range of social systems. The social diversity seen on Shackleford Banks is mirrored in other feral equid populations. Female horses
associating with territorial males are found in Exmoor (S. Gates, 1979), on
Cumberland Island (Lenarz, 1982), and in the New Forest (Pollack, 1980),
and female asses living in closed membership harem groups occur on Ossabaw Island off the southeastern United States. Although few detailed
measurements of the pertinent phenotypic and environmental constraints
outlined in the general model (Fig. 13.1) were made in these studies, the
basic ecological descriptions suggest that for each population exhibiting a
deviation from the "typical" pattern, one of the external constraining variables is somewhat atypical. On Ossabaw Island lush vegetaion is fairly

evenly distributed (NRC, 1982; McCort, 1979), a situation not commonly encountered by asses. Amicable behaviors outnumber aggressive ones and permanent harems occur. In the New Forest, the sex ratio is extremely biased by the commoners, who prefer to rear females. With a shortage of males, many females wander unassociated with, and unhindered by, males, who limit their range to territories encompassing the best grazing areas (Pollack, 1980). And at Exmoor, a river bisects the range making it easier for groups to remain apart (S. Gates, 1979). On Cumberland Island, both the extent to which a male can exclusively control an area and the quality of the vegetation he controls depend on his dominance status (Lenarz, 1982). Only the most dominant males can control the areas of highest quality vegetation on a year-round basis, and Lenarz suggests that in doing so they assist the reproductive success of their females. Unfortunately, he provides no data on group size or fertility of dominant versus subordinate males. The same correlation of male dominance status and territorial quality occurs on Shackleford (Rubenstein and Balbo, in prep.), and on Shackleford females associating with dominant males derive a significant fitness advantage (Fig. 13.4).

This social diversity results from a variety of factors. As these examples demonstrate, the distribution of resources affects the potential for females to form groups. Except in extreme environments, like on the western end of Shackleford, or in the steppe deserts of northern Kenya, closed membership groups of females are permitted because feeding competition is low. Whether or not they form is dependent on other factors and only indirectly on food. Predation is one factor that appears to provide a sufficient advantage. By forming permanent membership groups mutual reduction in one's personal vulnerability is assured. Another advantage derives from the sexual activities of males. At least on Shackleford Banks, females, by forming closed membership groups, assist males in becoming effective protectors of their own reproductive interests as well as those of females.

This is a novel advantage that seems to have evolved in a number of phylogenetically distinct harem-living groups of primates—the gorillas (Harcourt, 1979b), the hamadryas baboons (Kummer, 1968), and the gelada baboons (Dunbar, 1980a). The details of the social organization of these species are similar but not identical. One difference concerns the role of breeding-nonbreeding male relationships. In all, the nonbreeders are repulsed by the harem males, but the success is varied. In both gelada baboons and gorillas, harems often contain secondary males who attempt to steal copulations, or females. This rarely occurs in horses (cf. Miller and Denniston, 1979) or hamadryas baboons. Another difference involves the nature of the female-female bonds. In the gorillas and hamadryas baboons few relationships exist among females so that when the resident male disappears the female group fragments. In horses and gelada baboons the disappearance of a male leaves the female relationship unchanged; new males take over the entire group (Klingel, 1974; Dunbar, 1984a). Thus despite a common need for a male shield or hired gun, other relationships in these social systems are different. Whether they are the result of phylogenetic differences or are the result of subtle differences in ecology is as yet poorly understood. Perhaps small differences in the levels of feeding competition among females are involved. Gelada baboon females can apparently stress other females and suppress reproduction (Dunbar, 1980a). And on Shackleford Banks female horses already in a group attempt to deny strange females entry. Hints of density-dependent reproductive suppression can be gleaned from Figure 13.4b, and suggest that competition among females may be as reproductively important in equids as is male-male intergroup cooperation.

In horses and zebras relationships among males are affected by features of both the social and physical environment. Predictability of female associations and movements are most important in structuring male associations. When females form cohesive groups, males defend them against incursions by other males. When females live in temporary assemblages, male strategy is varied. If daily female movements are predictable, then males compete for territories along these highways. This is typically the case in Grevy's zebra (Klingel, 1974) and in asses (Moehlman, 1979a). But even in the New Forest where interference by commoners limits permanency of female bonds (Pollock, 1980), some males limit their activities to the areas where unattached females prefer to graze. When even these daily movements of females are unpredictable, males have no alternative but to search for, and attend, individual females in estrus. Apart from the Shackleford population this rarely occurs in equids, because at least movements to and from water are somewhat predictable.

Ecological pressures have some effect on male behavior, but mostly by fine tuning the major male alternatives. Extreme topographical features, such as those found on parts of Shackleford Banks, Exmoor, the New Forest, and Cumberland Island, can make the simultaneous defense of females and territories economical. Similarly the presence or absence of predators seems to induce nonbreeding males to form either cohesive or transitory assemblages. For horses and zebras the number of combinations of particular ecological and social features is large, and as the general model shows, can account for the social diversity displayed by the family.

14. Marmot Polygyny Revisited: Determinants of Male and Female Reproductive Strategies

————————————— • —————————————

KENNETH B. ARMITAGE

MUCH OF THE FOCUS on the nature of mating systems has centered on determining the conditions under which polygyny evolves. One model, the polygyny threshold, states that a female should choose to mate with an already mated male when she can expect greater reproductive success than if she mated with a remaining unmated male (Verner and Willson, 1966; Orians, 1969). In this model, both males and females benefit.The underlying assumption in the study of mating systems is that individuals attempt to maximize fitness. An analysis of the mating system of the yellow-bellied marmot revealed that the fitness of individual females, measured as reproductive output, decreased as harem size increased (Downhower and Armitage, 1971). Downhower and Armitage emphasized that the reproductive interests of males and females conflict; such conflict is now widely regarded as fundamental (Wittenberger, 1979, p. 272). Additional studies of marmots have necessitated a revision of the Downhower-Armitage model. Before describing the current model, based on twenty-two years of data, some of the assumptions underlying the development of polygyny models must be examined. The assumptions led us astray in the past and I hope an examination of these assumptions and a clear statement of the assumptions underlying the present interpretation of the marmot mating system will at least clarify the issues.

Most studies of polygyny focused on birds. Because most bird species are monogamous, theory centered on why some birds are polygynous. But a student of mammals, noting that most mammals are polygynous, might try to find models to explain monogamy. Implicit in the polygyny threshold model is the assumption that females operate independently, make choices based on potential reproductive success, and do not interfere with the decisions of later-arriving females. Yet, among mammals, females may be bonded to other females and cooperation or competition among females may decide the size of the mating group. Two recent papers made major efforts to broaden perspectives on the evolution of polygyny. Greenwood (1980) emphasized the relationships among philopatry, dispersal, and mating systems. In many mammals, males defend females (mate defense); in many birds, males defend resources (resource defense). Female mammals

303

and male birds are philopatric. Futhermore, philopatry should favor the evolution of cooperative traits among members of the sedentary sex. Emlen and Oring (1977) integrated ecology, sexual selection, and mating systems. They argued that polygamous mating systems occur when multiple mates, or resources sufficient to attract multiple mates, are energetically defendable and when animals have the ability to exploit this potential. Females independently choose whether to become polygynous (e.g., Emlen and Oring, 1977; Wittenberger, 1979); in resource defense polygyny, males determine the dispersion of females. However, Greenwood (1980) rightly pointed out that females may determine the dispersion of males, especially in harem defense polygyny (Emlen and Oring, 1977).

Female mammals, because of gestation and lactation, invest more in reproduction than males. Therefore, females should selectively choose a mate (Trivers, 1972). But there may be realistic constraints on female choice. Because of the greater reproductive investment of female mammals, females must satisfy their physiological and ecological requirements and males must adjust their strategies to the realities of what females must do. I hope to demonstrate that for yellow-bellied marmots, and probably many species of mammals, females determine the spacing patterns that make polygyny possible and males attempt to associate with clumped females.

BIOLOGY OF THE YELLOW-BELLIED MARMOT

The yellow-bellied marmot (*Marmota flaviventris*) is one of six species of marmots living in North America. Marmots are typical ground squirrels, along with prairie dogs, spermophiles, and chipmunks (Moore, 1959), of the subfamily Marmotinae of the family Sciuridae, order Rodentia. The yellow-bellied marmot is widely distributed in the western United States, especially in forest clearings and the alpine of the Cascade, Rocky, and Sierra Mountains (Frase and Hoffmann, 1980).

M. flaviventris occupies open areas, dominated by perennial forbs and grasses (Svendsen, 1974; Kilgore and Armitage, 1978), in which rock outcrops, boulders, or talus occur. Marmots excavate burrows at what appears to be every suitable site; where rock and/or soil structure do not permit burrowing, marmots are absent (Svendsen, 1974). Areas occupied by marmots become snow-free earlier in the year than adjacent forest areas do.

Marmot populations are clumped on habitat patches that range widely in size. Larger patches (total open area, $\bar{x} = 58$ ha) that typically harbor three or more adults and variable numbers of yearlings are designated colonial sites. Smaller sites ($\bar{x} = 6.6$ ha), where typically only one or two animals live, are called satellite sites (Armitage and Downhower, 1974; Svendsen, 1974). Some satellite sites are within 100 meters of a colonial site, but most are more distant. Although marmot habitats may be classified into these two types, they form an almost continuous range of sizes from a single boulder in an area of about 0.01 hectare to a rocky slope and meadow of 70 hectares or more. Mean population density is correlated with habitat area (Svendsen, 1974).

The annual cycle of marmots consists of two phases: heterothermal and homeothermal (Morrison and Galster, 1975). The annual cycle is a circannual rhythm of metabolism, body mass, and food consumption (D. E. Davis, 1976) that persists in yellow-bellied marmots maintained in the laboratory (J. M. Ward and Armitage, 1981). Because the phases of the circannual rhythm are tightly locked, emergence, reproduction, growth, and maintenance and preparation for immergence are sequentially programmed. Emergence patterns follow an age-sex sequence. For example, at North Pole Basin in 1976, 50 percent of adult male marmots emerged by 16 May; 50 percent of adult females, by 25 May; 50 percent of yearling males, by 28 May; and 50 percent of yearling females, by 5 June. Immergence is reversed: adult males and nonreproductive females immerge first, usually between mid- and late August; reproductive females immerge in early September and juveniles immerge about mid–September. Two or more marmots commonly occupy the same hibernaculum (Johns and Armitage, 1979).

Emergence typically occurs through the snow. At colonial sites, there is no marmot activity into or out of a site; marmots are vulnerable to coyote predation when crossing snow. During the first two weeks there is little activity; females remain near their burrows. The adult male is most active and is especially attentive to females. As females become receptive, they are increasingly active and consort with the male from one to several hours. Reproductive activity ceases after two weeks (Armitage, 1965; Nee, 1969). Subsequent to mating, females may shift burrows; home ranges overlap less. The synchrony of reproduction into a short period early in the homeothermal period is critical for survival of the young; young weaned late in the period have a very low rate of survival through their first hibernation (Armitage et al., 1976). When lactation begins, marmot habitats are rich in food resources (Svendsen, 1974; Kilgore and Armitage, 1978; Frase, 1983). Marmots are generalist herbivores (Frase, 1983), feeding on leaves and flowers of a wide variety of forbs and grasses, but avoid eating plants with known defensive compounds (Armitage, 1979). Juveniles appear above ground between three and four weeks of age; weaning occurs at this time (Armitage, 1981). Marmots use less than 4 percent of the aboveground primary production available to them (Kilgore and Armitage, 1978). Although marmot population density apparently is not food-limited, the number of offspring weaned by a female was significantly associated with food resources in a high-altitude population (D. C. Andersen et al., 1976).

Yellow-bellied marmots live as solitary individuals or as members of social groups, consisting characteristically of one adult male, one or more adult females, yearlings of both sexes, and juveniles (Armitage, 1962; Armitage and Downhower, 1974; Svendsen, 1974; Johns and Armitage, 1979). Members of social groups are called colonial. Peripheral animals live near colonial animals, but their home ranges lie outside the home ranges of the colonial residents. Peripheral animals, especially males, occasionally venture into the colony. Female residents attempt to exclude female intruders, but are dominated by male intruders. Males rebuff male intruders, but accept females. Solitary females may be associated with juveniles, but rarely with yearlings. Virtually all male yearlings disperse (Armitage and Downhower, 1974; Downhower and Armitage, 1981), but 72 of 135 female yearlings became resident in their natal populations (Armitage, 1984). Home ranges of resident yearlings overlap by 50 percent or more those of the resident adult females (Armitage, 1975).

Although the age of first reproduction is two years for both sexes, most marmots do not breed before age three or older (Armitage and Downhower, 1974; Armitage, 1981). Yearlings have a greater probability of living to age three than juveniles (Armitage and Downhower, 1974); therefore, the number of yearlings produced by an adult is a more critical measurement of reproductive success than the number of young.

The adult male is territorial (Armitage, 1974). Many habitat patches support only one male, but larger patches may have two or more. Virtually all of the variation in the population density of adults results from changes in the number of resident females. Changes in population density are not density-dependent (Armitage and Downhower, 1974; Armitage, 1975).

Predation is rarely detected (Armitage, 1982a). Marmots of all age classes gain weight normally (Armitage et al., 1976); diseased animals have not been detected. The loss of residents, which occurs over winter, is assumed to be mortality. This assumption is supported by the higher survival of young and adults when winter terminated earlier and by the higher survival of females that produced a litter when the onset of winter was later or the length of winter was shorter (Armitage and Downhower, 1974).

Immigration is significantly more likely to produce residents when the previous residents fail to return (Armitage, 1984). Recruitment of yearling females is highly variable. A recruit is an animal who becomes resident in its natal population; recruitment is defined as retention of yearling offspring in the parent's population without implying any particular proximal mechanism whereby it occurs. Social structure and recruitment are highly philopatric among closely related ($r = 0.5$) females (Armitage, 1984). Adult females that are successful recruiters produce more yearling females than nonrecruiters (Armitage, 1984). Sociable females (Svendsen and Armitage, 1973) have greater lifetime reproductive success as measured by the number of female yearlings produced, the number of daughters recruited, and the number of two-year-old daughters who were residents (Armitage, 1983). Recruitment is not affected when a new adult male replaces a previous resident.

Social behavior is highly variable (Armitage, 1975, 1977). This variability is partly attributable to individuality (Armitage, 1982b, 1983) and to the age-sex composition of the population (Armitage and Johns, 1982). The major variable affecting social interactions is kinship: behavior among females related by 0.5 is amicable (greetings, allogrooming) but is agonistic (chase, fleeing) when relatedness is 0.25 or less (Armitage and Johns, 1982). Although adult females generally behave amicably to their daughters, they behave neutrally to their yearling sons. Adult males behave cohesively toward females and agonistically toward males (Armitage, 1974; Armitage and Johns, 1982).

METHODS

Yellow-bellied marmots were studied every year since 1962 in the East River Valley, Gunnison County, Colorado, at an elevation of 2,900 meters. Each year at each study site virtually all animals are trapped, sexed, and permanently marked for identification. Each time an animal is trapped, it is weighed and its reproductive status noted. At various periods during the study, blood samples were collected for genetic studies. Juveniles were trapped as soon as possible after their first appearance above ground. In this way, maternity could be assigned except in seven instances (16 of 158 litters) when adult females occupied the same burrow system and young intermingled from the time they were first observed.

Usually our field season extended from about the first of June to late August. Consequently, in most years we missed mating and the last two or three weeks prior to immergence. Paternity was assigned to the male resident in early June. In those years when we were present early in the season, no male was replaced by another during the reproductive period. Genetic analyses verified that the June resident was the probable father of juveniles produced at a particular study site (O. A. Schwartz and Armitage, 1980).

Marmot populations were observed in excess of 250 hours each summer. Because of low population densities, all animals at a study site are monitored. Observation hours are concentrated in the morning, late afternoon, and evening when marmots are most active (Armitage, 1962). All social interactions are recorded. At regular intervals the position of each animal is recorded from a transparent numbered grid overlying a map of the study site. These census data are plotted by the Surface II computer program (Sampson, 1975) as three-dimensional block diagrams.

Differences in reproductive success among groups were assessed by analysis of variance or linear regression; differences among means were tested

CLIFF COLONY

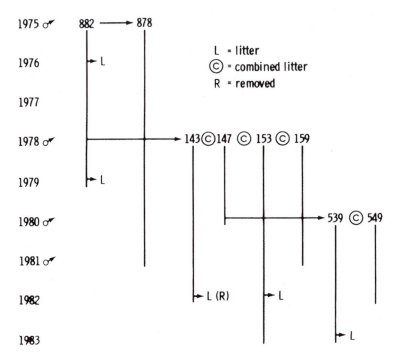

Fig. 14.1. An example of a matriline. The ♂ symbols indicate when a new male became resident. Note that not all litters (L) produce recruits, that females fail to breed in many years, and that some litters are intermingled (horizontal line crossing two or more vertical lines). An animal's number is recorded in the year of its birth (except ♀882, who was a resident adult in 1975). The vertical line indicates the years of residency.

by the *t* method of snk; and the distribution of social interactions was tested by χ^2 (Sokal and Rohlf, 1981).

RELATIONSHIPS AMONG ADULT FEMALES

Matrilines

Females form groups of one to five ($\bar{x} = 1.47$). A group of two or more animals consists of closely related animals who usually are full-sister or mother–daughter pairs (Fig. 14.1). Larger groups occur when a female recruits several daughters from the same litter; within these, average relatedness is 0.5. Females in these kin groups may produce litters in common, such that maternity cannot be assigned (Fig. 14.1). Larger groups also occur

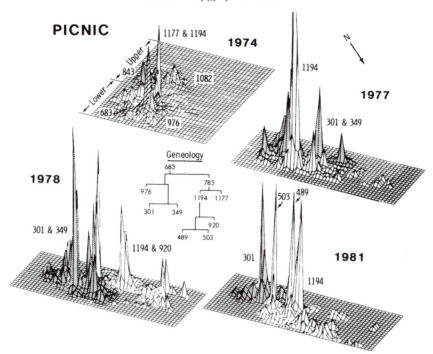

Fig. 14.2. Patterns of space use at Picnic Colony. The plots are oriented as if one were standing at the foot of the slope and looking uphill. The grids are 9 × 5 m. For 1974, the right side of the habitat was not plotted as no animals were using that area. For the other three years, only the lower half (Lower Picnic) is included. For each year, all resident adult females are shown. The height of each peak is a function of the number of times the animal was censused in that area. The genealogy of the residents of Lower Picnic begins with ♀ 683. The horizontal position of each animal on the chart indicates order of birth. Except for paired sisters, all births occurred in different years. In any year, different matrilines are distinguished by patterns of shading.

when nieces, granddaughters, and so on are added to the group and average relatedness decreases. Female kin-groups continue through space and time as matrilines (Armitage, 1984). Known lineages have persisted for eleven, twelve, and twenty-three years. Such lineages may undergo fission to form daughter matrilines. For purposes of analysis, I define a matriline as any number of females known to be related by descent that occupy a common home range. When home ranges diverge and females no longer share burrows or foraging areas, a single matriline is considered to have undergone fission.

Matrilines occupy space exclusively. There was no overlap of home ranges among members of three different matrilines at Picnic Colony in

TABLE 14.1
Frequency analysis of amicable and agonistic interactions by kinship for
adult females at Picnic Colony

Kinship group	Amicable		Agonistic		Amicable		Agonistic	
	1974				1977			
	E	O	E	O	E	O	E	O
Sisters	3.3	10*	1.2	0	3.3	10*	4	0*
Mother:daughter	3.3	2	1.2	0	—	—	—	—
Grandmother:granddaughter	6.7	6	2.3	1	—	—	—	—
Aunt:nieces	6.7	2*	2.3	6*	6.7	0*	8	12*
χ^2	17.5		8.9		20.3		6.0	
p	<0.001		<0.05		<0.001		<0.02	

	1978				1981			
	E	O	E	O	E	O	E	O
Sisters	3.5	13*	1.8	3	2.5	10*	—	0
Mother:daughter	3.5	8	1.8	0	7.5	5	—	1
Grandmother:granddaughter	—	—	—	—	—	—	—	—
Aunt:nieces	14	0*	7.3	8	5.0	0*	—	0
χ^2	45.6		2.7		28.3			
p	<0.001		>0.2		<0.001			

NOTE: Although there is no test of significance for individual rows, those values that contributed markedly to the total χ^2 are marked with an asterisk. O = observed interactions, E = expected interactions, — indicates the kinship group was not possible. In some instances, aunt:nieces also included grandnieces.

1974 (Fig. 14.2). At Upper Picnic, ♀843 represented the sole survivor of a matriline initiated at Lower Picnic in 1967; ♀1082 was descended from a matriline initiated at North Picnic, 300 meters distant. These females illustrate that matrilines may persist across space. Both animals died by 1980; thus the matrilines became extinct, the fate of most matrilines (Armitage, 1984). At Lower Picnic in 1974, the home ranges of the kin group overlapped (Fig. 14.2), but matrilineal fission was beginning. The peaks of major activity of ♀1177 and ♀1194 overlapped directly and were displaced from those of their grandmother and aunt.

Social Behavior

Social behavior is strongly related to matrilineal structure. Both amicable and agonistic behavior occur nonrandomly (Table 14.1; Armitage and Johns, 1982). Mother:daughter and grandmother:granddaughter amicable behavior occurs at a frequency predicted from the frequency of these kin pairs in the population. Amicable behavior between sisters occurs more frequently then predicted, and between aunts and nieces, much less frequently than predicted (Table 14.1). By contrast, agonistic behavior between sisters, mother and daughters, and grandmother and granddaughters occurs less than or about as expected whereas agonistic behavior between aunts and nieces occurs more frequently than expected.

Social and Spatial Dynamics of Matrilines

Sister pairs in 1974, 1977, 1978, and 1981 behaved amicably (Table 14.1) and had similar patterns of space use (Fig. 14.2). Social behavior of aunts and nieces was either more agonistic or less amicable than expected (Table 14.1). Home ranges of aunts and nieces (or grandnieces) either overlapped slightly (e.g., 976 vs. 1177, 1194 in 1974) or not at all (e.g., 301, 349 vs. 1194, 920 in 1978; 301 vs. 503, 489, 1194 in 1981). Both home range patterns and social interactions must be considered because social interactions between individuals who have lived in the same colony for more than one year typically decline over the period of joint residency (Armitage, 1977). For example, agonistic behavior was frequent between females 1194 and 301 in 1977, but never observed in 1981.

The shifting home range patterns suggest that a matriline attempts to exclude members of other matrilines from resources; e.g., preferred burrows, foraging areas, and rocks used for sitting or lying. In 1977, ♀1194 was dominant to either ♀301 or ♀349 alone, but when both 301 and 349 were present simultaneously ♀1194 retreated. As a result, ♀1194 moved downslope away from a primary burrow site which 301 and 349 used and extended her foraging area to the northwest (Fig. 14.2). In 1978, ♀s 301 and 349 occupied the prime talus and foraging areas of Lower Picnic, ♀1194 and her two-year-old daughter 920 maintained residency, but avoided space used by the other matriline. By 1981, ♀s 349 and 920 had died, but ♀1194 had two new two-year-old daughters. The matriline of three now occupied the prime resource area and 301 was displaced to the southeast edge of the habitat (Fig. 14.2). The space-use patterns of 1981 suggest that an animal in a subordinate position may avoid contact with dominant animals of a different matriline. This avoidance pattern may be a consequence of the behavioral interactions in previous years. For example, 301 was alone in 1979. She was subordinate in all encounters with 1194 and 920. Her space-use pattern of 1981 resembled that of 1979 (Armitage, 1984), despite no observed agonistic behavior between ♀301 and ♀1194 and her daughters. These results suggest that a major advantage of a matriline of two or more females is that resources are more readily obtained and defended and that the presence of several matrilineal associates may deter potential competitors.

Why does not one matriline with clear advantage drive out all members of another matriline? Two possible reasons come to mind. One, they may not be capable. Two, they may have garnered the necessary resources for

TABLE 14.2
The number of amicable behaviors between adult female and
juvenile yellow-bellied marmots

	Marmot meadow					Picnic	
	1978	1979	1980	1981		1974	1976
♀911	24*	1	6*	15*	♀1177	4	20*
♀918	18	6*	9*	20*	♀1194	8*	54*
♀179	—	—	3*	0*			
χ^2	0.8	3.6	3.0	18.5		1.3	17.0
p	>0.3	>0.05	>0.2	<0.001		>0.2	<0.001

NOTE: Females with litters are indicated with an asterisk. Because females and juveniles shared the same burrow system, it was assumed that social interactions occurred with equal probability between any adult female and any juvenile. A dash indicates an animal was not present. Sister pairs are 1177–1194 and 911–918; ♀179 is the daughter of ♀911.

growth and reproduction and the costs of attempting to gain more space may exceed any potential benefits. The benefits would come in future years when new offspring are produced to move into and occupy the space. I never observed any group of females cooperatively seek out and attempt to drive out competitors.

Cooperative Breeding

"Cooperative" and "communal" are applied to breeding units in which some members in a social unit behave as parents but are not the genetic parents of the juveniles they aid (J. L. Brown, 1978). W. D. Hamilton (1964) used "cooperative" for those situations in which the direct fitness of both actor and recipient is increased. In the discussion that follows, I will use "communal" in the sense of Brown and "cooperative" only in the sense of Hamilton. However, I will describe behaviors that appear to be cooperative, but measurements of fitness benefits are lacking.

Among rodents, only the naked mole rat breeds communally (Jarvis, 1981). Among the Marmotini, black-tailed prairie dogs breed cooperatively in the sense that members of the colony share territorial defense (Hoogland, 1981). Among yellow-bellied marmots, females may live in the same burrow system. If more than one female has a litter, the juveniles at weaning intermingle such that maternity cannot be assigned (Fig. 14.1; Armitage, 1984). Although some females forgo breeding, there is no evidence that these females act as helpers (Brown, 1978); thus, communal breeding is not known to occur. Cooperative breeding occurs because all females assist in the detection of predators and in the defense of the matrilinal space against conspecific intruders. If cohesive behaviors with juveniles are considered postweaning parental care, females share such care.

TABLE 14.3
The number of occurrences of amicable and agonistic behavior
between adult females at Marmot Meadow

	1978	1979	1980	1981	
	♀918:♀911*	♀911:♀918*	♀911*:♀918*	♀911*:♀918*	♀179*:♀918*
Amicable	14	1	7	4	0
Agonistic	1	6	8	5	18
χ^2	11.2	3.6	0.06	0.1	18.0
p	<0.001	$0.1 > p > 0.05$	>0.8	>0.7	<0.001

NOTE: Expected values were calculated on the basis that a female had equal probability of behaving amicably or agonistically in any encounter. Reproductive females are indicated with an asterisk. 911 and 918 are sisters; 179 is the daughter of 911 and the niece of 918.

At both Marmot Meadow and Picnic, females sharing burrow systems usually interacted with all juveniles even if the female was not a mother (Table 14.2). Only amicable behaviors were observed between juveniles and adults. In two instances, the frequency of amicable behaviors of individual females was nonrandom; in 1981, ♀179 at Marmot Meadow was not observed to interact with any juvenile and ♀1177 at Picnic interacted with juveniles less frequently than expected. Social interactions were tested to determine if an adult female demonstrated preferential treatment to any young when litters intermingled. Generally no such preference was detected; ♀1177, $\chi^2 = 5.9, p > 0.3$; ♀1194, $\chi^2 = 6.6, p > 0.3$; ♀301 (Picnic, 1978), $\chi^2 = 9.9, p > 0.2$, ♀911 (1981), $\chi^2 = 21.7, p > 0.2$; ♀918 (1981), $\chi^2 = 35.1, p < 0.01$. Female 918 interacted three times more often with one of nineteen young than with any other. Except for this one instance, there is no evidence that adult-female juvenile amicable behavior (mostly greetings) is biased toward the mother's own offspring.

Although adult-juvenile postweaning, above-ground social behaviors suggest cooperative breeding, other behaviors indicate that adult females attempt to maximize direct fitness (J. L. Brown, 1980) and maximize indirect fitness as a second choice. In 1978, reproductive ♀911 interacted amicably with her nonreproductive sister ♀918 (Table 14.3). However, when ♀918 reproduced in 1979, she treated nonreproductive ♀911 agonistically. Social interactions were fewer because ♀911 extended her home range to avoid contact with ♀918 and occupied a different burrow system for part of the summer (Frase and Armitage, 1984). In 1980 and 1981, when both females had litters, social interactions were few even though they lived in the same burrow system, had similar home ranges, and often fed within a few meters of one another (Frase and Armitage, 1984). By contrast, ♀179 occupied a different burrow system and utilized a foraging area generally distinct from that of her mother and aunt (Frase and Armitage, 1984). Each year she

Contingency table analyses of the number of occurrences of
amicable and agonistic behavior between adult female and
yearling yellow-bellied marmots at Marmot Meadow

	1978		1979	
	Amicable	Agonistic	Amicable	Agonistic
♀911	0	42	8	1
♀918	5	5	4	28
χ^2	23.2		19.8	
p	< 0.001		< 0.001	

NOTE: Female 911 was reproductive in 1978 and ♀918. in 1979.

moved with her litter just prior to weaning to the burrow system occupied
by her kin. Although when ♀918 was reproductive her behavior toward the
other adults was agonistic, she was unable to exclude either of them. Thus,
her sharing of resources and cooperative breeding may be interpreted as obtaining the best fitness possible under the circumstances.

The behavior of adult females toward yearlings is affected by kinship
(Armitage and Johns, 1982) and reproductive status. The patterns suggest
females attempt to maximize the direct component of inclusive fitness
(Brown, 1980). When reproductive, females 911 and 918 behaved agonistically toward yearlings (Table 14.4). In 1978, the yearlings were maternal
sibs of the females, but a year younger. In 1979, ♀918 was agonistic to her
niece and nephews, but 911 was amicable to her yearling offspring (Table
14.4). The daughter (♀179) of ♀911 was recruited into the population. In
1981, when the yearlings were descended from all the adult females, there
was no difference in the frequency of amicable or agonistic behavior between the adult females and the yearlings ($\chi^2 = 0.3, p > 0.5$). Because all
adult females were reproductive in 1981, reproductive status is insufficient
to explain behavioral patterns. Presumably the females could not distinguish among the yearlings and may have opted under those conditions to
maximize inclusive fitness. Thus, six of the thirteen yearlings were recruits.

Individual variability characterizes marmot behavior and complicates the
interpretation of behavioral patterns (Svendsen and Armitage, 1973; Svendsen, 1974; Armitage, 1983). When amicable versus agonistic behavior is
summed over all years of pregnancy, ♀179 was generally noninteractive,
even with juveniles (Table 14.2); ♀911 was amicable ($\chi^2 = 3.9, p < 0.05$);
and ♀918 was so agonistic ($\chi^2 = 12.7, p < 0.001$) that she contributed to
the significant level of agonistic behavior ($\chi^2 = 7.0, p < 0.01$) directed by
the adults toward her yearlings in 1980. At Picnic in 1976, the behavior between sisters 1177 and 1194 was entirely amicable, but ♀1177 interacted

TABLE 14.5

Frequency analysis of amicable and agonistic interactions
between the adult females and the yearlings
at Marmot Meadow in 1981

	Amicable		Agonistic	
	E	O	E	O
♀179	15	3*	13.3	3*
♀911	15	18	13.3	24*
♀918	15	24	13.3	13
χ^2	15.6		16.5	
p	<0.001		<0.001	

NOTE: All females were reproductive. O = observed interactions, E = expected interactions. Individual row values that are probably biologically significant are marked with an asterisk.

significantly less with juveniles than ♀1194 (Table 14.2). Adult females differed in their frequency of social interactions with yearlings. Female 179 had fewer amicable and agonistic interactions than predicted and ♀911 was more agonistic than expected (Table 14.5). These individual differences may be part of a strategy that emphasizes phenotypic plasticity in a variable and unpredictable environment (Armitage, 1983). Briefly, both the social, physical, and other biological environments of marmots vary across space and time. Any particular behavioral pattern may not be adaptive to all sets of environmental conditions, but a diversity of behavioral characteristics among an animals's descendants may produce a range of individuals adaptive to a range of environmental conditions.

Chance, such as unpredictable mass mortality, may significantly determine the outcome of female competition. By early August of 1981 the population at Marmot Meadow consisted of one adult male, three adult females, four yearling females, two yearling males, twelve juvenile males, and seven juvenile females. The three adult females moved off and were not seen thereafter. In 1982, only two of the female yearlings and one adult female were recovered. Similar mass disappearances occurred previously at Marmot Meadow and at two other colonies. Mortality is the suspected cause of the disappearance because extensive searches over wide areas failed to find any of the missing animals. Causes of the presumed mortality are unknown.

In summary, both cooperative and competitive elements occur in the social behavior of closely related yellow-bellied marmots. These results suggest that the recent controversey of whether black-tailed prairie dogs are cooperative breeders (Hoogland, 1983; Michener and Murie, 1983) was misdirected. Both yellow-bellied marmots and black-tailed prairie dogs are

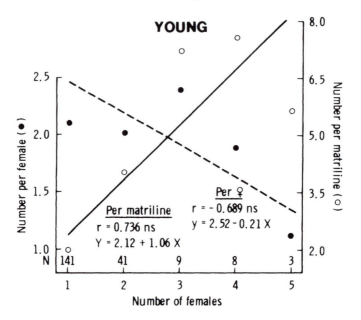

Fig. 14.3. The relationship between the production of young and the size of a matriline.

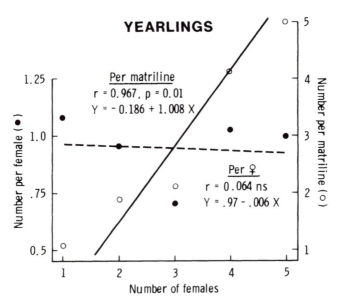

Fig. 14.4. The relationship between the production of year-lings and the size of a matriline.

social and both have cooperative and competitive components in their social systems (Hoogland, 1981, 1983; Armitage and Johns, 1982). What is important is not debating to what degree a species may be categorized as a communal breeder (e.g., Michener and Murie, 1983), but determining how animals try to maximize direct fitness. We expect that individuals should opt to maximize direct fitness because investing in the kin of other individuals may not be reciprocated (Rubenstein and Wrangham, 1980).

Reproductive Success

One possible consequence of living in groups is that direct fitness, measured as the per capita reproductive output of females, may decrease as group size increases. Such potential loss was reported for yellow-bellied marmots (Downhower and Armitage, 1971), black-tailed prairie dogs (Hoogland, 1981), and red deer (Clutton-Brock et al., 1982, p. 92).

The per capita production of young decreased and the total production of young increased with increased size of matrilines, but the relationships were not statistically significant (Fig. 14.3). There also is virtually no relationship between the per capita production of yearlings and size of matrilines (Fig. 14.4). Thus, for each additional adult female added to the matriline, one additional yearling is produced. If yearling output is used as an index of fitness, females lose no direct fitness (per capita production) by being members of matrilines and gain in indirect fitness (total output of matriline). However, inclusive fitness could decline if descendant kin do less well than collateral kin in subsequent years.

The analysis of matrilineal size and reproductive success lumped data from all colonies and years. Data are insufficient to do a similar analysis within years or within colonies. Different habitat quality might affect the matrilineal size:per capita production relationship. Large matrilines would be expected in better-quality habitat. The possible effects of habitat (= colony) differences were examined by testing for variation in the reproductive output of individual females per year and in the reproductive output per female per matriline per year. The first analysis examines the lifetime reproductive output of individual females, expressed as the average production per year, in order to account for differences in length of residency. A nested ANOVA tested for differences among matrilines. The second analysis ignores individual females and examines the yearly per capita production of matrilines. For both analyses, one objective is to partition the variance among groups (colonies), subgroups (matrilines), and within groups (individuals within matrilines or between years).

The production of young per female per year differed significantly among colonies ($F_{5,44} = 5.3$, $p < 0.001$), but not among matrilines ($F_{44,250} = 0.6$, $p > 0.75$). However, the differences among colonies explain only 5.3 percent of the variance; 94.7 percent of the variance occurs among individual females. When only those females that produced young are considered, the production of yearlings differed significantly among colonies ($F_{5,30} = 168.8$, $p < 0.001$), but not among matrilines ($F_{30,105} = 0.1$, $p > 0.75$). Again, all the variance is explained by variation among colonies (39.5%) or among individual females (60.5%). There is no relationship between the rank order of the size of the area the colony inhabits and the mean number of yearlings per female ($r_s = 0.03$, $n = 6$, $p >> 0.05$). Because our studies have not detected differences in habitat quality (Svendsen, 1974; Kilgore and Armitage, 1978), the differences among colonies may reflect sampling differences and differences in female residents. For example, Cliff Colony, with a mean significantly higher than all other means, was sampled only from 1976 to 1982 (Fig. 14.1). Unusually good success during that period could have biased the mean upward.

The yearly per capita production of young by matrilines does not differ among colonies ($F_{5,30} = 1.9$, $p > 0.1$) nor among matrilines ($F_{30,146} = 0.7$, $p > 0.75$). Only 1.4 percent of the variance is explained by differences among colonies; 98.6 percent of the variance is attributable to differences among years. The yearly per capita production of yearlings by matrilines does differ among colonies ($F_{5,30} = 3.8$, $p < 0.01$), but not among matrilines ($F_{30,145} = 0.8$, $p > 0.75$). Only 6.6 percent of the variance is explained by differences among colonies; 93.4 percent of the variance, by differences among years. In this analysis, the differences among years is best interpreted as differences among females. There are no significant differences in mean litter size associated with size of matrilines ($F_{4,140} = 1.8$, $p > 0.1$) or with colonies ($F_{5,139} = 2.2$, $p > 0.05$). However, the mean number of litters per female is 0.48. This value is not affected by the size of a matriline ($F_{4,199} = 0.2$, $p > 0.75$). Thus, on average, any female has a likelihood of producing a litter every other year. If she produces a litter, it is likely to consist of 4.2 juveniles. Some females produce a litter nearly every year of residency; others may skip several years in succession (e.g., Fig. 14.1). The variation in reproduction from year to year by individual females accounts for the variation among years in the reproductive output of matrilines. I conclude that the major source of variation in reproductive output is unequal reproductive success among females within matrilines. The lack of a significant effect of matrilineal size on the production of young and yearlings cannot be attributed to differences among colonies.

MALE REPRODUCTIVE STRATEGY

The production of young per female decreases significantly but the number of young per male increases significantly as harem size (= number of females two years old or older) increases (Fig. 14.5). Similarly, the number of yearlings per female decreases but the number of yearlings per male in-

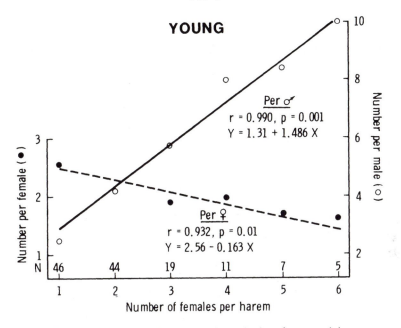

Fig. 14.5. The relationship between the production of young and the size of a harem.

creases with increased harem size (Fig. 14.6). This disadvantage to females in large harems arises because the number of litters per female decreases (Fig. 14.7). Harem size does not affect litter size ($F_{5,136} = 0.6, p > 0.5$). However, the number of litters per male increases significantly as harem size increases. Therefore, a male increases his direct fitness by mating with as many females as possible. A male should never choose to be monogamous, but monogamy is better than not breeding. On average, an adult, territorial male has a harem for 2.24 years. If monogamous, he would produce (2.24 years × 2.54 young/year) 5.7 young. However, if bigamous, he would produce (2.24 × 4.34) 9.7 young. The average harem size is 2.27 adult females. Thus, the average male can expect a lifetime production of (2.24 × 2.27 × 2.19) 11.1 young. Monogamy cannot be a male evolutionarily stable strategy (Maynard Smith and Price, 1973) provided that resources, and thus females, are clumped and that either the resources, the females, or both can be defended (Emlen and Oring, 1977).

The loss in annual reproductive output of females living in large harems could be compensated by increased survival. Additional years of reproduction may enable longer-lived females to have a lifetime reproductive output equal to or greater than those females living in small harems (P. F. Elliott,

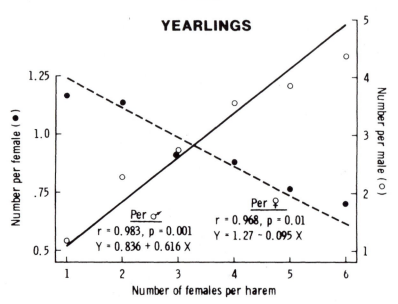

Fig. 14.6. The relationship between the production of yearlings and harem size.

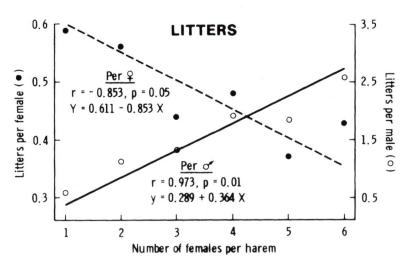

Fig. 14.7. The relationship between the number of litters per resident female and harem size.

1975; Wittenberger, 1979, p. 288). The survivorship of fifty-two known-aged females was compared to the mean harem size in which each female lived. There was no relationship ($r = -0.18, p > 0.1; Y = 4.7 - 0.34X$, where Y = survivorship and X = mean harem size in which each female lived). Similarly, lifetime production of young and yearlings was unrelated to the mean harem size in which each female lived (young: $r = -0.21, p > 0.1; Y = 11.2 - 1.16X$, where Y = number of young; yearlings: $r = -0.19, p > 0.1; Y = 5.6 - 0.61X$, where Y = the number of yearlings). These results are consistent with the previous analysis of lifetime reproductive output in matrilines in which most of the variation occurred among individual females.

COMPARISON OF MATRILINES AND HAREMS

The different effects of increased size of matrilines and harems on per capita female reproductive success suggest that reproductive costs are not a function of groups per se but of the kind of group. Harems may consist of one or more matrilines. Therefore, the reduced per capita reproductive success of females in harems > 1.0 may result from competition between matrilines. In a given location, the density of females in a matriline may be reduced to one for one or more years. Although the mechanism by which competition acts is unknown, evidence for a competitive effect is apparent from an analysis of reproductive success of females living alone in matrilines when harems of different sizes consist of more than one matriline.

Harem size affects the mean number of young produced by females ($F_{3,172} = 2.4, 0.1 > p > 0.05$). A female living singly in a harem > 2.0 produced fewer young than females living singly in a harem consisting of one or two females or than two females associating in a matriline (Table 14.6). The production of yearlings is affected similarly ($F_{3,172} = 2.7, p < 0.05$, Table 14.6). The lower reproductive success of females who are the sole members of matrilines in harems > 2.0 occurs because they produce fewer litters ($F_{3,172} = 2.2, 0.1 > p > 0.05$). The number of young ($F_{3,87} = 1.3, p > 0.25$) or yearlings ($F_{3,87} = 1.5, p > 0.25$) per litter does not differ among the four groups of females. These data suggest that the presence of two or more females in a matriline adjoining a female living singly inhibits her ability to reproduce; but if she reproduces, she is as successful in rearing young to one year of age as females living in more favorable social situations. Although these group effects are biologically significant, they account for only about 3 percent of the variance in the reproductive success of females living singly. Thus 97 percent of the variance is attributable to variation among females.

Harem size affects the number of recruits retained in the natal area ($F_{3,172} = 3.9, p < 0.025$). Significantly more yearling females are recruited into

TABLE 14.6
Reproductive success of females of matrilines in harems of different sizes

	Social groups			
	Only 1 female present, both matriline and harem = 1.0	2 females present, each is a matriline of 1.0; harem = 2.0	1 female in a matriline of 1.0; other matriline(s) total 2 or more females; harem > 2.0	2 females in a matriline; other matriline of 1 or more females; harem ≥ 3.0
Young	2.54	2.38	1.48	2.71
Yearlings	1.17	1.36	0.56*	0.71
Litters	0.54	0.61	0.37	0.57
Recruits	0.13	0.58*	0.14	0.29
Young/litter	4.20	3.91	4.00	4.75
Yearlings/litter	2.3	2.20	1.50	1.30

NOTE: All values are means per female. Means that differ statistically from other means in each row are marked with an asterisk. Means whose values are of probable biological significance are underlined.

matrilines when harem size is 2.0 (Table 14.6). This group effect explains 10 percent of the variance; the remaining 90 percent of the variance stems from individual variation among females. Two females associating in a matriline when harem size ≥ 3.0 produce more young, more young per litter, and as many litters as females living alone, and produce more recruits than females living in harems of 1.0 or harems > 2.0. Apparently there are reproductive benefits in matrilineal living beyond those obtained from suppressing reproduction in nonmatrilineal competitors. However, recruitment is lower than in harems of 2.0; why recruitment is more successful when harem size is two requires further analysis.

The causes of variation in reproductive success are largely unknown, but at least some variation is associated with differences in behavioral phenotypes (Armitage, 1983). Two lines of evidence suggest that social repression of reproduction occurs. Although two-year-old females may reproduce, only 32 percent of seventy-four did so. Because half of the adult females produce a litter, on average, I assumed that two-year-old females should have a 50 percent chance of reproducing. The presence or absence of adults significantly affected the reproductive success of two-year-olds. When adult females were present, reproduction was significantly reduced (fourteen of forty-eight two-year-olds produced litters; $\chi^2 = 8.3, p < 0.01$). If the two-year-old's mother was present, only eleven of thirty-six two-year-olds reproduced ($\chi^2 = 13.3, p < 0.001$). None of the seven two-year-olds reproduced when more distant kin were present ($\chi^2 = 7.0, p < 0.001$) but three of five reproduced when only unrelated adults in different matri-

TABLE 14.7

Contingency table analysis of the effects of living solitarily
or in proximity to other adult females on weaning success of
female yellow-bellied marmots

	Solitary	In proximity to others
Litter weaned	14	20
No litter weaned	2	20

NOTE: Subjects were females aged three or older who were recorded as pregnant. The "proximity" group included females of the same matriline or of adjacent matrilines when the home ranges of the females were contiguous. $\chi^2 = 6.7$, $p < 0.01$

lines were present. When no adults were present, ten of twenty-six two-year-olds reproduced. This proportion does not differ significantly from the expected 50 percent reproductive ($\chi^2 = 1.38$, $p > 0.1$). I identified from trapping records fifty-six females three-years-old or older who were recorded as pregnant or lactating. Whether females lived solitarily or in proximity to other adult females significantly affected their success at weaning a litter of juveniles (Table 14.7).

Females may reduce the reproductive success of other females if they can prevent their access to resources, especially food. Competition for resources would be most effective during early pregnancy when much of the area may be snow-covered, food resources are scarce, and females may be forced to utilize fat reserves for maintenance rather than for reproduction. Snow cover varies from year to year (Armitage and Downhower, 1974); perhaps only in those years of late snow cover does competition for food affect reproductive success. At our high altitude study site in North Pole Basin, females with inadequate foraging areas during gestation and lactation either failed to produce young or produced litters of lower than average size (D. C. Andersen et al., 1976). These females did not encroach on foraging areas of other females, which suggests that female competition, either active or potential, affects reproductive success of nearby females who are likely members of the same harem and may be kin. In ten of the twenty instances in which females failed to wean litters when they were presumed to be pregnant and living in proximity to other females, the other females were kin. In seven of the ten, all females were members of the same matriline.

DISCUSSION

Mate Choice

There is no evidence that females choose among males. Females are far more philopatric than males; sixty-three of ninety-one resident females were born in their colony of residency; only two of fifty-two males became resident in their colony of birth. The mean lengths of residency of males (2.24 years), females (3.26 years), and matrilines (4.35 years) differ significantly from one another ($F_{2,15} = 16.1$, $p < 0.001$). Therefore, most females during their residency can expect to mate with two males. Males disperse and settle, when possible, with resident females. Males may live peripherally to a population of females, but are prevented from residing with the females by the territorial defense of the resident male (Armitage, 1974). The strategy of living peripherally and waiting for an opportunity to become colonial is often successful; thirty of fifty-two colonial males were known to live peripherally for one or more years before becoming colonial. Social interactions between resident females and a strange male initially occur at a high level and then decrease (Armitage, 1974). Males always establish dominance; we have no evidence that females can exclude a potential colonial male. A female could refuse to mate, but she would pay a high direct fitness cost in that, on average, she would lose nearly one-third of her breeding potential. It is not practical for a female to travel to find a different male because she would be forced to travel across snow and be highly vulnerable to predation and exposure. In those instances when no male was present with a group of females, females did not emigrate and failed to breed. Thus, the lengths of residency, pattern of philopatry, and social behavior all support the interpretation that males seek out and attempt to associate with already resident females.

The Yellow-Bellied Marmot and the Polygyny Threshold Model

The polygyny threshold model does not apply to marmots. The fundamental assumption underlying the model is stated in the following quotation (Wittenberger, 1979, p. 288): "In territorial systems, polygyny can evolve only when already mated males attract additional females to their territories, so its evolution is clearly dependent on the choices made by unmated females." Marmots do not meet that requirement. Rather than females being attracted to males, males are attracted to females. Whether a male is monogamous or polygynous depends on the number of females residing in a habitat patch, not on the ability of the male to attract females. Furthermore, the organization of female yellow-bellied marmots into matrilines does not meet the implicit assumption that females act independently. The density of females on a habitat patch is determined by females (Armitage, 1975); I have no evidence that colonial males affect the density of resident females. In effect, females, by regulating the number of resident females in a habitat patch, determine the reproductive success of colonial males. The strategy available to a resident male is to increase his territory size to include additional females. In some habitats, males do expand territories when the opportunity arises (Armitage, 1974).

The question then arises, should a male remain in a habitat patch with only one female? We have little direct evidence that bears on this question, but several lines of evidence suggest moving is risky. First, a male must be able to add sufficient fat and locate a hibernaculum in order to survive until the next breeding season. If he emigrates, he is leaving a habitat already proven successful and with no assurance that he will find the necessary resources. Second, if he finds the resources, he may also find a resident male. More importantly, he needs to locate females. If he finds females, most likely a resident male will be present. Then our questing male must risk conflict in order to drive out the resident, or live as a peripheral and forgo reproduction until he can replace the resident. We have never observed a resident male lose an encounter with a potential usurper. Therefore, our male most likely will forgo reproduction and would likely be less fit than if he had remained in his original habitat. We do not know to what degree males move about seeking females. We currently are investigating this problem by radio-tagging males and tracking their movements.

Marmots were considered a mammalian example of resource defense polygyny (Emlen and Oring, 1977). However, Greenwood (1980) suggested that because of male-biased dispersal in mammals, males defend mates. This mate defense is similar to the harem-defense polygyny of Emlen and Oring. One source of confusion is whether females congregate to resources which males defend or whether males attach themselves to females. Clearly males attach themselves to females. I conclude that male territoriality in yellow-bellied marmots evolved its present form in response to the patchy distribution of females and the need to maintain residency with the females as the mechanism with the highest probability of producing future matings.

THE MATING SYSTEM OF GROUND SQUIRRELS

Mating systems may be viewed as attributes of grades of sociality (Michener, 1983). This proximate description of social structure has an underlying assumption that the various grades of sociality represent an evolutionary sequence; sociality is an end-point of evolutionary mechanisms rather than a means for increasing fitness. By classifying some societies as egalitarian, the intense female competition and the conflict between the reproductive interests of males and females is devalued. Finally, grades of sociality imply that males and females form a single social unit whereas it is more likely that social systems should be treated as two separate units; a male system, focusing on mate acquisition, and a female system, focusing on kinship (Armitage, 1984). Mating systems should be viewed as one of the life-history tactics comprising a species' reproductive strategy (Armitage, 1981). In this view, sociality is a consequence of the convergence of individual life-history strategies, which leads to the formation of a social unit characterized

TABLE 14.8

Selected characteristics of the mating systems of the Marmotini

	Breeding sex ratio as no. ♀♀/♂	% of adult ♀♀ breeding	Dispersing sex	Type of kin group	Infanticide committed by	Sex expressing territoriality, or IA	Kin-biased behavior	References
Spermophilus *tridecemlineatus*	2.9–3.3	>95	♂	C		Breeding adults IA	Alarm-calling	McCarley, 1966, 1970; Rongstad, 1965; Schwagmeyer, 1980; Schwagmeyer and Brown, 1983; Wistrand, 1974
beldingi	1.3–3.0	>95	♂	C	Unrelated ♂♂ or ♀♀	Parous ♀. Breeding ♂ IA. "resource-based lek"	Avoid fighting with close kin, close relative codefended, alarm-calling	Hanken and Sherman, 1981; Morton and Parmer, 1975; Sherman, 1977, 1980, 1981; Sherman and Morton, 1984
tereticaudus	2.6	>90	♂	C		Parous ♀	Reduced aggressiveness toward ♀ kin, alarm-calling	Dunford, 1977a, b, c
richardsonii	2.6	>95	♂	C	—	Parous ♀ Breeding ♂	Mother: Young, fewer chase-flee and more appeasement behaviors among uterine kin, alarm-calling	L. S. Davis, 1983, 1984; Michener, 1973, 1979a, b, 1980; Michener and Michener, 1977; Yeaton, 1972
armatus	3.3	>90	♂	CP	—	Parous ♀		Balph and Stokes, 1963; Slade and Balph, 1974
beecheyi	1.2–1.9	>90	♂	CP		Parous ♀ IA Breeding ♂		F. C. Evans and Holdenreid, 1943; Dobson, 1979, 1983; Fitch, 1948; Owings et al., 1977
parryii	2.0	>95	♂	C	Adult ♂♂	Breeding and prehibernation ♂	More amicable among ♀♀	Carl, 1971; McLean, 1982, 1983

			♂	C	Adult ♀♀	Breeding ♂ Parous ♀	Behavior among ♀♀	
columbianus	1.27	65–84	♂					Balfour, 1983; Boag and Murie, 1981; Festa-Bianchet, 1981; Festa-Bianchet and Boag, 1982; Festa-Bianchet and King, 1984; Murie and Harris, 1978, 1982
Cynomys leucurus	2.2	88	♂	CP	—	Parous ♀	Litters of "family groups" intermingle	Clark, 1977; Stockard, 1929
gunnisoni	1.6	52–80	♂	C, MP	—	Parous ♀ Harem ♂	Among yearlings	Fitzgerald and Lechleitner, 1974; Longhurst, 1944; Rayor, pers. comm.
ludovicianus	1.5	66	♂	M	Related ♀♀	Parous ♀ Harem ♂	Amicable behaviors, space sharing and defense	J. A. King, 1955; Hoogland, 1979, pers. comm.
Marmota monax	1.0–2.0	>95	♂?	CP	—	Parous ♂?	Alarm-calling Mother: Young, possible with ♀ yearling	Grizzell, 1955; Snyder, 1962; Svendsen, pers. comm.
flaviventris	3.1	45–50	♂	M	Adult ♀♀	Parous ♀ IA Harem ♂	Amicable behaviors, defense of shared space	Armitage, 1965, 1974, 1975, 1984; Armitage and Downhower, 1974; Armitage and Johns, 1982; Armitage et al., 1979; Brody and Melcher, pers. comm.; S. E. Thompson, 1979
caligata	1.0–1.8	40–50	♂?	MP	—	Parous ♀ Harem ♂	Probable	Barash, 1974, 1980, 1981; Holmes, 1984
olympus	1.6	40–50	♂	MP	—	Parous ♀ Harem ♂	Probable	Barash, 1973

KEY: C = kin clusters, M = matrilineal group, P = probable, IA = increased aggressiveness. Kin clusters occur when home ranges of closely related females are contiguous; M species form social groups.

327

by sharing resources (at least to some degree) and kin-biased social behavior. In effect, the life-history tactic model emphasizes looking at mating systems as the result of individuals' attempting to maximize inclusive fitness, especially direct fitness.

In the well-studied Marmotini, dispersal is male-biased (Holekamp, 1984); breeding sex-ratios are female-biased (Table 14.8). Philopatric females form kin clusters. In *M. flaviventris* and *C. ludovicianus*, the female kin-groups persist over time as closed matriarchal societies in the same area. In these species, and probably in *C. gunnisoni*, *M. olympus*, and *M. caligata*, the female group cooperates to exclude conspecifics and shares in kin-based behaviors, such as alarm-calling and amicable behavior (Table 14.8). However, kin-based behavior is not limited to matriarchal groups. Alarm-calling is kin-directed in *S. tridecemlineatus*, *S. beldingi*, *S. richardsonii*, and *S. tereticaudus*. Other kin-directed behavior may be characterized as reduced fighting and/or increased amicable behavior among closely related females (Table 14.8). Adult female *S. columbianus* may relinquish nest-sites to their yearling daughters (M. A. Harris and Murie, 1984).

Males and females of spermophiles are either territorial or are more aggressive during breeding (Dobson, 1983, 1984). Male territories or increased agonistic behavior by the spermophiles (Table 14.8) and *C. leucurus* usually are concentrated on a core area around their burrows. Male *S. beldingi* defend small mating territories on snow-free ridges near females' hibernacula (Sherman and Morton, 1984). Because these territories usually do not include the nest burrows of females, the function of territorial behavior is unclear, but probably assures the male of residency in the breeding population. Males unable to maintain territories are forced to live peripherally, usually in poorer habitat (e.g., Armitage, 1974; Carl, 1971; Morton and Parmer, 1975). Male *S. tridecemlineatus* wander widely during the breeding season and attempt to mate with estrous females in a scramblelike competition (Schwagmeyer and Brown, 1983). Males may guard females following copulations (e.g., *S. beecheyi*, Dobson, 1983). Male territories may overlap the core areas of several females with whom the males may have a higher probability of mating (e.g., *S. beecheyi*, Owings et al., 1977; *S. columbianus*, Murie and Harris, 1978). Where females are clumped on patchy habitat, territoriality is well developed, (e.g., *S. parryii*, Carl, 1971; McLean, 1983; *M. olympus*, Barash, 1973; *M. caligata*, Barash, 1981; and *M. flaviventris*, Armitage, 1974). In the prairie dogs *C. ludovicianus* and *C. gunnisoni*, clumped females apparently occur as kinship groups in an otherwise continuous population. The female groups form as a consequence of postweaning reproductive investment in which juveniles are provided space, food, and a social environment in which to mature (Armitage, 1981). These female groups are defended by adult males. In general, male territoriality prolonged beyond the breeding season occurs in those species in which the females are clumped and, hence, are defendable. Territoriality in these species is best interpreted as defense of females, i.e., mate guarding (Barash, 1981).

Female *S. beldingi* (Hanken and Sherman, 1981) and *S. tridecemlineatus* (Schwagmeyer and Brown, 1983), species in which males do not defend females, may mate with two or more males. In the territorial *M. flaviventris* (O. A. Schwartz and Armitage, 1980) and *C. ludovicianus* (Foltz and Hoogland, 1981), genetic analysis indicated that the territorial male most likely fathered all offspring of females in his harem. Thus, territoriality may also assure paternity and, in addition, may protect young from infanticide by other males (McLean, 1983).

Female defense of burrows or increased aggressiveness during pregnancy and lactation apparently is universal among the Marmotini (Table 14.8). Defense seems not to be related to resources, i.e., food or burrows, but to young (e.g., Festa-Bianchet and Boag, 1982). Although protection against predators likely is partially responsible for female defensiveness, the major selective factor may be defense against infanticide. Infanticide occurs in several species (Table 14.8) and indirect evidence suggests infanticide or the potential for infanticide may be widespread among the Marmotini (Sherman, 1981, 1982).

Mating systems of the Marmotini are consistent with the model that females are philopatric, that philopatry leads to the close association of female kin, and that males attach to female groups. Polygyny occurs when males achieve differential access to females either by scramble competition or by contest competition in which clumped females (harem polygyny) are defended. Although most marmot species are harem polygynous, polygyny is not obligate. Slightly more than one-third of yellow-bellied marmot mating associations were monogamous (Fig. 14.5). One of twelve woodchuck mating groups was monogamous (Svendsen, pers. comm.) and 19 percent of the social groups of the Olympic marmot were monogamous (Barash, 1973). In some populations, all hoary marmots are monogamous (Holmes, 1984). Holmes concluded that resources would not support more than one female at a site with a hibernaculum and that hibernacula were too far apart for a male to control more than one. Hoary marmots support the interpretation that harem polygyny occurs when resources permit the clumping of females.

Problems and Perspectives

A major problem is to determine what limits the size of a matriline. Although resources doubtlessly are the ultimate limiting factor, they may not be the proximal factor determining reproductive strategies. In most of our study areas hibernacula and other burrows and food appear to be capable of supporting more females than are typically present. Females may share burrows; when only a few burrows are available, more females could be resident than usually are. Resources may be critical in years when spring begins later than usual. Reduced breeding occurs in years of late spring in *S. beldingi* (Morton and Sherman, 1978), *S. columbianus* (Murie and Harris, 1982), *S. lateralis* (Bronson, 1979), and *M. flaviventris* (Nee, 1969; Armitage and Downhower, 1974). Fat stores are inadequate to support pregnancy and lactation; tissue stores contributed only 31 percent of the calories required to produce a litter of *S. parryii* (Kiell and Millar, 1980). Thus, females must increase ingestion; if food is not available, reproduction fails (D. C. Andersen et al., 1976). Female *S. elegans* occupying certain burrow sites produced significantly more young than females in other sites (Pfeiffer, 1982). The more successful sites occurred in areas of early snowmelt, which suggests that access to green vegetation early in pregnancy may affect reproductive output. Yearling female *S. columbianus* normally do not breed; some yearling females living on a site apparently richer in food resources produced litters (Festa-Bianchet, 1981). Further studies on the nutritional ecology of ground squirrels are essential if the relative importance of food in shaping mating strategies is to be determined.

There is widespread evidence that social suppression of reproduction occurs. In those species in which individuals live by themselves within aggregations on favored habitat, more than 90 percent of the females breed each year (Table 14.8). In those species that form social groups, from 40 to 84 percent of the females breed annually. Although some decrease in the number of reproductive females may be attributed to late springs or biennial breeding (Barash, 1973, 1974), reproductive failure in *M. flaviventris* and the production of fewer young per female by subordinate *M. caligata* (Wasser and Barash, 1983) clearly implicate social suppression. Furthermore, when the population of *S. armatus* was reduced by removal of up to 60 percent of the residents, yearling males became scrotal and some bred, and a larger percentage of yearling females bred than prior to reduction (Slade and Balph, 1974). Although increased aggression and parasitism were suggested as costs of sociality (Hoogland, 1979), reproductive suppression may be the major cost.

If sociality imposes a high reproductive cost, why do these animals form social groups? All social species are those that mature reproductively at age two or later. They live in confined environments; either the habitat patch is limited (e.g., marmots), or the space available for settlement is restricted (e.g., black-tailed prairie dogs). Fitness is higher when the probability of leaving reproductive descendants is increased by retaining direct descendants in their natal area. Furthermore, the association of females provides for defense against acquisitive conspecifics. However, because females strive to maximize direct fitness, conflict increases as subsequent matings reduce average relatedness. Therefore, matrilines should remain small and divide into daughter matrilines. Females should stive to sequester resources for use by their direct descendants; hence they should react to the presence of other females as the proximal factor limiting matrilineal size. Therefore, agonistic behavior, matrilineal fission, and dispersal should be largely independent of resources. Females may not be successful in excluding conspecfics; therefore, they should prefer to share resources with closely related females. In summary, there should be an ongoing waxing and waning of cooperative and competitive behavior determined by the kin structure of the population, the abilities of individual females, the location and abundance of resources, and the prospects for reproductive success.

Better measures of fitness are required. Although reproductive output is useful, fitness requires that descendants reproduce. Because all males and many females disperse from their natal areas, we must determine the reproductive success of dispersers in order to understand the significance of life-history strategies. Dispersal may be a tactic for increasing the probability of future reproductive success when settlement in the natal area is unlikely to lead to reproduction. Dispersal could be imposed on juveniles by adults attempting to increase their reproductive success or could represent a decision by the dispersers who perceive little likelihood of reproducing in their natal area.

Finally, the reproductive success of males not associated with clumps of females needs to be determined. We know little about the reproductive success of isolated females. These animals could play an important role in continuing matrilines across space and time. Males who mate with isolated females may be as reproductively successful as those associated with clumped females. Are these males who mate with widely spaced females owners of exploded harems? Are they at greater risk of predation? Do these males follow a viable strategy or do they represent the losers who are making the best of a bad job?

Ground squirrel mating systems appear to be adaptable to prevailing conditions. They offer an excellent opportunity to test the relationship between environment and mating system plasticity (Emlen and Oring, 1977). Long-term studies should be planned so that lifetime reproductive success and measures of fitness may be obtained. The paucity of measures of lifetime reproductive success may be the most serious shortcoming in studies of mating systems.

15. The Social Ecology of Gelada Baboons

———————————————— • ————————————————

R . I . M . D U N B A R

DURING THE 1960s, one of the main preoccupations of field workers was to relate the range of social systems they observed in nature to underlying environmental determinants. This endeavor, known generally as socio-ecology, was particularly associated with the name of John Crook (see Crook, 1970; Crook and Gartlan, 1966). Subsequently, the rise of socio-biology in the later 1970s altered the emphasis from group-level phenomena to the behavior of the individual. There was a consequent shift in the problems that interested field workers, with species-based studies giving way to problem-based studies in which the identity of the species was often incidental.

Although an important reaction against a creeping group-selectionism in traditional socio-ecology, this shift to problem-oriented studies has tended to obscure the fact that an individual's behavior is constrained by the biological system within which it exists. An animal is a set of compromises cobbled together in a somewhat haphazard fashion over evolutionary time, and the solutions to any one of its problems of survival and reproduction are constrained by the fact that it also has to solve the other problems (Goss-Custard et al., 1972; Dunbar 1983a, 1984a). Consequently, in order to be able to understand why an animal behaves in the way it does, in the final analysis we must understand the whole biological complex within which that behavior is being performed. This is not to say that there are no general rules that apply to all species: the point is, rather, that the way in which these rules function in a specific case depends on the demands made on the individual by other aspects of its biology—in other words, the same universally valid rules may predict contrasting optimal solutions to the same problem in different circumstances.

In this chapter, I try to examine the various levels of the gelada social system in order to show how the various components relate to each other and to identify the selective forces that have given them their shape.

THE GELADA

The gelada baboon, *Theropithecus gelada*, is the only surviving member of a genus that, during Plio-Pleistocene times, occurred widely throughout the savanna grasslands of sub-Saharan Africa. This genus was an offshoot of the common African baboon lineage (genus *Papio*) that left the confines of its ancestral wooded habitats and invaded the savanna grasslands as these opened up during the late Miocene and early Pliocene (see Jolly, 1972; Dunbar, 1983b). Some fifty thousand years ago, the genus all but went extinct as a result of what has come to be known as the "Pleistocene overkill" (P. S. Martin, 1966), leaving the gelada on the remote high-altitude grasslands of northern Ethiopia as its sole surviving representative.

The gelada are morphologically primitive to the genus, but they clearly share with their congeners a way of life that is (and was) totally adapted to an open country grazing niche (Jolly, 1972). The theropithecines are the primate equivalent of the equids, eating little other than grasses and relying on bulk feeding to solve the nutritional problems associated with a high-cellulose diet (Dunbar, 1983b). They are almost totally terrestrial, being poor tree climbers, and are found only in the vicinity of the deep gorges that dissect the Ethiopian plateau at altitudes of between 2,000 and 4,000 meters above sea level. They are dependent on these gorges for safe sleeping sites and for refuges from predators, and rarely stray more than half a kilometer inland from the gorge edge.

The gelada social system is relatively complex, since it consists of a hierarchical series of relationships among individuals (see Kawai et al., 1983). The basic social group is the one-male unit, which consists of a single breeding male and one to ten reproductive females, together with their dependent young. About 20 percent of these units contain an additional nonbreeding adult male, termed a "follower" (some units may in fact contain several followers). Males that do not belong to a reproductive unit form stable all-male groups. The reproductive and all-male groups share a common home range.

Figure 15.1 shows groupings of the thirty-nine reproductive units found at Sankaber in 1974. The figure uses a single-link clustering procedure to illustrate the similarities in the habitat usage of the units, based on the percentage overlap in the distribution of sightings of individual units in four gross sectors of the study area. Four primary clusterings emerge of units that have similar ranging patterns. Each of these groups is termed a "band." The measure used is relatively crude, and other indices such as the frequency of association in the same herd (see below) would yield a more ragged clustering pattern, though the same overall picture would be preserved. Bands range in size from 30 to 270 individuals; on average, a band contains about 115 animals distributed among approximately ten reproductive groups and one all-male group. Electrophoretic studies by Shotake (1980) suggest that bands are relatively homogenous genetically, with migration rates between them as low as 5 percent per generation (a figure that is in very close agreement with the observed migration rates of individual animals: Dunbar, 1984a).

Fig. 15.1. Single-link cluster analysis of the percentage overlap in range use between pairs of reproductive units in the Sankaber area. Four distinct clusters of units that have similar ranging patterns are clearly evident. Each cluster is a different band.

Kawai et al. (1983) have suggested that these tendencies for reproductive units to associate together reflect their ontogenetic history: the units of a band are thought to represent the product of repeated fissions over time of units descended from a single parent unity (see Dunbar, 1984a). Bands and other higher-level groupings listed by Kawai et al. (1983) can usually be identified only stochastically, though they undoubtedly have considerable reality to the animals themselves. Prolonged observation of known units is necessary for an observer to be able to identify the regularities in the grouping and ranging patterns of individual units. This stochasticity reflects the looseness of the bonds between individual units, which allows them to move and forage independently or together as circumstances require.

Because of the looseness of the relationships between the individual reproductive units, we need another term to refer to the aggregation of units that happens to be together in one place at any given moment: the term "herd" is used for this purpose. A herd implies no connotations of band membership: it is simply a collection of units foraging together and may consist of some or all of the units of one or more different bands. Most herds, however, consist of units of the same band. The ranging areas of the bands overlap extensively, although each band has a core area in which it spends most of its time. All-male groups are more labile in their attachment to their parent bands; some may spend up to a third of their time wandering alone or in the company of units from neighboring bands (though individual all-male groups vary widely in this respect).

The one-male unit is the reproductive and social unit in gelada society, whereas the band is the ecological and genetic unit and the herd is the foraging unit. Further details of the overall structure of gelada society and its component units can be found in Dunbar and Dunbar (1975), Kawai (1979), Kawai et al. (1983), and Dunbar (1984a).

STUDIES AND METHODS

The summaries that follow are based on an extensive series of field studies carried out by myself and my wife at Sankaber in the Simen Mountains of northern Ethiopia and at Bole (some 500 km to the south) and by Masao Kawai and his colleagues (Hideyuki Ohsawa, Umeyo Mori, and Toshitaka Iwamoto) in the Gich area of the Simen. On the whole, the methodologies of the two sets of studies were very similar. A series of detailed comparisons of data from the studies suggests that all the observed differences can be accounted for by differences in environmental and demographic variables (see Kawai et al., 1983; Iwamoto and Dunbar, 1983; Ohsawa and Dunbar, 1984; Mori and Dunbar, 1984). Most of the information on the detailed structure of interpersonal relationships derives from our studies at Sankaber, but comparisons with the Gich data suggest that precisely the same principles are at work there—and indeed in captive groups that have been studied by various researchers (Dunbar, 1982a).

Both sets of field studies were designed to be comprehensive. At Sankaber, for example, we carried out detailed studies of demographic structure and life history variables, behavioral ecology, reproductive and social behavior, vocal communication, and infant development. Our aim was to obtain as broad an understanding of gelada biology as possible within the limitations of the time and manpower available; inevitably, some aspects were studied in much greater detail than others. In particular, we lack any data on the biomedical aspects of the gelada study populations; as a result, we have had to infer certain aspects of the genetic and physiological characteristics of the animals using behavioral and morphological indicators. Ideally, we would also have needed to carry out a twenty-year study to follow through the lifetime consequences of individual behavior. We have, however, been able to circumvent some of the disadvantages of shorter-term studies by the use of within-population comparative analyses and by simulation. It is

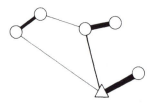

Fig. 15.2. Schematic representation of the pattern of social relationships among the adult members of a gelada reproductive unit. Thickness of the line joining any two individuals is proportional to the amount of time they spend interacting. Triangle, adult male; circles, reproductive females. Examples of sociograms for specific units can be found in Dunbar and Dunbar (1975) and Dunbar (1979a).

worth stressing that many insights were gained by interhabitat comparisons between the Gich, Sankaber, and Bole studies: these allowed a number of previously anomalous findings to fall into place by showing how certain aspects of behavior were influenced by environmental variables. These insights could never have been gained from a study of a single population because the range in environmental parameters over time is not sufficiently great at any one place.

RESULTS

The data on social relationships and reproductive strategies that are summarized here have been presented in full in the various quoted sources. The data are in general too complex to present in full here, and I shall present only token data to support or illustrate key points. In the final section, I shall look beyond the level of the individual to the social system's emergent properties. Because our data bearing on the adaptive significance of these aspects of gelada society have yet to be analyzed in any detail, I shall present a more speculative account aimed mainly at sketching the general outlines that qualitative impressions suggest that the final picture will fill in.

Relationships between Females

The adult females constitute the core of gelada society. A close examination of the pattern of social relationships within a unit reveals that the females tend to separate out into small clusters, often dyads, that interact particularly frequently (Fig. 15.2). These clusters turn out to consist of closely related females (usually mothers and their mature daughters: see Dunbar, 1979a, 1982a). Females who frequently groom together also tend to support each other in agonistic encounters, both against other members of their unit and against the members of other units. This mutual support in part determines the structure of dominance relationships among the females of the unit. The female dominance hierarchy is determined by what amounts to a two-step rule. First, all the females are assorted into a rank order on the basis of their natural aggressiveness (this being in part determined by their age and size); then those individuals who have close female kin of higher rank are inserted into the hierarchy immediately below that individual as a consequence of her support (Dunbar, 1980a).

The females of a given unit are, of course, all related to each other to some extent, and it seems likely that they all descend from a recent common ancestor. Units undergo fission at intervals (roughly equivalent to the time it takes the average unit to double its size). Demographic modeling suggests that the observed distribution of harem sizes is a consequence of a fission process that is random with respect to harem size (Dunbar, 1984a). Although it is not known for certain, it is probable that units split along genealogical lines when they undergo fission. Females who groom regularly together always stay in the same half (Dunbar, 1984a); in addition, it is known that the members of a given one-male unit are genetically more homogeneous than is the band from which they come (Shotake, 1980).

Relationships between individual females within a unit differ in a number of ways depending on the closeness of their genealogical relationships. Thus, closely related females (by which I mean mothers, daughters, and sisters) groom each other more per interaction than less closely related females (Dunbar, 1983c). The distribution of the number of interactees that each female has corresponds precisely to what would be expected if females interacted only with members of their own matrilines (where matrilines are defined as the descendants of a single *living* female), given the life-history parameters of the study population (Dunbar, 1983c). Most females groom to any significant extent with only one other female, however (hence the tendency for units to appear to consist of dyads). This is mainly because the female's time budget limits the amount of time she has available for social interaction (Dunbar, 1983c). The gelada's grass-based diet requires a relatively high proportion of time to be devoted to foraging if the animal is to meet its daily nutritional requirements (Iwamoto and Dunbar, 1983). Analysis of the distribution of social time suggests that, on average, after devoting time to grooming with her immature offspring, a female has sufficient time remaining to groom with only one other individual to any significant extent (Dunbar, 1983c).

Although females never give coalitionary support to females with whom they do not groom, they do support those females with whom they groom only occasionally. The fact that females groom only with close relatives suggests that they are well aware who their close relatives are and do not require strong grooming relationships as a condition of coalitionary support (Dunbar, 1980a, 1983c).

Some fairly extensive analyses suggest that females form coalitions with close relatives in order to keep their dominance ranks as high as possible over the course of their lives (see Dunbar, 1980a). Other things being equal, a female would normally rise in rank as she matured, reaching a peak during her physical prime, and then decline as she became older. By forming coalitions with other females, she can greatly reduce the rate of decline in old age: old females who had grooming/coalition partners held significantly higher ranks than those that did not (Dunbar, 1980a). Simulation analyses have suggested that the optimum coalition partner is a daughter (preferably a firstborn offspring) because this individual will be reaching her peak rank-holding potential just as the mother goes into her decline. In addition, the mother will be able to give the daughter rank support while the latter is a subadult (and would otherwise hold lower rank). Full details of these analyses can be found in Dunbar (1984a, chap. 8).

The functional importance of coalitions lies in the fact that the female birthrate is a function of dominance rank (Table 15.1) as well as of age (Dunbar, 1980b). Each unit loss of rank is worth about 0.04 births per year (on a mean birthrate for the Sankaber population of only 0.45 births per year: see Dunbar, 1984a). A detailed analysis of the cause of this loss of reproductive output (see Dunbar, 1980a) points to an increasing frequency of anovulatory cycles as rank declines. The duration of the period of postpartum amennorrhea is not related to dominance rank (analysis of variance of data given in Dunbar, 1980a: $F_{3,24} = 0.645$, ns), but high-ranking females tended to conceive again sooner than low-ranking females (fourteen out of eighteen independent dyads of simultaneously cycling females: binomial test, $p = 0.03$). In other words, high- and low-ranking females came back into estrus at about the same time postpartum, but it took low-ranking females longer to conceive.

Bowman et al. (1978) have demonstrated experimentally in the talapoin that stress due to harassment by higher-ranking individuals can block the LH-surge that triggers ovulation. Stress due to psychological and/or physical trauma is known to suppress reproduction in a number of species, including humans (for a recent review, see Wasser and Barash, 1983). Suppression of the LH-surge leads to extended estrous cycles because the system fails to "switch off" in the absence of the progesterone secreted by the corpus luteum following ovulation. Rowell (1972b) has observed that estrous cycles are longer following traumatic injury and when animals are kept in groups rather than alone in baboons, rhesus macaques, and chimpanzees. In the gelada, lower-ranking females have significantly longer estrous components to their menstrual cycles than do higher-ranking females (means of 14.3 vs. 10.7 days, respectively; Mann Whitney test, $n_1 = 3$, $n_2 = 4$, $p = 0.05$ 2-tailed). The amount of harassment received is known to be a cumulative function of declining rank (Dunbar, 1984a) and there is evidence that the frequency of harassment increases when a female comes into estrus (Dun-

TABLE 15.1
Rank-specific birthrate for gelada females

Dominance ranks	Infants born	No. of females	Birthrate (9 mos.)	Annual birthrate
1–2	7	21	0.333	0.444
3–4	5	18	0.278	0.371
5–10	3	16	0.188	0.251

SOURCE: Eleven reproductive units at Sankaber sampled over nine months in 1974–1975.

bar, 1980a). The weight of evidence clearly suggests that low-ranking females are being reproductively suppressed by harassment from higher-ranking individuals.

One consequence of the poor reproductive performance of low-ranking females (especially those that lack close female relatives to support them) is that they will become increasingly likely to desert their male in favor of another male with whom they might do better. However, females will not leave their units in search of a new male (for reasons that are not as yet clearly understood), so that "desertion" can occur only when another male joins the unit (either following a takeover or as a submissive follower). In general, the loss of reproductive output by the lowest-ranking female(s) is quite small in units with fewer than five females: as a result, the females of such small units are apparently unwilling to desert their incumbent male. Because desertion during a takeover is on an all-or-none basis by what amounts to a majority decision, males are only able to take over units with more than four females, and the ease with which they can do so increases steadily with harem size (Dunbar, 1984a).

It is worth observing that females seem to be more willing to desert those males that are less active in defending them against harassment by other units. The median frequency with which the harem males of two units that were later taken over were involved in agonistic encounters with other reproductive units was 0.44 times per hour (range 0.38–0.51), whereas the median for seven males that were not taken over was 0.99 times per hour (range 0.17–2.85). The difference, however, is not statistically significant (Mann-Whitney test, $p = 0.157$ 2-tailed). Males who were taken over were also less likely to engage in the ritualized encounters with all-male groups (Median rates of 0.04 and 0.19 times per hour, respectively; Mann-Whitney test, $p = 0.028$).

Relationships between a Harem Male and His Females

The male's relationships with his females are largely determined by the females' social preferences. Because a female's long-term reproductive objectives are best served by forming coalitions with close female relatives, the female's interest in interacting with her male is generally limited to his function as a source of sperm. The male's residence in the unit is relatively short (on average four to five years) compared with the average female's ten-year reproductive life span. Consequently, even though the male is invariably the most dominant member of the unit, his value as a coalition partner is rather limited. The problem is that he is likely to be replaced just at the point where he is becoming of most use to the female (i.e., in the later stages of her life when her own rank-determining abilities are on the decline). Females are therefore likely to use the male as a coalition partner only as a second-best solution in the event of not having a close female relative available (Dunbar, 1983d, 1984a). This accounts for the characteristic structure of gelada one-male units in which the females are formed into grooming dyads, and the male spends most of his time grooming with just one female, who herself has no female grooming partners.

Comparison of the relationships between a harem male and his main female partner and between the male and his other females reveals marked differences. With nonpartner females, grooming interactions are typically short, usually initiated by the male and terminated by the female, require a high rate of presenting by the male to elicit grooming, and involve many refusals to groom on the part of the female. His relationship with his partner female, on the other hand, closely resembles a typical female-female relationship, and it seems clear that the female treats the male much as she would any female partner. The male is also more likely to support his grooming partner than he is any other female (Dunbar, 1983d).

Interestingly, this relationship continues even after the male's replacement as harem-holder by another male. The only difference is that the female now mates with the new male while grooming and seeking coalitionary aid from the old male. This highlights a further feature of male-female relationships in the gelada, namely that sexual relationships are low-key, peremptory, and do not disrupt either the female's normal pattern of relationships or the social life of the unit as a whole (Dunbar, 1978a). In other words, there is a clear separation between social and sexual relationships.

Relationships between Males

Because of the highly structured nature of gelada society, males have a number of qualitatively rather different relationships with other males. We therefore need to distinguish the relationship that a harem-holder has with his follower, with another harem-holder, and with a male from an all-male group. We would also, of course, need to consider the relationships among the members of an all-male group as well.

Harem males are, in general, very tolerant of their followers. Followers come in two kinds, old males and young ones. An old follower is the previous harem-holder. (Dispossessed harem-holders always remain in their units as followers.) Once the relationship between them has been regularized, following the trauma of the takeover, the two males can in fact form a strong bond, which may on occasion extend to grooming. The old male usually supports the new male in the defense of the unit and, indeed, will often act alone in that capacity (Dunbar, 1984b). Young followers, on the other hand, are males who have opted not to try for a takeover and instead have joined a reproductive unit in a subordinate capacity. Once in the unit, the young follower will set about building up a limited set of relationships with the unit's more peripheral females, with a view to establishing an independent unit through harem fission at a later date. Although harem-holders may be described as being suspicious of the motives of any male who attempts to associate with his unit, a submissive response by such a male usually denotes a willingness to enter the unit as a follower rather than taking it over. From the harem-holder's point of view, a young follower can be an advantage since he may take over some of the females whose "loyalty" may be in doubt. This may diminish the females' collective willingness to desert by reducing the harem's effective size (i.e., the number of females bonded to any one male) below the critical threshold of five females. Harem males do seem to benefit from the presence of young followers insofar as large units that have followers are less likely to be taken over than those that do not (see Dunbar, 1984a).

A harem male's relationships with other harem males is best described as neutral—unless their respective females become involved in an encounter over infringements of social space, they generally ignore each other. A harem male will usually support his females if an encounter (many of which are initiated by the females) escalates to include most of the females in the unit (Dunbar 1983e). If both males become involved, they may end up threatening each other vigorously, but only come to blows in exceptional cases.

In contrast, a harem male's attitude toward the males of the all-male groups is decidedly antagonistic. These groups, of course, provide the reservoir from which future challengers of his hegemony over his harem will come. Harem males are noticeably worried by all-male groups and engage in ritualized interactions ("yelping chases") with their members. These interactions may help to reduce the likelihood of a takeover attempt in the immediate future, perhaps by implying something about the harem-holder's capacity to withstand a challenge; alternatively, or in addition, it may encourage the females to think that they have a superior quality harem male and therefore to be less willing to desert him (see Dunbar, 1984a).

All-male groups consist of adult and subadult males and are stable in composition over time (that is to say, males do not move from one group to another, but, having joined one, they remain a member of it until such time as they leave to enter a reproductive unit in search of their own harem). All-male groups are very cohesive and the members will commonly support each other in encounters with other all-male groups and reproductive units. As with the reproductive units, there is a tendency for grooming cliques to form, often consisting of males of different ages (Dunbar and Dunbar, 1975). This may suggest that males join all-male groups in which they already have relatives, a conclusion that is given some support by the finding that the variance in age structure of all-male groups is significantly greater than that in juvenile male peer groups (Dunbar, 1984a). Relationships among the males of an all-male group are generally friendly, though there is a great deal of ritualized greeting, which suggests the existence of an underlying tension in their relationships. These "greeting ceremonies" involve the presentation of the rear or the ventrum to another male, who then reaches through to touch the presenter's penis.

Individual Relationships and Society

In this section, I shall consider two emergent aspects of the social system: the influence of a population's demographic structure on male reproductive strategies and the functional significance of the higher levels of society (bands and herds).

Males, as we have already noted, can acquire their harems in one of two ways: either by taking over an entire unit intact or by joining a unit as a follower with a view to building up an incipient harem with one or two of the more peripheral females.

In discussing female–female relationships, I pointed out that low-ranking females are likely to be doing rather badly in reproductive terms and that this may incline them to desert their harem-holder should another male present himself. In practice, it seems that females only begin to do sufficiently badly to be willing to desert from about ranks four or five downwards. Consequently, units with four or fewer females seem to be above a critical "loyalty" threshold and are virtually impossible for outside males to take over. This is in part a consequence of the fact that the outcome of a takeover bid is determined not so much by the relative fighting abilities of the two males, but by the females' collective willingness to desert (see Dunbar and Dunbar, 1975; Dunbar, 1984a). The probability of being taken over is therefore zero up to a harem size of four females, and increases linearly thereafter (Dunbar, 1984a).

One obvious consequence of this is that the overall frequency of takeovers will depend on the distribution of harem sizes in the population: in other words, the more large harems there are, the more males will be able to effect a takeover. This is indeed the case (Fig. 15.3). In view of this, it is not therefore surprising to find that the proportion of males entering units as

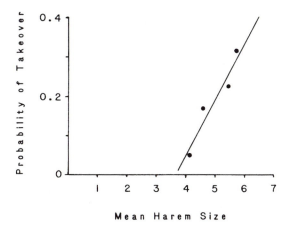

Fig. 15.3. Probability of takeover per unit per year plotted against mean harem size for four different bands (reproduced from Dunbar, 1984a).

followers is inversely proportional to the number effecting takeovers (see Dunbar, 1984a).

The final question I want to raise is why the gelada have the social system they do. Most of this section will be speculative since the relevant data have not yet been analyzed. However, the question is obviously of fundamental importance and, since it has not been given serious consideration since Crook's original formulations (Crook and Gartlan, 1966), it seems worthwhile sketching in the outline of what I think the data will eventually show.

I shall begin by reversing what would seem to be the logical order and first ask why the gelada form herds. I do this in order to be able to argue (at least as a preliminary hypothesis) that the tendency for females to group together in matrilineal groups (i.e., harems) is primarily a consequence of the reproductive costs they suffer by forming large herds. Herds disband whenever the opportunity to do so exists; this implies that females must have very strong reasons for forming large herds.

There would seem to be only two primary candidates as to why an animal like the gelada should form large groups: one is to defend food resources (Wrangham 1980) and the other is for defense against predators (Bertram, 1978; Rubenstein, 1978). It is hard to see how food resource defense could be a serious candidate in this case: grazers do not usually find it economical to defend territories because grass is one of the most abundant and evenly distributed resources. Nor are access to water, refuges, or sleeping sites

TABLE 15.2
Test of various hypotheses that might explain variations in herd size

	Escarpment	Gorge	Ridge	Fit to prediction
Herds with single units (%)	72.3	31.5	12.4	
1. Food availability[a]				
Grass cover (%)	58.3	49.0	55.0	N
Ground cover (%)	91.4	79.3	88.9	N
Total vegetation (%)[b]	99.2	98.1	102.0	N
Green grass in dry season (%)	41.0	4.2	4.6	N
2. Visibility[c]				
(median in m)	9.0	8.0	20.0	N
3. Predation risk				
Escapability rank[d]	1	2	3	Y
Mean slope	49	33	10	Y

[a] Estimated from point samples 1.2 m apart along a line transect of 450 points (two transects each on escarpment and gorge, one on ridge top).

[b] Sum of all points at ground level and bush level cover, but excluding all records of *Erica arborea* (which gelada did not eat).

[c] Measured as the greatest distance at which a sitting baboon could clearly see another sitting baboon in each of the cardinal compass directions (based on 20 points along the escarpment and gorge vegetation transects and 10 points on the ridge top transect).

[d] Measured as the relative availability of rock faces (1 = most, 3 = least).

likely to be worth defending, since these are all superabundant in the gelada's present habitat. Predation pressure, however, is a common cause of herd formation in open country bovids (Estes, 1974) and might provide a plausible reason why gelada form herds.

In order to justify this suggestion, it may be helpful to ask how herd sizes are influenced by various environmental variables that might plausibly be determinants of group size. The three obvious factors to consider are food availability, visibility conditions, and predation pressure. Each factor can be expected to influence herd sizes in different ways: 1) herd size will correlate with food availability because the animals will be forced to form smaller herds as grass density declines and becomes patchier in its distribution; 2) herd sizes will decline as visibility declines because it becomes increasingly difficult for a large group to stay together when visual contact cannot easily be maintained; and 3) herd sizes will correlate with the habitat's intrinsic risk of predation. Hypotheses (2) and (3) may be related if visibility affects the animals' confidence of detecting predators; however, if predation risk is negligible, visibility alone may impose a restriction on the animals' ability to maintain coordinated movement in large herds. (With the relatively loose relationships between gelada units, this is likely to be a serious problem.)

The hypotheses are put to the test in Table 15.2 using data on herd formation from three different habitats at Sankaber (the steep wooded escarpment face, the less steep thicketed gorge side, and the relatively open ridge top). As a measure of herd size, I have used the percentage of herds sighted that contained a single reproductive unit. This is a more reliable index than actual herd size since it was often difficult to get accurate counts of the large herds (200–600 animals: see Dunbar and Dunbar, 1975, table 28); on the other hand, it was easy to tell whether a herd consisted of just one reproductive unit or contained several of them. The differences between the habitats in the observed frequency of single-unit herds are significant: the ridge top had lower values than the gorge side in seventeen out of eighteen months (binomial test: $p < 0.001$) and the gorge side had lower values than the escarpment in thirteen out of thirteen months ($p < 0.001$).

Several different measures of food availability are given. These are all based on vegetation transects in which the presence or absence of cover was recorded at points spaced 1.2 meters apart (see Dunbar, 1978b). Visibility was estimated at every fiftieth point along these transects by determining the distance at which a sitting baboon would be able to see another sitting baboon in each of the four cardinal compass directions. The median distance is given for each habitat. Data for the escarpment and gorge habitats are based on two 450-point transects, that for the ridge top on one 450-point transect. Each habitat's predation risk was assessed visually in terms of the ease with which gelada could escape from a pursuing predator: this is largely a function of the relative abundance of sheer cliffs down which the gelada could go without a cursorial predator being able to follow. Two measures are given in the table, namely the rank of cliff abundance (assessed by eye) and the mean slope of the ground from the horizontal (measured off the topographical map prepared by Stähli and Zurbuchen, 1979).

It is fairly clear from the comparisons in Table 15.2 that only predation risk (i.e., gross topography) is correlated with the frequency of single unit herds. In other words, the likelihood of a reproductive unit being found alone seems to be correlated with the ease with which it can escape from predators. Food availability and visibility conditions appear to be irrelevant. We can use Bayes' Theorem to assess the relative likelihood of predation risk being the correct cause, given the observed distribution. To do so, we assign each outcome a likelihood of occurring given the validity of each of the three hypothesized causes. The most conservative solution is to assign each outcome a likelihood equivalent to its probability of being a random deviation from the correct order. Thus, if the predicted order is 1-2-3, an observed order of 1-2-3 will have a likelihood of $(1-1/6) = 0.833$ of being a correct estimate, while 3-2-1 will have a likelihood $(1-6/6) = 0.000$. With three possible and equally likely causes, the probability of predation risk being the right one, given the observed distribution of herd sizes, is 0.625

by Bayes' Theorem (see, e.g., Meyer, 1970); the probability that visibility is the right explanation is 0.250; and the probability that food availability is correct is 0.126. Predation risk is given further support by the fact that herds formed most readily on the relatively open plateau top in all three study areas (Gich, Sankaber, and Bole) and seemed to break up when they went down onto the steep slopes.

The main predators present in the Simen today are native dogs, jackals, hyena, and humans: gelada invariably responded to all of these with alarm barks, clustering of units, and/or flight to the nearest steep cliff. It is difficult to estimate the frequency of encounters with predators in retrospect since detailed records were not kept. However, dogs and humans, which are encountered by gelada on a daily basis throughout their geographical range, are probably the most serious. Bands that live in the vicinity of settlements commonly show clear evidence of this: the Gich band, for example, whose core range was closest to the only settlement within the park, had a very high frequency of maiming and injury. One instance of predation was recorded on a duiker (a small solitary antelope about the size of a female gelada that frequents the thicketed areas on the ridge top at Sankaber). In the recent past, gelada would also have been subject to predation by leopard and, at lower altitudes, by lion, though neither of these are common today. Leopard have been sighted in the Simen; caracal and serval occur at Sankaber.

At this point, a reconstruction of the likely course of theropithecine social evolution might be helpful. The ancestral theropithecines separated off from the ancestral woodland-base baboon stock and invaded the grassland habitats that were opening up during the late Miocene and early Pliocene, some four million or so years ago (Dunbar, 1983c). There seems to be no good reason to doubt that, at this point, their social system was in any significant way different to that of modern *Papio* baboons (i.e., multimale troops). On the open plains, predation is a more serious problem than it is in wooded habitats, since trees in which to take refuge are scarce. Open country antelope species live in significantly larger groups than those that inhabit thicketed or wood habitats (Estes, 1974), while *Papio* baboons live in significantly larger troops in open habitats than in wooded ones (Sharman and Dunbar, unpubl.). I suggest that this tendency toward increasing group size as the habitat becomes more open is a response to increasing predation risk: predators are less willing to attack large groups than small ones (see also van Schaik, 1983).

An alternative solution to the predator problem is simply to become much larger in body size, and it is interesting to note that the later species of extinct theropithecines that lived on the open East African savannas were very large indeed. Rough calculations based on the ecology of modern gelada suggest that the largest of these species, the 60–65 kg *T. oswaldi*, would

TABLE 15.3
Mean spread of reproductive units in relation to herd
size (*N* in parentheses)

Herd size	Mean spread		
	Escarpment	Ridge top	Gorge
<20	34.4 (19)	28.5 (22)	34.5 (10)
20–100	30.1 (8)	22.5 (28)	20.4 (19)
>100	20.1 (1)	14.5 (13)	—

SOURCE: Reproductive units sampled in feeding herds at Sankaber in 1975 during the dry season.

probably have been obliged to live in dispersed groups of no more than twenty individuals. Modern gelada are only a third as big and live in habitats that receive two or three times as much rainfall each year.

Because grass is a relatively evenly distributed resource, grazers are often able to maintain very high densities when feeding. Gelada herds of 300–400 animals are commonly compressed into an area no more than 200 meters in diameter under normal grazing conditions. In contrast, troops of the more frugivorous *Papio* baboons become much more dispersed: bands of fifty hamadryas baboons are commonly spread over one kilometer when foraging (Kummer, 1968; Stolba, 1979), while D.R. Rasmussen (1979) describes his troop of 120 *P. cynocephalus* as often being dispersed over three kilometers in the line of travel. In the case of the gelada, the reproductive units become increasingly bunched as herd size increases (Table 15.3). Bunching invariably leads to higher levels of aggression, and increased aggression will inevitably affect reproduction adversely.

My hypothesis is that, having begun with a *Papio*-like multimale social group, the increased aggression generated as group size grew resulted in the females tending to band together into matrilineal groups for mutual protection in order to buffer themselves against the physiological side effects of crowding. Matrilineal coalitions are in fact found within *Papio* troops and thus provide the precursor for gelada harems.

The gelada's present habitat on the Ethiopian plateau has some unusual features. The flat plateau top often provides substantial areas of good grazing, but these cannot be used efficiently by reproductive units unless they form herds. On the gorge sides, where they are safe from predators, there is no advantage in forming large herds, and their disadvantages therefore become more prominent. The gelada social system in fact has all the hallmarks of a flexible foraging system, allowing independent units to congregate in large numbers where this is advantageous, but to disperse again where it is disadvantageous.

One final problem remains: How do bands fit into this scheme and what determines their size? Iwamoto and Dunbar (1983) examined the likely correlates of band size and found that a band's size correlated most closely with the absolute size of its core ranging area. The mean density of animals was 72.5 per square kilometer, with rather little variation around this figure (n = 7 bands in three study areas). Intrahabitat variation in band size was considerably greater than interhabitat variation in mean band size, suggesting that general environmental variables were probably not relevant. Individual units were known to have the same core ranges over three-year periods in all three study areas and it seems likely that ranging areas had become traditionalized. Once band size exceeded the carrying capacity of the range, the band underwent fission (or individual units moved out). There was evidence (see Dunbar and Dunbar, 1975) that, when bands did undergo fission, they did so along lines of least association (and thus presumably minimum genealogical relatedness). In the light of this, I think the correct interpretation of bands is that they are incipient herds: in other words, ontogenetically related units are more familiar with each other, and therefore tend to form herds more readily. As a result, they tend to range in the same areas and therefore to constitute a band.

This account has been necessarily sketchy and tentative, and a number of inconsistencies remain to be ironed out. Nonetheless, being able to provide a convincing explanation for the historical sequence in the evolution of behavior that makes sense in terms of palaeo-environmental conditions seems to me to be an important step toward generating convincing explanations as to why animals behave in the way they do. In any case, the exercise has heuristic value in that it will often highlight important aspects of the animals' behavior that are as yet poorly understood.

DISCUSSION

The key points to note about the gelada's social system are: 1) the fundamental importance of the females in determining the overt appearance of society (and the consequent way in which the male's behavior is constrained by the females' social strategies) and 2) the way in which environmental variables interact with the female's primary physiological problems to impose limits on particular dimensions of the social system. Note also how the different levels of society reflect different functional problems and the selective forces relevant to those problems. Unlike most species, in which a single social grouping subserves all functions, the different functional units of society (predator defense, feeding, social, mating, genetic) are clearly segregated in the gelada.

Gelada obviously contrast with the common baboons of Africa (genus *Papio*) in this respect. Among *Papio* baboons, all these functions take place within the same social setting, the multimale troop. This is a cohesive group averaging some forty to fifty individuals of all ages and both sexes, with several breeding males. The social structure of these groups bears some similarities to the gelada insofar as females do show tendencies to form matrilineally based coalitions, and there is evidence that these coalitions may serve to buffer the females against reproductive suppression caused by harassment from other females (Wasser and Barash, 1983). Some females may also form particularly close bonds with individual males, as in the gelada (see Ransom and Ransom, 1971; Dunbar, 1973; Packer, 1979a; Altmann, 1980); these relationships too may serve to buffer females against excessive harassment (Dunbar and Sharman, 1983). There are, however, a number of differences between the gelada and the *Papio* baboons. Gelada bands, for example, are apparently less permeable than *Papio* troops. Almost all *Papio* males migrate from their natal troops to breed elsewhere (Packer, 1979b), whereas in the gelada it has been estimated that only about 30 percent of males acquire their harems in bands other than the ones they were born into (Dunbar, 1984a).

The hamadryas baboon (*Papio hamadryas*) is of particular interest in the present case because, like the gelada, it has a multilevel society based on one-male reproductive units that are grouped into bands (Kummer, 1968). Hamadryas one-male units are smaller than those of the gelada: on average, they contain only about two females. However, they do have male followers like those of the gelada. Hamadryas do not have all-male groups; instead, the excess males form a loose attachment on the band's periphery. One striking contrast between hamadryas and gelada reproductive units is that, while gelada females never leave their natal units, hamadryas females commonly do so and some females may transfer from one male to another several times during the course of a lifetime (Sigg et al., 1982). It is important to note, however, that most of these transfers are between units of the same band; like other *Papio* females, hamadryas females do not often migrate out of their natal bands/troops (Sigg et al., 1982).

Hamadryas bands are almost certainly the homolog of *Papio* troops. Hamadryas bands are more cohesive than those of the gelada: the constituent units normally maintain a coordinated route of daily travel even though they may become widely scattered in the process (Sigg and Stolba, 1981). There is also an intermediate level of social organization between the band and the reproductive units which are termed "clans" (Sigg et al., 1982). Clans are patrilineally organized and harems are commonly "passed on" from one male to another within clans. There appears to be no clear equivalent in the gelada social system to the hamadryas clan.

So, despite the superficial similarity, the hamadryas reproductive units differ in a number of important structural respects from their gelada counterparts. Studies by Kummer (1968), Stammbach (1978), and Sigg (1980) have shown that the structure of relationships among adult hamadryas is starlike, with the male at the focus of the group. The females all groom with the male and do not normally groom to any significant extent with each other, whereas the converse is the case in the gelada. Dominance is an important factor determining a hamadryas female's access to her male. The male's own herding behavior (which can involve vicious bites to the neck) may be an important factor preventing grooming between females. In general, a female's loyalty to her male seems to be largely a function of the extent to which he herds her: studies of hybrid *anubis* × *hamadryas* males showed that hybrid males had a less well developed herding instinct and had less stable harems as a result (Nagel, 1971). The female's tendency to follow her male in response to his herding seems to be entirely learned: *P. anubis* females released into hamadryas bands very quickly learned to behave like "good" hamadryas, while hamadryas females released into *P. anubis* troops soon gave up any tendency to follow males (Kummer et al., 1970).

The gross similarity between the hamadryas and gelada social systems has often led commentators to assume that they are the products of similar selection pressures (see, e.g., Crook and Gartlan, 1966; Jolly, 1970). Hamadryas society is generally presumed to be the consequence of the typical *Papio* troop becoming dispersed in an arid, poor-quality habitat. Recently, Popp (1983) has suggested that the hamadryas social system is a case in which male–male aggression has been minimized as a result of the effects of a poor environment on body size and sexual dimorphism: herding of one or two females is an energetically inexpensive alternative to the intense male–male competition that characterizes the other *Papio* species.

Although the gelada has been assumed to be the result of similar selection pressures (mainly on the grounds of the Crook and Gartlan [1966] claim that they live in a seasonally arid habitat), it does not in fact fit easily into any of these schemes. Their habitat lies at the upper end of the rainfall distribution for baboon habitats (see Popp, 1983, table 1): the Ethiopian plateau receives 1,100-1,500 millimeters of rainfall per year, which is more than double that for most African savanna habitats. (Rainfall is known to be a good index of primary productivity in most habitats: Rosenzweig, 1977; Coe et al., 1976.) Both Crook and Popp would presumably have to argue that the gelada social system was a hangover from Pliocene times when the species' ancestors roamed the more arid lowland savannas of eastern Africa. This seems a poor argument from an evolutionary point of view.

The account of the gelada social system's evolution that I have proposed differs radically from that given for the hamadryas. It seems likely that the two species have evolved in quite different directions from the same starting point under the influence of different sets of environmental pressures. The hamadryas seems to involve the dispersal of the basic *Papio* troop under conditions of a poor food supply and low predator pressure, while the gelada seems to be a case of increasing the size of the basic troop in response to high predation pressure where a dramatic change of diet facilitated high local densities. The apparent similarities in the two species' social systems seem to me to have more to say about the often rather superficial ways in which we tend to classify social systems.

16. Ecology and Social Relationships in Two Species of Chimpanzee

———————————— • ————————————

RICHARD W. WRANGHAM

FOR A FAMILY containing only four species, the social systems of the great apes are strikingly diverse. Orangutans (*Pongo pygmaeus*) are essentially solitary. Gorillas (*Gorilla gorilla*) form stable bisexual groups with rarely more than two males or six females. Chimpanzees (*Pan troglodytes*) and bonobos (or pygmy chimpanzees) (*P. paniscus*) live in closed social networks (communities) with as many as one hundred or more individuals. Within communities the mating systems and association patterns of the two chimpanzee species differ substantially.

These social variations present an attractive problem because the ecology of the four species is similar in many ways. All are confined to habitats that include at least a small amount of tropical rain forest. They breed all year and the young travel with their mothers from birth. They sleep in nests throughout their habitats, and have no permanent sleeping sites. There are no obvious differences in their vulnerability to predation, disease, or bad weather. Their main ecological differences appear to lie in foraging patterns, and even these have striking similarities, because all species feed from rather discrete patches on easily digested food (even the gorilla: Waterman et al., 1983).

Previously I have suggested that foraging patterns are primarily responsible for differences in great ape social organization. Fig. 16.1 shows that group size varies among the four species. Differences in group size are not explained by differences in population density: at any density bonobos form the largest parties and orangutans the smallest. Therefore food density, which presumably controls population density, is unlikely to be a strong influence on group size. Food distribution varies, however, and the nature of each species' food type and foraging style suggests that differences in average group size may be caused partly by differences in the intensity of feeding competition. For example, orangutans have the smallest group size (mean 1.8 individuals). This is apparently because their diet is restricted to rare food items distributed in discrete sources, such that two animals feeding together would reduce each other's food supply or be forced to travel farther. Mountain gorillas, on the other hand, form larger groups. They appear able to do so because they exploit relatively abundant forage with low

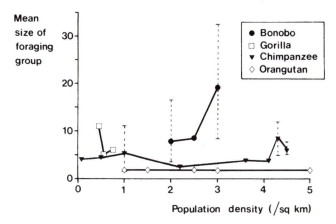

Fig. 16.1. Foraging group size and population density in the great apes.

Vertical dashed line shows range of monthly group (party) sizes. Data sources are given in order of increasing population density, together with sample size (number of parties), and code showing: A, all groups counted; T, only traveling groups counted; I, infants and juveniles included; *, infants and juveniles excluded; #, recalculated from original data.

Orangutan: Ranun, Sumatra: MacKinnon, 1974 (45, TI); Ulu Segama, Borneo: MacKinnon, 1974 (146, TI); Tanjung Puting, Borneo: Galdikas, 1979 (?, —); Kutai, Borneo: Rodman, 1973, 1977 (—, AI); Ketambe, Sumatra: Rijksen, 1978 (434, AI).

Chimpanzee: Mt. Assirik, Senegal: Tutin et al., 1983 (267, AI); Kasakati, Tanzania: Izawa, 1970 (78, —); Okorobiko, Equatorial Guinea: Sabater-Pi, 1977 (60,—); Kibale, Uganda: Ghiglieri, 1979, 1984 (597, T*); Gombe, Tanzania: Goodall, 1968, Wrangham, 1975 (498, T*); Budongo, Uganda: Sugiyama, 1968; Suzuki, 1975 (514, —); Mahale, Tanzania: Nishida and Kawanaka, 1972; Nishida, 1979 (987, AI); Bossou, Guinea: Sugiyama and Koman, 1979 (284, A*).

Gorilla: all data taken from Harcourt et al., 1981: Kahuzi/Bwindi/Tshiaberimu, East Africa, 33; Mt. Alen/Abuminzok-Aninzok, Equatorial Guinea, 29; Virunga, Rwanda, 31.

Bonobo: Lomako, Zaire: Badrian and Badrian, 1984 (268, AI); Yalosidi, Zaire: Kano, 1983 (36, AI); Wamba, Zaire: Kuroda, 1979, Kano, 1982 (172, AI).

variance in the quality of food items; hence, competition for food has less effect than in orangutans (Wrangham, 1979a).

The focus on feeding competition has been helpful in indicating constraints on great ape social evolution, but there are no data available to allow quantitative tests of its importance in different species. A qualitative test is made possible, however, by new data on bonobos, which have recently been studied in the wild for the first time. They have been shown to have larger and more stable parties than chimpanzees (Fig. 16.1), as well as other differences in social organization. If the size of chimpanzee parties is constrained by feeding competition, what explains the species difference? Why are bonobos able to form larger parties at equivalent population densities?

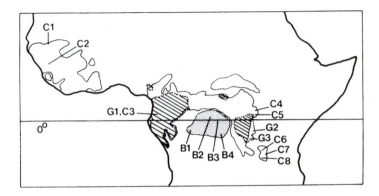

Fig. 16.2. Distribution of African apes.

Key: Chimpanzee: solid outline, empty. Gorilla: dotted outline, empty. Chimpanzee and Gorilla sympatric: hatched. Bonobo: solid outline, dotted.

Distributions are taken from Kortlandt (1962) (chimpanzee), Maple and Hoff (1982) (gorilla), Kano (1983) (bonobo). Accurate distribution maps have not been made for any of the species.

Study sites are coded by species. Bonobos: B1, Lac Tumba; B2, Lomako; B3, Wamba; B4, Yalosidi. Chimpanzees: C1, Mt. Assirik; C2, Bossou; C3, Equatorial Guinea; C4, Budongo; C5, Kibale; C6, Gombe; C7, Kasakati; C8, Mahale. Gorillas: G1, Equatorial Guinea; G2, Virunga; G3, Kahuzi.

There are three possible solutions. First, feeding competition among bonobos may be less severe because they are adapted to using different food types from chimpanzees (e.g., foods occurring in larger patches). In support of this idea, Kinzey (1984) found morphological evidence of differences in feeding adaptations. Bonobo molars have a longer shearing surface than chimpanzee molars, suggesting more adaptation to leaf-eating by bonobos. No other difference in feeding adaptations has been proposed. Because there are few data from the wild it is still unclear whether there is any consistent difference in body size, despite the common assumption that bonobos are smaller than chimpanzees. Probably, however, bonobos are the same size as some chimpanzees and smaller than others. Fifty-three measurements assembled by Jungers and Susman (1984) show no difference between the body weights of bonobos (females 33 kg, males 45 kg) and East African chimpanzees *P. t. schweinfurthii* (females 33 kg, males 43 kg), while the Central African chimpanzees *P. t. troglodytes* appear to be larger (females 47 kg, males 60 kg). Bonobos have a shorter mouth, smaller teeth, and more gracile arms and trunk than chimpanzees, but the effects of these differences on foraging strategy are not understood (S. C. Johnson, 1981).

Second, the environments of the two species may differ systematically in ways that lead to reduced feeding competition among bonobos. However, no evidence of differences has yet been found in their forest habitats, which lie either side of the Zaïre River (Fig. 16.2). The altitudinal and latitudinal limits of bonobo habitats fall entirely within the range of chimpanzee habitats. The most obvious difference in their ecology is that unlike bonobos, chimpanzees occupy not only moist forests but also drier and more open habitats (to the north and east of the Zaïre Basin, Fig. 16.2). But this does not explain the differences in social behavior.

Third, grouping patterns may be unrelated to food distribution. For instance, they may be determined by the threat of predators. This would have important implications, because it would weaken the idea that feeding competition is primarily responsible for differences in social organization of the other apes.

In this chapter, therefore, I compare chimpanzees and bonobos as a test of the hypothesis that differences in social organization are explicable by differences in food distribution. Ultimately this should involve demonstrating not only that there is a difference in feeding ecology, but also that it accounts for the particular ways in which social relationships differ in the two species. Because the social relationships of bonobos are still not well known the focus here will be on grouping patterns.

I begin by describing the social relationships of chimpanzees and how they are affected by foraging constraints and social competition. I then review bonobo behavioral ecology, and propose that the species differences in group size and composition are indeed explicable in terms of differences in food distribution. However, ecological factors explain only why bonobos are able to form larger groups than chimpanzees, not why larger groups are favored. By extending an argument from the chimpanzee social system, I suggest that the reasons why bonobos form large groups lie in social behavior rather than ecology.

CHIMPANZEES

Study Sites and Methods

Chimpanzees are the best studied great ape. At Gombe (Goodall, 1983) and Mahale (Kasoge) (Nishida, 1983a), both in Tanzania, habituated individuals have been under continuous observation for at least twenty years. Other socio-ecological studies with at least partially habituated individuals have been carried out in Guinea (Bossou: Sugiyama, 1981), Senegal (Mt. Assirik: Tutin et al., 1983), and Uganda (Budongo: Suzuki, 1975; Kibale: Ghiglieri, 1984). Unhabituated populations have also been observed, and these provide data on home range size, population density, diet, and grouping patterns (Equatorial Guinea: Sabater-Pi, 1977; Tanzania: Suzuki, 1969; Izawa, 1970).

TABLE 16.1
Chimpanzee ecology: Variation among study sites

	Mt. Assirik	Kibale	Gombe	Budongo	Kasoge	Bossou
Population density (km²)	0.1	2.2	3.6	4.1	4.3	4.5
Daytime range (km)						
Male	—	—	2.8–3.5	—	—	—
Female	—	—	2.6–2.8	—	—	—
Mean	—	—	—	—	>1.8	—
Community range (km²)	~300	23–30	10–13	19	10–17	5–6
Mean/median party size	4	2.6	4	3.9	8.6	6
Community size	~25	>44	19; 36	>80	27; 106	23
Community sex ratio (M/F)	—	0.86	1.5; .67	.67	.61; .64	.57
Activity budget						
%feed	—	52–62	56	—	—	—
%travel	—	10–12	14	—	—	—
%rest/groom	—	25–37	30	—	—	—
Time eating						
%fruit/seed	—	89	59	—	57	—
%leaf	—	7	21	—	7	—

SOURCES: Mt. Assirik: Baldwin et al., 1982; Tutin et al., 1983; Kibale: Ghiglieri, 1979, 1984; Gombe: Goodall, 1968; Wrangham, 1977, 1979b; Budongo: Sugiyama, 1968; Suzuki, 1975; Mahale: Nishida, 1974, 1979; Hasegawa and Hiraiwa-Hasegawa, 1983; Bossou: Sugiyama and Koman, 1979; Sugiyama, 1981.

For three reasons the data from Gombe and Mahlae may be unrepresentative of social relationships in other areas. First, the Gombe and Mahale populations (*P. t. schweinfurthii*) have rather similar body size, ecology, and behavior, whereas some other populations have obvious differences such as more forested habitats or substantially lower population density (Table 16.1). Differences such as these could affect social organization. Second, they both occur near the extreme southeastern edge of the species range (Fig. 16.1), and may thus be atypical. Third, chimpanzees are provisioned with agricultural foods both at Gombe and Mahale, so their behavior may be unnatural in some ways (Reynolds, 1975; Ghiglieri, 1984). Provisioning has certainly had some effects at Gombe, including larger parties and increased rates of aggression at the feeding site (Wrangham, 1974), although not apparently outside it (Wrangham and Smuts, 1980). For these reasons generalizations about chimpanzee social ecology are not yet legitimate. It is quite clear that chimpanzees do not exhibit uniform behavior throughout their range, yet the nature of the variation is still uncertain. When chimpanzees have been studied in detail in west and central Africa we will be able to expand on the descriptions that follow.

Major distortions of natural behavior through provisioning are possible but unlikely. First, during the time that individual ranging patterns were being recorded, the amount of food provided to each individual was small, two or three bananas once a week at Gombe (Wrangham, 1974), or ten centimeters of sugar cane per day at Mahale (Hasegawa and Hiraiwa-Hasegawa, 1983). Second, after the Kasekela community at Gombe fissioned in 1972 only one of the daughter communities visited the feeding site, although both were observed. There was no indication that the availability of bananas caused any differences in ranging and grouping patterns between the provisioned and unprovisioned communities (Wrangham and Smuts, 1980). Third, at both sites unprovisioned chimpanzees have been observed regularly, although not in detail, since before provisioning began. Again, no differences in grouping and ranging patterns are apparent. Fourth, at least some of the putative effects of provisioning are improbable, such as a major rise in party size. Since at least 98 percent of foraging occurs in the natural environment it is difficult to see how provisioning could increase mean party size significantly. Goodall (1983) argued similarly that it is unlikely to have generated the observed pattern of intercommunity aggression. Fifth, data on grouping, ranging, feeding, and social behavior are collected primarily or entirely on chimpanzees in the natural habitat rather than at the artificial feeding areas. I therefore assume that long-term ranging, grouping, and feeding patterns have not been affected significantly by provisioning, even though short-term patterns were undoubtedly influenced during earlier periods (Wrangham, 1974).

Ecology

Though chimpanzees use habitats varying from closed forest to open grassland, their ranges invariably include at least 1 percent of forest, which provides critical food supplies either in the wet season (Suzuki, 1969) or the dry season (Baldwin et al., 1982). Their diets are dominated by ripe tree-fruit (Table 16.1), and food sources are reached mainly by traveling on the ground (88-96 percent of time in different seasons: Wrangham, 1977). Humans are the only known predators, but large carnivores are considered an important threat in some open areas (Tutin et al., 1981, 1983).

Community Organization

All individuals live in communities ("unit-groups"), which are closed social networks containing from 19 to 106 individuals (median 36, see Table 16.1). Community members share a common home range within which they forage and sleep alone or in temporary parties. No permanent associations are found among adults.

Both at Gombe and Mahale most but not all adult females are immigrants from other communities. During fifteen years at Gombe thirteen females immigrated into the Kasekela community, two natal females emigrated, while five natal females remained (Goodall, 1983): in this sample 72 percent of the adult females recruited into the community were immigrants. At least four of the five nonemigrating females became mature while their mothers were still living in the community. Similarly at Mahale two females that gave birth within their natal community (M-group) associated frequently with their mothers (Hiraiwa-Hasegawa et al., 1984). This suggests that females are more likely to breed in their natal communities if they can continue to associate with their mothers as adults.

Adult males, by contrast, normally breed in their natal communities. At Gombe, for example, all of the nine males who reached adulthood between 1968 and 1980 remained in the community where they grew up (Goodall, 1983). However, only one individual has been followed from birth to adulthood, and it is possible that some adult males are not close relatives since a low rate of migration by juvenile males has been seen both at Gombe ($n = 2$) and Mahale ($n = 2$). In each case the juvenile immigrated in the company of a migrating female presumed to be a mother or sister (Goodall, 1983; Nishida et al., 1985). Adult males have also been suspected of intercommunity migration at Bossou, but this awaits confirmation (Sugiyama, 1981).

Resident communities at Gombe and Mahale are thus composed largely of related adult males and unrelated adult females. In these sites population density is high (3–5 per km²), community ranges shift only slightly during the year, and intercommunity interactions are frequent. At low density (below 0.5 per km²), communities migrate as units between discrete seasonal ranges. Nothing is known about intercommunity encounters. In such areas the kinship structure is unknown (Izawa, 1970; Tutin et al., 1983).

Relationships among Females

Both at Gombe and Mahale adult females spend much of their time alone with their immature offspring. From 1972 to 1975, for instance, forty-seven all-day observations spread throughout the year on fourteen females showed that Gombe mothers spent 65 percent of their time alone, and their mean party size consisted of 1.6 adults, with accompanying offspring (Halperin, 1979; Wrangham and Smuts, 1980). This solitary tendency is correlated with a marked lack of affiliative social interactions among females (Table 16.2). Nishida (1979) characterized female groups as "passive homogeneous aggregations," and this is supported by Gombe data showing that females in parties were attracted to rich food sources rather than to each other (Wrangham and Smuts, 1980).

Though females are rarely seen to seek out each other's company, regular companionships do occur, in three main patterns. 1) Six cases are known of adult females whose mothers were in the same community. In five, mother and daughter traveled together frequently and sometimes supported each

TABLE 16.2
Frequency of grooming combinations in chimpanzees and bonobos

	Female–Female	Female–Male	Male–Male	N
Chimpanzees				
Budongo (Sugiyama 1968)	2%	10%	55%	135
Gombe (Goodall, 1968)	3	9	29	*
Mahale (Nishida, 1979)	10	39	46	477
Kibale (Ghiglieri, 1979)	28	17	24	92
Bonobos				
Yalosidi (Kano, 1980)	7	17	0	99
Lomako (Badrian and Badrian 1984)	17	25	7	72
Wamba 1974–1975 E-group (Kano, 1980)	18	29	6	295
Wamba 1976–1977 B-group (Kuroda, 1980)	36	49	11	143

NOTE: Cells show the proportion of all adult grooming pairs observed, or (at Gombe) the proportion of time spent grooming, which is closely equivalent (Ghiglieri, 1979). Studies vary widely in sample method and observation biases, so results should be taken only as an indication of the direction of trends. Partner availability depended on sex ratio in the community (Tables 16.1, 16.3), sex differences in associations, and sample method. Note that among bonobos, males consistently groom each other less than chimpanzees do despite having a higher sex ratio.
* Data from two months at a provisioning site.

other in interactions with others (Goodall, 1971, 1977, 1983; Hiraiwa-Hasegawa et al., 1984). The sixth was a pair who had been separated for a year, and they treated each other almost as if they were unrelated (Nishida, 1979). 2) Nulliparous females and unrelated mothers sometimes associate strongly. Here there is no evidence that the adults are drawn to each other, however: the young female behaves as an alloparent, and is thought to be attracted by the infant rather than the mother (Nishida, 1983a). 3) In Kibale (Ngogo), where mean party size for females was similar to Gombe (2.4 including immatures), mothers with infants associated with each other more than with males or nulliparous females (Ghiglieri, 1984). This pattern has been reported in other studies of only partially habituated groups (Beni: Kortlandt, 1962; Budongo: Suzuki, 1975), and its significance is still unclear. It implies that mothers benefit by traveling with other mothers, yet in the only study of alloparental behavior, Nishida (1983a) found that lactating females were either uninterested in or abusive to the infants of others. Possibly the nature and significance of female grouping patterns vary between study sites.

At Gombe and Mahale, however, the solitary and asocial nature of females is clear, and is reflected in their distribution. At Gombe all mothers travel throughout the community range, but each has a preferred core area where she spends the majority of her time. Thus for five females 80 percent of each individual's observations occurred within a stable core area whose size, over at least eight days (median 2.1 km²), was less than 20 percent of the community's annual range (13 km²). The core areas of different females were scattered evenly throughout the community range. This has been interpreted as being due to female-female competition for feeding ranges (Wrangham 1979b; Wrangham and Smuts, 1980). Comparable data are not available for Mahale, but females are reported to be dispersed in their own individual ranges (Nishida, 1979). Hasegawa and Hiraiwa-Hasegawa (1983) discuss several recent immigrant females who spent most of their time in a third of the community range.

Although resident females are asocial, they are rarely aggressive to each other. It is not yet clear if they have consistent dominance relationships but there is some evidence that they do, because in greeting interactions older females are generally higher-ranking (Bygott, 1979; Nishida, 1979). Since there is competition for core areas within the community range, nulliparous immigrant females elicit aggression from resident females (Pusey, 1980; Hasegawa and Hiraiwa-Hasegawa, 1983). The most extreme case of aggression by a female was infanticide, committed on at least three occasions (Goodall, 1977) by a mother whose adolescent daughter was in the process of establishing her own core area. Possibly this benefited the killer by repelling other mothers and helping to establish a core area for her daughter, but more evidence is needed (Goodall, 1983).

The low frequency of interaction between most community females suggests that they obtain little benefit from each other. There are no observations of females cooperating in community interactions (Goodall et al., 1979; Nishida, 1979) and there is therefore no indication that they assist each other in defense of the shared range.

Male–Female Relationships

In contrast to the dispersal of sexually inactive females in individual core areas, males range widely and comparatively evenly over the community range (Wrangham, 1979b). They travel farther per day (e.g., at Gombe, 48 percent farther, Wrangham and Smuts, 1980), in larger parties, and encounter more individuals of each sex than females do (Ghiglieri, 1979; Nishida, 1979; Reynolds and Reynolds, 1965; Wrangham and Smuts, 1980). Their greater time with others occurs partly because they give long-distance calls, including characteristic food-calls at sources capable of feeding several individuals. These food-calls, which are not given by females, attract both sexes in equal numbers (Wrangham, 1977; Ghiglieri, 1984). Males also display or run between parties in apparent attempts to bring them together (Sugiyama and Koman, 1979, pers. obs.).

Male sociability is almost wholly promiscuous so far as females are concerned. Males pay little attention to sexually inactive females, grooming them less than they are groomed in return, and the only evidence of disproportionate association over the long term is among close kin (mother-son pairs, or maternal siblings). Even siblings do not always have a special relationship (Nishida, 1979). In mixed-sex parties males decide the direction and timing of travel, while females follow passively (pers. obs.). This is particularly obvious in the case of immigrant nulliparous females, who spend most of their first two years in a community following males, apparently for protection against resident females (Nishida, 1979; Pusey, 1980). Short-term associations also occur as a result of recent mothers following males, possibly for protection against infanticide (Goodall, 1977).

Females with sexual swellings, by contrast, are clearly attractive to males: large parties gather around them. Females become tumescent (i.e., develop sexual swellings) throughout the year. They travel widely throughout the community range, and males follow them, grooming them more than they are groomed (Nishida, 1979; Tutin, 1979). Over 95 percent of copulations occur during the nine-to-twelve day period of maximal tumescence, and at this time females copulate at high rates, i.e., 0.5 copulations per hour for each male present (Tutin, 1979; Tutin and McGinnis, 1981; Hasegawa and Hiraiwa-Hasegawa, 1983). During two years in a Gombe community with twelve adult females, at least one adult female was sexually receptive on 57 percent of days (Tutin, 1979). Thus mating opportunities occur frequently, even if unpredictably.

There is no evidence of long-term relationships between particular mating pairs. Instead, mating partnerships are determined principally by behavior during the period of female receptivity. Ovulation occurs on the last day of maximal tumescence ($\pm$ 1 day) (Graham, 1981). As this approaches, aggressive competition or consorting attempts by males become increasingly frequent, and individual males succeed in restricting others from mating. The outcome of aggressive competition has not been shown to be affected by female choice. Males attempting to initiate a consortship, however, are more likely to succeed in doing so if they are friendly to females in general, i.e., likely to groom them or share food with them (Tutin, 1979; Tutin and McGinnis, 1981; Hasegawa and Hiraiwa-Hasegawa, 1983). This suggests that female discrimination occurs, and that short-term relationships can influence mating success. Nevertheless all females are willing to mate any male, so the dispersed distribution of females favors males who search for mates throughout the community range.

Relationships among Males

Strong social bonds among adult males have been reported from all studies which have discussed male relationships, based on patterns of association, alliance, and grooming (Bygott, 1979; Ghiglieri, 1984; Goodall,

1975; Nishida, 1979, 1983b; Riss and Goodall, 1977; Sugiyama, 1968; Sugiyama and Koman, 1979; see Table 16.2). Male–male cooperation is expressed in two main contexts within communities.

First, male allies assist each other by using threat or aggression against rivals during competition for status or resources. In some cases allied males are maternal kin, but not always (Nishida, 1983b). The advantage of high rank appears to lie primarily in its effect on mating competition. Thus in three different communities it has been shown that the highest ranking male achieved an overwhelming proportion of matings on days of probable ovulation (Tutin, 1979; Hasegawa and Hiraiwa-Hasegawa, 1983; Nishida, 1983b). The highest-ranking male also achieved the most copulations at Bossou (Sugiyama and Koman, 1979).

Second, males cooperate in food-getting, especially in giving food-calls and catching vertebrate prey. Neither activity seems likely to account for the presence of male bonds, however. Little food-calling occurs when food is scarce, and there is no evidence that it increases foraging efficiency (Wrangham, 1977). The frequency of predation by chimpanzees is trivially low in many areas (W.C. McGrew, 1983), and no benefit has been found from cooperation, because meat intake per individual falls as party size increases (Busse, 1978). Indeed, it is not even clear whether chimpanzee predatory behavior should be termed "cooperative" at all, because the joint action seen in hunts is easily interpreted as directly selfish (Busse, 1978).

Intercommunity Relationships

Interactions between communities are almost invariably hostile, and fall into two main types, based on observations among four communities at Gombe and two at Mahale (Goodall et al., 1979; Nishida, 1979). First, territoriality involves a party of resident males collaborating to supplant neighbors encountered during foraging. This has been observed regularly at both sites following full habituation, and can involve supplants of both males and females. Territoriality occurred particularly clearly for several years in Mahale, where a large community in search of ripe fruit annually supplanted its smaller neighbor for a season, and then vacated the area of overlap when its food supply dwindled (Nishida, 1979). Nothing is known of intercommunity relationships in other populations, but it is clear that the ranges of low-density populations are too big to be defended as territories (Table 16.1).

Second, aggressive raiding occurs by males who patrol borders of their own range in search for members of neighboring communities (Goodall et al., 1979). When neighbors are heard or seen in a party of several individuals, the patrollers do not advance. When lone neighbors are found, however, the patrollers may become raiders, advancing deep into the neighboring range to stalk and physically attack an adult male or old female. Such violent attacks are infrequent, but are known to be responsible for the deaths of one female and one male at Gombe, and are thought to account for a further two female and six male deaths at Gombe (Goodall, 1983, pers. comm.), and an unknown proportion of seven male deaths at Mahale (Nishida et al., 1985). Members of two communities at Gombe and one at Mahale have been attacked, and in each case attacks occurred during a sequence of hostilities directed by a larger to a smaller community over a period of months or years (up to at least four years). Chimpanzees attacked violently during intercommunity encounters were alone in seven out of eight cases (Goodall et al., 1979). (In the exception, the victim appeared to be weak from illness. His companion ran away without assisting him.) Together with numerous observations of larger parties calling or displaying at each other without physical aggression, this strongly suggests that attacks are made selectively on lone individuals. It therefore means that lone individuals are particularly vulnerable.

Consistent losses in intercommunity encounters have at least three possible results. 1) The community range size may be reduced either temporarily (Nishida, 1979) or permanently (Goodall, 1983; Nishida et al., 1985). 2) The community may become extinct as a result of the loss of all adult males (Goodall et al., 1979; Goodall, 1983; Nishida et al., 1985). 3) Females may emigrate to the aggressor community (Uehara, 1981; Goodall, 1983; Nishida et al., 1985). Consequently the aggressor communities may gain benefits through increased access to resources and/or a higher ratio of reproductively active females to males.

Ecological Influences on Chimpanzee Social Organization

It is remarkable that in spite of a wide range of habitats, range sizes, and population densities, certain aspects of chimpanzee social organization are rather uniform. The mean size and composition of parties, in particular, varies little among six different study sites (Fig. 16.1; Tutin et al., 1983), indicating that substantial differences in habitat and predation pressure have little influence on grouping patterns, at least at a relatively gross level of analysis. The threat of predation has been held to be responsible for the occurrence of large parties migrating occasionally between forest patches in open country near Mt. Assirik (median 19, Tutin et al., 1983). However, it did not appear to affect the size of their daily foraging parties. No other effects of predation risk on chimpanzee grouping patterns have been proposed.

All studies have found that party size fluctuates over the year (Fig. 16.3), and that food sources are often too small to satisfy all members of a party. Smaller parties have been attributed to feeding competition for three reasons. First, parties are smallest when overall food density appears least (as indicated by fruit availability and body weights) (Sugiyama, 1973; Nishida, 1974; Wrangham, 1977). Second, party size increases significantly at large

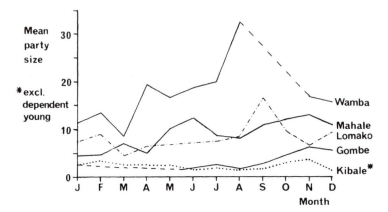

Fig. 16.3. Annual variation in party size of chimpanzees and bonobos.

Data sources are: Wamba, Kuroda, 1979; Mahale, Nishida, 1974; Lomako, Badrian and Badrian, 1984; Gombe, Wrangham, 1975; Kibale, Ghiglieri, 1979.

food sources (Wrangham, 1977, Ghiglieri, 1984), whereas parties break up when individuals spend time together without feeding (Wrangham and Smuts, 1980). Third, more time is spent feeding in smaller than larger parties. Gombe females spent 67 percent of their time feeding when alone, for example, compared to 50 percent of their time feeding when in parties ($p <$ 0.01, Wrangham and Smuts, 1980); Gombe males similarly fed significantly longer when alone than when in parties (Bygott, 1974). For both sexes the relationship between percentage of time spent feeding (f) and party size (n) in parties containing from one to seventeen individuals was $f =$ 59.4 $-$ 1.74n ($r = -0.79$, $p < 0.02$) (recalculated from Wrangham, 1977). Fourth, females fed for longer when in parties for a short time than for a long time (71 percent vs. 28 percent, $p < 0.01$; "short" $= < 1.5$ hr) (Wrangham and Smuts, 1980). Together with evidence that females spent only little time traveling when in parties, this suggested that when they spent time in parties they tended to do so at large food sources.

These observations are all consistent with the hypothesis that competition, whether indirect or direct, reduces foraging success in parties. However, it has not yet been shown how closely the rate of energy gain is related to feeding time. Rates of intake are rather consistent within food types, but vary between food types partly as a function of the ease or difficulty of harvesting (Hladik, 1977; Riss and Busse, 1977; Wrangham, 1977). The competition hypothesis can therefore be tested in the future by direct measurements of the rate of nutrient gain. Aggressive competition for food occurs rarely, presumably because food patches are not economically defendable (J. L. Brown, 1964).

Even by the competition hypothesis, aggregation of dispersed feeders is expected at large food patches. However, this does not explain why social bonds should occur, as they do among males in particular. Instead, reproductive competition has been argued to be ultimately responsible. Males attempt to defend access to females from other males. Effective protection depends on cooperation between males, because lone individuals are defeated by coalitions. Therefore, according to this hypothesis, selection has favored males that invest in alliance relationships. Sets of allied males compete for dominance over a local area or community, so that all males are members of a male-bonded community (Ghiglieri, 1984; Goodall, 1983; Nishida, 1979; Wrangham, 1979b).

Chimpanzee social organization thus appears to depend on a delicate socio-ecological balance. If the competition hypothesis is correct, party foraging reduces feeding efficiency significantly. This means that the best strategy for females, whose reproduction is presumed to be limited by food intake, is to spend most of their time alone. On the other hand, party foraging allows sufficient feeding time that males, whose reproductive interests depend critically on male bonds, are able to spend much of their time in parties. Sexually active females, who are not yet investing in a fetus or nursing offspring, are likewise able to tolerate the costs of party foraging. The fact that party size varies in different seasons indicates that constraints on grouping can change rapidly. In a similar way, local differences in food supply occur between neighboring communities (e.g., Nishida, 1979), and may be responsible for differences in party size. This could explain why individuals in small communities sometimes forage alone despite the danger of being found and attacked by parties from a larger community, as they sometimes are.

If these ideas are right, a systematic shift between species in the intensity of feeding competition would also have important effects. Orangutans are more solitary than chimpanzees, possibly because the costs of party foraging are greater (Rodman, 1984). Bonobos, on the other hand, are less solitary than chimpanzees. This implies that the costs of party foraging are less for bonobos than for chimpanzees. Why this might be is examined below.

BONOBOS

Study Sites and Methods

Bonobos are confined to Zaïre (Fig. 16.2), and are known well from two sites. At Wamba they have been studied since 1974, initially without provisioning (Kuroda, 1979, 1980), and, since 1977, with limited provisioning of two communities (Kano, 1982). At Lomako a single community has been

TABLE 16.3
Bonobo ecology: Variation among study sites

	Lomako	Yalosidi	Wamba
Population density (no./km^2)	2	>2	3
Daytime range (km)	>2	—	2.4
Community range (km^2)	22	—	30–60
Mean party size	7.6	8.5	16.9
Community size			
Range	∼50	—	59–∼120
Mean	—	—	80 ($n = 4$)
Community sex ratio (M/F)	∼0.61	—	.77–.94
Activity budget			
%feed	—	—	<30
%travel	—	—	13
%rest/groom	—	—	>43
Time eating			
%fruit/speed	—	—	—
%leaf	—	—	—

SOURCES: Lomako: Susman et al., 1980; Badrian and Badrian, 1984; Yalosidi: Kano, 1983; Wamba: Kuroda, 1979; Kano, 1982; Kano and Mulavwa, 1984.

observed irregularly since 1974 without being provisioned (Badrian and Badrian, 1984). Unhabituated animals have been studied briefly at Yalosidi (Kano, 1983) and Lac Tumba (A.D. Horn, 1980). Individual relationships have not been described in detail at any site.

Ecology

Bonobos differ from chimpanzees in being restricted to continuous evergreen forest (both primary and secondary). Like chimpanzees their main food is ripe fruit from trees or vines. Data from Wamba suggest that they feed for a lower proportion of the day than chimpanzees (Tables 16.1, 16.3) (Kano and Mulavwa, 1984). They travel both on the ground and in trees, and it is not yet clear if they are more terrestrial (A.D. Horn, 1979) or arboreal (Susman et al., 1980) than chimpanzees. No predation has been recorded except by humans, who are responsible for low bonobo densities in some areas (e.g., Lac Tumba, Horn, 1980).

Community Organization

Communities of bonobos are similar in many ways to those of chimpanzees (Table 16.3), and in the following sections only the major differences are reviewed. Patterns of recruitment into communities appear similar for two reasons. First, two nulliparous female bonobos are known to have associated many times with at least two communities, and temporary visits have been made by two further nulliparous females and one mature female

(Kano, 1982). Second, putative adolescent sons are more likely to travel with their mothers than daughters are (Kano, 1982). Female migration between communities is therefore expected to be typical (Badrian and Badrian, 1984).

The five bonobo communities which have been counted are large (median 63, Table 16.3) compared to those of chimpanzees (median 36, Table 16.1). Since four of these come from one study site (Wamba), it is not yet clear if this is a species difference. Bonobo community ranges are larger than those of chimpanzees at equivalent population density, possibly because their communities are bigger. The sex ratio at Wamba is higher than in most chimpanzee communities, but at Lomako it may not be (Tables 16.1 and 16.3).

Relationships among Females

Female bonobos have more frequent affiliative interactions than female chimpanzees do. First, they have high rates of homoerotic behavior (genito-genital rubbing, i.e., mutual rubbing of sexual swellings), which is almost unknown in chimpanzees (Kuroda, 1980; Badrian and Badrian, 1984). Second, they are believed to groom each other more than female chimpanzees do (Kuroda, 1980), although on the basis of present evidence (Table 16.2) it is not yet clear if this is always the case. Third, food-sharing has been found to be more common among females than males. Kano (1980) found that 61 percent of eighty-nine food-sharing interactions were among females, whereas in chimpanzees most food-sharing by adults is among males (Goodall, 1968). Tolerant relationships among females were also inferred to exist by Kano (1980, 1983), because females are more likely to feed close together than males are. Aggressive episodes between females are reported to be infrequent compared to those among males (Kano and Mulavwa, 1984).

Badrian and Badrian (1984) have suggested from such observations that female bonobos differ from chimpanzees in being "strongly bonded" to each other. Females have not yet been shown to have differentiated relationships, however. Female dominance relationships have been looked for without success, and females do not support each other in competitive interactions (Kuroda, 1980). Genito-genital rubbing occurs principally when females meet, and has been argued to reduce tension (Kuroda, 1980). It is therefore possible that affiliative behavior among female bonobos is a method of reducing competition rather than maintaining bonds.

Male–Female Relationships

Male-female relationships differ in at least four ways from those of chimpanzees. First, males and females have differentiated patterns of association, forming "subgroups" which have consistent membership for at least two years (Kano, 1982; Kitamura, 1983; Badrian and Badrian, 1984). Subgroups are defined as sets of individuals who regularly form parties together over a long period of time, and may be thought of as "subcommunities." The composition of five subgroups at Wamba and Lomako varied from two to eight males, and one to eight females. Three subgroups were shown by Kano (1982) and Kitamura (1983) to have core areas in different parts of the community range, while a fourth, which contained two females (mothers) and the five highest-ranking males of the community, covered the whole area. Thus many, perhaps all, females have close long-term relationships with particular males within the community. Female chimpanzees, by contrast, have equally close relationships with all males in the community, except for kin.

Second, no sex differences have been found in association patterns. In particular, the sex ratio in small mixed parties, which favors males heavily in chimpanzees, is the same as in big parties and equal to the community sex ratio in bonobos. This pattern holds both at Wamba and Lomako, unlike Gombe or Mahale where the sex ratio in parties consistently favors males (Kuroda, 1979; Badrian and Badrian, 1984).

Third, party size is not changed by the presence of tumescent females (Kano, 1982), unlike the results from all studies of chimpanzees (e.g., Tutin et al., 1983). Furthermore almost all parties contain tumescent females, for example, 98 percent of 172 (Kano, 1982; cf. Badrian and Badrian, 1984). Female bonobos apparently have sexual swellings for the majority of their adult lives, as suggested by Kano's data (1982) showing that in 1,170 observations of females, 67 percent of the females were tumescent. The difference from chimpanzees (tumescent for only 3.8 percent of their adult lives, Tutin, 1975) occurs because bonobos have sexual swellings sooner after a birth, for longer during pregnancy, and for a longer period during the menstrual cycle (Kano, 1982).

Fourth, among sexually cycling females bonobos mate at a higher frequency than chimpanzees outside the period of maximal tumescence. Thus the majority of copulations at Lomako were by females with submaximal swellings (Badrian and Badrian, 1984). At Gombe, by contrast, copulations during maximal tumescence occurred at seven times the rate found during other phases of the cycle.

Relationships among Males

Two measures show that relations among males are less affiliative in bonobos than chimpanzees. First, male-male grooming is reported to be uncommon, and in comparison to male-female grooming it is clearly less frequent than in chimpanzees (Table 16.2). Second, male coalitions have not been seen, and occasional massed branch-dragging is the only kind of cooperative display that has been observed (Kuroda, 1980). Patterns of food-sharing and spacing in feeding trees also suggest that bonobo males are less cooperative than chimpanzee males (Kano, 1980, 1983).

In addition to having weaker male bonds, bonobos apparently have less severe aggression than chimpanzees. Kuroda (1980) observed no biting, stamping, or rolling in fifty-six attacks by males, whereas these occur routinely in attacks among chimpanzees (Bygott, 1979). Although dominance relations exist, tension between males is said by Kuroda (1980) to be at a low level even when they meet, unlike chimpanzees (Bauer, 1979; Bygott, 1979).

Intercommunity Relationships

Few intercommunity interactions have been seen. They are known to cause tension and normally involve small parties avoiding larger parties (Badrian and Badrian, 1984; Kano, 1982; Kano and Mulavwa, 1984). This suggests that large parties bring benefits in intercommunity competition, despite the increase in intracommunity competition. One violent intercommunity interaction has been observed, causing "serious injuries to several individuals," but it has not been described in detail (Kano and Mulavwa, 1984).

ECOLOGICAL DIFFERENCES BETWEEN CHIMPANZEES AND BONOBOS

The difference in grouping patterns between chimpanzees and bonobos is partly explicable if bonobos suffer less disadvantage in feeding together than chimpanzees do. This means that bonobos must have regular access to a food source that: 1) allows individuals to feed together with less competition than occurs at chimpanzee foods; 2) occurs widely throughout the geographical range of the bonobo; 3) is used more by bonobos than by chimpanzees.

All feeding studies show clearly that the preferred food of both species is ripe tree-fruit. One possibility, therefore, is that the fruit patches of bonobos are systematically larger than those of chimpanzees, as suggested by Badrian and Badrian (1984). Unfortunately there is no evidence that such a difference exists. Until relevant data are available this hypothesis cannot be pursued.

An alternative hypothesis, however, is partly testable. This proposes that a key food resource differentiating bonobos and chimpanzees is the diet eaten during periods when preferred foods (ripe tree-fruit) are not available. Diets are normally classified according to the part of the plant, that is, fruit, leaf, etc. This classification indicates nothing in bonobo diets that satisfies the above requirements. However, if foods eaten are grouped by their pattern of distribution, the literature suggests that such a food source exists,

TABLE 16.4

Arboreal and terrestrial foods in the diets of chimpanzees and bonobos

	No. food species	% monthly food intake
Chimpanzee (Gombe)		
ARB	21	93
THV*	4	7
Chimpanzee (Equatorial Guinea)		
ARB	23	62
THV	3	21
Bonobo (Wamba)		
ARB	16	67
THV	3	33

SOURCES: Wrangham, 1975; Sabater-Pi, 1977, figs. 4, 5; Kuroda, 1979, p. 167. Data given only for "major foods," i.e., all those given by Sabater-Pi and Kuroda, and all foods at Gombe eaten for more than 10 percent of feeding time in any month.

KEY: ARB = foods from trees, even if eaten on the ground.

THV = terrestrial, herbaceous foods (entirely Zingiberaceae and Marantaceae for Equatorial Guinea and Wamba; includes Acanthaceae at Gombe).

* Includes one terrestrial woody shrub, *Monanthotaxis poggei.*

although its importance in bonobo diets has not previously been reviewed. The food source is terrestrial herbaceous vegetation, particularly Zingiber-aceae (ginger) and Marantaceae (arrowroot).

These plants occur throughout the humid evergreen forests of central Africa, especially in swampy areas, secondary forests, and disturbed growth in primary forest, but not in open woodland or grassland. Many species often grow in large patches, or "fields" (Sabater-Pi, 1979; Kuroda, 1979), and their fruits, pith, and shoots are eaten by all the African great apes. Evidence of the particular importance of terrestrial, herbaceous vegetation (THV) for bonobos is as follows.

1) Bonobos use THV extensively. THV has been reported to be eaten regularly in all bonobo study sites, (Horn, 1980; Kuroda, 1979; Kano, 1979, 1983; Kano and Mulavwa, 1984; Badrian and Badrian, 1984; Badrian and Malenky, 1984). Ecological studies have been most intensive at Wamba, where Kuroda (1979) recorded all the major food species ($n = 19$) over ten months. Though only three (16 percent) species were in the Zingiberaceae (*Aframomum* sp.) or Marantaceae (*Megaphrynium macrostachyum* and *Haumania liebrechtsiana*), these contributed over 30 percent of the major monthly foods, including at least two in each month (Table 16.4).

Equivalent data are not available elsewhere, but reports are similar. At Yalosidi, Kano (1983) reported that during the three months of study "shoots, herbaceous stems, and piths from ground food plants . . . occupied as important a place in the diet as did tree-fruits." At Lomako a preliminary study indicated that "important staples in the bonobos' diet were found at ground level in the herbaceous layers of the forest . . . the majority of these foods were leaf shoots or pithy stems'' (Badrian et al., 1981). Subsequent data showed that seven herbaceous species comprised the second most frequently eaten food type (after ripe fruit), and that one species (*H. liebrechtsiana*) was ''the single most frequently eaten food in the Lomako area'' (Badrian and Malenky, 1984).

2) THV is eaten when preferred foods are scarce. Kuroda (1979) and Badrian et al. (1981) refer to THV as a staple eaten when arboreal fruits (including vine-fruit) are scarce. Badrian and Badrian (1984) state that ''an important part of [bonobos'] diet consists of herbaceous foods that are both ubiquitous and non-seasonal.'' Kano (1983) was explicit: ''. . . these fibrous foods, which are characterized by year-round reproduction and lack of notable seasonality in the amount of production, perform the important function of maintaining the diet level of the pygmy chimpanzee during periods when there is insufficient fruit.'' This appears to be true in all areas.

3) THV allows animals to feed together. Badrian and Badrian (1984) note: ''feeding competition . . . may also be reduced through utilization of the evenly distributed and continuously available herbaceous foods. Thus bonobo females may associate together (and with males) without reducing their foraging efficiency.'' Kuroda (1979) found that bonobos ''split into small groups when feeding in the *Megaphrynium* field or the African ginger field.'' ''Small'' here means a monthly average of at least 8.5 (Fig. 16.1). It is therefore large compared to chimpanzee parties during periods of food scarcity (fewer than 2: Wrangham, 1977).

There is a second reason for thinking that a diet of THV allows animals to feed together. Gorilla diets are primarily THV, and gorillas form stable bisexual groups. The gorilla diet is known best from the Virunga Volcanoes, where terrestrial leaves, shoots, and stems account for 86 percent of all feeding records (Fossey and Harcourt, 1977). Less detailed studies in West African lowland forest show similar reliance on THV (Tutin and Fernandez, 1985). Thus three species of *Aframomum* (Zingiberaceae) make up 80–90 percent of the gorilla diet in Equatorial Guinea (Sabater-Pi, 1979). Among eastern lowland gorillas, herbaceous vegetation has also been described as a critical food which determines the gorilla distribution (Schaller, 1963). All gorillas have stable groups, larger than those of chimpanzees but smaller than the mean party size of bonobos (Fig. 16.1).

4) THV is used more by bonobos than by chimpanzees. Some of these herbaceous foods are eaten year-round by forest-living chimpanzees in West Africa, especially the fruits, shoots, and leaves of *Aframomum*. Thus Sabater-Pi (1979) estimated that three species of *Aframomum* contributed about 15 percent of the annual food for chimpanzees in Okorobiko, Equatorial Guinea (see also Table 16.4), and he pointed to their importance as ''a reserve nutrient that can be exploited throughout the year.'' In Gabon the

371

Zingiberaceae and Marantaceae are important chimpanzee foods in the wild (Tutin and Fernandez, 1985); and a heavily provisioned group of chimpanzees, reintroduced to a forest on a Gabonese island, ate ten species from the two families, for an average of about 8 percent of the diet (Hladik, 1973). Thus forest-living chimpanzees do exploit THV.

In drier areas, such as western Tanzania, *Aframomum* spp. can also be important staples for part of the year (May to July in Filabanga: Kano, 1971; March to April, Kasakati: Suzuki, 1969). Mahale is more forested than other Tanzanian sites, and *Aframomum* spp. provide up to 10 percent of the diet and are available all year (Nishida, 1974; Nishida and Uehara, 1983). Such foods are used little by Gombe and Kibale chimpanzees, however (Goodall, 1968; Ghiglieri, 1979; see Table 16.4), because they are uncommon (pers. obs.).

Thus chimpanzees clearly eat THV when it is available, but it is not always common. Two reasons are suggested. First, many chimpanzee habitats are drier than those of bonobos, and THV does not grow well in seasonally dry areas. Second, where THV is abundant, that is, in forest habitats, gorillas are often present also (Fig. 16.2). THV is the principal food source of gorillas, and in Equatorial Guinea their biomass (35-55 kg/km^2: Harcourt et al., 1981) is comparable to that of sympatric chimpanzees (30-80 kg/km^2: calculated from data in Sabater-Pi, 1979, following Harcourt et al., 1981). This means that gorillas eat more THV than chimpanzees do, and are possibly important as competitors. The importance of competition from other species is unclear. Kano (1983) reported that neither monkeys nor any other mammal fed on the THV foods of the bonobo, in contrast to their tree foods, which were shared with many species. It seems likely that elephants and other mammals may be significant competitors in some areas.

Bonobos thus exploit an important food source that is usually less abundant for chimpanzees. As Kano (1979) concluded from a survey of several populations in the eastern part of their range, bonobo diets include "both the frugivorous foods of (chimpanzees) and the fibrous foods of gorillas." These "fibrous foods" are thought to allow gorillas to live in groups. I propose that their importance in bonobo diets explains why bonobos can have larger parties and more stable associations than chimpanzees.

BENEFITS OF SOCIAL BONDS IN THE GREAT APES

Even if feeding competition among bonobos is relaxed by their using THV during periods of fruit shortage, this does not explain why females form larger parties than chimpanzees. Present data on bonobos suggest no ecological benefits to grouping, for example, in terms of predator avoidance or cooperative food-getting. No social benefits have yet been indicated either. However, in the other great apes the benefits of grouping are explained more easily in terms of their social than their ecological functions. Since this raises the possibility that social benefits underlie bonobo grouping also, the possible functions of grouping are briefly reviewed.

Among chimpanzees social bonds often serve to reduce intraspecific aggression by a third party. Such bonds have therefore been argued to function as protective mechanisms. Thus it has been argued for chimpanzees that nulliparous immigrant females travel with males because they are thereby protected from aggression by resident females (Pusey, 1980); that females with young infants travel with males because this protects their offspring from infanticidal attacks by other females (Goodall, 1977); that males travel with close allies within the community to protect or enhance their dominance status when meeting rival males (Nishida and Hiraiwa-Hasegawa, in press); and that older females who travel with males, or males who travel together, obtain protection against lethal attacks by males in neighboring communities, because they are less likely to be attacked when in a party. These relationships alone can account for the gross pattern of chimpanzee grouping, that is, sociable males sometimes accompanied by females. By contrast, predator defense or other ecological benefits are an improbable source of chimpanzee sociality, particularly because males are more gregarious than females.

In orangutans sexual consortships are the only common form of group among adults. Females who travel with dominant adult males experience reduced rates of forced copulations, which otherwise occur at high rates despite intense female resistance (Mitani, 1985). No ecological function for orangutan consortships has been proposed.

In chimpanzees and orangutans, therefore, grouping certainly appears to be promoted by social rather than ecological benefits. Social benefits may similarly be important for gorillas, but here the case is less clear because predator defense could be important also. The social benefits of grouping are seen when adult males defend their groups against potentially lethal attacks by other males. Females who travel with the most dominant males therefore receive the most effective protection for themselves and their infants (Harcourt, 1978; Fossey, 1984). Since lone females would be courted by bachelor males it has been argued that female choice of protective males can account for the existence of groups (as well as their intragroup and intergroup relationships) (Wrangham, 1982). Males also defend against predators, however, especially by displaying at humans. Defense against conspecifics and defense against predators are therefore both possible functions of grouping. No data yet exist to show which hypothesis is preferable. (See Wittenberger, 1980b, and Harcourt et al., 1981, for discussions of infant mortality in relation to predation, and Stewart and Harcourt, in press, for arguments in favor of the predation hypothesis.)

An alternative ecological hypothesis has been proposed to account for gorilla groups. Greenwood (1980) suggested that in birds, male philopatry helps males to obtain and defend the resources needed to attract a female, and that females disperse to avoid inbreeding. Like many birds, gorilla females disperse farther than males, so Greenwood suggested that females form groups so as to obtain resources controlled by males. There is no evidence, however, that adult male gorillas control access to important resources. Home ranges overlap extensively and are not defended. Fossey (1983) indicates that intergroup interactions only rarely occur at favorite food patches, whereas their frequency appears to depend on the defensibility of a group's females (Harcourt, 1978). For instance, males may lead their groups in pursuit of other groups for several days before attacking (Fossey, 1983); and the aggressiveness of intergroup relationships depends on the number of females in each group, males with few or no females being the most aggressive (Harcourt, 1978). Unless evidence emerges that gorillas defend access to food sources the hypothesis applied to birds is not appropriate for explaining gorilla groups.

While firm tests remain to be conducted, therefore, the great apes offer cases where social benefits of grouping seem at least as important as ecological advantages. In each case, the social benefit is protection from harassment by a conspecific. The threat is constant, because there are individuals ready to harass, attack, or force an unwanted copulation whenever unprotected individuals are found. So long as individuals travel alone or only in small parties, evidence suggests that they are vulnerable, whereas they obtain a significant measure of protection by traveling in groups, particularly with adult males.

Bonobo Sociality: A Hypothesis

Since chimpanzees travel in larger parties when more food is available, apparently for social benefits, an obvious hypothesis is that bonobos form larger parties than chimpanzees for similar reasons. If so, we can expect to find reduced levels of aggression in bonobos, that is, improved protection from intraspecific harassment. Present data indicate that this occurs, but quantitative comparisons are not available.

Bonobos differ from chimpanzees in a variety of social patterns in addition to party size, and ultimately these complex differences must be explained. Although a detailed analysis is not yet possible, there are several aspects of bonobo sociality that fit the idea that social benefits have accrued, especially to females, in the evolution of bonobos from a hypothetical chimpanzeelike ancestor.

First, the exaggeration of female sexuality in bonobos, together with the fact that males accompany females and mate them almost daily, is a clear indication that selection has favored increased attractiveness in females. Their elaborate sexual behavior, which includes highly variable vocalizations, facial expressions, and posturing (Savage-Rumbaugh and Wilkerson, 1978), may also be a method of attracting males. If THV enables female bonobos to forage in parties without significant feeding costs, it may be advantageous to females to attract males as social protectors, while remaining in small core areas to reduce travel costs (cf. Wrangham, 1979a).

This does not explain, however, why females have evolved special mechanisms to attract males, as opposed to merely choosing a male companion (as gorillas do). I suggest that the answer lies in the history of bonobo adaptation from a chimpanzeelike ancestor. The fact that female bonobos have evolved special attraction mechanisms indicates that males were not willing to be attracted; i.e., that female and male strategies are in conflict. Such a conflict occurs in chimpanzees: females occupy small core areas while males travel widely to find females who are sexually receptive. If the bonobo ancestor was similar to modern chimpanzees, selection could favor females who obtain male companions by being as attractive as possible to males. Hence their prolonged sexual receptivity and elaborate sexual behavior.

A problem here is the relationship between proximate and ultimate mechanisms of attracting males. Ultimately males should find sexier females attractive only if they thereby achieve greater fitness. Among bonobos this could happen if the effect of long-term relationships between females and males is to give low-ranking males higher reproductive success than they would achieve under the chimpanzee system. This would contribute to explaining why male-male competition is reduced in bonobos compared to chimpanzees.

Second, bonobo subgroups differ from chimpanzee parties in having a more equal sex ratio. Other things being equal, the sex ratio within a subgroup should not exceed the community sex ratio, since otherwise its males should move to other subgroups with a more favorable sex ratio. At the same time, if males are favored as protectors, two are significantly better than one. Hence, females should prefer to travel with at least two males, in which case there should be at least two females. By this argument females increase the number of males they travel with by being willing, rather than preferring, to travel with other females. It therefore expects that female relationships are tolerant but not affiliative. This issue has not yet been examined.

This argument fails to explain why the majority of gorilla groups have only one breeding male (64-77 percent, Harcourt et al., 1981). Since adult male gorillas in the same group have been observed to cooperate effectively during intergroup aggression (Fossey, 1983), and females tend to move to groups with a high ratio of silverback males to females (Baker, 1978), it is puzzling why male-male cooperation is not more common in gorillas.

Third, bonobos show clearer differences in the behavior of adult males than chimpanzees do. Kano (1982) found that the five highest-ranking males formed a subgroup which covered the community range, unlike three other subgroups that traveled in small core areas. This suggests that dominant males pursue a chimpanzeelike strategy. Subordinate males, by contrast, may increase their probability of fertilizing particular females (i.e., those in their own subgroup) while reducing their chances of fertilizing other females. Thus, their strategy is more like that of gorillas.

DISCUSSION

Field studies show clearly that bonobo social behavior has been reorganized compared to that of the chimpanzee. Female bonobos spend more time with each other, have friendlier interactions, and have even evolved a new pattern of sexually stimulating greeting behavior. They spent more time with males than female chimpanzees do, are sexually receptive for longer, and have more elaborate sexual interactions. Males are less inclined to fight, groom each other less, and less given to traveling all over the community range than male chimpanzees are. These changes appear to have occurred when a presumed chimpanzeelike ancestor of the bonobo (S. C. Johnson, 1981) became geographically isolated south of the Zaïre River.

How can the changes be characterized? The two principal studies have prompted the same comparison. Kuroda (1979) concluded that Wamba bonobos have "some intermediate [traits] between the common chimpanzee and the gorilla." Badrian and Badrian (1984) found that Lomako bonobos "combine elements of the cohesiveness of gorilla society with the flexibility of chimpanzee social organization." I have argued that one reason why the social organization of bonobos combines elements of gorilla and chimpanzee is that their food sources are equally mixed. They combine the THV of the gorilla and the tree-fruits of the chimpanzee.

Thus, as in gorillas, use of THV allowed larger and more stable groups in which protective relationships could flourish. Females stopped traveling on their own, and they welcomed the presence of a male. Unlike gorillas, females prolonged their attractiveness, males congregated around them, and for unknown reasons male-male competition was reduced. Possibly female bonobos synchronize, or conceal, their ovulation date more effectively than chimpanzees do (Turke, 1984).

The importance of THV has not yet been established, and other ecological differences may therefore prove to be equally or more important. At present the only alternative hypothesis was proposed by Badrian and Badrian (1984), who suggested that tree-fruit patches are significantly larger for bonobos than chimpanzees. This will be testable with further data. It appears to me less attractive than the THV hypothesis because it does not explain why bonobos can form large parties when food is scarce. Nor is there any explanation for why tree-fruits should occur in larger patches south of the Zaïre River than north of it.

Why do bonobos rely on THV more than chimpanzees? The only morphological difference known in bonobo and chimpanzee feeding adaptations is that bonobo molars function better for shearing, and therefore for eating vegetative growth (Kinzey, 1984). Since there is no evidence that bonobos eat more tree-leaves than chimpanzees, this suggests that bonobos have an evolutionary history of dependence on THV. A likely reason is that bonobos have evolved in a THV-rich habitat (i.e., their present range south of the Zaïre River) without competition from gorillas. Fig. 16.1 shows that gorillas and bonobos straddle the equator, where they are confined to areas of high rainfall. The simplest hypothesis to explain the absence of gorillas from bonobo habitat is that following a period of forest contraction (A. C. Hamilton, 1981) there were no apes south of the Zaïre River. By chance the chimpanzeelike ancestors of bonobos subsequently crossed the river (S. C. Johnson, 1981) while gorillas failed to do so. Bonobos then took advantage of the vacant gorilla niche.

Accordingly, there are two reasons for differences in the feeding behavior of bonobos and chimpanzees. First, THV is more abundant and available in bonobo habitats. This is because unlike chimpanzees, bonobos occupy only THV-rich habitats. Furthermore where chimpanzees occupy moist forest they frequently compete for THV with gorillas. Second, bonobos use THV extensively because they are adapted for eating it efficiently. It will be interesting to find out if bonobos have evolved any feeding adaptations in addition to the dental shearing ability described by Kinzey (1984).

Tests of the hypothesis that great ape grouping patterns are constrained by feeding competition will be possible from further field study. For instance, it is expected that chimpanzees with abundant THV will be relatively bonobolike in their social behavior, and that bonobos with little THV will be more chimpanzeelike in their behavior. The social organization of the Central African chimpanzee, *P. t. troglodytes*, will be particularly interesting, because its range includes areas where THV can be expected to be most abundant. On morphological and behavioral grounds it has been considered to be the subspecies most closely related to bonobos (E. Schwartz, 1934, quoted in Groves, 1981).

Comparison of closely related species can help in understanding the behavioral ecology of unusual social systems, which occur in too few animals to allow a statistically satisfactory correlation of ecology and behavior. The present comparison ties a small change in ecology to a set of related changes in behavior and physiology. The ideas thus generated are testable by further information on feeding competition, food distribution, and the social benefits of grouping.

Similar efforts may generate rules by which relationships respond to ecological change, and will show whether rules are consistent within taxonomic groups. In the great apes the rule looks consistent: vulnerability to conspecifics is the principal source of social bonds, and defensive groups are therefore formed when foraging constraints permit.

17. Male and Female Mating Strategies
on Sage Grouse Leks

•

ROBERT M. GIBSON

AND JACK W. BRADBURY

LEKS, MALE ASSEMBLIES that females visit for mating, are a spectacular, but uncommon promiscuous mating system that occurs in a diverse set of vertebrate and invertebrate taxa (Bradbury, 1981; Oring, 1982; Thornhill and Alcock, 1983). Lek systems are commonly defined by four criteria: absence of paternal care, clusters of displaying males, location of mating aggregations away from resources required by females, and apparent freedom of females to choose mates. Additional, but not defining, characteristics of most vertebrate lek species are highly skewed male mating success and, presumably as a result, pronounced sexual dimorphism.

These features are classically illustrated by sage grouse. Leks form for a few hours each day at dawn (and at dusk and on moonlit nights in peak season) between early March and mid-May. Typically lek sites host 20–100 males each and are located at 1–3-kilometer intervals throughout suitable habitat; some sites are used annually while others are occupied only when population densities are high. Females visit leks singly or in groups from late March to early May and mating is concentrated in a ten-to-fourteen-day window within this period (R. L. Patterson, 1952). Most females apparently mate once and then lay a clutch of seven or eight eggs that is reared without male assistance (Patterson, 1952; Lumsden, 1968). The sexes are strongly dimorphic: males are twice as heavy as females and possess striking specializations for display, including a long tail and a three-liter esophageal air sac. There are also sex differences in group size, ranging patterns, gross food intake, and annual survival (Patterson, 1952; Dalke et al., 1963; Beck and Braun, 1978; T. Remington, pers. comm.).

In this chapter we review lek mating in sage grouse in relation to two topics of current interest in mating system evolution. First, what are the links between ecology, female dispersion, and lek mating? Second, how does sexual selection operate on leks? Because relevant data are only just beginning to emerge, we shall only sketch issues related to the former topic and devote more space to the latter. Our analysis will indicate that some revision of established thinking on lek mating is necessary and point to areas where further data are needed to resolve outstanding issues.

METHODS

Data reviewed here come mainly from our own ongoing study of female ranging and lek behavior of sage grouse at three sites in eastern California and previous intensive studies of lek behavior by Wiley (1973a) and Hartzler (1972) in Montana and Wyoming.

Female ranging data were obtained by radio telemetry of birds captured by spotlighting and fitted with battery- or solar-powered transmitters attached to a poncho harness. From 1982 to 1984 six females were tracked daily (or more frequently in some cases) from late winter into May and a further four were followed through April (the mating period).

Our lek studies have entailed monitoring ten lek sites, with from 5 to 100 males per site, at one-to-four day intervals from early March to mid-May annually since 1980. Since 1981 one lek has been the focus of intensive daily attention in each year. Observations were made from distances of 200–300 meters using high-power telescopes. This method has the advantage that activities of individual males can be monitored over the entire lek without disturbances. Individual males are recognized by color bands (fitted to birds spotlighted at night) or natural differences in tail shape and markings. By contrast, Wiley and Hartzler's data were collected from observation blinds sited on leks and refer to only a subset of the total males present. Wiley's data were collected during the mating peak at three different leks (one per year), while Hartzler's spanned the entire season on one lek over three years.

Sampling methods and variables sampled were similar in all three lek studies. Variables measured included numbers and locations of copulations, display repetition rates, fighting activity, and spatial ranges of individual males. These data were obtained by focal animal samples (display and fighting) taken directly in the field (Hartzler, our study) or from time-lapse film records (Wiley), focal and ad lib observations (copulations), and regular scan samples (ranging data). We also recorded movements of individual females on the lek from arrival to departure. In addition we took a number of morphological measures from birds captured for marking. Nonstandard procedures used in data analysis are described in relevant sections of the following account.

ECOLOGY, FEMALE DISPERSION, AND LEKS

Study of links between ecology, female dispersion, and lek mating is in its infancy. Pioneering eco-correlate studies of birds identified broad associations between gross diet and both promiscuity and absence of paternal care (Lack, 1968). However, perhaps because finer-grained ecological differences are important, the correlations fail within some taxa (e.g., the grouse:

Wiley, 1974; cf. Wittenberger, 1978) and do not explain variation in mating systems among promiscuous species.

Recently, attention directed toward the relationship between female dispersion and mating system has led to the recognition of a positive association between the degree of male clustering and female range size (Bradbury, 1981). This correlation has generated hypotheses linking female movements with male mating strategies, though the links with paternal care remain comparatively unexplored. Male self-advertisement, contrasted with resource or mate defense, has been suggested to be a "default" mating strategy adopted when high female mobility makes these more common alternatives untenable (Bradbury, 1981; Emlen and Oring, 1977). Beyond this, variation in male dispersion from even (e.g., ruffed grouse) to highly clustered (sage grouse) has usually been attributed to male-specific factors, e.g., habitat requirements or predation on displaying males (reviewed by Bradbury and Gibson, 1983). However, two recent models indicate qualitatively how increased female home range size alone could intensify the degree of male clustering independently of male-specific factors.

In the first, "hotspot," model males settle on a female encounter surface produced by the spatial utilization distribution of the female population. The height of the encounter surface at any point reflects both the numbers of females passing through that point and the frequency with which each female does so. Males are assumed to sample a number of locations and then to settle according to ideal free rules at the site giving the highest expected mating success. The model differs from previous resource settlement models (Fretwell and Lucas, 1970; Parker, 1978) in that settlement of any given site not only devalues that site for later settlers, but also all surrounding areas visited by females encountered at the settled location. This feature causes male clustering at hotspots created by female home range overlap/contiguity to intensify as female home range size increases (Bradbury and Gibson, 1983; Bradbury et al., submitted). An alternative "female choice" model adds to hotspots a female mating preference for clustered over solitary males. This further lowers the expected mating success of a male settling within one female home range diameter of a settled site. As a result males resettle and a stable dispersion is reached when clusters of males (leks) are spaced at least one female home range diameter apart (Bradbury, 1981). By contrast, the hotspot model allows leks to be both more distant and closer than this criterion.

Both female ranging and male settlement patterns in sage grouse appear to fit the premises of the models. Prior to nesting, females move in ranges of around 1,000 hectares per month (100–1,000 times the values for non-lekking grouse species) that overlap extensively with those of other individuals (Eng and Schladweiler, 1972; Pyrah et al., Wallestad and Pyrah, 1974; pers. obs.) Further range increases during the mating season result from movements between disjunct wintering and nesting areas by some females in nonmigratory populations (three or six females radiotracked in our study) and presumably all females in populations where birds migrate between winter and summer ranges (Dalke et al., 1963).

Male sampling that could allow assessment of both male and female densities at different locations (as required by both models) is suggested by the behavior of yearling males, which attend leks in large numbers from the mating peak on (while females are attending), visit several different lek sites during the season, and often establish lek territories toward the end of this period (Patterson, 1952; Wiley, 1973a; Emmons, 1980). Occasional inter-lek movements by adult males (Emmons, 1980; pers. obs.) indicate some resettlement in older birds also.

Female ranging patterns provide further evidence favoring the hotspot over the female choice model. Female ranges during the mating period frequently encompass more than one lek, and some radiotagged females have visited two or more leks (pers. obs.; Dunn, 1983). Thus range size data provide a poor fit to the female choice model but are consistent with the hotspot alternative. Conceivably, hotspots might occur where females spend most time and/or on major traffic routes between such areas, for instance, between daily feeding and roosting areas or between wintering and nesting ranges. Data on both the paths and timing of female movements (dawn and dusk) versus the location and timing of lek display are consistent with the traffic route idea. However, a larger sample of female ranges will be necessary to evaluate this conclusion critically.

Given a preliminary qualitative fit to the hotspot model, a residual problem is the degree to which this model alone is sufficient to explain the actual degree of male clustering, or whether additional factors are necessary. Both habitat restriction (leks are usually in open areas, e.g., meadows within sagebrush-dominated plant communities) and predation by golden eagles (which take up to 5 percent of displaying males per year in our study areas) are candidates; however, the former's effect is unlikely to be large since leks occupy only a small fraction of available open areas.

SEXUAL SELECTION ON LEKS

The processes by which variance in male mating success on leks is generated are of particular interest for theories of sexual selection. Although a role for female mate choice is widely accepted in species where males provide resources or parental care, its occurrence in lek species that lack both is controversial (Halliday, 1983). For example, while some authors assume that female choice for particular male traits occurs in lek species (S. J. Arnold, 1983), others maintain that female choice is preempted by intermale interactions (LeCroy et al., 1980). Even among those who accept female

Sage Grouse Leks

TABLE 17.1
Normalized variances in male mating success on sage grouse leks

No. of males in sample	No. of copulations	Mating success variance index (H)	Lek size	References
12	51	0.110	58	Lumsden (1968)
8	107	0.293	158	Wiley (1973a)
19	82	0.314	30	Wiley (1973a)
21	42	0.344	260	Wiley (1973a)
12	78	0.191	60	Hartzler (1972)
31	216	0.541	75	Hartzler (1972)
27	178	0.255	80	Hartzler (1972)
12	19	0.552	12	This study (1983)

choice there is considerable disagreement over the extent to which choice is active or passive (sensu Parker, 1983; e.g., Arak, 1983) and the degree to which it is seen as merely reinforcing the effects of male combat or as an independent process (Wiley, 1973a; Hartzler, 1972). The possibility that interactions among females, in addition to male–male and male–female interactions, may play an additional role has been largely ignored.

The following sections present data on mating skew on sage grouse leks and then critically review the roles of interactions both within and between the sexes in determining the allocation of matings.

Variance in Male Mating Success

Table 17.1 compares the normalized variance in the fraction of matings per male (Bradbury et al., 1985) on sage grouse leks for eight data sets culled from the literature and our own study. This measure should not be confused with the Shannon-Weiner information index or diversity. $H = (n \times \Sigma [c_i - 1/n]^2)/(n - 1)$ where n is the number of males and c_i is the fraction of matings performed by the ith male. H takes values from zero if matings are equitably distributed to one if all matings are by one male; random values are around 0.01 (varying slightly with n and number of copulations). Alternative measures of mating success variance applied to avian lek mating systems have recently been discussed by Payne (1984).

Two points emerge from Table 17.1. First, in all data sets H indices are at least an order of magnitude higher than expected with random mating. Second, the degree of mating skew is quite variable both between leks and between different years on the same lek. The following sections examine some factors that may contribute both to mating skew and to its variability.

Interactions between Males and Mating Success

The possibilities for male interactions to influence matings depend jointly

383

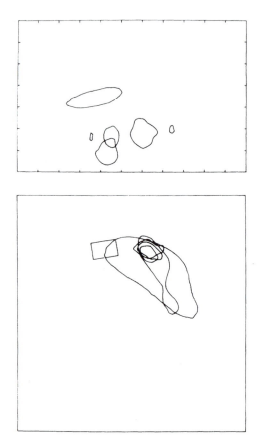

Fig. 17.1. Superimposed contour plots of 50 percent areas of use for male sage grouse lek ranges at Wheeler Flat in two years. The upper plot delimits 50 percent areas of use for four neighboring territorial males in 1982. The bounded area measures 110 × 70 meters and was used by 10–25 males throughout the season. The two smallest peaks are outlying parts of the ranges of the two central males. The lower plot shows equivalent data for five males in 1983. The area shown is a 1,050 × 1,050 meter square with the location and dimensions of the 1982 plot indicated by the enclosed rectangle.

on the extent to which different males interact, the nature of interactions that occur, and the effects on female mating behavior of intermale encounters. Since ranging patterns of males during lek display are quite variable, determine encounter opportunities, and are correlated with patterns of dominance, our discussion is focused around male spatial behavior on leks. For convenience we consider separately two extremes of a continuum that we term stable and unstable leks.

Stable leks. On many sage grouse leks most males occupy small display ranges (territories) that include a small core (ca. 0.01 ha) of exclusive use and broad margins that overlap the ranges of neighbors. Occasional males occupy larger areas that may overlap more extensively with those of others, or move between a number of small noncontiguous display ranges, on the same lek or occasionally on two different leks, throughout the breeding season (Hartzler, 1972; Wiley, 1973a; pers. obs., Fig. 17.1, upper plot). This situation limits the extent of encounters both socially and spatially to areas of range overlap between territorial neighbors.

With rare exceptions (that result in the eviction of a male from its territory), agonistic interactions between neighboring territorial males have no clear winners or losers, as indicated by subsequent approach and avoidance. By contrast males invariably win interactions with intruders into their core areas but may (without consequences for their territorial status) lose interactions outside their usual display territories (Hartzler, 1972; Wiley, 1973a; pers. obs.). Because interactions between males are relatively symmetrical, any effects on relative mating success must be more subtle than mere global exertion of dominance. Male interaction patterns when females are on leks suggest two potential routes for a male effect.

First, copulations initiated in territory overlap zones are regularly interrupted by neighbors who knock the copulating male off the female's back, apparently before sperm transfer in 70 percent of cases (Hartzler, 1972); interrupted copulations accounted for 15.4 percent of copulation attempts recorded annually in Wiley and Hartzler's studies and a similar, though more variable, proportion in our own data from smaller leks. Females falling victim occasionally remate with the same male, but may alternatively move to another male (sometimes the attacker) or even leave the lek without remating (pers. obs.; J. W. Scott, 1942). Thus copulation interference can reshuffle mating opportunities among males. However, for it to increase variance in mating success, attacks must be directed relatively more toward males with fewer initial opportunities to mate. Figure 17.2 shows that although the predicted trend occurs, it is not significant ($r = -0.302$, $n = 40$); it may arise because males with more mating opportunities are more likely to interrupt neighbors' matings (per opportunity to do so) than are less successful males (data combined from Hartzler and Wiley's studies: $r = 0.361$, $n = 78$, $p < 0.01$). In combination these data suggest that copulation interference is at best only a minor contributor to variance in male mating success.

A second possible effect of male interactions is suggested by an increase in the intensity and frequency of fighting by males involved in mating, attributed to increased rates of intrusion by other males (Hartzler, 1972; Wiley, 1973a). To the extent that this is likely to retard male tenure by increasing the probability of eviction or rate of energy reserve depletion, it

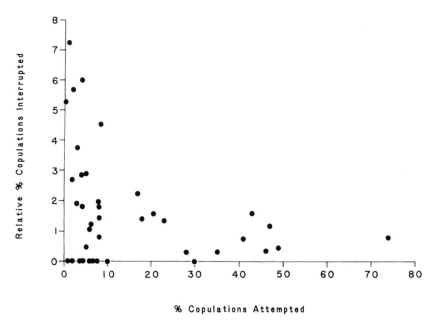

Fig. 17.2. Copulation interference in relation to numbers of copulations attempted. Each point is a value for a male over a single mating season. Numbers of attempted copulations are expressed as a percentage of the total copulation attempts recorded on the lek in that season. Copulation interference is measured by the fraction of the male's copulation attempts that were interrupted divided by the fraction of all attempts on the lek that were interrupted in that season. Data are from Hartzler (1972) and Wiley (1973a).

might both cause matings to be more equitably distributed and reduce the relative mating success of males more susceptible to these effects. However, although eviction and "burnout" occur, their frequency is low (Hartzler, 1972; Wiley, 1973a). Thus, as with copulation interference, the magnitude of any resulting effects is likely to be small.

Before leaving this topic, we note that intensification of fighting in successful males has sometimes been cited as evidence that males compete more intensely for sites at which females mate, with the presumed result that only more dominant males achieve breeding status (Wiley, 1973a). Leaving aside the issue of spatial mating preferences (discussed later), the evidence (though necessary) is insufficient to clinch the point; doing so would require showing that unsuccessful males were less well able to withstand an equivalent increase in intruder pressure. An additional prediction of the same hypothesis, that males should compete more intensely for future "mating centers" prior to the mating period is not supported by data on early season fighting (Hartzler, 1972).

Unstable leks. Though played down in earlier literature (Scott, 1942; Lumsden, 1968; Wiley, 1973a), strikingly different male ranging patterns in which entire aggregations shift within a display session, from day to day, or both, are a regular feature of sage grouse leks. In our study areas such movements occur regularly both before females attend and during the mating period (at times when females move from the lek to adjoining scrub) even in years when male spatial organization is otherwise stable. For example, in the most recent year of our study most males on one site moved 50-200 meters from their initial locations on a meadow to adjoining scrub during 40 percent of thirty-five display sessions. The frequency of such movements appears to vary both among sites and between years. Data from another year in which leks remained unstable throughout the season, coincident with unusually heavy snowpack and low population levels, showed two shifts in male behavior on unstable leks that increase the potential influence of male interactions on mating success.

First, male lek display ranges were significantly larger and overlapped more extensively than in years when leks were stable (Fig. 17.1). Individual 50 percent areas of use (measured by D. J. Anderson's [1982] MAP index) increased from 0.018 to 2.830 hectares while 95 percent MAPs increased from 0.040 to 8.357 hectares (Mann-Whitney tests: $p < 0.01$ in each case; $n = 4$ and 7). As a result most males could interact freely with most others.

Second, fighting changed in ways suggesting relatively clear-cut dominance relationships among males. Compared to previous years fights were shorter (Mann-Whitney test: $p < 0.02$), more frequently escalated to wing-beating (G-test: $p < 0.01$), and were more frequently decisive in outcome (G-test: $p < 0.001$); the analyses are based on focal samples of seven males each from 1981-1982 (stable leks) and 1983 (unstable leks). Consistent dominance relationships within pairs of males are suggested by data from ten dyads for which more than one interaction was recorded. In eight cases one male consistently beat the other; in one of the remaining pairs one male won two fights and four were tied, while in the last dyad wins were evenly split (Table 17.2). A ranking of individuals by the proportion of opponents beaten suggests substantial differences in fighting ability among males. This was tested by contrasting observed numbers of circular versus transitive triadic dominance relationships with those expected if dominance was determined independently within each dyad. Estimates of the probability of independence ranged from 0.01, obtained by simulation, to 0.063, using Kendall's method (see Appleby, 1983).

To see if the enlarged potential for male interference to affect mating success was realized we computed an index of fighting success for each male (see Gibson and Guinness, 1980) and correlated it with numbers of copulations. The result suggests that dominance could account for around 25 percent of mating success variance in this sample ($r = 0.503$, $n = 8$). However, the correlation is not significant, indicating either a type two error due to small sample size or that dominance and matings are unrelated. The former seems more likely since winners of fights often appeared to gain or maintain access to females at the expense of losers.

Summarizing, both the scope and routes for male interactions to influence matings directly are quite variable, reflecting variation in male ranging patterns between sites and seasons. In the most favorable situation male interactions may account for as much as 25 percent of the variance in mating success. However, more typically their influence is probably far smaller.

Interactions between Males and Females

Because females initiate copulation (by soliciting), they appear to have considerable control over the allocation of matings. The relegation of male interactions to a minor influence on mating success is often taken as evidence that female choice has a major effect on the mating distribution. However, the details of how choices are made and the identity of the salient cues are more controversial. We review evidence relevant to these issues below.

Female movements and mate choice. Any choosing process implies sampling. Female movements provide evidence suggesting both that females sample males prior to mating and that the sampling process may entail active rather than passive choice (sensu Parker, 1983).

Observations of small numbers of banded or radiotagged females plus comparisons of numbers of females attending and mating on intensively monitored leks (Lumsden, 1968; Dunn, 1983; pers. obs.) suggest that most females visit leks for two to three days prior to mating, mate once, and then lay a clutch without additional matings. Some females visit the same lek repeatedly while others may visit two or more leks on different visits. Thus, across days opportunities exist for extensive and repeated sampling of males before mating.

To investigate the choice process further we have mapped female movements from arrival to departure for forty-two female lek visits. Typically females flew onto the lek and then walked along winding paths that took them through the territories (50 percent use areas) of ten or eleven displaying males. This pattern was characteristic both of females that did not mate during a visit and of the majority that did. Of the latter, most mated with males that they had passed earlier in their visit; some of these moved between the chosen male and one or two others repeatedly before mating, suggesting sequential comparison of males. By contrast, other females moved directly from their arrival point to a male and mated without preliminary "sampling," suggesting either relatively instantaneous choice or that choice was based on an earlier visit.

TABLE 17.2
Dominance status and mating success of eight male sage grouse at Wheeler Flat lek,
April–May 1983

Winner	Loser								Fighting success	Copulations
	BT	Q	WBW	R	N	D	YYY	OG		
BT	—	4	1	2	1	1	3	0	.713	0
Q	4	—	3	3	3	5	8	1	.703	14
WBW	0	0	—	0	1	1	1	0	.550	0
R	0	0	0	—	3	0	1	0	.517	1
N	0	0	0	0	—	1	1	0	.428	4
D	0	0	0	0	0	—	0	1	.366	0
YYY	0	0	0	0	0	0	—	0	.291	0
OG	0	0	0	0	0	0	0	—	.267	0

NOTE: The fighting success index was calculated by a method that weights each male's rank by the ranks of each of his opponents (Gibson and Guinness, 1980).

The preceding data are more consistent with a process of active choice in which females compare males under equivalent conditions and reject some, rather than with passive attraction. Under the latter process a female will accept any male but is most likely to mate with the one providing the strongest intensity stimulus; the outcome of the process will depend on the disposition of males relative to the female's initial location as well as their relative stimulus strengths (Parker, 1983). A more rigorous test will require knowledge of the cues used in mate choice, which are discussed in the following section.

Mate choice cues. Mate choice cues can conveniently be bracketed into two classes whose relative importance has been controversial: spatial and phenotypic. Wiley (1973a) argued that females preferred to mate at particular central locations within the lek and that the monopolization of these sites by a few males was responsible for mating skew. In contrast, Hartzler (1972) suggested that phenotypic differences among males were the cues. We first reassess the evidence for spatial versus phenotypic cues and then evaluate current evidence on the importance of particular phenotypic traits.

1) Spatial versus phenotypic cues. Three aspects of the spatial distribution of matings have been used in support of Wiley's model: central location of successful males, stability of mating centers, and clustering of successful males within the lek. However, the possibility that the phenotype cue model or other processes could generate the same phenomena has been largely ignored. In investigating this issue we have assumed two explicit versions of the spatial cue argument: in one females choose the geometric center of the

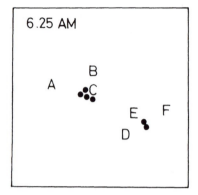 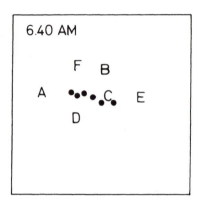

Fig. 17.3. Male relocation around females on Wheeler Flat lek, 1 May 1983. Locations of six males (A–E) and six females within an 80 × 80 meter area are shown for two scans fifteen minutes apart. Between 0625 and 0640 two females at the lower right moved up to join four others around male C. Subsequent moves by the other males generated the conformation at 0640 with C centrally located within the group.

male aggregation while in the second they use some topographic feature within the lek identified, post hoc, by the location of the most successful male.

A recurrent observation on sage grouse leks has been that successful males are centrally located (Scott, 1942; Wiley, 1973a). If male locations are fixed before mating begins, this is a necessary prediction of the aggregation center version of the spatial cue argument. The phenotypic cue model would not predict consistent central location of successful males without some additional process causing the pattern. In fact, the latter condition is met: once females cluster around a particular male others attempt to approach as closely as possible (Hartzler, 1972; Wiley, 1973a). The way in which this can generate central successful male location is particularly clear in unstable leks (Fig. 17.3). Until data controlling for male relocation are collected, the observed spatial pattern is insufficient either to critically support the spatial cue position or to discriminate between this and the phenotype model.

Stability of mating center locations through time is a clear prediction of the topographic spatial cue idea. In contrast, the phenotypic cue model would predict this only if correlations between phenotypic cue values and location remained constant through time. Available data (Lumsden, 1968; Wiley, 1973a; Hartzler, 1972; pers. obs.) show that shifts in locations where females mate are commonplace within seasons on sage grouse leks, occurring in two or three years in both Wiley's and Hartzler's studies and in each year of the others. Consistent with the phenotype model, Hartzler was

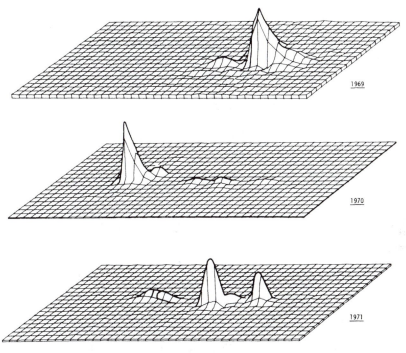

Fig. 17.4. Smoothed three-dimensional plots of copulation distributions Fords Creek lek in three years. Data are from Hartzler (1972). The area mapped is 90 × 180 meters.

able to correlate some shifts with the disappearance of successful males. The spatial stability of mating centers across years was carefully investigated by Hartzler who found substantial shifts from year to year on the same lek (Fig. 17.4). Thus female choice using topographic cues appears unlikely.

Finally, both spatial cue models predict that mating success should fall off monotonically with distance from the lek center. The phenotypic cue model does not make this prediction, though some correlation would be expected to result from male relocation (see above). Thus while spatial cue models require consistent correlation between male location and mating success, the phenotype model can accommodate this pattern, or no correlation, or some intermediate outcome depending on the strength of the relocation effect.

To investigate this issue we took male territory and mating data from each year of Wiley's and Hartzler's studies and computed correlations between mating success and distances from a male's territory center to each of two

391

TABLE 17.3

Correlations between mating success and distance from the lek center in
male sage grouse

Location	Center of male aggregation		Location of top male		No. of males
	r	*p*	*r*	*p*	
Muddy Springs 4/11–4/17/67	−0.227	0.221	−0.802	0.269	7
Muddy Springs 4/18–4/25/67	−0.183	0.377	−0.866	0.056	7
Fords Creek 3/27–4/5/68	−0.204	0.227	−0.623	0.275	12
Fords Creek 4/9–4/25/68	−0.281	0.135	−0.687	0.014	14
Dry Sandy 1969	−0.349	0.046	−0.463	0.140	21
Fords Creek 1969	−0.532	0.037	−0.751	0.026	12
Fords Creek 1970	−0.057	0.470	−0.287	0.986	21
Fords Creek 1971	0.062	0.616	−0.195	0.679	21

NOTE: All distances are from the center of the male's territory to the central reference point. The center of the aggregation is the average of the x and y coordinates of all male territory centers. One-tailed *p*-values were obtained by simulation (see text).
SOURCES: For 1967–1969, Wiley, 1973a; for 1969–1971, Hartzler, 1972.

lek centers: the average *x* and *y* coordinates of the pooled male territory centers (i.e., the aggregation center) and the location of the most successful male's territory. Data from the first two years of Wiley's study were each split into two periods to accommodate mid-season shifts in male territories. Probabilities of obtaining correlations as small or smaller than those observed were obtained by 1,000 simulations in which we held territory location and per male mating success constant but randomly reallocated males among territories. Table 17.3 shows that although correlations were generally negative, as predicted under the spatial cue models, they are significant in only two of eight samples for either spatial referent.

Summarizing, arguments cited previously in support of the spatial cue position are either inconclusive or unsupported by available data. The latter do, by default, support the idea that females choose males using phenotypic cues. We next turn to stronger evidence favoring this conclusion.

2) Phenotypic cues and female choice. Because of the large number of potential candidate traits, identifying phenotypic cues used in mate choice is a challenging task. To narrow down the search we have used computer simulation to investigate what degree of variation in male traits is required to generate the amount of mating skew observed on leks, assuming reasonable discrimination abilities by females and a single cue distributed a specified way among males (details in Bradbury et al., 1985). The most relevant model to the present context was one in which the cue was normally distributed with a specified coefficient of variation that could be varied between

TABLE 17.4

Coefficients of variation for phenotypic measures of
adult male sage grouse

Measure	CV (%)	N	References
Wing length	3.3	32	1
Tarsus length	5.1	10	1
Bill (culmen) length	6.5	32	1
Tail length	8.1	32	1
Body weight	8.3	32	1
Body fat (as % of weight)	21.1	9	2
Strut rate	12.2–22.2	12–31	1,3

SOURCES: 1) this study; 2) Hupp (1983); 3) Hartzler (1972).
NOTE: Strut rates are corrected for proximity to females.

runs. The results indicated that quite large coefficients of variation (10 percent or more) were necessary to generate the degree of mating skew observed on sage grouse leks, suggesting that the most likely candidate cues will have coefficients of variation at least as large as this.

Coefficients of variation for some morphological and behavioral traits of adult male sage grouse are listed in Table 17.4. It is clear that most linear morphological measures (perhaps excepting tail length) are less variable than the simulations deem necessary. However, both behavioral measures and energy reserves on which behavioral differences presumably depend fall within the acceptable range of values.

The most promising behavioral cue is the highly stereotyped strut display that contains complex visual and acoustic features (Wiley, 1973b). Its repetition rate is highly correlated with the presence of females and their proximity both within display sessions and over the mating season (Hartzler, 1972). Several authors have correlated individual strut rates with mating success with similar results: strut rates and mating success are positively correlated but high strut rate is not sufficient to ensure mating success (Wiley, 1973a; Hartzler, 1972; pers. obs.). Although Wiley suggested that the correlations might be a spurious product of female proximity, the same relationship is found when strut rates are corrected for this factor (Fig. 17.5). Statistically significant relationships between strut rate and mating success are present in all the larger data sets, with strut rate accounting for up to 37% of the variance in matings in bivariate regression analyses.

If strutting is associated with female choice, as the preceding data suggest, there is evidently more involved than just display repetition rate. At least four classes of variable might account for the relatively poor fit observed so far.

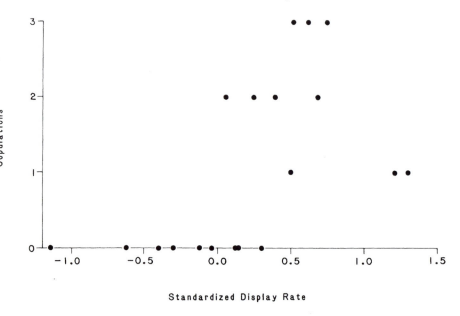

Fig. 17.5. The relationship between mating success and strut display rate for a lek at Lake Crowley in 1984. Display rate was corrected for bias due to female proximity and time of day by regressing measured strut rate on both of these variables simultaneously and taking standardized residuals as a corrected measure.

First, average measures of strut rate may obscure differences in the time at which the most active males hit their display peak. Since most matings occur only within a narrow time window and only a subset of males would be expected to peak at this time, such differences might lower the correlation between matings and strut rate.

Second, there may be qualitative differences in display that are obvious to females but have been ignored so far. Wiley (1973b) examined temporal properties of the display in a small sample and found them relatively invariant both within and between males. He also found no relationship between mating success and temporal measures among four males. However, negative evidence based on such a small sample is not persuasive. Recent analyses have demonstrated individual variation in several acoustic display components, one of which is significantly correlated with mating success after controlling for display rate (Gibson and Bradbury, 1985).

Third, other behavioral or (despite the arguments advanced previously) even spatial factors might affect choice. To examine this we subjected Hartzler's largest sample (eighteen males) to a multiple regression analysis, using matings as the dependent variable and strut rate (corrected for female proximity), a composite fighting index, and two measures of distance from

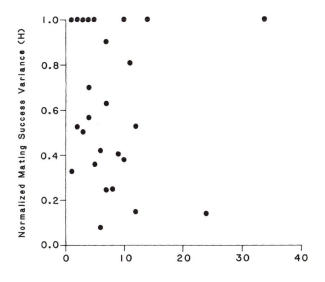

Fig. 17.6. The relationship between the normalized variance in male mating success (*H*) and the number of females copulating during the observation period. Data from Wiley (1973a) and Wheeler Flat leks in 1983.

the lek center as independent variables. The results indicated that strut rate retains its predictive ability when entered with either or both of the other variables in the regression. Partial regressions were not significant for either fighting or spatial indices.

Finally, interactions either among males (see above) or females may affect the mating distribution. We turn to the latter possibility next.

Interactions between Females

Two tyes of interactions between females have been suggested to affect the allocation of matings in lek species: interactions such as mate choice copying that increase the probability that two females will mate with the same male, and aggressive interactions that have the reverse effect.

Copying other females' choices. The possibility of choice copying arises from synchrony in female lek visits. This should potentially allow females to be influenced by matings occurring both during their nonmating visits and on the day they mate. Some approximate estimates of the opportunity to observe other females mating, based on our female lek attendance and mating data from one intensively monitored lek, suggest that 30–59 percent of females had opportunities to observe matings on previous lek visits, 64 percent had opportunities to observe earlier matings on the day they themselves mated, while only 18–27 percent of females could not have observed other females mate.

Despite these opportunities, however, we have been unable to find supporting evidence that choice copying occurs. An indirect test for same-day copying can be based on the following argument. If each female chooses her mate independently of other females who mate within the same session, then the distribution of matings should be unaffected by the number of females mating per session. Alternatively, if females copy the choices of others when given the opportunity, mating skew within sessions should increase with the number of females mating. To test this we took data from each year of Wiley's study and our own for 1983. For each year we pooled data for days when the same number of females mated and calculated the normalized mating success variance (H, see above) for each type of day. The resulting plot of H versus numbers of females mating does not support the copying hypothesis (Fig. 17.6). Data that could demonstrate between-day copying are not yet available.

Interference between females. Two types of apparently disruptive interactions occur regularly between females on leks: peck fights that often result in withdrawals by one female and mounting attempts on soliciting females that lead to fight or withdrawal of the victim (J. W. Scott, 1942; Lumsden, 1968; Wiley, 1973a; pers. obs.) These interactions are closely correlated with mating activity (G-test on association between interactions and mating activity in a sample of 122 days; $p < 0.001$) and nearly always involve at least one and often two females that solicited to males during the same lek visit (pers. obs.). Thus disruptive interactions involve sexually receptive females in situations where females may be competing for access to the same male.

Several consequences of interference between females are possible. For example, females that are harassed by others might be unable to mate with a chosen male during a display session, forcing mating with a second choice alternative or a delay in mating to a later lek visit. While the latter would be unlikely to alter the mating distribution, since successful males usually attend daily and have a low turnover rate, it might affect female fitness if the timing of breeding is critical to brood success. Rates of involvement in disruptive interactions appear to be relatively low, averaging 0.21 per female visit on one intensively studied lek ($n = 42$; pers. obs.). However, further data on the rates, social distribution, and consequences of female interactions are needed before a clear picture can emerge.

Two emergent themes of the preceding account are the need for further data in certain key areas and the necessity to revise some well-established notions about the social dynamics of lek mating. Bearing these in mind, some further comments on our conclusions are appropriate.

The first and most obvious area of ignorance is in the mapping of female dispersion onto ecological variables in lek species. The failure of cross-specific correlations between diet and mating systems in lek taxa indicates that future one-species studies will need to relate female ranging patterns to much finer differences in food availability, and also consider constraints on ranging imposed by predation/parasitism and energy allocation. For example, analyses of the large-size sage grouse winter ranges will need to consider the birds' selectivity for particular plants based on protein and monoterpene levels in leaves (T. Remington, pers. comm.), divergent cover requirements during the day versus at night plus the need to avoid predictable use of the same sites (the "MX effect") imposed by predation, and the periodic need to shelter from the worst effects of winter storms (pers. obs.). Winter-to-summer range shifts must further take into account nest-site habitat requirements (Wallestad and Pyrah, 1974) and a diet shift from sagebrush leaves to forbs (on which the young are also dependent) that occurs at this time (R. L. Patterson, 1952; J. G. Peterson, 1970; Wallestad et al., 1975). There are encouraging signs that data with which to assess and rank these alternatives may soon be available.

Although recent progress in developing and testing models that link female dispersion to mating strategies is encouraging, some problems remain. On the theoretical side there is a need to consider links between female ranging and parental care patterns, and to add to the basic models factors such as habitat restrictions and additional male fitness components that may constrain male settlement. On the empirical side, three weaknesses are evident. First is a need for more data on female ranging and male dispersion with which to test whether leks are in fact on "hotspots." Second is the desirability of making tests of the model quantitative. Finally, since tests that seek matches between female and male dispersion are unable to test divergent underlying assumptions of the models, for example, the extent to which the match is due to males responding to female dispersion (hotspot model), females responding to males (male-factor models), or some combination of the two (female choice model), it is possible that a good fit to one of the models could arise for the wrong reasons. This is an important limitation of current tests that could in principle be resolved by experiments that perturb the dispersion of each sex in turn while monitoring the other sex's response.

The analysis of male and female mating strategies on leks and their effect on mating allocation is clearly a multivariate problem. One weakness of our treatment, therefore, is that the data did not always allow us to use appropriate multivariate methods. Despite this four conclusions seem reasonably firm.

First, in contrast to earlier accounts that stress the spatial stability of lek organization, male ranging and interaction patterns are quite variable within and between seasons and lek sites. As a result the relative strength of factors affecting mating allocation, especially male interactions, can be expected to vary. Differences between estimates of mating success variance on leks from different years and sites point to the same conclusion.

Second, despite interactions both between males and between females, female choice clearly plays a major role in allocating matings. This conclusion is based on the failure of male interactions to explain a large fraction of the mating success variance even under the most favorable conditions (unstable leks), female behavior indicating prechoice sampling of prospective mates, and the absence of evidence for a female choice-copying process. More information is needed to assess the impact of disruptive interactions among females, but it is clear that any effect these have in equalizing mating differentials is insufficient to offset factors generating mating skew.

Third, the prevalent notion that females choose males on the basis of their location within the lek (Wiley, 1973a) is not supported. Spatial patterns of mating, particularly shifts in mating locations within and between seasons and lack of consistent clustering of successful males, are hard to reconcile with this idea, though they are consistent with choice for phenotypic cues. As a corollary, there is no basis for supposing that a positive correlation between the effects of male competition and female choice arises because females choose certain sites for which males then compete. Instead, female choice and male competition should be regarded as potentially independent processes until correlations between their effects on particular traits have been unravelled empirically.

Finally, there is encouraging preliminary evidence that female choice may be based on differences in male display. This conclusion fits with recent data from both black grouse leks, which show some display components to be highly correlated with mating success (Kruijt et al., unpubl. data). Although here we have focused on display rate, other display variables are the subject of recent analyses that are reported elsewhere (Gibson and Bradbury, 1985).

18. Grouping, Associations, and Reproductive Strategies in Eastern Grey Kangaroos

———————————— • ————————————

PETER J. JARMAN

AND COLIN J. SOUTHWELL

THE EASTERN grey kangaroo *Macropus giganteus* is one of the largest of the Macropodidae, a family of Australasian marsupial herbivores whose species occupy niches equivalent to those of the medium and small ungulates on other continents. Eastern grey kangaroos are among the most social macropodid species, yet their groups are usually small, despite their preference for mesic, lightly wooded, or savanna habitats. To a biologist familiar with the social organization of ungulates of similar size, organization of kangaroo society is remarkably obscure. Overtly defined and defended territories, leks, temporary or permanent monopolization of a group of females by one adult male, frequent all-male groups, or partial segregation of the sexes for even part of the year, are all absent. Instead, almost all classes seem to mingle freely, at all times, in open-membership groups, without spatial constraints.

Eastern grey kangaroos are also among the most strongly sexually dimorphic of the macropodids (and of all terrestrial mammals), suggesting great intrasexual competition among males for matings. This chapter looks at aspects of that competition and its effect upon the population's organization, and at features of the reproductive strategies of the two sexes.

STUDY AREA AND ANIMALS

Eastern grey kangaroos are widespread in relatively open habitats in the higher rainfall (300–1,000 mm per annum) areas of eastern Australia from Tasmania to northern Queensland. They avoid dense forest. In semi-arid shrublands they congregate near treed watercourses. They are predominantly grass-eaters, moving and feeding during the night and for two to four hours after dawn and before dusk. They prefer leaves of soft, green, and therefore often short, grasses. When not feeding they lie or rest crouched; they prefer to rest in shade in hot, sunny weather. They are usually found in groups both when feeding and resting, and the mean size of groups is from two to five animals (Caughley, 1964; Kirkpatrick, 1966; Southwell, 1984a;

R. J. Taylor, 1982), although groups of over fifty can be found in some areas. Superficially groups appear to contain any combination of sex- and age-classes, but this observation is discussed in more detail below. Group membership is open. Kaufmann (1975) has suggested that the population is organized into "mobs," fairly discrete subpopulations of regularly inter-acting individuals.

Kangaroos epitomize the peculiarities of marsupial reproduction. An eastern grey kangaroo is born after a gestation of thirty-six days, weighing under one gram. Blind and embryonic in appearance, it finds its way into the pouch and attaches itself to a nipple. After four to five months it begins to poke its head out of the pouch, and may start to leave the pouch for short periods a month later, usually while the mother is resting (Stuart-Dick, pers. comm.). At about 10½ months old it leaves the pouch permanently. In some other macropodid species this happens just before the female next gives birth, but in eastern grey kangaroos there is usually a delay of some weeks before the next juvenile is born. The fully emerged juvenile, now called a young-at-foot, continues to suck, putting its head into the pouch to do so. The female may suckle two juveniles at once, a small pouch-young attached to a teat, and a young-at-foot. Weaning can occur at any stage from fifteen to eighteen months (Stuart-Dick, pers. comm.). Breeding occurs year-round in the wetter parts of the species' range, but with a peak of births in summer leading to juveniles permanently vacating the pouch in spring and early summer (pers. obs.).

Grey kangaroos finally leave the pouch weighing about 5.5 kilograms. Growth continues throughout life in both sexes, but at very different rates, resulting in old males that are several times the weight of females of the same age. Females over four years old average about twenty-four kilograms (Taylor, 1981), and individuals may reach thirty kilograms. They can first conceive at about two years old, weighing about eighteen kilograms. Males are physiologically mature by 2½ years and twenty-five kilograms, and a few eventually reach as much as ninety kilograms (Jarman, pers. obs.). Males in some populations are slightly browner grey than the silver-grey fe-males, but there is otherwise little difference in coloring or marking be-tween the sexes, unlike two other large kangaroos, the red kangaroo *Macro-pus rufus*, and the wallaroo or euro *M. robustus*, in which the sexes differ strongly in coloring. Grey kangaroo males develop longer and more heavily muscled forearms, and more muscular shoulders and necks, than do females (Jarman, pers. obs., 1983). Their hands and claws are also larger. Large males grow thicker skin over the shoulders and belly. These aspects of di-morphism represent an exaggeration of the weaponry and protection of the male, which increases with age; the older and larger males have relatively as well as absolutely the longest and heaviest forelimbs. There is thus the physical basis for a male hierarchy related to individual size and age.

Strategies of Eastern Grey Kangaroos

Information used here comes from studies made of kangaroos on the New England tablelands of northern New South Wales. In particular, we describe a population at Wallaby Creek (near Tooloom) living on partly tree-cleared, cattle-grazed pasture and adjacent logged forest. The study area covers about five square kilometers of the valley of the Wallaby Creek, which has wet eucalypt and rain forest on one side, and on the other rough pasture of short grasses and tall tussocks, with remnant stands of eucalypts, casuarinas, and evergreen forest trees. Altitude ranges from 400 to 700 meters above sea level; annual rainfall is about 1,000 millimeters, falling mainly in summer. Frosts occur in winter. Nine other macropodoid species occur in the valley, and there is a numerous population of dingoes (*Canis familiaris dingo*), which prey upon the kangaroos.

The rough pasture provides a variable food resource for the kangaroos, their preferred short, green, low-fiber grasses being most abundantly available and continuously distributed on the lower slopes, which are the most heavily grazed by cattle and cleared of trees. On the steeper slopes there are either remnant areas of forest with little grass cover, or pastures dominated by tall, tussock-forming grasses, which the kangaroos find less palatable. The short, green grasses are most nutritious and accessible in spring and summer. Studies elsewhere (R. J. Taylor, 1981) have shown no appreciable differences in the diets of eastern grey kangaroos of different sex or age classes. These kangaroos appear not to have special habitat requirements associated with any particular life stage or reproductive situation. Although mothers restrict their ranging when their juvenile leaves the pouch permanently, they do not select a particularly cover-rich habitat for this, as do some species that leave the juvenile lying out at this stage (e.g., red-necked wallaby *Macropus rufogriseus*: C. N. Johnson, pers. comm.). When not feeding the kangaroos may seek shelter from wind, rain, frost, or sun. Appropriate shelter is more available in the remnant forest areas, but is still adequately available in or adjacent to the cleared pasture. Free water is rarely drunk in the study area but is always available in farm dams or streams throughout the area. Grey kangaroos are preyed upon by dingoes, and infrequently wedge-tailed eagles (*Aquila audax*) attempt to take juveniles. The kangaroos avoid dingoes by surveillance and then flight. We cannot yet evaluate habitats for the relative advantages they offer against predators but think that the kangaroos would be less easily taken unawares in the open habitats of the tree-cleared pasture. Thus in this study area the eastern grey kangaroo population has well-dispersed resources equally available to all its sex and age classes. Of these, food is the most heterogeneously distributed in space and time, being most abundant and of highest quality in spring and summer on the lower slopes of the pastureland (the area described below as the high-density stratum).

In a given season the kangaroos may prefer to feed on specific patches of pasture that provide relatively greener, leafier, or more continuously dispersed palatable grass. Such patches are never the only food sources and usually cover several hectares, which is large relative to the typical individual's home range (30 to 90 ha). Preference for patches changes with maturation of grasses, state of weather, and incidence of fires; and this, together with their size and marginal superiority to other pasture, would make them unprofitable to defend perennially or even temporarily as exclusive food sources. We have seen no sign of any defense of them. A kangaroo may spend most of one day's feeding in one patch, or it may feed and move entirely between patches. Rates of movement while feeding are slow: often under one hundred meters per hour and much less when feeding intensively on productive, short, green, leafy grass.

The study area carries a pegged one-hectare grid, used to record the locations of all sighted groups, and as a basis for habitat evaluation. Nearly all kangaroos using the core of the study area can be recognized by natural features and are well habituated to the observers. Individual males are classified by size and development as either juvenile, subadult, or small, medium, or large adult. Females are classified as juveniles, subadult, or adult, and the latter by their current reproductive state into six classes. These, and their usual duration in the development of a juvenile, are: female with discernible, small pouch-young, 2.4 months; with medium pouch-young, 1.7 months; with large pouch-young, 1.2 months; with young-at-foot, about 4 months; with young-at-foot and a small pouch-young simultaneously (no estimate of duration); and female without any discernible young (referred to hereafter as without young or with no young), usually 2 to 4 months. It must be emphasized that females in the class "without young" may actually be carrying a very small pouch-young, too small to make the pouch bulge perceptibly, and that a proportion of the young-at-foot may still be returning to the pouch.

Data on occupancy, group size and composition, and ranges and associations of individuals were collected during monthly patrols of transects through the study area, from April 1982 to March 1984. Monthly field trips lasted eight to twelve days. The data on occupancy have been used to divide the greater part of the study area into two strata, carrying adult kangaroos at mean densities of about 40 km^{-2} and 16 km^{-2} respectively. There were only small seasonal shifts in occupancy and the two strata represent areas of contrasting density at all times of year (Fig. 18.1). Although defined on differences in our records of occupancy by the kangaroos, the two strata also differ in the resources they offer. The high-density stratum has less tree cover, fewer steep slopes, and more short, green grass than has the low-density stratum; it therefore gives the kangaroos more abundant, higher-quality food in more open habitat where surveillance for predators is easier. The

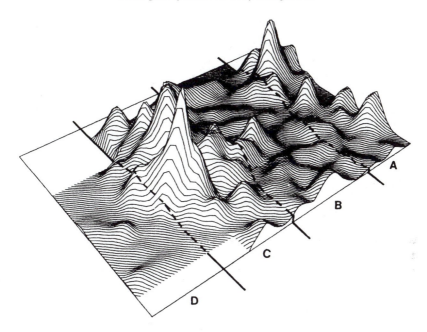

Fig. 18.1. Year-round occupancy (vertical scale) values for eastern grey kangaroos in the Wallaby Creek study area; data collected on systematic transects from April 1982 to February 1984. The marked zones are: A, a hilltop zone of mixed kangaroo densities (not discussed here); B, the low-density zone of the mid-slopes; C, the high-density zone of the lower slopes and riverine flats; and D, a generally unoccupied zone of dense forest (not discussed here). The river called Wallaby Creek runs from north to south (left to right in this projection) along the boundary between zones C and D. Compare this projection with Figure 18.9 for scale.

strata are not occupied by totally distinct populations of kangaroos. Most adult individuals occupying the high-density stratum have also been recorded in the low. Other adults from beyond the two strata also enter the low-density stratum without entering the high. We interpret the high-density stratum as being central to the ranges of one or two mobs while the low-density stratum is peripheral to the ranges of those and another mob. Records of kangaroos seen in the two strata are contrasted in some of the following analyses to investigate whether there is any evident displacement of classes of individuals from the preferred high-density stratum.

The objective of the study has been to elucidate the reproductive strategies of male and female eastern grey kangaroos. Clues to these strategies, provided by analyses of grouping and association between classes of adults, are presented in this chapter.

GROUP SIZE

Two earlier studies of eastern grey kangaroos on the New England table-lands (Southwell, 1981; R. J. Taylor, 1981) investigated grouping, comparing kangaroos at sites contrasting in their available resources. Neither found significnt seasonal variation in mean group size. At only one of the five studied sites did group size vary significantly with time of day, being smallest in the early morning (Taylor, 1981). Both studies found that group sizes were significantly correlated with population density, as: $\log_n$ mean group size $= 0.95 + 0.01$ density (animals/km^2) (Southwell, 1981); $\log_n$ mean group size $= 1.02 + 0.01$ density (animals/km^2) (Taylor, 1981). Even the changes in group size with time of day were best explained by changes of density between habitats used for feeding and resting.

Southwell also found a highly significant correlation between group density (number of groups per unit area) and population density. This relationship was logarithmic, while that between group size and population density was exponential. Consequently, at low population density a small increase in density leads to a greater increase in number of groups than in size of groups, while at high population density the reverse is true and an increment in density is reflected in larger, rather than more numerous, groups (Southwell, 1984a). These studies tend to confirm Caughley's (1964) suggestion that group size in large kangaroos was determined by the rates at which animals randomly (in time) joined or left groups. Southwell (1984b) showed that this rate rose with population density and group size, and that there were no socially induced upper or lower limits to the sizes of groups. At all study sites the majority of grey kangaroos were in company with at least one other.

Data from Wallaby Creek agree with these findings. Group size is not limited by tight and closed associations of individuals (other than those between mothers and dependent young-at-foot). Although frequencies of association vary between pairs of individuals, even the most frequently associating kangaroos are not always found together. At Wallaby Creek the mean group sizes, three to six animals (Table 18.1), are similar to those of most eastern grey kangaroo populations in similar climatic regimes. There are suggestions of seasonal variation in mean group size in the high-density stratum, the two summer samples (November to March) showing slightly larger mean group sizes than samples from other seasons. However, the difference is small enough to have been caused by permanent emergence in summer of juveniles from the pouch; seasonally regular variation in group size of adults is not apparent. Nor is there an obvious difference between the first year of data-gathering, when the area received very low rainfall, and the second year, of above-average rainfall. Grouping, then, does not appear to be sensitive to strong seasonal and annual variations in quality and quan-

TABLE 18.1
Mean group sizes within high- and low-density strata in time periods
for the first two years of the study

Period	High-density stratum		Low-density stratum	
	All animals	Adults only	All animals	Adults only
Apr.–June 1982	4.99	4.29	3.89	3.63
July–Oct. 1982	4.86	3.68	3.65	3.06
Nov. 1982–Feb. 1983	5.91	4.19	3.97	3.21
Apr.–July 1963	5.03	3.90	3.65	3.17
Aug.–Oct. 1983	5.03	3.81	4.21	3.47
Nov. 1983–Mar. 1984	5.56	4.86	3.31	3.02

NOTE: Summer samples are from November to February or March.

tity of pasture. Nor does there appear to be any socially induced, seasonally regular variation in adult group size. Consequently, there is no season when larger groups provide any male with a better than usual chance to monopolize access to females.

POPULATION COMPOSITION AND DENSITY STRATA

In 15,012 sightings in which the classes of all individuals in the group could be determined, 23 percent of the individuals were juveniles or subadults. This proportion varied seasonally, rising as high as 32 percent in summer 1982-1983 in the high-density stratum, and falling as low as 8 percent in the low-density stratum in autumn and early winter 1982. There was a significant overall difference between the high- and low-density strata in the proportion that subadults and juveniles formed of their sampled populations; these were 0.252 and 0.185 respectively ($\chi^2 = 69.07$, $df = 1$, $p < 0.001$). The sampled adult population had an overall sex ratio of 0.49 males to one female, indicating mortality or dispersion biased heavily against males, which in this population outnumber females 1.5:1 when they leave the pouch (Stuart-Dick, pers. comm.).

An adult sex ratio so heavily biased against males indicates a polygynous mating system. Species of large, terrestrial, herbivorous mammals that display polygyny and are found in open-membership groups generally show a hierarchical organization among their males. One consequence of male-imposed organization to look for in such a society would be a displacement, from the area densely occupied by females, of some males because of the aggression of others. Such displacement was not evident in a comparison of the adult sex ratios of samples from the high- and low-density strata, nor did variations in the ratio show any clear seasonal pattern. Thus there was no

TABLE 18.2

Ratios indicating the relative proportions of classes of adult male
and female kangaroos in the high and low-density strata in
samples from all months combined, and summer samples only

Ratio of classes of adults	Density strata		
	High	Low	Significance
All year			
Medium:large males	1.79:1	1.33:1	$p < 0.005$
Small:large males	1.24:1	1.16:1	ns
Large males:females	0.11:1	0.14:1	$p < 0.01$
Medium males:females	0.21:1	0.19:1	ns
Small males:females	0.14:1	0.16:1	$p < 0.05$
Summer samples			
Medium:large males	2.32:1	1.39:1	$p < 0.005$
Small:large males	1.49:1	1.10:1	$p < 0.005$
Large males:females	0.09:1	0.14:1	$p < 0.01$
Small males:females	0.19:1	0.22:1	$p < 0.001$

sign of a perennial or seasonal male strategy of monopolization of access to females by spatial exclusion of a large fraction of the whole adult male population. However, when the composition of the male population is looked at in finer detail, some patterns of apparent relative displacement do emerge.

The sampled populations of adult males recorded in the two density strata differed in the proportions of their classes (Table 18.2). There were significantly more medium males to each large male in the high-density stratum than in the low, and this difference was greatest in summer. The ratio of small to large males did not differ between strata but differed seasonally. Because of the timing of their maturation from subadults, there were more small adult males to each large male in summer (1.486:1) than at other times (mean 1.096:1). This seasonal variability may have masked other changes in the ratio of small to large adult males.

If classes of adult males are considered in relation to females, further differences between the strata emerge (Table 18.2). Relative to adult females, both small and large adult males were less abundant in the high-density stratum, especially in summer, but medium males showed no differences between strata or seasons. It should be remembered that the absolute occupancy of the high-density stratum was 2.5 times that of the low, so that, for example, large adult males were still recorded absolutely more frequently in the high-density stratum despite being recorded at a lower rate relative to females there than in the low-density stratum.

There seem, therefore, to be two signs in these data of spatial organization in the adult male population. A proportion of the large adult males appear to decrease their use of the high-density stratum in summer. This movement is not apparent among medium males, but small males also increased in frequency relative to females in the low-density stratum in summer. We emphasize that the differences are slight. They come nowhere near a total exclusion of one class of males from the area most densely occupied by females.

Female classes are classified by reproductive state, not size or age, and an analysis of their relative distribution between density classes seeks signs of choice of one stratum or the other by females with young at a certain stage of development. None of the three classes of females with pouch-young was significantly more common, relative to all females, in either stratum; but females with young-at-foot or with young-at-foot plus a small pouch-young were significantly more common (Table 18.3), and females with no young were less common, in the high-density than the low-density stratum. The difference between the strata in the proportions of the female population formed by females with young-at-foot remains significant even when females without young are removed from the sample. The significance of the differences between strata in the proportion of females without young was similarly high in summer and nonsummer samples (summer: 0.286 high; 0.385 low; $\chi^2 = 20.95$; nonsummer: 0.235 high; 0.300 low; $\chi^2 = 22.48$; $df = 1$; $p < 0.001$ in each case). The same was true of females with young-at-foot (summer: 0.274 high; 0.219 low; $\chi^2 = 7.43$; nonsummer: 0.179 high; 0.152 low; $\chi^2 = 5.20$; $df = 1$; $p < 0.025$).

Several interpretations of these data are possible. Either females tend to move out of the high-density areas when they are temporarily without young; or females who live permanently in the low-density areas are reproductively less active, spending longer between births; or females in the low-density areas are losing young-at-foot sooner than are those in high-density areas while maintaining a similar interbirth interval. Predation may have played a part. The study site carries a variable population of dingoes which the landholders occasionally reduce by shooting or trapping. During spring and summer of 1983-1984 dingoes were particularly active in the high-density stratum and nearly every juvenile of known females disappeared soon after each emerged from the pouch. At that time the proportion of females with young-at-foot fell lower, and of females with no young rose higher, in the high-density than in the low-density stratum. Table 18.3 shows that trends in distributions of female classes between density strata reversed between the first and second years of sampling. Differential predation in the two areas is as likely an explanation of this as is any suggestion of changes in preference by the females. Preliminary analysis of the ranging of some known individuals shows no consistent shift in a female's occupancy from one density stratum to the other as her reproductive class changes. Our data

TABLE 18.3

Proportions of the adult female population sampled on transects in time periods in the high- and low-density strata and in the whole two years of sampling

Period	High-density stratum						Low-density stratum					
	FNY	FSY	FMY	FLY	FYF	FYS	FNY	FSY	FMY	FLY	FYF	FYS
Apr.–June 1982	.375	.312	.179	.046	.087	0	.463	.297	.140	.064	.036	0
July–Oct. 1982	.124	.202	.228	.163	.267	.015	.223	.170	.238	.152	.218	0
Nov. 1982–Feb. 1983	.216	.239	.092	.053	.374	.025	.373	.231	.051	.048	.270	.027
Apr.–July 1983	.237	.348	.190	.100	.088	.038	.210	.329	.256	.073	.110	.023
Aug.–Oct. 1983	.321	.187	.173	.094	.225	0	.290	.198	.184	.092	.235	0
Nov. 1983–Mar. 1984	.435	.320	.120	.041	.063	.021	.409	.303	.136	.035	.111	.005
Whole two years	.253	.261	.164	.090	.213	.018	.325	.240	.167	.088	.171	.008

KEY: Female classes: FNY = females with no young; FSY = females with small pouch-young; FMY = females with medium pouch-young; FLY = females with large pouch-young; FYF = females with young-at-foot; and FYS = females with young-at-foot and a small pouch-young.

so far do not, therefore, show that differences in composition of the female records in the two density strata were the result of predictable movements by the females of particular classes, but may have resulted from unpredictable predation.

The strong differences, which existed in most seasons, between the compositions of female populations in adjacent parts of the study area seem to have affected the distribution of male classes little, perhaps because those differences least involved the frequencies of females with small pouch-young, the class most likely to enter estrus. They formed similar proportions of the female population in each stratum (see below). Thus males, when optimizing their reproductive tactics, need respond to the absolute abundance of this class (which was more abundant in the high-density stratum) but not to its relative abundance (which did not differ between strata). Table 18.3 shows females with small pouch-young to have been most plentiful in summer and autumn.

COMPOSITION OF GROUPS

Frequency of Single-Sex Groups

There appears to be only a small degree of broad spatial organization, affecting the adult males and subadults and juveniles mostly, in the Wallaby Creek population. Further evidence of organization by one class, in pursuit of its mating strategy, might be found in the finer distribution of classes of adults among groups. A negligible proportion of groups contained no adults. Of the rest, 9 percent contained adult males but no adult females and 40 percent contained adult females but no adult male. Since females outnumbered males two to one, this is a glaring disparity in distribution of the sexes between groups. An analysis of 3,419 groups in the two density strata shows that all-male groups were relatively more frequent in the low-density stratum (0.076 of all groups in the high-density stratum versus 0.110 in the low; $\chi^2 = 11.12$, $df = 1$, $p < 0.002$); and the proportion of groups without adult males was greater in the high-density stratum (0.428 of all groups in the high-density stratum versus 0.352 in the low; $\chi^2 = 19.55$, $df = 1$, $p < 0.001$). The difference between the strata in their frequencies of no-male groups remained constant between seasons (summer samples: high 0.42, low 0.33; nonsummer samples: high 0.43, low 0.36); but the difference in relative frequencies of all-male groups was exaggerated in summer samples (summer: high 0.07, low 0.15; nonsummer: high 0.08, low 0.09).

Single-sex groups might contain one or several adults. The ratio of single-female to multifemale no-male groups did not vary between density strata, but did vary with the seasons. There were 1.04 single-female groups to each multifemale no-male group in autumn, 1.45:1 in winter and spring,

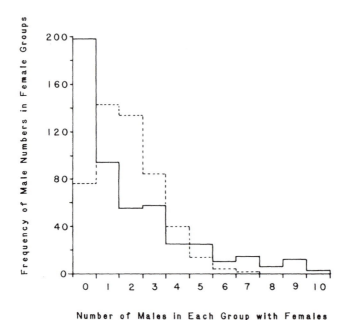

Fig. 18.2. Distribution of the frequencies of numbers of males sighted in 500 groups containing females. Solid line = observed distribution; broken line = distribution expected if males had been allocated to groups randomly. Observed mean number of males per group (when any males were present) = 3.1; expected mean = 2.2.

and 1.68:1 in summer ($\chi^2 = 11.61$, $df = 2$, $p < 0.005$). The ratio of single-male to multimale no-female groups did not differ between seasons, but did differ between density strata. In the high-density stratum there were 2.4 single males for every multimale no-female group, compared with 4.4:1 in the low-density stratum ($\chi^2 = 4.88$, $df = 1$, $p < 0.05$).

These figures show single-sex groups to consist most commonly of females; 23 percent of all groups contained just one adult female and no adult male; another 17 percent of groups contained several adult females and no male. By contrast single males form only 7 percent of all groups, and multimale groups without females a mere 2 percent. Single-sex grouping is a seasonally variable but common phenomenon among females, and a rare but density-related occurrence in males. The reasons for single-sex grouping appear likely to be different for each sex. The 40 percent of females found without any accompanying male was not for lack of males. When a female was in a group containing males, the number of males averaged between 2.8 and 3.2 (varying with the class of focal female), more than

TABLE 18.4

Classes of females and the types of groups in which they were
found in the first year of the study, taken from sightings
of individual females of known reproductive class

Female classes	Alone	With no male	With only one male	Alone with males	N
FNY	4.0%	28.8%	28.0%	4.8%	125
FSY	8.9	35.2	20.1	6.1	179
FMY	9.6	39.4	21.2	1.0	104
FLY	9.7	31.9	13.9	2.8	72
FYF	22.0*	50.8*	16.9	5.1	177

KEY: Female classes as in Table 18.3.

* Differs significantly ($p < 0.001$) from other classes combined.

enough adult males to have provided at least one to each female group if
they had been evenly distributed. Either some females actively avoided
males, or males were clustering. The latter is supported by Figure 18.2,
which compares the observed numbers of adult males with each of 500 female groups with the number which would have been expected if the males
had been randomly allocated to groups. Significantly too many female
groups lacked males, and at the other extreme there were more groups containing large numbers of males than would have been expected.

Female Classes and Group Composition

Records of the sightings of known individuals in the first year of the study
have been used to examine the types of groups in which individual females
occurred when in particular reproductive classes (Table 18.4). Females with
no young or with the three stages of pouch-young differed little in the frequencies with which they were seen alone, without males, with only one
male, or as the only female with males. However, females with young-at-
foot were significantly more likely than all the others to be alone or without
adult male company. Another study at Wallaby Creek (Stuart-Dick, pers.
comm.) has shown that a female whose juvenile is at the stage of permanently leaving the pouch tends to avoid other kangaroos, and to confine her
ranging to a small area. Our records confirm this, and it is these solitary females who caused the summer peak in ratio of single-female to multifemale
no-male groups mentioned above. Juveniles most commonly vacate the
pouch permanently in spring or early summer.

When females were seen with other females, there was a weak tendency
for some classes to associate. Thus females with no young, with young-at-
foot, and with small pouch-young were most often accompanied by females
of their own respective classes. Females with large pouch-young tended to

TABLE 18.5

Classes of females, and females in the field trip when
they showed signs of being in estrus or proestrus, and the
average numbers of adult males within their group when
any males were present (i.e., excluding no-male sightings)

Female classes	Large males	Medium males	Small males	Total adult males
FNY	0.98	1.27	0.87	3.12
FSY	0.73	1.40	1.06	3.19
FMY	0.76	1.25	0.78	2.79
FLY	0.69	1.39	1.14	3.22
FYF	0.67	1.33	0.97	2.97
Estrous or proestrous females	1.12	2.09	0.73	3.94

KEY: Female classes as in Table 18.3.

avoid like company and to be accompanied by females without young or
with young-at-foot.

When accompanied by males, the classes of females differed little in
either the average number of males or the proportions of the three adult
classes in the group (Table 18.5). These proportions closely reflected the
composition of the adult male population at that time. Therefore no class of
female appears to associate with a distinct set of males.

Estrous Females and Group Composition

There was a subset of females, however, that differed significantly from
the rest in the number and composition of the males accompanying them.
These were the females approaching or in estrus. For several days, sometimes more than a week, before her brief estrus, a female kangaroo is attractive to males, who try to associate closely with her, inspect her pouch
and cloaca if they can, sniff her urine, follow her as she moves, and eventually mount her if permitted to do so. Close consorting, close inspection,
courtship, and mating are the prerogatives of the most dominant male currently with the female. A more dominant male can supplant a subordinate
at any stage, but subordinates can still remain in the group containing the
proestrous or estrous female, provided they keep their distance from the female and her dominant consort. The classic picture of a kangaroo approaching estrus is of a small female moving slowly, followed by half-a-dozen
males in descending order of size, from a huge dominant to a small adult,
little bigger than the female herself. (All these statements are based on our
observations in this, and other studies.)

But this associating group of males is not just a random selection from the
available population. Analyses of the compositions of groups containing

proestrous or estrous females show that, compared with groups accompanying all other females, there were more males, and that significantly fewer were small males and more were medium and large ($\chi^2 = 9.50$, $df = 2$, $p < 0.01$; see Table 18.5). The ratio of medium to large males is no different from that within groups of males recorded with nonestrous females. These differences are conservatively estimated since all records for estrous females for the whole ten- to-twelve-day field trip in which they were seen to be closely followed have been used to describe their proestrous and estrous associations (since the precise beginning and end of estrus were not usually known).

As might be expected, in the observation period in which a female was estrous or proestrous she was alone in less than 3 percent of sightings (cf. 8 percent for other females except those with young-at-foot). More strikingly, she was the only female in a group with adult males in 31 percent of sightings compared with 4 percent for other females (except those with young-at-foot. $\chi^2 = 42.9$, $df = 1$, $p < 0.001$). Estrus was nearly twice as common in summer (November to March) as in the rest of the year (4.2 females scored as in estrus or proestrus per summer field trip, cf. 2.2 for other months).

Thus females who were approaching or in estrus clearly received more attention from adult males, especially the medium and large males. Most estrous females belonged to the class of females with small pouch-young. Of thirty observed or deduced matings, 70 percent involved females with small pouch-young, and 80 percent of fifty cases of females being closely consorted with or intently courted by males involved females with small pouch-young. Yet this class as a whole was not singled out for special treatment by males, reflecting the relatively small proportion of her total time in that class for which a female would be in or approaching estrus.

Male Classes and Group Composition

Dominance in the male society is closely related to size and determines access to estrous and proestrous females. Consequently one might expect to find a differentiation among males, based on relative size, in their association with females. Such differentiation occurs (Fig. 18.3) but, rather surprisingly, large males were found alone more often, and with females less often, than were medium or small males. Large males were also found as the only adult male with females significantly less often than were small males. Large males certainly do not seem to inhibit the other two classes' extent of contact with females in general, and appear on these data to have reduced their own.

Large males might limit access by the others to certain classes of females; but this appears not to have happened (Fig. 18.4). The three classes of males accompanied an almost identical mix of classes of females. Yet that mix was

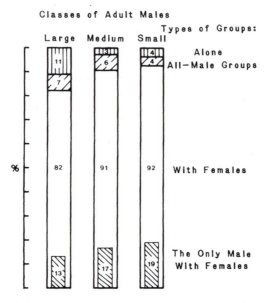

Fig. 18.3. The relative proportions of the types of groups in which the three classes of adult male kangaroos were recorded. Data are from sightings of known individual males at Wallaby Creek in 1982–1983. Sample sizes were 414 groups for large males, 656 for medium, and 335 for small. Values on the bar graphs are percentages and the vertical scale is marked at 10% intervals.

not a simple reflection of the composition of the female population, differing from it highly significantly ($\chi^2 = 37.14$, $df = 4$, $p < 0.001$) in containing fewer females with young-at-foot and more with small pouch-young than would be expected of unselective association. Even if the females with young-at-foot are excluded from the analysis because of their tendency to isolate themselves, the composition of the females seen with adult males was significantly different from that of the population at large ($\chi^2 = 17.65$, $df = 3$, $p < 0.001$).

Nor did the classes of females accompanying single males differ significantly between classes of males, and this subset of females was no different from that accompanied by several males. Again, large males do not appear to have monopolized sole access to any class of females.

In a small proportion of the records an adult male accompanied a single adult female. This could be a promising prelude to exclusive courtship and hence a time for exercised dominance to be detectable in the selection of the class of escorted females. But again the male classes differed little. Individuals of each class were seen alone with a single female equally frequently, and the mix of females with which they occurred differed nonsignificantly.

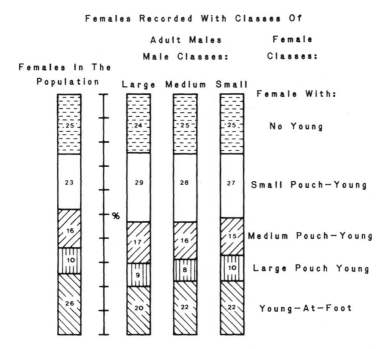

Fig. 18.4. The relative proportions of the reproductive classes among the adult female kangaroos recorded with each adult male class, and in the population at large. Samples were 341 groups with known large males, 594 with medium, and 310 with small, and the population composition of the females was taken from 4,575 records of females on systematic transects. Values on the bar graphs are percentages and the vertical scale is marked at 10% intervals.

Classes of adult males appear to behave similarly toward females, enjoying similarly free access and associating with similar proportions of classes of females. Strangely, large adult males had least contact with females and were most often solitary.

Dominance in Large Adult Males, and Association with Females

The interactions between individual adult males at Wallaby Creek clearly showed which males were dominant and which were subordinate. The following analyses concern kangaroos in the central part of the study area. These individuals were the best known and most observed, and were chosen because the whole of their ranges fell within the study area. They probably constitute what Kaufmann (1975) called a "mob," a subpopulation of mutually interacting individuals which interact far less with members of adjacent mobs. At any one time there were four to seven large adult males in this mob, and one of these was clearly dominant. The occupier of this most

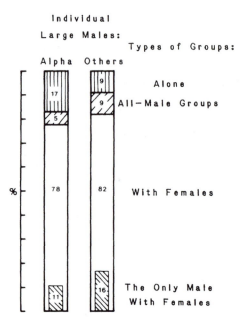

Fig. 18.5. The relative proportions of the types of groups in which the currently alpha male kangaroo and other large males were recorded. Sample sizes were 151 sightings for the alpha male and 349 sightings for other known large males. Values on the bar graphs are percentages and the vertical scale is marked at 10% intervals.

dominant position (who we will now call the alpha male) changed twice in the first two years of the study. In the following analyses we contrast the current alpha male, an amalgam of the three holders of that position, with the other large males.

The alpha male was seen alone much more frequently than were the other large males (Fig. 18.5), was in multimale plus multifemale groups no more frequently, and was in all-male groups or the only male with females less frequently than other large males ($\chi^2 = 10.96$, $df = 3$, $p < 0.025$). Indeed, the frequency of solitary records for the alpha male accounted for much of the difference between large males and other classes of males reported earlier. Other larger males were solitary little more often than medium or small males.

In records summed over the two years of the study the alpha male did not differ from the other large males in the composition of the males or females with which he was recorded ($\chi^2 = 1.98$, $df = 5$, $p > 0.05$), being recorded nonsignificantly more often than were other large males with females with small pouch-young (Fig. 18.6), especially when the female was the only fe-

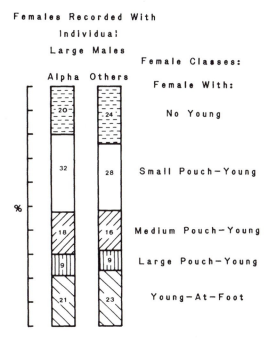

Fig. 18.6. The relative proportions of the reproductive classes among the adult female kangaroos recorded with the currently alpha male and with other known large males. Sample sizes were 299 recorded females for the alpha male and 659 for the other large males. Values on the bar graphs are percentages and the vertical scale is marked at 10% intervals.

male in the group. When a large male was the only male with several females, the alpha male was significantly more likely to be with females with small or medium pouch-young than were other large males ($\chi^2 = 7.38$, *df* $= 2, p < 0.025$). He was highly unlikely to be the only male with females without young. Thus there is some evidence that the alpha male achieved a small degree of exclusivity in his overall association with classes of females.

Figure 18.7 (a-e) presents month by month the proportion of the classes of adult females seen with the alpha male and other large males over the two years of the study, and compares those with the proportions of the female classes in the population. When the records for this two-year period are summed (Table 18.6), both alpha and subordinate large males show only low selectivity for the company of particular classes. The subordinate males need have selected for or against only 6.6 percent of the females, and the alpha males 8.4 percent, to have produced the observed frequencies of association. Nevertheless, the preferences expressed by both ranks of large

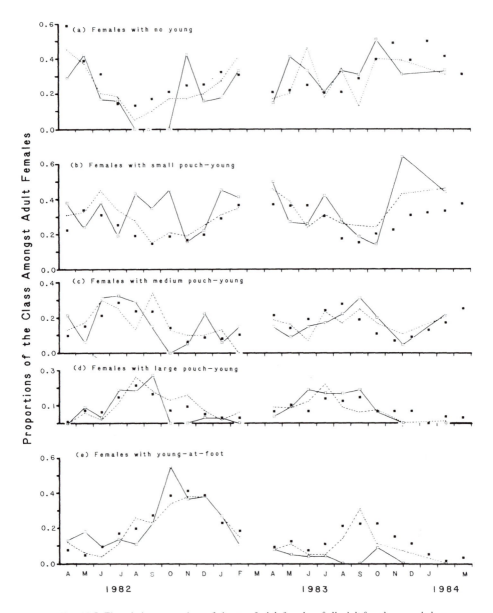

Fig. 18.7. The relative proportions of classes of adult females of all adult females recorded on transects in the two years of the study (black squares), recorded with the alpha male (white squares and solid line), and recorded with other large males (broken line). There were no samples in March 1983, and samples with the males have been combined for November and December 1983, and January through March 1984.

418

TABLE 18.6

Females recorded with large males, distinguished as the currently alpha male and other large males, for the two years of the study

Large males	Female classes						Totals
	FNY	FSY	FMY	FLY	FYF	FYS	
Alpha male							
Observed number	134	179	90	45	71	5	524
Expected number	152.7	139.9	85.0	45.2	93.1	8.1	524.0
Difference as % of records	− 3.6	+ 7.5	+ 0.9	0	− 4.2	− 0.6	± 8.4
Other males							
Observed number	368	444	245	125	218	16	1,416
Expected number	416.0	382.0	227.1	112.3	256.1	22.6	1,416.1
Difference as % of records	− 3.4	+ 4.4	+ 1.3	+ 0.9	− 2.7	− 0.5	± 6.6

NOTE: Female categories as in Table 18.3. Expected occurrence with males derived from proportions in the population.

males were similar, being against females with young-at-foot or no young, neutral or slightly for females with medium and large pouch-young, and most strongly in favor of females with small pouch-young. The alpha male's preference for the latter class appeared to be more emphatic than that shown by subordinate males.

Figure 18.8 shows that there were seasonal variations in the apparent selection by both ranks of large males for their female associates, so the overall measure of preference (Table 18.6) is a poor reflection of instantaneous preference. The strength of preference for association with each female class can be expressed as the proportional shift from the expected distribution of sightings among classes in each month, thus:

$$\frac{N_{i(obs)} - N_{i(exp)}}{\Sigma N_{1-i(obs)}}$$

where $N_{i(obs)}$ = the number of females in class i seen with the male in that month

$N_{i(exp)}$ = the number of females which would have been seen with the male if he had accompanied females unselectively

$\Sigma N_{1-i(obs)}$ = the total of females of all classes seen with the male in that month.

The monthly expression of total preference by that male would be the sum of absolute values of this statistic for each class of females, divided by two (because positive and negative preferences arise from one shift). Measured in this way, monthly preferences expressed by the alpha male were on average higher (22.8 percent) than those by subordinate males (14.2 percent).

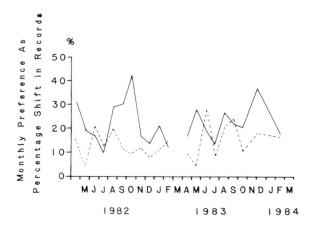

Fig. 18.8. The monthly extent of preference for association with classes of females shown by the alpha male (solid line) and other large males (broken line), calculated as the percentage departure from random association that would have been needed to produce the observed composition of females seen with the male(s).

There were two periods of high preference shown by the alpha male (Fig. 18.8), spring 1982 and summer 1983-1984. Preferences by the alpha and subordinate males were not correlated.

The six classes of adult females and twenty periods of time used in the analysis (November and December 1983, and January to March 1984 were each combined into single periods) gave 120 class × time cells in the matrix of expressed male preference. Alpha males expressed positive or negative preference exceeding 10 percent for a class of female in thirty-three of these cells, while subordinate large males did so in only seventeen. There were only two of the twenty time-periods in which alpha males did not exceed a 10 percent preference for or against some class of female, while subordinates failed to show that level of preference in nine periods.

There were suggestions in the data of month-to-month sequencing of the alpha male's preferences for classes of females. Females typically progress from having small, to medium, to large pouch-young, then to having young-at-foot, and back to having no young or small pouch-young again. This progression can be seen in the movement of the seasonal peaks of frequency of the classes of females in Figure 18.7. The alpha male's changing monthly preferences occasionally seemed to follow this same sequence of development of females' reproductive classes. Only twice did the alpha male show strong, positive preference for females with young-at-foot, in May and October 1982. In June 1982 he associated positively with females with small and medium pouch-young, and in July with females with medium and large pouch-young. In November 1982 he associated strongly with females without young, followed in December and January by association with females with small and medium pouch-young. Instances of strong association with females without young, in May, August, and October 1983, were each followed in the next one or two months by strong association with females with small or medium pouch-young.

Such sequencing was not evident in the records of preferences for female classes shown by subordinate males; nor, indeed, was it simple, regular, and predictable for the alpha male. His preference values for females with young-at-foot in one month were significantly correlated ($r = 0.610$, $df = 10$, $p < 0.05$) with those for females with no young in the next month for the first year of sampling, but the relationship was not significant in the second year. In the second year, what had formerly been a mildly expressed preference against females with young-at-foot turned to strong avoidance of them. Instead, from June to September 1983 the alpha male showed positive preference for females with large pouch-young, a class he had avoided for the previous eight months. This change probably related to differences between the years in the patterns of female reproduction, which may have been better discerned by the resident alpha male than by us infrequently present human observers.

THE REPRODUCTIVE STRATEGIES

The preceding descriptions of grouping and association of adult kangaroos at Wallaby Creek confirm that eastern grey kangaroos are a species with only a lightly organized society: organized, that is, in showing clear spatial or temporal patterns in the distributions of classes of individuals relative to each other. This lightness of organization is unusual among similarly sized species of sedentary, savanna-dwelling, mammalian herbivores, and calls for explanation.

The only major difference between reproductive classes of females in their grouping or association is brought about when females with young permanently emerging from the pouch isolate themselves for some time. Stuart-Dick (pers. comm.) argues that the relative fragility of the bond between mother and newly emerged young-at-foot necessitates social seclusion (although not physical hiding in this species) until the young kangaroo has accepted being denied reentry into the pouch despite still being allowed to put its head in to suck. The juvenile has to learn to move with the mother rather than be transported. Young have been seen to follow the wrong female when in a group. Such confusion in which female to follow would make a newly emerged young-at-foot very vulnerable if it were in an easily detected group surprised by a predator. Thus females of the class with young-at-foot were found alone and without males more than were any others. The classes of females resembled each other closely in their grouping and associations. The form of care for the young for its first ten months alters its mother's maintenance activities very little. A female without young and females with young at different stages of pouch-life choose the same habitats, move at similar rates, require qualitatively similar resources of food, shelter, and opportunity to avoid predators, and can lead sufficiently similar lives for there to be no need to differentiate among themselves in grouping and associations to any great extent.

The exception is the female in or approaching estrus. She is distinguished by the number and mix of males with her, although that may be involuntary, and by being found in a group where she is the only female with males much more often than is any other class of female. She is by no means always alone with males; it is not an obligatory part of her mating strategy. Indeed, estrous females may be closely followed, courted, and mated in the midst of multifemale groups. The exceptional proportion of sightings of an estrous or proestrous female alone with males reflects her greater rate of movement while in this state, averaging over three times the normal diurnal rate. She gives the appearance of moving because she is chivied by importunate males, but the persistent movement often takes her out of a group and far across her range, although she looks free to double back and keep moving within the group if she wished to. The males do not appear to direct her movements. We have no evidence, however, that a proestrous female begins wide-ranging movements *before* picking up a male escort, so the question of who initiates and induces her movement remains open. As she is attractive to males for about a week, during which time she moves widely, being courted or followed by an ever more noticeable train of males, it is unusual for any except the alpha male finally to mate with her (dominant males accounted for twenty-seven of thirty-five observed mountings). As a mating strategy it contains an element of conspicuous resistence to premature close inspection and mating, coupled with wide-ranging advertising of her state. These grey kangaroos occur in small, impermanent groups and have a male society so strongly hierarchical that, without her moving and advertising, the female might have only a one-in-ten chance of being, for the few hours of her estrus, in a group containing the locally indisputably dominant male. That male offers the female nothing in the way of protection or privileged access to resources. If the female reproductive strategy has been selected because it leads to mating with the locally most dominant male, the advantage presumably lies in linking the female's genes to heritable characteristics that have led to dominance in that male. Synchrony between females poses a minor risk to this strategy. The central mob contained about thirty adult females, and for the two years of study we estimate that there was, at the seasonal worst, a one-in-ten chance of one female's estrus occurring on the same day as another's. When two females are simultaneously approachng estrus, the alpha male monitors both, moving rapidly between them if they are in different groups.

The male's strategy complements the female one by his not only inspecting females for himself but also watching the reactions of other males to females. A male is attracted to a closely followed and courted female, and will himself approach her more or less closely depending on his rank relative to those of the males already with her. Relative rank is all-important since she will eventually mate with the highest-ranking male present. Thus the general male strategy contains two major elements: to gain high relative rank, and to detect estrous females.

Gaining rank involves long-term tactics such as survival (since relative rank is related to age) and maintenance of growth rate (since growth appears to be individually variable, persistent, and decelerates with age), and possibly even such immutables as being born to a high-quality mother with a resource-rich home range (although it will be some years before we can investigate this). It also involves short-term tactics of perpetual interaction with peers and assessment through sparring and more remote interactions. These occur in all situations and seasons but may intensify in the vicinity of a courted female, even among males too low in the hierarchy to approach the female and her escorts. The alpha male, once he has assumed his rank, has few sparring interactions (until he is finally challenged), maintaining his dominance by less direct intimidation.

Within one mob the male hierarchy is continuous and culminates in a single male; there is not a stratum of coequal, spatially separated, territory-holding dominant males at the top. Once that alpha male is displaced he is unlikely to reenter the hierarchy at a lower rank and climb to the top again. Two alpha males were defeated during this study; both deteriorated rapidly in condition, one soon dying, the other taking to a solitary existence on the periphery of his former range for some months before disappearing. A third is deteriorating fast at the time of writing. They suggest a tenure of alpha-rank in this population of about one year, reached at an age of about ten years. Their deterioration in condition could result from their greater daily rates of movement, larger ranges, more time alert, and lower time feeding than other large males.

A male might gain in relative rank by moving to an area where there are few larger males. The shift by some large males of some of their occupancy from the high- to the low-density stratum in summer might exemplify this, but since the alpha male was among those who did this (Fig. 18.9) the phenomenon is more probably a result of seasonal changes in the monitoring of females. It is interesting that small adult males consistently form a higher proportion of the male population in the low- than the high-density stratum, and that the disparity increases in summer, while medium males do not show such differential distribution. The value of changing tactics of spatial

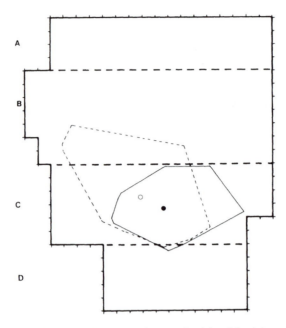

Fig. 18.9. Seasonal changes in the range and center of activity of the alpha male eastern grey kangaroo in the central, lower part of the Wallaby Creek study area in 1982–1983. Ranges shown are for winter (July to October 1982; solid line and solid dot; 39 sightings; 41 ha) and summer (November 1982 to March 1983; broken line and open circle; 32 sightings; 66 ha). The zones are labeled as in Figure 18.1, zone B being the low-density and zone C the high-density zones. This shows the summer expansion of his range taking in part of the low-density zone. The outline of the study area (bold line) is marked at 100 m intervals.

occupancy by males as they develop has yet to be explored, but medium males tend to use smaller ranges than do small or large males. The larger ranges and apparent seasonal changes in occupancy of small males may reflect a continuation of their process of dispersing out of natal ranges. Some even disperse away from and then return temporarily or permanently to their natal range (Stuart-Dick, pers. comm.).

The similarity between all classes of males in their choice of female company, and the tendency for males to be clustered rather than spread evenly between female groups, suggest either that all males are responding to the same cues of female attractiveness or that some males (presumably younger ones) are following the lead given by the others. It is unlikely to be the alpha male's lead that is being followed, since he was solitary more often than any other males and managed to express a preference for a class of females more strongly than other males. Yet he was least often the only male with a group of females, which could indicate that other males gravitated to him.

Perhaps the ways in which alpha and subordinate males selected their female company explain these findings. Mating opportunities were most likely to be offered by females with small pouch-young, but females of other classes might also enter estrus, so almost all female classes were worth checking. Moreover, with the exception of some females with young-at-foot, the female classes mingled fairly freely in groups, so even a strong association with one class resulted in coincidental contact with the others. With this in mind, the general male tactic of slight preferential association with females with small pouch-young, but contact at the same time with all other classes, appears to be an optimal way of monitoring the female population for mating opportunities while at the same time feeding, resting, and moving normally.

The alpha male improves upon this almost passive monitoring by wide-ranging, active searching and checking. He is seen alone more often than the other males because he moves between groups frequently. He spends very little time in all-male groups, perhaps because he spends little time in prolonged agonistic interactions, being able to depend, once his status is established, on brief threats which he can administer in any circumstances. He selects the classes of females with whom he is found more than do other large males, and may change this selected class from month to month. Does this mean that the class most likely to contain estrous females changes from month to month, or that the alpha male is monitoring over several months a segment of the female population which contains those females most likely next to come into estrus? Either explanation involves belief in the alpha male's ability to predict either the class or the individuals likely to produce the next estrous females. Such prediction is quite plausible, and, coupled with each female's apparently lifelong residence within a small home range and the alpha male's extensive ranging and checking of females, would make knowledge of the probable future location of estrous females a major component of the alpha male's mating strategy.

The price for this knowledge is paid in greater daily movement and reduced feeding and resting time, leading eventually to his collapse. To reach his dominant position he has had to grow faster than, or outlive, all his peers. For a decade he has sparred and fought his way up through the male hierarchy. But the rewards are great. For a year he fathers almost all the offspring conceived within his subpopulation. He monopolizes reproduction in an area far larger than he could possibly patrol as an exclusive territory. His absolute dominance allows him to interrupt any courtship, no matter how far advanced it may be. The risk he runs is not finding an estrous female or of being diverted from her. And that residual risk explains the presence of the train of attendant, subordinate, but still hopeful males.

COMPARISON WITH OTHER MACROPODS

The species of Macropodoidea show the same general trends in correlations between size, aspects of their behavioral ecology, and sexual dimorphism as do the Bovidae or Cervidae (Jarman, 1974, 1983; Kaufmann, 1974). The smallest species, weighing one or two kilograms, feed highly selectively on scarce foods, many of which have to be unearthed. These species are usually solitary, living well dispersed, at low density. Many spend the day in a nest, emerging at dusk to feed. We have observed one such species, the rufous rat-kangaroo, or bettong (*Aepyprymnus rufescens*), at Wallaby Creek and in captivity. It is homomorphic and females are well able to repulse importunate males. The male reproductive strategy involves monitoring the female's estrous state by inspecting her nest; he also tries to inspect the female herself but risks violent rebuff. The nest is also used as a predictor of the female's occurrence. The male emerges from his nest ten to twenty minutes earlier at dusk than do females, and he uses that time to move between nest-sites checking on the presence and status of the females. As a female approaches estrus he will visit her nest soonest and try to spend increasing time with her, at least in the early night; this is equivalent to the consortship of grey kangaroos. In captivity most mating occurs immediately when the now-estrous female emerges from her nest, is brief, and is terminated by the female driving off the male. The brevity of estrus makes predicting where and when the female will be in estrus very important in the male strategy. Unfortunately we know little as yet about the spacing behavior of these fascinating small macropods in the field.

A large genus of small to medium-sized macropods, the rock-wallabies *Petrogale*, are even more predictably spatially organized by the distribution of the caves and rocky refuges in which they shelter from predators and the weather during the day. Refuges must have specific characteristics of aspect, ingress, and egress, and are not usually abundant. A female takes up permanent residence in a refuge, perhaps sharing it with two or three female relatives. One study of the bush-tailed rock-wallaby *P. penicillata* (K. P. Joblin, pers. comm.) has indicated that size-related dominance among these females may be linked to reproductive success, the largest, dominant female producing most of the refuge's surviving juveniles. Males compete to control a small number of refuges and hence access, during the day, to the females in them. A male moves among his refuges, checking females, and will consort with a female approaching or in estrus. Unfortunately nothing is known about the spatial relationships of these rock-wallables at night when they leave the refuges to forage.

Males of other macropod species appear to use clues to the probable spatial occurrence of free-ranging females to enhance their chances of detecting estrus. Several medium-sized macropods live in forest or scrub but come out into clearings to graze at night. In some of these, such as the red-necked pademelon *Thylogale thetis* (K. A. Johnson, 1977) and swamp wallaby (pers. obs.), large males emerge sooner and range farther in the evening and may patrol more of the forest/pasture boundary than do smaller males. These large males can then inspect the females, which fairly predictably come to feed in the pasture or clearing. Another species, the red-legged pademelon *T. stigmatica*, lives in rain forest where it browses and eats fallen fruits. Brief observations at Wallaby Creek suggest that males may use patches of fallen fruit and the species' daytime resting sites between buttresses of forest trees as points of greater predictability to aid them in monitoring females.

Among the larger wallabies and kangaroos, which do not use nests or refuges and whose food is not so patchily distributed, there may be few spatial clues for a male to follow to find females. However, not all are as clueless as the eastern grey kangaroos described in this chapter. For example, red kangaroos *M. rufus* can show marked segregation of population classes. C. N. Johnson and Bayliss (1981) showed that, in samples taken across habitats occupied at different densities by a perennially breeding population of red kangaroos, the proportions of males classified as large and of females classified as having large pouch-young or young-at-foot correlated significantly. The correlation persisted despite seasonal changes in habitat preference. Those females were the ones most likely to be in estrus which, in red kangaroos, follows within days of the permanent emergence of a juvenile from the pouch (in contrast to the eastern grey kangaroos described above). Their segregation from other females could represent the same tendency to isolation at this critical stage in the juvenile's development as occurs in eastern grey kangaroos, accompanied by selection of a favorable habitat. If so, then the differential distribution of the large males between habitats is being induced by the potentially estrous females' resource needs. Moreover, the male red kangaroos's mating strategy appears to differ from that of grey kangaroos in the segregation of large from other classes of males. Perhaps this is a necessary additional measure brought about by the large red male to protect his chances of exclusive access to estrous females in a situation where those potentially estrous females are scattered singly and have reduced their ranging.

For all the diversity of their species-typical mating strategies and spatial organizations, the macropods do not show some of the forms of social behavior common in eutherian terrestrial herbivores such as the bovids or cervids. Defense of resource-sufficient territories has yet to be demonstrated for any macropod species, and when space is defended (as in rock-wallabies) it lacks the sharp boundaries and overt signals given by so many antelope or deer. While skin secretions and urine still convey social information in macropods, their use is less obvious and less "ritualized" than in

427

many ruminants. Vocalizations are used socially but at short range; no extant macropod bellows like a rutting red deer stag *Cervus elaphus* or roars like a territorial impala male *Aepyceros melampus*. There are no leks and, despite some social clustering of males in species like the eastern grey kangaroos described here, no discrete behavior herds in macropods. Macropod social behavior appears so muted largely because of the lack of strongly marked and defended territoriality as part of the male reproductive strategy. On the other hand, persistent growth and relative sizes of males, and of females in some species, play a greater part in macropod reproductive strategies than in those of all except the largest terrestrial eutherian herbivores. As we have suggested above, the two phenomena, persistent growth and absence of mating territories, are probably linked, although why the relatively small macropods should pursue a strategy which is confined to the largest species among eutherian herbivores (Jarman, 1983) remains to be analyzed.

19. The Ecology of Sociality in Felids

———————————— • ————————————

CRAIG PACKER

PERHAPS no mammals are as conspicuously solitary as members of the Felidae, yet the felids include one of the most remarkably social of all mammalian species: the African lion. Because almost all cat species are strictly carnivorous and females are solitary in all species except lions, comparison of the ecology of female lions with that of other felids should reveal the conditions that have resulted in lion sociality.[1] Until now, most reviews of felid sociality have ascribed group living in lions to the "advantages" of cooperative hunting of large prey (e.g., Schaller, 1972; Kruuk, 1972; Bertram, 1978, 1979; Gittleman, 1984; Macdonald, 1983). However, there has not been a convincing attempt to explain why cooperative hunting would be advantageous in lions but not in any other felid species. Furthermore, as shown below, the available data on hunting success in lions show that individual lions hunting in groups do not gain greater amounts of food than do solitary hunters.

In this chapter I briefly contrast lion social organization with that of other felids, compare preference for prey of large body weight across species, and test previous hypotheses about the advantages of group foraging in lions. I show that although females in moderate-size prides have higher reproductive rates (Packer et al., in press), group foraging does not confer obvious advantages in lions, and may even be disadvantageous under certain circumstances. I describe group dynamics in lions and suggest an alternative explanation for lion sociality. Since female lions do most of the hunting in the pride (Schaller, 1972) and it is the gregariousness of *female* lions that is so unusual among felids. I will be primarily concerned with the behavior and ecology of females. I briefly review gregariousness in male felids and contrast group formation in male lions with that of females.

FELID SOCIAL STRUCTURE

Since most solitary cat species show a similar social organization, I will first summarize the main features of the solitary species and then describe lion social organization in detail. Data on lions are based primarily on studies of

---

[1] I exclude domestic cats, which occasionally form social groups in households (Dards, 1978; Liberg, 1981), because the ancestral *Felis sylvestris* is always solitary, and because gregariousness may be a trait that has been favored by cat fanciers.

twenty prides in the Serengeti and Ngorongoro Crater, Tanzania. Two of these prides have been studied continuously since 1966 and the rest since 1974 (Schaller, 1972; Bertram, 1975a; Packer et al., in press). A. E. Pusey and I have studied these lions since 1978.

In all of the solitary cat species, females are intolerant of any conspecifics except their dependent offspring or males during the females' periods of sexual receptivity. In some species females have exclusive home ranges (tigers: Sunquist, 1981; bobcats: Bailey, 1974; lynx: Berrie, 1973; European wildcat: Corbett, 1979; servals: Geertsema, in prep.; and leopards: Bertram, 1982), while in others females have ranges too large to defend and thus their ranges overlap considerably (cougars: Seidensticker et al., 1973; jaguars: Schaller and Crawshaw, 1980; and cheetah: G. W. Frame, 1980). Although adult females may assist their maturing daughters to acquire a range near their own in some species (tigers: Sunquist, 1981; and possibly cheetah: Frame, 1980), no persistent associations between a mother and her mature daughter have been observed.

Female lions live in permanent "prides" consisting of two to eighteen related females, their dependent offspring, and a coalition of one to seven adult males that have entered the pride from elsewhere (Schaller, 1972; Bertram, 1975a; Packer and Pusey, 1982). Average pride size is similar across Africa (Van Orsdol, 1981). Male offspring are usually evicted with their fathers when a new male coalition takes over the pride, and sexually immature females are often evicted with them (Packer and Pusey, 1983a, 1984). Most females are recruited into their mothers' pride, but about 25 percent leave their natal pride at male takeovers or when their mothers give birth to a subsequent litter (Pusey and Packer, in press). Some females are solitary, being either the sole survivor of a pride (N = 2 cases in our study) or an individual evicted by incoming males before she had reached sexual maturity (four cases). If a solitary female successfully rears daughters, these may be recruited and the female returns to group living (Hanby and Bygott, 1979). However, there are no cases of unrelated females joining together to form a pride in our study areas (see also Table 19.3 below).

There is no apparent dominance hierarchy among females of a pride (Schaller, 1972; Bertram, 1979; Packer and Pusey, 1985). Unlike many other social carnivores, there is no reproductive suppression within the pride and all adult females typically breed at a similar rate (Packer and Pusey, 1983c, 1984). The number of surviving offspring per female is significantly higher in prides of three to ten adult females than in either smaller or larger prides (Packer et al., in press).

Pride females are often scattered in small groups throughout the pride's range and it is rare to find all pride members together. On average any two females of the same pride spend only 20-30 percent of their time together (Schaller, 1972). The fission-fusion nature of groups in a lion pride is illus-

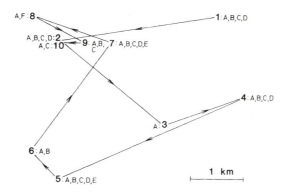

Fig. 19.1. Daily grouping and ranging patterns of Female A of the Gol Pride, Serengeti, 21–30 April 1980. Numbers refer to sightings on consecutive days and straight lines indicate changes in location. Letters B–F refer to the other five females in the pride. Females that were within 200 meters of Female A each day are listed next to that day's number.

trated in Figure 19.1, which shows the ranging and association patterns of an adult female over ten consecutive days. In this chapter the term "pride" refers to a set of individuals that regularly associates with each other (see Schaller, 1972), and "group" refers to a temporary aggregation of individuals from the same pride. Hence the female in Figure 19.1 was a member of a pride comprising six females and she was in groups ranging from one to five females over the ten-day period. Although their total range is too large to defend successfully, female lions do defend the portion of their range they currently occupy (Schaller, 1972). Pride ranges are stable over generations, and when prides split, the original range is subdivided (Pusey and Packer, in press). Solitary females also show stable ranges, but are often peripheral to their natal pride and thus range more widely.

Males are also solitary in most species, but typically have larger home ranges that overlap the ranges of several females (tigers: Sunquist, 1981; jaguars: Schaller and Crawshaw, 1980; cougars: Seidensticker et al., 1973; bobcats: Bailey, 1974; wildcats: Corbett, 1979). In contrast, male cheetah form stable coalitions of one to three males that defend territories much smaller than the ranges of females (Frame, 1980). Female cheetah are somewhat nomadic, ranging over very large areas, and male territories are found in areas utilized by a number of females (Frame, 1980). Coalitions of male lions defend female prides rather than a specific territory and large coalitions often "control" several adjacent prides simultaneously (Bygott et al., 1979; Pusey and Packer, 1983). Per capita reproductive success is higher in larger coalitions (Bygott et al., 1979; Packer et al., in press). Male-male relations are discussed in more detail in the final section below.

Schaller (1972) divided the Serengeti lion population into two categories: residents and nomads. Further observations in the Serengeti have shown that while most males have a nomadic phase, and may even remain nomadic for their entire lives if they never gain access to a pride (Bygott et al., 1979; Pusey and Packer, 1983), almost no females are truly nomadic. Subadult females may be evicted from their pride with subadult males, but the females either return to or settle near their natal pride (Pusey and Packer, in press). Adult females may leave their pride range temporarily either during extreme prey scarcity or while accompanying subadult offspring that have recently been evicted by new males (Packer and Pusey, 1983a, 1984), but they eventually return to their former range. Among Schaller's "nomadic females" were a number of females that moved over enormous areas of sub-optimal habitat (the Serengeti Plains). Subsequent observations showed that when seasonal extremes in prey availability became less severe on the plains, these females were resident in the same areas (Hanby and Bygott, 1979). Therefore, when considering lion social organization, all females should be considered as showing a high degree of philopatry, although they may temporarily leave their usual range or expand their range under certain circumstances.

FELID ECOLOGY AND SOCIALITY: THE COOPERATIVE HUNTING HYPOTHESIS

Kleiman and Eisenberg (1973) pointed out that the stealthy hunting style of most felids virtually requires solitary living. Therefore the evolution of group living in lions requires an explanation. Bourliere (1963) noted that solitary carnivores typically prey upon animals smaller than themselves whereas social carnivores often capture relatively large prey. Kruuk (1972, 1975) developed this observation more fully and suggested that sociality itself may be an adaptation to capturing large prey. Predators would be better able to subdue large prey if they hunted cooperatively. Kruuk (1975) tested this hypothesis in each family of the Carnivora, but found that the predicted association between group living and a preference for large prey held well only in the canids and hyenids. Nevertheless, the lion's preference for large prey is the most commonly cited explanation for lion sociality.

Table 19.1 shows the ratio of *maximum* prey size to female body weight across all felids. Compared to other species, lions do occasionally capture prey very much larger than themselves (e.g., buffalo, eland, and giraffe). However, such very large prey comprise only a minor proportion of the lion's diet in many areas (Schaller, 1972, tables 36 and 37), and are typically captured by males rather than by females (Schaller, 1972, table 62; see also below). Therefore, a more relevant ratio would be based on the prey

TABLE 19.1
Relative prey size in felids

Species	Female weight (kg)	Prey size/ Female weight		References
		Maximum	Mode	
Lion *Panthera leo*	141.0	7.45	1.06	1, 2
Tiger *Panthera tigris*	143.8	2.78	0.52	3, 4
Jaguar *Panthera onca*	77.6	2.26	0.39	5
Cheetah *Acinonyx jubatis*	60.5	4.13	0.26	1, 6, 7
Cougar *Felis concolor*	46.4	6.17	2.44	8
Leopard *Panthera pardus*	32.7	3.98	1.08	1, 2, 4
Snow leopard *Panthera uncia*	31.7	2.46	1.26*	9
Lynx *Felis lynx*	12.7	2.25	0.12	10, 11, 12, 13
Caracal *Felis caracal*	9.7	3.40	<0.10	14
Serval *Felis serval*	8.8	5.23	<0.10	14, 15
Bobcat *Felis rufus*	6.2	7.25	0.30	16, 17
European wildcat *Felis sylvestris*	5.1	0.49	<0.10	18
African wildcat *Felis lybica*	3.9	0.64	<0.10	14, 19, 20

SOURCES: 1) Schaller, 1972; 2) Bertram, 1982; 3) Schaller, 1967; 4) Sunquist, 1981; 5) Schaller and Vasconcelos, 1978; 6) McVittie, 1979; 7) Frame and Wagner, 1981; 8) Hornocker, 1970; 9) Schaller, 1977; 10) Saunders, 1963; 11) Saunders, 1964; 12) Haglund, 1966; 13) Nellis and Keith, 1968; 14) Smithers, 1971; 15) Kingdon, 1977; 16) Gashweiler et al., 1960; 17) Gittleman, 1984; 18) Corbet and Southern, 1977; 19) Stuart, 1977; 20) Rowe-Rowe, 1978

NOTE: Ratio of maximum prey size to female weight excludes domestic prey species (e.g., water buffalo for tigers and reindeer for lynx) since these are unlikely to have been typical prey during the recent evolution of each cat species. Modal prey size is the size category most often taken by each species, and wherever possible is based on direct observations of prey capture, data from radio-collared animals, or from stomach/scat samples. Ad lib data are biased toward large prey (Bertram, 1979) and snow leopards are asterisked since all data were based on ad lib observations.

size preferred by females and Table 19.1 also shows the ratio of modal prey size to female body size. Cougars, leopards, and possibly snow leopards prefer prey that are even larger relative to their own body weight than do female lions. Thus, although lions are certainly at the high end of the spectrum, several other species would also be expected to be social if a preference for large prey was the *only* cause of sociality.

Even if lions did prefer the relatively largest prey, it would not be proof that lion sociality evolved as a consequence of the advantages of cooperative hunting: cooperative hunting may be an adaptation to group living, rather than the evolutionary force resulting in group living (see Alexander, 1974). Furthermore, the association would not necessarily implicate cooperative *hunting* (Kruuk, 1975). Groups may form in order to defend large carcasses either against conspecifics (e.g., coyotes: Bekoff and Wells,

FOOD INTAKE PER LION PER DAY (kg)

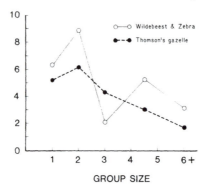

Fig. 19.2. Rates of daily food intake for individuals in different size groups while hunting either Thomson's gazelle or wildebeest/zebra, as calculated by Caraco and Wolf (1975). I have combined their results from wildebeest and zebra because Schaller's original data on hunting success did not differentiate between the two species. I have also included data on wildebeest and zebra for hunting groups of three: Caraco and Wolf intentionally excluded this point in their paper.

1980) or against other species (e.g., hyenas versus lions: Lamprecht, 1978b; but see below). Therefore it is essential to examine the available data on cooperative hunting in lions before accepting or rejecting the applicability of Kruuk's (1972) hypothesis to the felids.

COOPERATIVE HUNTING IN LIONS: THE CARACO-WOLF MODEL

Caraco and Wolf (1975) analyzed Schaller's (1972) data to estimate the optimal group size for hunting lions. Schaller had found that hunting success was twice as high for pairs as it was for solitary hunters, but that hunting success of three or more was no higher than that of pairs. Caraco and Wolf's analysis pointed out that it would be disadvantageous to be in too large a group because the captured prey would have to be divided into ever smaller portions while capture rate remained the same. However, groups do gain an added benefit since they can sometimes capture several prey simultaneously. Caraco and Wolf included a multiple-kill correction factor and determined the rate of food intake for each group size. According to their calculations, the hunting group size that maximized rate of food intake per individual per day would be two when hunting either Thomson's gazelle or medium-sized prey (wildebeest and zebra) (Fig. 19.2). In both circumstances their findings suggest that group hunting is optimal in lions since the optimal hunting group size is greater than one.

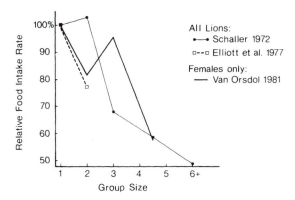

Fig. 19.3. Rates of individual food intake relative to that of solitary hunters, as calculated by the method described in the text. *Relative* rates of food intake are presented because it is the shape rather than the elevation of each curve that is of interest here and because it facilitates comparisons across studies. I have combined Schaller's 1972 (thin line) data on hunting success across all prey sizes. Hunting success does not vary significantly with prey size and thus the effect of hunting different-size prey affects only the elevation rather than the shape of the curve. Ideally, the data should be based on the number of females in each group (since they do most of the hunting), but only Van Orsdol, 1981 (heavy line) has presented such data. I have excluded Caraco and Wolf's "hyena loss" factor (which would award 10 percent greater levels of relative food intake to members of groups of > 4), since hyenas rarely steal *meat* from lions (see text).

Caraco and Wolf's model has been a very important and influential contribution, but unfortunately Schaller's data were not collected to test these hypotheses, and their model includes two assumptions that may invalidate the conclusion that group hunting is optimal in lions.

1). The multiple-kill correction factor. Caraco and Wolf awarded all groups of two or more the same proportion of multiple kills, although Schaller (1972, p. 254) stressed that the chances of making multiple kills were highest for groups of four or more. Consequently, the model overestimates the rate of food intake for groups of two and three, which is significant since the model shows the optimal group size to be two.

I have reanalyzed Schaller's data (as well as more recent data from other studies) using a multiple-kill factor that more closely approximates Schaller's findings. Schaller did not report the precise relationship between group size and proportion of multiple kills, but did give the average proportion of kills that included multiple carcasses (20.5 percent, excluding buffalo: 1972, table 42) and stated that most were made by groups of four or more. Thus I have assumed that the proportion of multiple kills increases linearly with increasing group size from 0 percent for solitaries to 33.3 percent for groups of six and more, which has the same mean across group sizes of two or more as given by Schaller. Note that this still overestimates the proportion of multiple kills made by groups of two and three, but not as much as in Figure 19.2.

By this method, the estimated rate of food intake per individual per day is generally highest for lions hunting alone (Fig. 19.3). In some areas, pairs or trios may do as well as a solitary hunter, but overall it appears that lions would enjoy the highest rates of food intake if they were as asocial as any other felid!

2) Number of hunts per day is independent of food intake and of group size. Caraco and Wolf based their calculations on the assumption that a lion could hunt only three times per day, regardless of the levels of recent food intake or of the number of other lions in its group. J. P. Elliott et al. (1977) and Van Orsdol (1981) have since published data showing that hunting frequency increases with decreasing levels of food intake: hungry lions hunt more often. This finding essentially invalidates the results presented in Figures 19.2 and 19.3, since lions suffering from reduced food intake per *hunt* may be able to compensate merely by hunting more frequently each day (note that Figs. 19.2 and 19.3 actually indicate rates of food intake per *hunt* since it is assumed that all group sizes hunt with the same frquency). Unsuccessful hunts often involve exhausting chases and if only a few group members become exhausted during cooperative hunts, it would be especially likely that large hunting groups could compensate by hunting more frequently. However, data on hunting rates of different group sizes are not available.

Rather than end this section with the conclusion that solitary hunting maximizes rate of food intake in lions, I wish instead to emphasize that two types of data must be collected before Caraco and Wolf's model can be properly utilized. First, the average biomass of kills made by groups of different sizes (to replace the extrapolated multiple-kill factor), and second, the hunting rates of different-size groups. Note that both will have to confer very strong advantages to large groups if they are to account for the higher per capita reproductive success of females in prides of three to ten.

GROUP FORAGING AND FOOD INTAKE: ALTERNATIVE DATA

Although our own lion studies have not yet focused on hunting behavior, we do have extensive data on group dynamics and rates of food intake. We regularly censused twenty prides and made daily observations of feeding and grouping patterns from July 1978–May 1981, February–May 1982, November 1982–January 1983, and July–October 1983. Whenever we locate a lion or group of lions we note the identity of each individual, their "belly size" if they are standing (see below), and the age, sex, and species of any carcass upon which they are feeding. These observations are referred to as

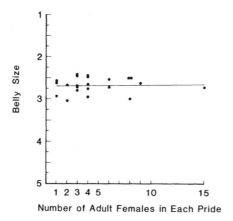

Fig. 19.4. Average belly size of females living in different sized prides. Each point is the mean across females of each pride. "Belly size" is based on a scale developed by Bertram (1975b) with 1.0 being the fattest possible and 5.0 the thinnest. When belly size is regressed against pride size, $r^2 = 0.0044$, $y = 2.6942 - 0.0036x$. Data are taken from sightings made on non-consecutive days and the belly sizes of each female were measured at least four times (range = 4–27). Data are from all study prides in the Serengeti and Ngorongoro Crater (1978-1981). Two prides contribute two points each since they were each observed at times when pride size differed markedly.

"sightings." We also collected detailed data on feeding behavior during watches of approximately two hours each at nearly one hundred kills.

The weight of stomach contents can be estimated visually from the profile of a standing lion's belly (Bertram, 1975b), and thus the levels of recent food intake of different individuals can be compared. Unfortunately "belly size" data have two important limitations. First, lions scavenge as well as hunt (Schaller, 1972; Kruuk, 1972), therefore without direct observations of how the lion acquired its food, belly size cannot be related to *hunting* success. Nevertheless, belly size does indicate *foraging* success and thus these data can be discussed in terms of group foraging. Second, belly size cannot be related to current group size. Figure 19.1 illustrates the ephemeral nature of group sizes within a pride, but belly size declines only gradually after a large meal. Therefore it is never certain whether a well-fed or poorly fed lion has joined a group before or after it has eaten.

However, certain aspects of lion social organization can be used to estimate the effects of group size on levels of food intake. First, our study includes prides ranging in size from two to eighteen adult females as well as a number of solitary females. Whereas females in large prides are able to form large groups, solitary females and those in very small prides can never do so. Therefore, if forming large foraging groups increases levels of food

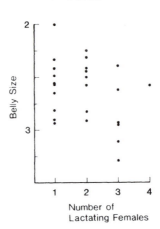

Fig. 19.5. Average belly size of lactating females rearing their cubs alone or communally. Each point is the mean across females of a set when all were lactating simultaneously (minimum of three measurements per female, range = 3–16). Age of cub does not have a significant effect on mother's average belly size. Singletons and pairs had significantly larger belly sizes than sets of three and four ($U = 26.5$, $n_1 = 18$, $n_2 = 7$, $p < 0.05$, two-tailed). Data are from all prides (1978-1983).

intake, belly size would be expected to increase with increasing pride size. However, there is no relationship between pride size and average belly size (Fig. 19.4), and thus solitary females and those in small prides do not suffer from reduced levels of food intake.

Second, females do not always alter group size as rapidly as the female in Figure 19.1. Females from the same pride with small cubs of similar age rear them cooperatively, and thus lactating females spend almost all of their time together (Schaller, 1972; Rudnai, 1974). Consequently, associations between lactating pridemates are much less variable than those between females in other reproductive states. As a result, average group size for mothers is set by the number of pridemates with similarly aged cubs rather than by characteristics of the food supply. In contrast to a nonlactating female in a large pride that might forage solitarily when prey size is typically small (Schaller, 1972, also see below), a lactating female in the same pride always remains in association with the other mothers. Thus lactating females are constrained to remain in groups with lower expected food intake, as predicted by Figure 19.2 and 19.3. Figure 19.5 shows that compared to singletons and pairs, sets of three and four mothers apparently suffer from lower levels of food intake.

The consequences of lowered food intake during lactation can be profound. The belly size of cubs is closely related to the belly size of their mothers ($r_s = 0.672$, $n = 15$ sets of mothers and cubs, $p < 0.01$) and poor nutrition in cubs increases their mortality (cubs' average belly size versus percentage survival to twelve months: $r_s = 0.663$, $n = 13$ sets of cubs, $p < 0.05$).

In summary, females that are solitary or are members of small prides achieve rates of food intake similar to those of females in larger prides. Lactating females that are constrained by the presence of cubs to remain habitually in groups of three or four suffer from lower food intake than mothers that rear cubs alone or in pairs. Taken together, these findings suggest that group foraging decreases rates of food intake in female lions, but that the fission-fusion nature of the pride allows females to forage in smaller groups when necessary and thus females in large prides do not suffer reduced food intake.

Rather than a mechanism that improves food intake, communal cub rearing appears to be an adaptation against infanticide: groups of females are more successful than solitaries in defending their cubs against alien males (Packer and Pusey 1983a, c, 1984, in prep.). Small cubs are killed whenever new males enter the pride and moderate-size prides suffer lower frequencies of male takeovers than do smaller or larger prides (Packer et al., in press).[2] Although communal defense against infanticidal males may be an important advantage of group living in female lions, by itself it cannot explain the distribution of sociality across species: infanticide by males also occurs in tigers and cougars (see review in Packer and Pusey, 1984), but females in these species are nevertheless solitary. Communal rearing strategies will be examined in detail elsewhere.

OTHER POSSIBLE ADVANTAGES OF GROUP FORAGING

Maximizing the rate of food intake is not the only possible advantage of group foraging. Three other possibilities may confer an advantage to moderate-size prides over smaller prides.

1) Cooperative hunting minimizes risk of starvation by minimizing variance in food intake (see Caraco et al., 1980; Rubenstein, 1982). Schaller's data suggest that a single female has to hunt six times to ensure a single success, whereas groups need only hunt three times. When prey is so scarce that only one or two hunts can be attempted per day, a run of bad luck could be fatal to a solitary female. However, the data used in Figure 19.4 show *no* relationship between pride size and variance in belly size. Variance in belly size was calculated for each female and the average variance in each pride was regressed against the number of females in that pride ($n = 22$, $r^2 = 0.0002$). Nevertheless, if such a life-and-death situation occurred only once every year or two it would still be of great importance, though too rare to be

---

[2] The reason for the higher rate of male takeovers in very large prides is not clear, but may result from a far greater attraction of male coalitions to the range of such large numbers of females (see Packer et al., in press).

documented. The above data were collected over all times of year and seasonal variations in prey availability may have swamped any effect from group foraging. To test this hypothesis properly, extensive observations should be made of different size groups during periods of extreme prey scarcity.

2) Cooperative hunting decreases risk of injury during prey capture (Schaller, 1972; Kruuk, 1975). Schaller suggested that by hunting in groups, lions could prey on very large animals that would be too dangerous to a single hunter. In some areas (e.g.. Manyara Park: Schaller, 1972; Kafue Park: Mitchell et al., 1965). large prey such as buffalo do comprise an important part of the lions' diet. However, there are no data on the success rate of different size groups in capturing buffalo, nor are there good data on risk of injury. Limited data on fatalities due to buffalo are available from our study. Between 1979 and 1983 one adult female in Ngorongoro Crater was observed being killed by buffalo (S. Trevor, pers. comm.) and three others were found dead and were assumed to have been killed by buffalo on the basis of their wounds and the trampling of nearby vegetation (pers. obs.; S J. Cairns and A. Geertsema, pers. comm.). These females were members of prides of two, three, four, and seven females respectively, suggesting that females in smaller than average prides have a somewhat higher risk of being killed by buffalo. However, the relationship is not statistically significant (proportion of Ngorongoro females killed in small prides [two out of ten] versus proportion killed in large prides [two of twenty-seven] $p > 0.20$, Fisher test), nor is it known if the females were killed while attempting to capture the buffalo. Buffalo will sometimes chase lions without provocation (pers. obs.).

As stated earlier, there are many areas where large, dangerous prey comprise only an insignificant part of the lions' diet, and thus I am skeptical of the importance of their safe capture in the evolution of lion sociality. However, even small to medium-size prey such as warthog, zebra, and wildebeest can occasionally injure a lion, and relatively minor wounds may reduce a female's subsequent hunting success (Bertram, 1978). Data are needed on wounding rates based on direct observation of prey capture, as well as data on the hunting success of wounded females. Also, group hunts may typically require shorter chases and thus not only be less energetically expensive but also result in less wear and tear to each hunter.

3) Group foraging increases success in competition with spotted hyenas (Schaller, 1972; Lamprecht, 1978b; Eaton, 1979). Lions and spotted hyenas have very similar prey preferences, and often compete over carcasses (Kruuk, 1972; Schaller, 1972). Lions are dominant to hyenas, except in encounters between few (1–2) lions and many (20-40) hyenas. Schaller found that 44 percent of carcasses fed on by lions are eventually lost to hyenas. But are these losses substantial enough to account for group living

in lions? Schaller (p. 272) found no striking relationship between the number of lions feeding at a carcass and the probability of losing it to hyenas (and Fig. 19.4 suggests that solitary female lions forage as successfully as group-living females). Furthermore, almost all published observations, as well as our own, show that lions surrender only the remnants of a kill (i.e., skin and bones) after they have gorged themselves on the meat. The crucial question is whether a female lion feeding solitarily loses more meat to hyenas than a lion feeding socially loses to her companions. Data presented in the next section show that lions lose considerable quantities of meat to conspecifics. In contrast, the quantity of *meat* lost to hyenas was negligible. During our detailed observations of lions feeding at ninety-one carcasses, the lions lost only two carcasses to hyenas when more than 10 percent of the meat was remaining, and Van Orsdol (1981) reported a similar finding. Although primarily meateaters, hyenas are also adapted to eating large bones (Kruuk, 1972), whereas lions will only eat the bones of small or immature prey. However, lions will sometimes spend hours gnawing at bones, removing the last traces of meat, and it is then that hyenas become boldest and try to take them away from the lions. This is presumably because the remains have become more valuable to the hyenas than to the lions.

In summary, the minimization of risks of starvation and injury may eventually prove to be important advantages of cooperative hunting in lions, but supporting data are not yet available. Competition with spotted hyenas does not appear to be an important advantage of group foraging in lions. However, even if both minimizing the risks of starvation and injury prove advantageous, detailed data from other felids would be required to show why these factors have resulted in sociality only in the lion.

LION GROUPING PATTERNS AND PREY SIZE: AN ALTERNATIVE VIEW

Prey size appears to be one of the most important variables determining group sizes within a lion pride (Schaller, 1972; Caraco and Wolf, 1975). However, this is primarily because a large carcass attracts a large number of lions. It is difficult to estimate *hunting* group size without direct observations during the moment of prey capture. Most published data are on *feeding* group size. We likewise lack good data on hunting group size and in the following analysis, I attempt to provide reasonable estimates of hunting group size as well as direct measures of feeding group size in order to measure the precise effect of prey size on group size.

Where prey size varies seasonally, consecutive kills will tend to be of the same prey species and hence of similarly sized prey. Therefore, group size prior to prey capture can be assumed to reflect preferred hunting group size for prey of a particular body size since the group will be likely to encounter

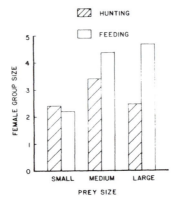

Fig. 19.6. Effect of different-size prey on female group size in the Masai Pride. Small prey include all species with adult weight < 100 kg (e.g., gazelle and warthog); medium-size prey, 100–250 kg (e.g., wildebeest and zebra); large prey, > 300 kg (e.g., buffalo, eland, and giraffe). Hatched bars refer to group sizes within two days before or after a kill of a particular size, but not in the presence of a carcass. These are assumed to reflect hunting group sizes. Open bars are group sizes in the presence of a carcass and thus show feeding group sizes. There was significant heterogeneity ($p < 0.01$, Kruskal-Wallis) both in hunting ($n = 276$ sightings) and feeding ($n = 86$) group sizes of different-size prey species. Feeding group sizes are significantly larger than hunting group sizes ($p < 0.01$) for both medium-size and large prey. There is no significant difference between hunting and feeding group sizes for small prey ($p > 0.20$). Data are from 1978–1981, during which time the Masai Pride included an average of fifteen females.

consecutive prey of the same species. Seasonality of prey size is particularly striking in the Masai pride, one of Schaller's original study prides and the one which much of the Caraco-Wolf analysis was based. This pride mostly captures medium-size prey from November to May, when the migratory herds are present in their range, and females mostly take small prey during the remaining months (the males occasionally catch buffalo during this same period). Figure 19.6 shows the average group size of the females in the Masai pride during periods when prey of different sizes were available, and the influence of the presence of carcasses of different sized prey on group size. Although females in this particular pride were in the largest hunting groups when medium-size prey were the primary food source, the increase following the capture of a large or medium-size prey was even more pronounced.

The increase in group size in the presence of a medium or large carcass occurs because many lions eventually arrive to feed at the carcass: the presence of a kill is the most common context in which female pridemates meet (Fig. 19.7). The larger the carcass, the longer it persists (edible biomass of

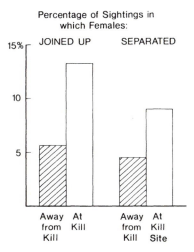

Percentage of Sightings in
which Females:

Fig. 19.7. Effects of the presence of a carcass on females moving into proximity of each other. A female is significantly more likely to move to within 200 meters of another female when the latter is at a kill (one sample $\chi^2 = 12.67$, 1 df, $p < 0.001$, $n = 70$ sightings in which females joined). Females also separate more frequently after feeding ($\chi^2 = 5.38$, $p < 0.05$, $n = 53$), emphasizing that feeding groups are often just temporary aggregations. Data are from all prides (1978-1981).

all kills at a site versus time from onset of feeding until 90 percent is consumed: $r_s = 0.792$, $n = 38$, $p < 0.001$, range $= < 5$ minutes for a 2-kg carcass to > 2 days for a 500-kg carcass). Consequently, more lions will be likely to locate and feed from a larger carcass. However, on the rare occasions when a carcass remains undiscovered by the remainder of the pride, one or two individuals may feed from a medium-size carcass for several days.

A female can often monopolize a small carcass (in competition with other females), but is much less successful in preventing other females from joining her on a medium or large carcass (Fig. 19.8). Therefore, a medium or large carcass is readily accessible to latecomers. It is important to recall that there is no dominance hierarchy among females at a kill. Instead, latecomers respect the first female's ''ownership'' of a small carcass or of a specific site at a larger carcass (Packer and Pusey, 1985). A larger carcass has many more acceptable feeding sites and the first female defends only her own site rather than the entire carcass.

Not only does a carcass attract pridemates, it also attracts individuals from other prides. Most interpride encounters occur at kills (Fig. 19.9). Females are intolerant of strange females within their range and interpride en-

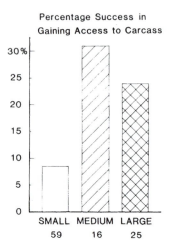

Percentage Success in
Gaining Access to Carcass

Fig. 19.8. Proportion of attempts by females to join a feeding female on a kill that were successful in spite of being threatened by the feeding female. Effect of size class of prey species is significant ($G = 6.26$, 2 df, $p < 0.05$). Numbers below size classes are the number of times a feeding female threatened a second female that attempted to join her at the kill. Data are from detailed observations of feeding lions in all prides (1978–1983).

counters usually end either with the larger group of females chasing away the smaller or the smaller group spontaneously avoiding the larger. However, if an interpride encounter occurs at a kill, the females finish feeding before chasing each other. Thus although females will defend their range against intruders, they do not cooperatively defend individual carcasses. On four occasions females from different prides were seen actually feeding together (on the carcasses of an elephant, giraffe, buffalo, and warthog). Similar large, temporary aggregations at very large carcasses have been observed in cougars (Seidensticker et al., 1973) and provisioned tigers (Schaller, 1967).

These data emphasize the importance of large prey size on gregariousness in lions: individuals aggregate at large kills. But as a result considerable meat is lost to conspecifics that were not present at the time of the kill—meat that could have been eaten by those who captured the prey if the latecomers could have been excluded.

The data in Figure 19.6 show that hunting group size is largest when females are hunting medium-size prey, the heaviest prey regularly captured by females of this particular pride (most of the large prey were believed to have been captured by the adult males). This finding can be interpreted in at least two ways. First, medium-size prey require cooperative hunting by the females, perhaps because of reduced risk of injury. Second, once they have

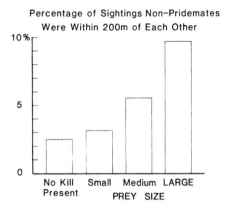

Fig. 19.9. Effect of the presence of carcasses of different sizes on the frequency of interpride encounters (one sample $\chi^2 = 14.91$, 3 df, $p < 0.01$, $n = 94$ interpride encounters). Data include observations involving all age-sex classes and all prides (1978–1981).

been captured, medium-size prey provide sufficient meat for more females than do small prey and thus being a member of a larger group is less costly than when hunting small prey (Caraco and Wolf, 1975; Kruuk, 1975). Large aggregations that had formed at larger carcasses would therefore be more likely to persist until the next prey is captured.

THE EVOLUTION OF GROUP LIVING IN LIONS

A large carcass is perhaps the ultimate large, ephemeral, and patchy resource. Enough meat is available to feed several lions, but is consumed within a day or two, and at any given time there may be only a few carcasses available over a wide area. The lion's diet therefore shows a fundamental similarity to that of many other social vertebrates (e.g., weaverbirds: Crook, 1964; primates: Clutton-Brock and Harvey, 1977; carnivores that specialize on small, patchily distributed prey: Kruuk, 1978b; Macdonald, 1983). Why among all the felids are only lions social?

I suggest that lion sociality results from the unique combination of three factors: preference for large prey, openness of habitat, and high population density. Table 19.2 shows that lions are the only felid to show extremes for all three of these variables. The consequences of a preference for relatively large prey are emphasized in Figures 19.6 and 19.8: large prey size allows several individuals to feed from the same carcass. How do an open habitat and high density relate to lion sociality? Open–country species live in larger groups than forest species in many mammals (antelopes: Jarman, 1974; primates: Clutton-Brock and Harvey, 1977; carnivores: Gittleman, 1983) and

TABLE 19.2
Relative prey size, density, and typical habitat of felids

Species	Modal prey size/Female wt	Female density (per 100 km²)	Habitat	References
Lion	1.06	7.9	Open savannah	1, 2
Tiger	0.52	2.7	Forest	3, 4, 5
Jaguar	0.39	3.1	Forest	6
Cheetah	0.26	2.9	Open grassland	1, 7, 8
Cougar	2.44	0.7	Woodland/brush	9, 10, 11
Leopard	1.08	2.9	Forest/brush	1, 12, 13
Snow leopard	1.26	1.0	Forest/brush	14
Lynx	0.12	5.9	Forest	15, 16
Bobcat	0.30	7.1	Forest/brush	17–22

SOURCES: 1) Schaller, 1972; 2) Van Orsdol et al., in prep.: 3) Schaller, 1967; 4) Borner, 1978; 5) Sunquist, 1981; 6) Schaller and Crawshaw, 1980; 7) Eaton, 1974; 8) Frame and Wagner, 1981; 9) Hornocker, 1969; 10) Hemker, 1982; 11) Hopkins et al., 1982; 12) Muckenhirn and Eisenberg, 1973; 13) Bertram, 1982; 14) Schaller, 1977; 15) Saunders, 1963; 16) Berrie, 1973; 17) Provost et al., 1973; 18) Bailey, 1974; 19) Fendley and Buie, 1982; 20) Lancia et al., 1982; 21) Lembeck, 1982; 22) Litvaitis et al., 1982

NOTE: Modal prey size is from Table 19.1. Female density is the median value across study sites and is based on the number of adult females over a given area, except in lynx and bobcat. Direct estimates of female density were unavailable for these species and are instead based on the inverse of home range size, since females have nonoverlapping home ranges. Thus, densities for these two species are probably too high. Estimates of density of cougars and snow leopards are probably too low since both suffer from intensive hunting pressure by humans except in marginal habitat.

this relationship is usually attributed to improved protection against predators—an unlikely explanation in the case of the lion. It has also been suggested that open habitat allows the coordination of group hunting by lions (Sunquist 1981, Gittleman, 1983). However, Van Orsdol (1981) found no relationship between degree of cover and hunting group size, which suggests that lions do not engage in cooperative hunting more often in open areas.

Openness of habitat has another consequence: a carcass is much more conspicuous to vultures and to mammalian scavengers, including lions. Lions watch for vultures, particularly when prey is scarce, and they will travel several kilometers to where vultures have landed. A lion at a kill can also be seen from far away, and lions will join feeding pridemates from several kilometers away. In contrast, Sunquist (1981) found it difficult to locate tiger kills visually, even from short distances, because of dense vegetation. Schaller (1972), Van Orsdol (1981), and ourselves have all noted that lions sometimes move carcasses from open areas in an apparent attempt to conceal them. Large carcass size ensures that food will still be available by the time scavenging lions arrive: a small carcass would have been consumed. Thus in species such as the cheetah, which also live in open country but specialize on relatively small prey, intraspecific scavenging would be less rewarding than in lions. In fact, cheetah almost never scavenge from each other (T. Caro, pers. comm.).

Throughout almost all of their range, lions live at higher densities than any of the other big cats. The lion's preferred prey species are all typically found in large concentrations and I suggest that high lion density is a prerequisite to sociality rather than its consequence. As a result of high density, lions are particularly likely to suffer losses of meat to conspecifics: a kill made by a single lion would be likely to attract many other lions simply because there are more lions in the vicinity. Therefore, there is greater scope for females to come into close proximity in lions than in any other large felid.

Furthermore, at high population density a dispersing subadult female might have more difficulty in establishing a new range (see Emlen 1982a, b). It will be impossible to determine whether costs of dispersal in lions are higher than in other species, since it is the costs of leaving relative to those of staying that are important and in no other felid species do females remain with their mothers. However, dispersing female lions do have lower fitness than females that are incorporated into their natal pride and the costs of dispersal are more severe in areas of higher population density (Pusey and Packer, in press). Dispersing females in Ngorongoro Crater suffer higher mortality than nondispersers; whereas dispersing females in the Serengeti have lower reproductive rates than nondispersers, but do not suffer higher mortality. Population density in the Crater is two to four times higher than that in the Serengeti.

The lion's preferences for large prey size and open habitat, and their high population density, may have resulted in sociality in the following manner. An ancestral female lion that was as solitary as any other felid was confronted with the ecological conditions of a modern lion. She typically captured prey that were similar in size to her own body weight and the carcasses regularly attracted other females to the site of the kill (Figs. 19.7-19.9). As her daughters approached maturity, their continued association with her slightly depressed her rate of food intake (Fig. 19.5). However, the costs of dispersal may have been so high relative to these losses that the mother would have increased her inclusive fitness by allowing the daughters to remain in her range (see Rodman, 1981). Furthermore, if the mother and her mature daughters subsequently associated as sporadically as do modern pridemates (Fig. 19.1), but their ranges overlapped so that the mother's contact with nonrelatives became less frequent (Waser and Jones, 1983), then the initial disadvantage of group foraging might have disappeared (Fig. 19.4). In addition, any loss of meat to her daughters would have been compensated by the fact that it was lost to close kin rather than to more distantly related neighbors (W. D. Hamilton, 1964). The mother could not dominate

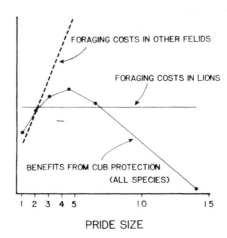

PRIDE SIZE

Fig. 19.10. Costs and benefits of sociality in lions and other felids. The benefits of cub defense are the inverse of the frequency of male takeovers for each pride size (from Packer et al., in press). The pride size with the lowest frequency of male takeovers suffers the lowest incidence of infanticide, and hence gains the greatest benefits from sociality. I assume that this pattern would be similar in other felids if they were also social. The "costs" of foraging are the losses of meat to conspecifics by females in different sized prides. In lions these costs are based on Figure 19.4 but are conjectural for other species. The elevations of these curves are drawn so that optimal pride sizes are given by the regions where the benefits of cub defense exceed the costs of foraging: 3–7 in lions (against an observed of 3–10) and 1 in other felids. Solitary females of other felids are also assumed to suffer lower foraging "costs" than do solitary female lions (see text).

her mature daughters (Packer and Pusey, 1985) and thus could not prevent them from breeding even if it was in her best interests to do so (see Vehrencamp, 1983). Once the mother and daughters reared their cubs communally in order to defend them against infanticidal males, then a modern pride had been achieved.

In contrast, female leopards or cougars (which often take large prey but live at much lower densities) gain greater benefits from living alone: they are unlikely to lose meat to unrelated conspecifics and would thus suffer a much greater reduction of food intake by allowing their daughters to remain within their ranges. Thus the costs of group foraging both to the mother and to the daughters may outweigh any advantage of cooperative cub defense.

In Figure 19.10 I have plotted the observed relationships of food intake rate and risk of infanticide across all pride sizes in lions, and have also plotted the hypothetical relationships for an ordinarily solitary felid. I assume that the advantages from cooperative defense against infanticidal males in most solitary species would be the same as in lions, but that food intake would decrease rapidly with increasing "pride" size in the solitary species.

I have also assumed that solitary females of the solitary species have a higher feeding efficiency than solitary female lions because they do not lose as much meat to conspecifics. Lions may be the only species in which females are able to gain a net advantage from sociality because they are neither able to benefit from solitary foraging, nor suffer prohibitive costs from group foraging.

There are three important predictions from this analysis for which we do not yet have adequate data. First, solitary females should lose comparable quantities of meat to other lions as do pride-living females. Second, losses to conspecifics by solitary female lions should also be higher than those suffered by females in other species. Third, female lions should show a greater tendency to be solitary in areas of low population density. All available data on the number of females per pride are from areas where the population density is nearly an order of magnitude higher than that of other large felids. However, the only areas where lions occur at very low density are extremely arid (e.g., parts of the Kalahari desert: Eloff, 1973), and thus lions may aggregate anyway at waterholes (see Macdonald, 1983).

Conversely, females of other species should show greater gregariousness in areas of high density. Although there are no good data on this point in other felid species, supporting data are found in other carnivore families. Spotted hyenas are found at higher densities and prefer larger prey than either striped or brown hyenas and are the most gregarious of the three species (Kruuk, 1972, 1976; Owens and Owens, 1978). Kruuk's (1972) study of spotted hyenas illustrated how huge aggregations could form at large kills and that cooperative hunting played a relatively minor role in prey capture. Female white-tailed mongooses forage solitarily for insect prey but their ranges overlap extensively with their mother's range in areas of high density, whereas females have nonoverlapping ranges in areas of low density (Waser and Waser, 1985).

Group territoriality may also be favored in situations where the presence of a few additional animals does not seriously deplete resources and where there is a moderately high level of intruder pressure (Davies and Houston, 1981; J. L. Brown, 1982). Although much of the data presented in this chapter are consistent with this idea and female lions do show cooperative range defense, I have not emphasized group territoriality for two reasons. First, since it does not directly involve defense of carcasses, we cannot easily quantify the precise consequences of such behavior. Second, there is no apparent relationship between rate of food intake and pride size (Fig. 19.4) as would be expected if group territoriality increased feeding efficiency. However, group defense may result in reduced hunting frequency and thus reduce injury rates and energy expenditure during prey capture. Studies are in progress both on cooperative territorial defense and on cooperative hunting.

449

TABLE 19.3
Kinship in male coalitions and female prides

	Always composed of close relatives	Ever include nonrelatives
Male coalitions	8	9
Female prides	19	0

NOTE: Based on male coalitions of known origins that gained access to female prides between 1978 and 1983, and on the composition of female prides in 1983 which had recruited members after 1975. $p < 0.001$, Fisher test.

SEX DIFFERENCES IN KINSHIP PATTERNS IN LIONS

In both sexes, individuals gain mutualistic reproductive advantages from group living: per capita reproductive success of male coalitions increases with increasing coalition size and is at a maximum in prides of three to ten adult females (Packer et al., in press). However, there is a striking difference between the sexes in their choice of companion in our study areas: whereas males frequently form coalitions with unrelated partners, females of the same pride are always close kin (Table 19.3). It is not surprising that solitary males and pairs of males join up with unrelated partners, since each male can thereby increase his reproductive success (Packer and Pusey, 1982, 1983b).[3] However, it is puzzling that solitary females do not form larger prides with other solitaries or pairs.

This sex difference may result from the differing consequences of philopatry and dispersal to each sex. Large male coalitions gain higher reproductive success because they are more likely to gain access to a pride, gain larger prides, and control prides for longer periods (Bygott et al., 1979). None of these factors depends critically on familiarity with an area: males maintain exclusive access to a set of females rather than to a range. Thus a solitary or pair of males can move over great areas in search of companions and settle in any pride they can take over (Pusey and Packer, in press). In contrast, females almost never move substantial distances from their natal range. Even those that are not incorporated into their mothers' pride nevertheless remain in a portion of their natal range or in an area immediately adjacent to that range. It is not known whether the lower reproductive success of these dispersing females results from a loss of access to all or part of their natal range, but it seems likely that they must learn the location of new safe denning sites and good hunting areas during critical times of year (see

[3] Coalitions of male cheetah are also often composed of nonrelatives (Frame, 1980), but the consequences of cooperation for each male are not yet known.

also Waser and Jones, 1983). Therefore, there may be real constraints that prevent females from leaving their natal area to find additional companions.

The only known cases of unrelated females forming new prides lend some support for this idea. After a severe drought, a number of females in the Kalahari Desert moved over forty kilometers to a new area and several solitaries formed new prides with nonrelatives (Owens and Owens, 1984). Once the females had been forced to disperse to a new area they were then willing to accept unrelated companions. In our study areas, we have twice seen pairs of unrelated solitary females in close and apparently amicable association outside their usual ranges. However, in both cases the two females returned to their respective ranges shortly thereafter and continued to live as solitaries.

CONCLUSIONS

The evolutionary causes of lion sociality are complex, but the distribution of sociality in felids shows many similarities to other taxa of higher vertebrates. Obviously, more data are needed to ascertain the effects of cooperative hunting on food intake, injury rate, and risk of starvation; but the available data suggest that group living has evolved in spite of disadvantages of group hunting. Although solitary foraging may yield the highest rate of food intake to *hunting* lions, scavenging is particularly profitable in this species and thus large groups often form at kills. Consequently, the average rate of food intake is independent of pride size, and across all felids only female lions can form stable subgroups to defend their cubs against infanticidal males. The kin-biased structure of female prides apparently results from high levels of natal philopatry.

20. Social Evolution in Birds and Mammals

———————————— • ————————————

RICHARD W. WRANGHAM

AND DANIEL I. RUBENSTEIN

ONE AIM of this book is to examine the extent to which social evolution in
different taxonomic groups can be understood through a series of common
principles. At the most general level the same fundamental rules presumably relate ecology and social organization in all animals: social behavior
evolves as an adaptation to maximize fitness, given a particular set of ecological pressures. However, the precise way in which ecological pressures
generate social organization in different taxa is not yet firmly known. Furthermore, the problem of understanding socio-ecological principles has in
some respects been made more difficult than it was twenty years ago by the
discovery of marked differences in social relationships in many species with
gross similarities in their grouping patterns (e.g., Dunbar, Chapter 15).
Long-term social relationships based on individual recognition are particularly prevalent in birds and mammals compared to other animals (Wrangham, 1983), probably partly because species in these taxa have relatively
advanced cognitive abilities. This means that in these groups socio-ecological studies must go beyond an analysis of grouping patterns to explain patterns of competitive and cooperative relationships. In this chapter, therefore, we consider how the analysis of social systems to date has paved the
way for a unified theory of social evolution capable of explaining social relationships both between and within the sexes. Such a theory needs to spec-
ify which behavioral variables are most directly subject to ecological pressures, and what ecological or other pressures are most influential in social
evolution. These problems are discussed in turn, illustrated with case studies from this volume. In the final section we examine briefly some general
problems raised by the comparative approach taken throughout the book.

BEHAVIORAL VARIABLES AND SOCIAL RELATIONSHIPS

The classical socio-ecological method is to analyze the ecological basis of
grouping patterns without paying attention to the types of social relationships within or between groups. Historically this was a natural starting point
because grouping patterns were documented before social relationships. For
the same reason it continues to be a useful method when species are first
studied because it provides correlations with heuristic value. For instance,

Leighton found that larger species of hornbills tend to live in smaller groups (Chapter 6). Although different species of hornbills eat similar foods, he showed that larger species eat more from a given patch. Since group size is presumably constrained by food dispersal, species differences in body size appear to be ultimately responsible for differences in group size. This is a helpful idea because it suggests an explanation for the species distribution of helpers, which occur only in the smaller species living in the larger groups: helping may be inhibited in the larger species because their group size is constrained by feeding pressures, putting a new twist on the theory that the distribution of helpers is best explained by the availability of breeding opportunities (Emlen, 1984) as was shown to be the case in jackals (Moehlman, Chapter 4), and scrub jays (Woolfenden and Fitzpatrick, Chapter 5). Among hornbills, species differences in body size may be ultimately responsible for the distribution of helping.

Valuable though this type of analysis is, it has limited power because it does not consider the nature of social relationships. Helpers tend to be males in some species (most birds, e.g., Florida scrub jays: Woolfenden and Fitzpatrick, Chapter 5), and can be either sex in others (canids: Moehlman, Chapter 4). Indeed, in some species the sexual distribution of helping can vary between populations, as occurs in bicolored wrens *Campylorhyncus griseus* (Austad and Rabenold, 1985). Again, male helpers do more work than females in many species, such as Florida scrub jays (Woolfenden and Fitzpatrick, Chapter 5), but less than females in acorn woodpeckers *Melanerpes formicivorus* (Koenig et al., 1983). The reasons for this variability between and within species are not understood (Koenig et al., 1983), but it means that an analysis of grouping patterns on its own is insufficient to explaining social relationships. Accordingly, other effects of ecology must be included in order to understand the full system.

Consideration of ecological influences that might explain social relationships can sometimes lead to new explanations for the pattern of grouping. For example, many nocturnal primates forage alone within individual territories. Their solitariness is sometimes ascribed to their antipredator tactics, because these are cryptic animals that would probably be more easily detected by predators if they foraged in groups (Clutton-Brock and Harvey, 1977; Schaik and van Hooff, 1983). Foraging alone, however, does not require exclusive, nonoverlapping home ranges. To account for territoriality in these species an additional explanation is needed. Territoriality in many birds and in nocturnal primates has been explained as occurring because food resources are economically defendable by lone individuals (e.g., Hladik, 1975). If this is shown to be the case for the nocturnal primates, it will account not only for territoriality but also for solitariness. Hence, it is not yet clear if the explanation in terms of predator avoidance (which has not been tested) is needed.

A similar example is provided by Packer (Chapter 19). The traditional explanation for lions living in social groups is that it makes cooperative hunting possible. Although this explains the grouping pattern, it does not explain why females form semi-closed groups that defend their carcasses from other groups. Packer argues that lions differ from other cats in having frequent access to easily locatable and defendable resources (large carcasses), and that the need for allies to help in defending those resources can explain why females form large closed groups. In this case Packer shows that the original explanation of the grouping pattern is not only unnecessary, but appears to be wrong: group hunting does not appear to be as efficient as previously thought. The need for joint defense of resources therefore seems to explain both the kinds of social relationships that are formed, and the fact that lions hunt in groups. A similar example is provided by Wrangham's (1980) analysis of primate groups, where it is suggested that the need for allies in defense of high-quality food patches generates semi-closed groups of females with cooperative relationships. It is therefore no longer clear whether predator pressure contributes significantly to explaining species differences in primate grouping, as has often been argued (Crook, 1970; Alexander, 1974; Schaik and van Hooff, 1983).

These examples imply that explanations derived solely for grouping patterns should be treated cautiously until social relationships have been accounted for also. Furthermore, it is important to note that in species with long-term social relationships, it is the relationships themselves that need to be explained, rather than merely the interactions, which can be cooperative or competitive. The distinction between social interactions and social relationships is critical because the costs and benefits of a given interaction are strongly influenced by the relationship within which it occurs. For example, an escalated fight started by competition over a trivial resource may be inexplicable unless one knows that a long-term dominance relationship is at stake (deWaal, 1983). Similarly, mutualistic relationships involving delayed reciprocity may look altruistic when viewed merely as interactions; and cooperative relationships may include agonistic components, which, if viewed as isolated interactions, may make them seem to be competitive. Often, therefore, the occurrence of particular types of social interactions cannot be explained without reference to other types of social interactions.

Recognition of the importance of analyzing social interactions in the context of long-term social relationships raises two questions. First, how should social relationships be described and classified? Second, which kinds of social relationship are most important for understanding the influence of ecology on social organization? The classification of social relationships is still at an early stage, but there is evidence that the overall patterns of social relationships are not necessarily well correlated with each other in different species even where some aspects are strikingly similar. For instance, female lions and female vervet monkeys *Cercopithecus aethiops* are similar in forming cooperative relationships to defend resources against other groups. Yet they appear to differ because intragroup dominance interactions among females, especially those involving alliances, occur more frequently in vervets than in lions (Wrangham, 1983). This kind of difference between intergroup and intragroup relationships, which also occurs between terrestrial and arboreal cercopithecines (Gouzoules, 1984), is important because it indicates that social networks must be described in terms of all four axes of interindividual interaction (cooperation, competition, spite, and altruism, W. D. Hamilton, 1964) rather than merely with respect to one of these.

Nevertheless, there are two ways in which an ecological analysis of social organization benefits from a focus on a limited set of social relationships. First, since the most fundamental problem in social evolution is understanding why animals act altruistically, cooperative and altruistic elements of social relationships are especially important. Intense forms of competition within semi-closed groups are not explicable without an account of the social glue responsible for the existence of the group in the first place.

Second, social relationships among breeding adults are critical because they have the most immediate effect on fitness, and are therefore the ones subject to the most intense selection pressures. Socio-ecological studies have often focused on mating relationships rather than including a specific analysis of relationships within sexes. This is because socio-ecology began and has been continued most forcefully among birds (Orians, 1961; Vehrencamp and Bradbury, 1984). A focus on intersexual aspects of mating relationships is appropriate in birds because cooperation among breeding adults of the same sex is rare, so that differences among species' social systems can be accounted for to a large extent in terms of male-female relationships.

In a few birds and many mammals, however, mating relationships are not the only, or even the major, source of variation in social systems. In birds, for example, the occurrence of helpers is not predicted by the type of mating system: monogamous pairs may or may not have helpers. Many mammals are polygynous, but the range of relationships within each sex is enormous and poorly correlated with mating types. This is most clearly illustrated when a variety of single male, multifemale mating systems (harems) are compared (Rubenstein, Chapter 13 and below). As in the case of ecological explanations of grouping patterns, therefore, this means that the social system cannot be accounted for purely in terms of the mating system.

This principle is not new, of course. It is responsible, for instance, for the distinction between resource defense polygyny (where female-female relationships are unimportant) and female defense polygyny (where the system depends critically on female sociality) (Emlen and Oring, 1977). There is still, however, a strong emphasis on mating relationships in the social ecology literature (cf. Vehrencamp and Bradbury, 1984). We regard explicit recognition of the need to analyze relationships both between and within sexes, as exemplified in this book, as a necessary precursor to a general theory of social ecology. Furthermore, as described in the introduction, the impact of ecological factors is generally expected to be greatest on female relationships and distribution, setting up the conditions which lead by a series of feedback loops to the eventual social system. This is diagrammed in Figure 13.1.

SOURCES OF SOCIALITY

In a global review of the origins of animal sociality, Alexander (1974) argued that the ecological factors favoring the evolution of social groups (as opposed to aggregations) fall into two main classes, predation pressure and resource distribution. The essence of the argument is that grouping invariably confers disadvantages in terms of increased competition and disease. Even where the resource distribution allows animals to form groups at low cost, therefore, positive benefits must exist to outweigh these inevitable costs. With rare exceptions (such as hyraxes huddling for warmth) all such benefits come from either escaping predators or acquiring resources. The major exception was the case of human groups, which were supposed to be favored as a response to the aggression of neighbors. In this case conspecifics were supposed to have analogous effects to predators, leading to the formation of groups despite their inherent disadvantages.

The idea that predation and resource distribution are the two principal factors responsible for variation in social behavior is widely accepted, and the preceding chapters give numerous examples of their effects. These chapters also support, however, the idea that it is useful to recognize a third major category of factors responsible for the formation of social groups, namely intraspecific competition. Intraspecific competition is routinely argued to have effects on social grouping, but it has not previously been elevated to the rank of a common cause of gregarious life. This is perhaps partly because it is a less independent variable than predation pressure or resource distribution. In fact, its intensity often reflects the strength of these ecological forces. It appears to us, however, to merit equal treatment with the two classical factors as a major source of sociality. Like predation and resource distribution, the effects of intraspecific competition are expressed in different ways in different species, are mediated by ecological constraints, and can account for the evolution of complex social groups even when no other factors favor grouping. This means that investigations of animal sociality should routinely examine all three influences. To substantiate these arguments we consider the three factors in turn. We examine both their effects on sociality and the ways their effects interact with each other.

Predator Pressure

High predation risk is normally argued to favor large groups, although in species relying on cryptic behavior to avoid predators it favors solitariness (Pulliam and Caraco, 1984). There are many mechanisms by which individuals achieve greater safety in groups, and their importance varies between species. Their effects on social relationships also vary. For instance, three types of mating systems are attributed to predation in this volume: polygyny, monogamy, and polyandry.

First, Robinson (Chapter 9) showed that fledging success in yellow-rumped caciques depends on the degree of protection from mammalian predators. Safe nesting sites occur in small, uncommon patches where coloniality is therefore favored. Yellow-rumped caciques cooperate in mobbing predators, but there is no evidence of differentiated social relationships between pairs. Predation has a strong indirect effect on mating relationships, however, because it forces females to cluster. This means that clumps of females are economically defendable by males, leading to polygyny.

In other species where predation favors groups, the result need not be polygyny, because one female may monopolize breeding by suppressing the reproductive effort of other females. This leads either to monogamy or polyandry. Thus in dwarf mongooses there is good evidence that individuals need to be in large groups in order to have a reasonable probability of survival. Rood (Chapter 7) showed that adult mortality is lower in large packs, despite a high rate of successful predator attacks (Rasa, 1983a). Elaborate antipredator behavior has also been described. The implication is that unmated adults choose to help rather than breed alone because breeding in small groups would bring a high risk of death. The only mammals in which nonbreeding adults are helpers also have reproductive suppression of females by a dominant female (carnivores: Emlen, 1984; Moehlman, Chapter 4; tamarins: Terborgh and Wilson Goldizen, 1985). This suggests that female–female competition is a critical feature for the development of helping, presumably because it constrains breeding opportunities.

Female–female competition appears to be equally important in the development of some forms of polyandry, in which the ecological basis is not yet generally understood (Erckmann, 1983). The chapters by Oring and Lank (Chapter 2) on spotted sandpipers and by Petrie (Chapter 3) on moorhens offer two examples. In both cases polyandrous relationships appear to be favored partly by a high predation risk, as is often suspected for polyandrous birds (Emlen and Oring, 1977). In Petrie's moorhens most pairs are monogamous, but there is a potential for polyandry because female egg production rate is not apparently limited by food availability, and the chicks are precocial. The fact that entire clutches are often lost means that it pays females to spread their eggs in many baskets and to be polyandrous provided that excess males in good condition are available to aid incubation. The same is true of spotted sandpipers, where again a low rate of polyandry is sometimes found. In Oring and Lank's study, however, the relationship between predation and polyandry is taken a step further, because populations that are exposed to high rates of predation are forced, like yellow-rumped caciques and dwarf mongooses, to live in dense clumps. Unlike caciques, female spotted sandpipers both nest and forage within individually defended areas. Female exclusion then sets a limit on the number of breeding females, similar to dwarf mongooses, and by skewing the sex ratio increases the incidence of polyandry. It is not yet clear what determines why females are able to defend an area sufficient for two nests, or why males do not exclude other males, but the association between a high predation rate, increased nest density, and increased frequency of polyandry is convincing.

In yellow-rumped caciques and dwarf mongooses, high risks of predation are argued to generate larger groups than would otherwise be favored. The same effect of predation is proposed for gelada baboons by Dunbar (Chapter 15), who shows that low-ranking females experience low reproductive rates, probably as a result of harassment by dominant females. Accordingly, there must be a compensating advantage for sociality. The fact that geladas form larger groups in more dangerous areas supports the idea that the benefits come from reduced costs of predation.

Predation has been argued to explain sociality not only in gelada baboons but also in many other group-living primates, where resource dispersal appears to impose constraints on the size of groups that form, and here the effects of ecological constraints on grouping look important (Crook, 1972; Schaik and van Hooff, 1983). Terborgh (1983), for example, showed how group size varies with food types in South American monkeys. Species dependent on large, concentrated resources, such as squirrel monkeys *Saimiri sciureus*, can afford to live in large groups, whereas those eating small, evenly distributed resources, such as tamarins *Saguinus* sp., are forced to live in small groups. Whether predation is indeed a universal impetus for sociality is a matter of debate (Cheney and Wrangham, 1986), but the importance of species differences in ecology as a constraint on grouping is widely accepted.

Although these examples illustrate how predator pressure can favor grouping in different ways depending on other pressures, models have yet to be developed to explain the kinds of social networks which different antipredator tactics generate within groups. Where antipredator tactics depend on cooperation, closed groups may be more likely, whereas if they depend on dilution, open and flexible groups might be expected (Helfman, 1984; Wrangham, 1982). This is a particularly important problem for understanding the effects of predation on primate groups, since social relationships differ extensively in species with similar group size and composition. In antelope little is yet known about female social relationships, and it is possible that the effect of predator pressure is commonly restricted to a general tendency for gregariousness, rather than for structured social relationships between females (Gosling, Chapter 12; Rubenstein, Chapter 13).

These examples demonstrate that the effects of predation on social behavior are often mediated through intervening variables such as whether and how females monopolize resources. The effects of intense predator pressure on social organization are therefore not easily predicted. A further problem is that the effects of predator pressure are not yet sufficiently well understood to explain why they lead sometimes to large groups and in other cases to small groups. For example, the simakobu *Nasalis concolor* is a leaf-eating monkey found only in the Mentawai Islands (Indonesia), where no mammalian or bird predators on monkeys are known to have occurred until 2,000 years ago. Since then some populations have been subject to intense hunting pressure by humans. Populations which are not hunted live at high density and form small polygynous groups, whereas heavily hunted populations live at low density and are monogamous (Watanabe, 1981). Thus in this case the effects of intense predator pressure are opposite to those normally expected for diurnal primates, causing smaller groups than normal. Why this happens is not clear, and will presumably be understood only when the ecological basis of polygyny is known for this species. It reminds us, however, that until the effects of predator pressure are understood better, the tendency for many prey species to live in large groups should not automatically be ascribed to predation.

Resource Distribution

There are numerous ways in which resource distribution shapes social behavior, and it is impossible here to attempt a complete review. We focus instead on the ways in which resource distribution and intraspecific competition interact to generate social systems. They may do so in two kinds of circumstance, when the distribution of resources either merely permits grouping, or actively favors it. Thus resource dispersal may affect either the costs of grouping, or both the costs and benefits.

First, different patterns of resource dispersal cause the costs of grouping to vary widely between species. Evidence that food distribution limits group size to different degrees in closely related species is presented for hornbills (Leighton, Chapter 6), cercopithecine monkeys (Andelman, Chapter 10), horses (Rubenstein, Chapter 13), and great apes (Wrangham, Chapter 16). Within these taxa, species living in smaller or less cohensive groups use food patches at which an increase in group size would apparently lead to a reduction in foraging efficiency. This means that even if animal groups are always disadvantageous, they are more costly for some species than for others. Accordingly, some species should be able to form groups which confer comparatively small benefits, while others should need much larger compensation. However, in all cases a benefit must be found. The fact that grouping costs are low is not itself adequate to explain the occurrence of permanent groups.

Second, the distribution of resources can favor sociality in both non-breeding and breeding animals. For nonbreeder sociality, Florida scrub jays provide strong support for the ecological constraints model of helping, because their habitat is sharply defined and offers few vacant territories (Woolfenden and Fitzpatrick, Chapter 5). Among canids also, a shortage of marginal habitat appears to be an important contributor to the costs of novel breeding attempts (Moehlman, Chapter 4), because of costs associated with the rearing of many extremely altricial young and the need for group hunting. In these cases groups occur not because individuals are thereby better able to exploit the habitat, but because intraspecific competition due to habitat limitation constrains breeding opportunities, sometimes in both the parents and the offspring, as Armitage shows in marmots (Chapter 14). The distribution of resources is a critical aspect of the development of groups, because it not only allows groups to occur, but also limits the formation of groups when optimal habitat is limited.

Reproductive opportunities are less clearly limited where breeding adults form groups, but the ecological constraints model may still be applicable. Packer (Chapter 19) shows that the traditional idea that lion groups function to allow cooperative hunting is poorly supported and fails to explain the distribution of felid sociality. He demonstrates that scavenging and high population density are better predictors of sociality. This forces a reexamination of the way ecology and group life are related in lions. Packer suggests that females allow their kin to share the foraging range because the costs of sharing large carcasses are low, whereas the costs of disperal are high if the habitat is saturated. This argument is clearly analogous to the ecological constraints model proposed by Emlen (1984), because it suggests that the key variable favoring group-living is the high cost of dispersal.

As Packer notes, however, it has yet to be shown that the costs of dispersal are indeed high (or would be if lions were not social). An alternative possibly to the idea that group-living is favored by the high costs of dispersal is that it is favored by the high costs of solitary life, because solitary females are unable to defend carcasses against groups of rivals. According to this hypothesis, grouping can become an evolutionarily stable strategy merely because it imposes high costs on solitary females, while generating lower costs to individuals in groups (Wrangham, 1982). Some support for this idea comes from data on lions in the Kalahari Desert (Owens and Owens, 1984). Subsistence on small prey during the dry season causes females to forage alone in larger home ranges than normal. At these times prides dissolve and females are known to disperse for long distances. By the time the rains begin and the large game return, new prides have formed that include many nonkin members. Thus, the formation of prides is clearly associated with the presence of large game, whereas there is no evidence that the costs of dispersal are particularly high at that time.

The relative importance of the costs of dispersal or the costs of solitary life in generating group-living is an important issue for many animals (Emlen, 1984). Either way, the lion data suggest that group-living is favored not only by the distribution of resources but also by the effect of intraspecific competitors on resource distribution. Furthermore, in lions an additional pressure from intraspecific competition affects female grouping: female lions benefit from each other's assistance in reducing the rate of infanticide by males (Packer, Chapter 19). Thus, the lion social system includes many complexities not explained by the traditional idea that groups are favored by the benefits of cooperative hunting.

In their discussion of human mating systems Flinn and Low (Chapter 11) also propose that resource distribution and intraspecific competition interact to generate certain types of sociality. Unlike Packer's, their chapter takes the basic grouping pattern of humans as given. Humans normally form communities (band, villages, nations, etc.), within which females rear their own offspring with the aid of one or more males. Within populations, mating and investment patterns vary widely and appear to be correlated with resource distribution. The association between the presence or absence of significant material resources and the preferred type of cousin marriage, for example, is difficult to explain in other ways. The relationships proposed by Flinn and Low between resource distribution and competition for mates are clearly far more elaborate than any system in other animals. In part this stems from the fact that females and males (as allies) are resources that can be manipulated and exist regardless of the distribution and abundance of ecological resources. Nevertheless, the underlying principle may be widely applicable: intricate social networks are generated by certain types of intraspecific competition, while the prevalent type of competition is itself determined by the nature and distribution of resources.

Intraspecific Competition

The analyses of felid and human social systems share the view that economic defendability of resources is a key characteristic determining the patterns of social relationship. However, felid and human systems differ because resource distribution is seen as determining social relationships among females in lions, but primarily among males in humans. The factors determining whether females or males (or both sexes) defend resources in different species have not in general been examined, although this is a key issue for understanding not only the basis of different forms of polygyny, but also the kinds of social relationship that occur within sexes (Bradbury and Vehrencamp, 1977a, b; Emlen and Oring, 1977). Emlen and Oring

(1977) argued that "when important resources are unevenly distributed or spatially clumped, certain males can defend areas containing a larger quantity or better quality of resources than others." This is doubtless valid, but it leaves open the question of why females cannot defend such resources in the way argued for males. A general answer, presumably, is that it pays males to defend resources against males where the resources tend to attract more fertile females than the male would be able to mate by alternative strategies; whereas it pays females to defend resources against other females only when they thereby increase their net rate of resolute gain. And since defense of a food resource often severely limits the rate of consumption, females may not generally find it economical to defend food resources. The patterns of resource dispersal that favor each strategy have yet to be established.

The preceding section has shown that ecological influences favoring sociality (either group life or the kinds of social relationships within groups) do not necessarily act alone: they may be mediated by and interact importantly with the pattern of intraspecific competition. In this section, by contrast, we review evidence that, when strong enough, certain forms of intraspecific competition can act alone to promote sociality.

This idea has been proposed for a variety of species. For instance, it applies to monogamous birds without male parental investment, where males sequester females, such as brood-parasitic ducks, cowbirds, and cuckoos, as well as a variety of nonterritorial migratory ducks (Wittenberger and Tilson, 1980; McKinney, Chapter 8). In these species males compete for mating rights to females and subsequently guard them against other males. There need be no advantages in terms of predator avoidance or resource gain. "Male guarding behavior protects females from harassment by unmated males, since harassment and attempted [forced copulation] are both commonplace" (Wittenberger and Tilson, 1980).

In ducks, pair formation occurs well in advance of the breeding season, as McKinney (Chapter 8) describes. An important component is female "inciting" behavior. Males respond to female displays by courtship before they pair, whereas after they are paired they respond to the same display by attacking male rivals. In nonterritorial species males guard the females only until the clutch is laid, at which point the female is sexually uninteresting to other males. Thus in the extreme form of this system the male contributes nothing except defense against other males throughout the period of sexual activity. This defense is necessary only because of male sexual activity, but it is valuable from the female's perspective because it allows her to continue feeding and raising young efficiently. Hence the social system emerges as a consequence of intraspecific competition, with the input of ecological variables reduced to permissiveness.

Wittenberger and Tilson (1980) discussed mate defense monogamy in the context of mating systems theory, and therefore argued that one or other sex is the controlling sex: in this case, males. As McKinney (Chapter 8) shows, however, the system is not imposed on females in a proximate sense. Each female works actively to establish a bond with a male, normally unmated, but not necessarily, if few males are available. (See also Lumpkin, 1983, who argues that prolonged female sexual responsiveness has evolved in a variety of birds because of benefits to the female of attracting a male who guards her against harassment.) The basis of female choice is still uncertain, but male dominance over others appears to be important in a variety of species. Thus it is clear that females behave as if they benefit from the presence of a mate.

In an ultimate sense, nevertheless, the social system can indeed be argued to have evolved as a result of male behavior (cf. Rubenstein, Chapter 13). If no males existed, females can be imagined to feed, lay, and rear their offspring as they do in the presence of males. Mated males are "hired guns," necessary only because guns have been invented.

The hired gun principle has been argued to have effects in other species where resources are undefendable and predation appears unimportant to the development of social relationships. Polygyny in horses (Rubenstein, Chapter 13) and gorillas is explicable because of the benefits of male protection from harassment by other males, as are the communities of chimpanzee and bonobos (Wrangham, Chapter 16). It may be a useful principle that applies to other groups, but has been obscured by searches for direct links between ecology and sociality.

The hired guns need not be male. Matrilines are argued to function as part of a strategy of defense against potential infanticide by other females in yellow-bellied marmots (Armitage, Chapter 14). Parallel arguments are given for matrilines in gelada and lions, except that harassment and competition by other females are viewed as more subtle strategies than infanticide (Dunbar, Chapter 15; Rubenstein, Chapter 13). Armitage stresses this point as he argues that female marmots, ground squirrels, and prairie dogs all repress the reproductive abilities of close kin, but by a variety of ways that fall along this continuum.

The classic situation where the behavior of one sex shapes the behavior of the other, independent of resources, occurs among lekking species. As Gibson and Bradbury show in sage grouse (Chapter 17), female movements determine male distributions, and even within leks female choosiness incites males to strut. It is still unclear why leks occur at all. However, in some species they occur only at high population densities (e.g., Uganda kob *Kobus kob*, Leuthold, 1966), or in habitats where predation pressure is extremely high (e.g., topi *Damiliscus korrigum*, Gosling, Chapter 12), where males abandon their normal strategy of territorial defense of food resources. This suggests that ecological pressures have a permissive effect, allowing leks to develop because resources are not defendable.

The effect of female choice in shaping social organization is not limited to leks or mate-guarding groups such as dabbling ducks and gorillas. In the eastern grey kangaroo (Jarman, Chapter 18), females in estrus roam widely advertising themselves to all the males in a neighborhood. Male hierarchies are strictly linear and are based on size. So by roaming over the entire habitat estrus females are virtually assured of mating with the largest and probably oldest males. The advantages of this strategy to females are not clear, but it does seem that they are able to adopt it because there are no stringent ecological pressures forcing them into particular kinds of groups.

These cases stress that intraspecific competition can promote social behavior even in the absence of any classical ecological benefit, just as Alexander (1977) argued for humans. It appears that vulnerability to conspecifics resembles predation risk, resource distribution, and other less common ecological factors in having effects either on its own or in combination with other factors.

EMERGENT CONSIDERATIONS

When a collection of essays focuses on how special features of ecology shape particular social relationships, it is possible to concentrate on details and miss the wood for the trees. In this collection a few important themes recur often enough to warrant further attention.

Biological versus Ecological Determinism

The apparent snug fit between ecology and sociality suggests that most animals come close to achieving optimal solutions to problems posed by nature. In a proximate sense behavior appears to be determined by ecological circumstance. But this should not be taken to imply that there is no genetic control of behavior. As Maynard Smith's (1977) game theory approach has shown, behavioral variation can be maintained in a population by two mechanisms. In one, individuals with different genetic predispositions can coexist at a frequency where each receives equal payoffs. In the other, all members of the population have the same genetic predisposition to adopt a variety of behaviors. As in the previous case, the point of payoff equality establishes the relative frequencies of each behavior. In most of the studies in this volume, behavioral variability appears to be maintained by behavioral flexibility. Thus selection in the past for a particular genetic constitution has freed individuals to behave according to the demands of the environment.

Phylogenetic Constraints

Behavioral flexibility is not infinite. Morphology and physiology often limit a species' behavioral options. For instance, animals without shearing carnassials make poor carnivores. Such limitations are usually the direct result of adaptations to other ecological circumstances, but it should not be assumed that they always are. For instance, within semi-closed groups of birds the pattern of giving aid can show remarkably little discrimination, with different individuals being treated with approximately equal generosity, regardless of kinship or the probability of future reciprocity (e.g., Mexican jays *Aphelocoma ultramarina*: Brown & Brown, 1981a; stripe-backed wrens *Campylorhyncus nuchalis*: Rabenold, 1985). In some social mammals, by contrast, intense discrimination based on kinship or reciprocity is routine (e.g., terrestrial cercopithecines: Hinde, 1983). Even within mammals the degree of differentiation of relationships within groups appears to vary; for instance, coalitions involving competitive elements have been reported more often in primates than in other mammals (Wrangham, 1983). It seems likely that differences in cognitive ability contribute to these taxon-specific differences. In the same way didelphid marsupials appear to have simpler social systems than those of eutherian mammals with similar ecology (Charles-Dominique, 1983). The ability to recognize other individuals and to evaluate probable benefits of particular acts could both be important. If so, the poor cognitive abilities of birds could be responsible for the fact that they form simpler social networks than those of many mammals.

Although this is a reasonable hypothesis, it cannot be properly evaluated until the economics of discriminatory behavior are better understood. For instance, Rabenold (1985) suggested that simple hard-wired behavior programs may be the most effective strategy in group-living birds, assuming that the payoff for a given behavior is highly predictable. Unfortunately, the predictability of payoffs probably depends on the flexibility of the behavior of conspecifics, so that predictability depends in part on the species' cognitive abilities. Nevertheless, it may yet be shown that the spatial and temporal distribution of resources favors greater discrimination in mammals than birds, just as Sherman and Holmes (1985) have indicated that kin recognition abilities in different species of ground squirrels appear to be adapted to the complexity of their social systems. Similarly, ecological hypotheses for species differences should always be investigated before phylogenetic constraints are invoked (Wilson, 1975).

Similarities and Differences in the Socio-Ecology of Birds and Mammals

Phylogenetic constraints may be responsible for some differences between avian and mammalian social systems, such as the degree of differentiation of social relationships. However, many aspects of social organization are similar in the two taxa. Both birds and mammals, for example, include species with monogamous, polygynous, or polyandrous social relationships, leks and cooperative breeding, indicating that, in general, mating relationships show similar variety in birds and mammals. In some cases where differences are known, it seems likely that they are the result of differences in ecology. For example, there is a higher frequency of territorial monogamy in birds than in mammals (Wittenberger and Tilson, 1980). Although the economics of territorial differences are understood well in only a few species (e.g., sunbirds: Gill and Wolf, 1975), this difference is probably attributable to territories being economically defendable by individuals or pairs more often among birds than among mammals.

Social relationships within sexes, by contrast to mating patterns, appear less diverse in birds than mammals. First, closed foraging groups with more than one breeding female are exceedingly rare in birds but common in mammals. Furthermore, where they occur in birds (e.g., anis, *Crotophaga sulcirostris*: Vehrencamp, 1978), affiliative relationships among breeding females are poorly developed. This is probably related to the fact that within closed groups complete reproductive suppression among females (as opposed to partial reproductive suppression: Wasser and Barash, 1983) is more extensive among birds than among mammals. The localization of breeding effort at a nest-site appears to be partially responsible for the effectiveness of reproductive suppression in birds (Robinson, Chapter 9). Aggressive competition among breeding females is effective at nest-sites, and many mammals that rear altricial young at nest-sites (e.g., dens) also show closed groups with only one breeding female (Emlen, 1984).

Second, competitive relationships among breeding mammals are more highly developed in mammals than birds. Polyandrous mating within cooperative groups has been demonstrated in one bird (acorn woodpeckers: Joste et al., 1985), whereas extensive mate sharing is known in a variety of primates and carnivores (Andelman, Chapter 10; Wrangham, Chapter 16; Moelhman, Chapter 5, and Parker, Chapter 19). Mammals form larger closed groups than birds do. Since in some mammals the number of males per group increases with group size, factors controlling group size may be responsible for the greater elaboration of both female–female and male–male relationships in mammals than birds. Pervasive ecological differences between birds and mammals therefore appear to be important in determining the size and structure of social groups. A systematic comparison of the socio-ecology of the two taxa has not yet been conducted, however.

The Problem with a Name

After the major features of a mating or a social system have been described, a shorthand label is often attached, and the system is placed into one of the many broad, well-established, and conventionally accepted categories. For example, all mating systems composed of one male and many females are usually termed harems. Yet as many of the chapters in this volume have shown, harems come in a variety of forms, each emerging from a unique set of environmental circumstances. The confusion is great since harems can contain females having affiliative relationships primarily with the male (hamadryas baboons), with both the male and unrelated females (horses), with the male and related females (gelada baboons), or with the male and both types of females (marmots). By lumping these species together the ecological rationale for each set of relationships is obscured. Moreover, even describing these systems as "harems" unwittingly creates the impression that in these groups males control females. As the above studies have shown, this rarely is the case. Similarly, monogamous systems include many variations in social relationships, such as in the division of labor and intersexual dominance (Bossema and Benus, 1985; Wrangham, 1986). The importance of analyzing systems according to the types of social relationship they include, rather than merely by group size and sex ratio, means that categories such as "harem," "monogamy," "polyandry," and other labels that describe principally mating relationships should not be treated as complete.

Different Glues for Different Sexes

The search for the causes of sociality often produces a single benefit that is thought to apply to all members of the society. For example, grouping in wild dogs might be attributed to increased benefits derived from cooperative hunting, or coloniality in caciques might be attributed to reduced vulnerability of nests to predators. But as many of the studies in this volume have shown, the benefits that bind together different classes of individuals within the same society can be very different. Moreover, they are often products of different kinds of glue. As Packer (Chapter 19) suggests, males may band together mutualistically to augment their mating prospects, while females may come together as kin and, by incurring a small cost, may reduce the likelihood of incurring the much greater cost of losing a large carcass to strangers. In addition, at the intersexual level lionesses appear to profit by mutualistically defending cubs against infanticide practiced by newly arriving males. Since different classes of individuals often have different needs, it would not be surprising if subgroups within a society associated for a variety of different reasons. In general, single causes of sociality are most likely to apply where ecological pressures affect both sexes equally, for example, where adult survival is threatened or where the sexes cooperate in obtaining resources. However, these circumstances are not expected to be common, because sex differences in reproductive strategy commonly lead to differences in the importance of particular ecological pressures.

Mating Systems, Intrasexual Cooperation, and Dispersal

One of the most important predictors of the pattern of social relationships within groups is the kinship system. This is not a universal relationship. For instance, in acorn woodpeckers males are more likely to breed in their natal groups than females, but females that stay tend to provide more help than males (Koenig et al., 1983). Generally, however, cooperation appears more extensive among individuals of the sex that tends to stay together as kin. For example, in solitary mammals females tend to be philopatric and have tolerant relationships with female kin, unlike males (Waser and Jones, 1983); in cooperatively breeding birds males are more likely to have kin within groups, and are normally more frequent helpers (Koenig et al., 1983); and in primates the amount of cooperation within sexes is closely correlated with the tendency for kin to live together. For instance, in hamadryas baboons females transfer between groups and males cooperate, while in gelada baboons males transfer between groups and females cooperate (Dunbar, Chapter 15; Struhsaker and Leland, 1979; Wrangham, 1980). This means that an explanation of the pattern of social relationships is tied closely to the problem of explaining dispersal patterns.

Greenwood (1980) proposed that species differences in dispersal patterns were explicable by the mating system. Using the classification of mating systems proposed by Emlen and Oring (1977) and Bradbury and Vehrencamp (1977), Greenwood argued that resource defense systems favor male philopatry and female dispersal, whereas mate defense systems favor female philopatry and male dispersal. Male philopatry was hypothesized to be favored in resource defense systems because it increases the ability of males to acquire resources. In mate defense systems, correspondingly, female philopatry was thought to occur because females invest more heavily in young than males, and therefore benefit by being sedentary. Thus Greenwood (1980) suggested that resource defense mating systems tend to be patrilineal, while mate defense systems tend to be matrilineal.

It is now clear that there are significant exceptions to these generalizations. Thus among group-living primates, most of the species in which male philopatry occurs show no resource defense by males (mountain gorilla *Gorilla gorilla berengei*, red colobus *Colobus badius*, hamadryas baboons *Papio hamadryas*: Pusey and Packer, 1986; Cheney, 1986). These species suggest that a more complete explanation of dispersal patterns can be obtained by including a consideration of intrasexual relationships. The spirit of Greenwood's argument was that sex biases in philopatry arise if there is a sex bias in the advantage of remaining in or close to the natal range. Greenwood suspected that the mating system would be a good predictor of sex biases in the benefit for philopatry. However, the primate data show that patterns of intergroup dispersal are correlated not with the presence or absence of resource defense, but instead with the tendency for females or males to cooperate in intrasexual competition (Pusey and Packer, 1986). Thus in mountain gorillas, red colobus, and hamadryas baboons, males within groups form alliances against males in neighboring groups. The fact that they cooperate appears to be more important in determining the benefits of philopatry than whether or not they attempt to defend resources. This observation retains the spirit of Greenwood's argument, while noting that the mating system as such is a less important variable than the way in which males or females compete. As socio-ecologists continue to broaden their analyses from a focus on mating patterns to a systematic consideration of relationships within sexes, it seems likely that more cases like this will emerge.

CONCLUSIONS AND SUMMARY

Although no predictive theory of social evolution exists, a number of general principles seem clear. First, social organization is the result of the interaction between proximate ecological pressures and "culture"—the social traditions already in place and shaped by local and phylogenetic history. Second, ecological pressures can operate rather directly on social relationships, which are formed in different ways depending on sex, breeding status, and other phenotypic attributes. This contrasts with the more traditional view that ecology first determines the size, composition, and dispersion of groups, and then through this context shapes relationships. Third, a suite of ecological pressures—the distribution of critical resources, the defendability of the resources, the intensity of predator pressure, and the intensity and nature of intraspecific competition—provide much of the force behind social evolution.

Unfortunately no crisp generalizations emerge as to how these forces shape social organization. For example, we are still puzzled as to: Why some societies have strong dominance structures whereas others do not (Vehrencamp, 1983); why in some societies the young adopt the rank of their parents, whereas in others they do not (Silk, 1986); and why such different ontogenies lead to differences in juvenile dispersal strategies (Altman, 1980). In part differences in phylogenetic history, which reflects patterns of sexual dimorphism, diet, and vulnerability to predation, as well as differences in demography, which reflects age structure and sex ratios, are responsible. But despite these particularities the following common themes are emerging.

Unravelling the evolution of any social system must begin with an understanding of the roots of female behavior, since the behavior of males is largely adapted to that of females. Finding these roots can be assisted by answering three queries. First, does the distribution of the critical resource (food, water, safe sites) force females to forage, travel, or live alone? If the resources are sparsely distributed, or intense predation can only be avoided by crypsis, then females will be forced to spend most of their time apart (Gosling, Chapter 12; Jarman, 1974). Second, does the nature of the critical resource facilitate its exclusive defense? If territoriality by lone individuals is economically feasible then females will rarely meet, otherwise they may aggregate when not contesting this critical resource, and a variety of fission-fusion type of social systems may develop (Robinson, Chapter 9; Gosling, Chapter 12; Rubenstein, Chapter 13; Wrangham, Chapter 16).

If females are not forced to forage, travel, or rest alone, then a third query must be answered—does the distribution of the resources or the overall structure of the habitat permit females to aggregate, or force them to do so? Permissive habitats are bountiful ones where resources are distributed fairly evenly, and competition among females is low. In such situations the risks of either predation or male harassment may be so high that females aggregate around males that can provide benefits sufficient to reduce these risks and offset the costs of competition (Rubenstein, Chapter 13; Wrangham, Chapter 16). Since competition for the prime males may be keen, permanent associations may develop. Conversely, in other habitats where resources are distributed in rich but scattered patches competition is intensified and may force females to aggregate, ensuring that others and not oneself, are excluded (Packer, Chapter 19). Even within groups coalitions may form to assist partners in garnering a disproportionate share of the resources that the group already controls (Wrangham, 1980; Dunbar, Chapter 15).

After an understanding of the nature and the determinants of female relationships have emerged, then the effects of these relationships on male behavior can be explored in a similar way. Since competition for elusive reproductive females is so intense, males usually avoid other males when searching for, and obtaining access to reproductive females. But the exceptions are striking (e.g., humans: Flinn and Low, Chapter 11; hamadryas baboons: Dunbar, Chapter 15; chimps: Wrangham, Chapter 16; and lions: Packer, Chapter 19) as is the fact that some also defend directly, or indirectly, one or more females whereas others do not. Moreover, this defense can be as brief as one reproductive episode or as long as a breeding lifetime. Ultimately female distributions and associations limit male options, but the demography of the male population also plays a major role by adjusting the costs of various male strategies. Consequently, the linkage between particular male and female strategies is obscured (e.g., antelope: Gosling, Chapter 12; Old World monkeys: Andelman, Chapter 10). Only with more long-term studies, like those in this volume, will these issues be brought into sharper focus.

Acknowledgments

———————————— • ————————————

L. W. Oring and D. B. Lank, Chapter 2. This chapter is based upon research conducted with the support of National Science Foundation grants GB 42255, DEB 77-11147, DEB 79-11147, and PCM 8315758. C. Gratto's help in production of the manuscript was invaluable. S. Haig kindly provided unpublished data; and D. Bosanko, R. Huddles, M. Lord, and D. Traun provided logistical help. Over the years, more than twenty people assisted with this work in the field. To all of them, a special thanks.

P. Moehlman, Chapter 4. I am grateful to friends at Ndutu Tented Camp and Olduvai Gorge for their support and assistance. Conversations and discussions over the years with colleagues studying behavioral ecology have all contributed to my understanding of these animals. D. Macdonald, J. Malcolm, J. Goodall, L. Frame, G. Frame, T. Collins, J. Gittleman, P. Waser, J. Rood, D. Rubenstein, R. Trivers, J. Downhower, and N. Sedransk have been particularly helpful. Financial support was provided by the National Geographical Society, The Harry Frank Guggenheim Foundation, and the Yale School of Forestry and Environmental Studies. S. Lowe ran the log-log linear regressions and L. Bowden, J. Mount, and D. Miller typed the manuscript. I thank the Tanzanian Scientific Council, Tanzanian National Parks, the Ngorongoro Conservation Area Authority, and the Serengeti Wildlife Research Institute for making this research possible. This is SWRI publication No. 241.

G. E. Woolfenden and J. W. Fitzpatrick, Chapter 5. We remain grateful to the Archbold Biological Station and its staff for providing an ideal working environment amidst the jays, and for their dedication to the protection of the Florida oak scrub. We thank Bobbie Kittleson and Jan Woolfenden for their help in the field and in data organization. We appreciate helpful comments and criticisms made by Wayne Hoffman, Jack Hailman, Kevin McGowan, and Ron Mumme on earlier drafts of this chapter. Field work has been supported by numerous institutions; we especially thank the National Geographic Society, American Philosophical Society, Chapman Memorial Fund, and the Conover Fund of the Field Museum of Natural History.

471

Acknowledgments

M. Leighton, Chapter 6. I wish to thank the Indonesian Institute of Sciences and the Subdirectorate of Forest Planning and Nature Conservation for permission to conduct research in Indonesia, and NSF for funding. An anonymous reviewer and the editors offered helpful comments on the manuscript.

J. P. Rood, Chapter 7. I thank the Tanzania National Scientific Research Council for permission to conduct research on mongooses at the Serengeti Wildlife Research Institute, and the National Geographic Society, the Max-Planck-Institut für Verhaltensphysiologie, and the Harry Frank Guggenheim Foundation for providing the funding that made the research possible. I am also grateful to J. Gittleman, D. Kleiman, D. Macdonald, H. Rood, D. Rubenstein, P. Waser, C. Wozencraft, and R. Wrangham for their constructive comments and criticism of an earlier draft of this chapter.

F. McKinney, Chapter 8. I am very grateful to M. G. Anderson, D. Rubenstein, P. Sherman, H. B. Tordoff, R. A. Wishart, R. Wrangham, and two anonymous reviewers for constructive criticisms of the manuscript. My research has been supported by the National Science Foundation (grants GB 36651x, BNS 76-02233, BNS 7924692, and BNS 83-17187), the National Geographic Society, and the Graduate School and Field Biology Program of the University of Minnesota.

S. K. Robinson, Chapter 9. I would especially like to thank the members of the Princeton population biology group, John Terborgh, Henry Horn, Daniel Rubenstein, John Bonner, John Hoogland, and Robert May for their support and advice throughout the project. John Terborgh provided the opportunity to work in the Manu National Park and was continuously helpful both in the field and at Princeton. John Hoogland first suggested looking at the mating system of the cacique and Daniel Rubenstein suggested a systematic study of female-female interactions. Henry Horn helped bring the entire project together with his comments and criticisms of oral and written presentations of the material. I also thank Diane Wiernasz, Charles Brown, and Charles Munn for their helpful comments. Nancy Burley and an anonymous reviewer made some excellent suggestions for improving the manuscript. I am grateful to the Peruvian Ministerio de Agricultura, Dirección General Forestal y de Fauna, and all officials of the Parque Nacional del Manu for their cooperation and for permission to work in the Manu Park. Financial support was provided by the Frank M. Chapman Memorial Fund of The American Museum of Natural History, the Society of Sigma XI, the Department of Biology of Princeton University, and the National Science Foundation.

Acknowledgments

S. J. Andelman, Chapter 10. I am grateful to Michael Beecher, Carolyn Crockett, and Gordon Orians for discussion and constructive criticism of the manuscript. Richard Wrangham and Daniel Rubenstein were extremely patient with my delays in preparing this manuscript, and also made several useful suggestions. Research on vervets was supported by a National Science Foundation Predoctoral Fellowship and a University of Washington Dissertation Fellowship, and by NSF grant BNS 80-08946 and a Harry Frank Guggenheim Foundation grant to D. L. Cheney and R. M. Seyfarth. I am grateful to the Office of the President and the Ministry of Tourism and Wildlife of the Republic of Kenya for permission to conduct research in Kenya.

M. V. Flinn and B. S. Low, Chapter 11. Many people were helpful foils in developing the ideas presented here. We are particularly grateful to Dick Alexander, Richard Wrangham, Bill Irons, Dan Rubenstein, and the members of the University of Michigan Human Behavior and Evolution Group for thoughtful criticism and discussion.

D. I. Rubenstein, Chapter 13. I thank H. S. Horn and R. W. Wrangham for comments on previous drafts of the chapter. The courtesies extended to me by the Duke University Marine Laboratory while I studied horses, and the assistance rendered by the Office of the President and the Government of Kenya while I worked on zebras, are much appreciated. This research was partially funded by National Institutes of Mental Health grant PSHMH 34890 and National Science Foundation grant BSR 8352137.

K. B. Armitage, Chapter 14. This research was supported by National Science Foundation grants G 16354, GB 1980, GB 6123, GB 8526, GB 22494, BMS 74-21193, DEB 78-07327, and DEB 81-21231. Field facilities were provided by the Rocky Mountain Biological Laboratory. Sharon Hagan prepared the figures and Jan Elder and Coletta Spencer typed the manuscript. I greatly appreciate the many students and colleagues who assisted in data collection and analysis and who provided intellectual stimulation and generously shared their ideas and insights. I also thank Richard Wrangham, Daniel Rubenstein, Paul Sherman, and an anonymous reviewer whose stimulating comments on an earlier draft of this chapter did much to improve its content and sharpen the developing ideas. However, they are not responsible for what readers may interpret as "flights of fancy."

R.I.M. Dunbar, Chapter 15. The fieldwork on which this chapter is based was supported by grants from The Science and Engineering Research Council (U.K.) and the Wenner-Gren Foundation for Anthropological Research. The Ethiopian Government Wildlife Conservation Organisation kindly gave permission to work in the Simen Mountains National Park and provided invaluable logistic support.

R. W. Wrangham, Chapter 16. D. L. Cheney, W. C. McGrew, T. Nishida, D. I. Rubenstein, R. M. Seyfarth, and B. B. Smuts made valuable comments on this chapter, which was prepared while I was a Fellow at the Center for Advanced Study in the Behavioral Sciences. I am grateful for financial support provided by the National Science Foundation (grant BNS 76-22943).

R. M. Gibson and J. W. Bradbury, Chapter 17. The work reported here was supported by National Science Foundation BNS 79-23524 and BNS 82-15426. We thank the California Department of Fish and Game and U.S. Forest Service for logistical assistance and Daniel Rubenstein for extensive comments on an early draft of this chapter.

P. J. Jarman and C. J. Southwell, Chapter 18. We are grateful to the landholders at Wallaby Creek, Messrs. E. and J. Hayes and V. Mulcahy, for their toleration of our use of their properties, and to the District Forester, Urbenville, for his cooperation. We also thank the landholders for their protection of the area and its wildlife. Several students have assisted in aspects of the fieldwork. We are especially grateful to our Wallaby Creek colleagues, Chris Johnson and Robyn Stuart-Dick, for their help and company in the field and for their comments on a draft of this chapter. This research was supported by Grant D18115308 from the Australian Research Grants Scheme.

C. Packer, Chapter 19. All the field data in this chapter were collected in collaboration with Anne Pusey, who also provided many important contributions to the ideas presented here. Data on cub survival include valuable observations made by Monique Bergerhoff-Mulder and Sara J. Cairns. I thank Thomas Caraco, Timothy Caro, David Macdonald, Anne Pusey, Daniel Rubenstein, Karl Van Orsdol, Larry Wolf, and Richard Wrangham for their many valuable comments on the manuscript. Fieldwork was supported by grants from the H. F. Guggenheim Foundation, National Geographic Society, American Philosophial Society, Sigma Xi, and NIMH grant MH15181. I am also grateful to the Government of Tanzania for permission to conduct the research and to the Serengeti Wildlife Research Institute for the use of its facilities.

Literature Cited

●

Aberle, D. F. 1961. Matrilineal descent in cross-cultural perspective. In D. M. Schneider and K. G. Gough, eds., *Matrilineal Kinship*. pp. 655–727. University of California Press.

Ables, E. D. 1975. Ecology of the red fox in America. In M. W. Fox, ed., *The Wild Canids*, pp. 216–36. Van Nostrand Reinhold, N.Y.

Ackerman, C. 1964. Structure and statistics: The Purum case. *American Anthropologist* 66:53–65.

Acosta, A. I. 1972. Hand-rearing a litter of maned wolves *Chrisocyon brachyrus* at Los Angeles Zoo. *Int. Zoo Yearb.* 12:170–74.

Afton, A. D. 1979. Time budget of breeding northern shovelers. *Wilson Bull.* 91:42–49.

Afton, A. D. 1980. Factors affecting incubation rhythms of northern shovelers. *Condor* 82:132–37.

Afton, A. D. 1985. Forced copulation as a reproductive strategy of male lesser scaup: A field test of some predictions. *Behaviour* 92:146–67.

Afton, A. D., and R. D. Sayler. 1982. Social courtship and pairbonding of common goldeneyes, *Bucephala clangula*, wintering in Minnesota. *Can. Field Nat.* 96:295–300.

Albignac, R. 1973. *Faune de Madagascar*, 36: Mammiferes Carnivores. O.R.S.T.O.M., Paris.

Albignac, R. 1976. L'ecologie de *Mungotictis decemlineata* dans les forets decidues de l'ouest de Madagascar. *La Terrre et la Vie* 30:347–76.

Alexander, R. D. 1971. The search for an evolutionary philosophy of man. *Proc. Roy. Soc. Victoria, Melbourne* 84:99–120.

Alexander, R. D. 1974. The evolution of social behavior. *Ann. Rev. Ecol. Syst.* 5:325–83.

Alexander, R. D. 1977. Natural selection and the analysis of human sociality. In C. E. Goulden, ed., *Changing Scenes in the Natural Sciences: 1776–1976*, pp. 283–337. Bicentennial Symposium Monograph, Phil. Acad. Nat. Sci. Special Publ. 12.

Alexander, R. D. 1979. *Darwinism and Human Affairs*. University of Washington Press.

Alexander, R. D. 1985. The biology of moral systems. Typescript.

Alexander, R. D., and G. Borgia. 1979. On the origin and basis of the male-female phenomenon. In M. S. Blum and N. A. Blum, eds., *Sexual Selection and Reproductive Competition in Insects*, pp. 417–40. Academic Press, N.Y.

Alexander, R. D., J. L. Hoogland, R. D. Howard, K. M. Noonan, and P. W. Sherman. 1979. Sexual dimorphisms and breeding systems in pinnipeds, ungulates, primates, and humans. In N. A. Chagnon and W. Irons, eds., *Evolutionary Biology and Human Social Behavior: An Anthropological Perspective*, pp. 402–435. Duxbury Press, N. Scituate, Mass.

Altmann, J. 1974. Observational study of behavior: Sampling methods. *Behaviour* 49:227–67.

Altmann, J. 1980. *Baboon Mothers and Infants*. Harvard University Press.

Altmann, J., S. A. Altmann, G. Hausfater, and S. A. McCuskey. 1977. Life history of yellow baboons: Physical development, reproductive parameters, and infant mortality. *Primates* 18:315–30.

Altmann, S. A. 1962. A field study of the sociobiology of rhesus monkeys. *Macaca mulatta. Annals of N.Y. Acad. Sci.* 102:338–435.

Altmann, S. A. 1974. Baboons, space, time, and energy. *Amer. Zool.* 14:221–48.

Altmann, S. A., and J. Altmann. 1970. *Baboon Ecology*. University of Chicago Press.

Literature Cited

Altmann, S. A., S. S. Wagner, and S. Lenington. 1977. Two models for the evolution of polygyny. *Behav. Ecol. Sociobiol.* 2:397–410.

American Ornithologists' Union. 1983. *Check-list of North American Birds.* 6th ed. Allen Press, Lawrence, Kansas.

Andelman, S. J. 1985. Ecology and reproductive strategies of vervet monkeys in Amboselli National Park, Kenya, Ph.D. diss. University of Washington.

Andelman, S. J., J. G. Else, J. P. Hearn, and J. K. Hodges. 1985. The non-invasive monitoring of reproductive events in wild Vervet monkeys (*Cercopithecus aethiops*) using urinary pregnanediol-3 -glucuronide and its correlation with behavioural observations. *J. Zool. London* 205:467–77.

Andelt, W. F. 1982. Behavioral ecology of coyotes on Welder Wildlife Refuge, South Texas. Ph.D. diss. Colorado State University, Fort Collins.

Andere, D. K. 1981. Wildebeest *Connochaetes taurinus* (Burchell) and its food supply in Amboseli basin. *Afr. J. Ecol.* 19:239–50.

Andersen, D. C., K. B. Armitage, and R. S. Hoffmann. 1976. Socioecology of marmots: Female reproductive strategies. *Ecology* 57:552–60.

Andersen, F. S. 1948. Contributions to the biology of the ruff (*Philomachus pugnax* L.) Pt. 2. *Dan. Ornithol. Foren. Tidaskr.* 42:125–48.

Anderson, D. J. 1982. Home range: A new non-parametric estimation technique. *Ecology* 63:103–112.

Anderson, D. R. 1975. Population ecology of the Mallard: V. Temporal and geographic estimates of survival, recovery, and harvest rates. *U.S. Fish Wild. Serv. Res. Publ.* 125.

Andersson, M. 1984. Brood parasitism within species. In C. J. Barnard, ed., *Producers and Scroungers: Strategies of Exploitation and Parasitism*, pp. 195–228. Croom Helm, London.

Ankney, C. D., and C. D. MacInnes. 1978. Nutrient reserves and reproductive performance of female lesser snow geese. *Auk* 95:459–71.

Ankney, C. D., and D. M. Scott. 1982. On the mating system of brown-headed cowbirds. *Wilson Bull.* 94:260–68.

Appleby, M. C. 1983. The probability of linearity in hierarchies. *Anim. Behav.* 31:600–608.

Arak, A. 1983. Male-male competition and mate choice in anuran amphibians. In P. Bateson, ed., *Mate Choice*, pp. 181–210. Cambridge University Press.

Armitage, K. B. 1962. Social behaviour of a colony of the yellow-bellied marmot (*Marmota flaviventris*). *Anim. Behav.* 10:319–31.

Armitage, K. B. 1965. Vernal behaviour of the yellow-bellied marmot (*Marmota flaviventris*). *Anim. Behav.* 13:59–68.

Armitage, K. B. 1974. Male behaviour and territoriality in the yellow-bellied marmot. *J. Zool. London* 172:233–65.

Armitage, K. B. 1975. Social behavior and population dynamics of marmots. *Oikos* 26:341–54.

Armitage, K. B. 1977. Social variety in the yellow-bellied marmot: A population-behavioural system. *Anim. Behav.* 25:585–93.

Armitage, K. B. 1979. Food selectivity by yellow-bellied marmots. *J. Mammal.* 60:628–29.

Armitage, K. B. 1981. Sociality as a life-history tactic of ground squirrels. *Oecologia* 48:36–49.

Armitage, K. B. 1982a. Marmots and coyotes: Behavior of prey and predator. *J. Mammal.* 63:503–505.

Armitage, K. B. 1982b. Social dynamics of juveniles marmots: Role of kinship and individual variability. *Behav. Ecol. Sociobiol.* 11:33–36.

Armitage, K. B. 1983. Individuality, social behavior, and reproductive success in yellow-bel-

Literature Cited

lied marmots. Paper presented at the symposium Origins and Significance of Individual Differences, Animal Behavior Society, Bucknell University.

Armitage, K. B. 1984. Recruitment in yellow-bellied marmot populations: Kinship, philopatry, and individual variability. In J. O. Murie and G. R. Michener, eds., *Biology of Ground-Dwelling Squirrels*, pp. 377–403. University of Nebraska Press.

Armitage, K. B., and J. F. Downhower. 1974. Demography of yellow-bellied marmot populations. *Ecology* 55:1233–45.

Armitage, K. B., J. F. Downhower, and G. E. Svendsen. 1976. Seasonal changes in weights of marmots. *Amer. Midl. Nat.* 96:36–51.

Armitage, K. B., and D. W. Johns. 1982. Kinship, reproductive strategies and social dynamics of yellow-bellied marmots. *Behav. Ecol. Sociobiol.* 11:55–63.

Armitage, K. B., D. W. Johns, and D. C. Andersen. 1979. Cannibalism among yellow-bellied marmots. *J. Mammal.* 60:205–207.

Arnold, K. A. 1983. *Quiscalus mexicanus*. In D. H. Janzen, ed., *Costa Rican Natural History*, pp. 601–603. University of Chicago Press.

Arnold, S. J. 1983. Sexual selection: The interface of theory and empiricism. In P.P.G. Bateson, ed., *Mate Choice*, pp. 67–107. Cambridge University Press.

Aschoff, J., and H. Pohl. 1970. Rhythmic variations in energy metabolism. *Fed. Proc.* 29:1541–52.

Ashcroft, R. E. 1976. A function of the pairbond in the common eider. *Wildfowl* 27:101–105.

Asdell, S. A. 1964. *Patterns of Mammalian Reproduction*. Cornell University Press.

Aswad, B. 1971. *Property Control and Social Strategies in Settlers in a Middle Eastern Plain*. Anthropological papers no. 44, Museum of Anthropology, University of Michigan, Ann Arbor.

Atwood, J. L. 1980. Social interactions in the Santa Cruz Island scrub jay. *Condor* 82:440–48.

Austad, S. N., and K. N. Rabenold. 1985. Reproductive enhancement by helpers and an experimental enquiry into its mechanism in the bicolored wren. *Behav. Ecol. Sociobiol.* 17:19–27.

Bachmann, C., and H. Kummer. 1980. Male assessment of female choice in hamadryas baboons. *Behav. Ecol. Sociobiol.* 6:315–21.

Backhaus, D. 1959. Beobachtungen uber das Freileben von Lelwel-Kuhantilopen (*Alcelaphus buselaphus lelwel*, Heuglin 1877) und Gelegenheitsbeobachtungen an Sennar-Pferdeantilopen (*Hippotragus equinus bakeri*, Heuglin 1863). *Z. Saugetierkde* 24:1–34.

Badrian, A., and N. Badrian. 1984. Social organization of *Pan paniscus* in the Lomako Forest, Zaïre. In R. L. Susman, ed., *Evolutionary Morphology and Behavior of the Pygmy Chimpanzee*, pp. 325–46. Plenum Press, N.Y.

Badrian, N., A. Badrian, and R. L. Susman. 1981. Preliminary observations on the feeding behavior of Pan paniscus in the Lomako Forest of central Zaïre. *Primates* 22:173–81.

Badrian, N., and R. K. Malenky. 1984. Feeding ecology of *Pan paniscus* in the Lomako Forest, Zaïre. In R. L. Susman, ed., *Evolutionary Morphology and Behavior of the Pygmy Chimpanzee*, pp. 275–99. Plenum Press, N.Y.

Bailey, R. 1985. The socio-ecology of Efe pygmies in the Ituri forest, Zaïre. Ph.D. diss., Harvard University.

Bailey, T. N. 1974. Social organization in a bobcat population. *J. Wildl. Mgmt.* 38:435–46.

Baker, R. R. 1978. *The Evolutionary Ecology of Animal Migration*. Hodder and Stoughton, London.

Baldwin, P. J., W. C. McGrew, and C.E.G. Tutin. 1982. Wide-ranging chimpanzees at Mt. Assirik, Senegal. *Int. J. Primatol.* 3:367–85.

Literature Cited

Baldwin, P. J., J. Sabater Pi, W. C. McGrew, and C.E.G. Tutin. 1981. Comparisons of nests made by different populations of chimpanzees. *Primates* 22(4):474–86.

Balfour, D. 1983. Infanticide in the Columbian ground squirrel, *Spermophilus columbianus*. *Anim. Behav.* 31:949–50.

Ball, I. J., D. S. Gilmer, L. M. Cowardin, and J. H. Riechmann. 1975. Survival of wood duck and mallard broods in north-central Minnesota. *J. Wildl. Mgmt.* 39:776–80.

Balph, D. F., and A. W. Stokes. 1963. On the ethology of a population of Uinta ground squirrels. *Amer. Midl. Nat.* 69:106–126.

Bancroft, G. T., and G. E. Woolfenden. 1982. *The Molt of Scrub Jays and Blue Jays in Florida*. Ornithol. Monogr. no. 29.

Barash, D. P. 1973. The social biology of the Olympic marmot. *Anim. Behav. Monogr.* 6:171–245.

Barash, D. P. 1974. The social behaviour of the hoary marmot (*Marmota caligata*). *Anim. Behav.* 22:256–61.

Barash, D. P. 1980. The influence of reproductive status on foraging by hoary marmots (*Marmota caligata*). *Behav. Ecol. Sociobiol.* 7:201–205.

Barash, D. P. 1981. Mate guarding and gallivanting by male hoary marmots (*Marmota caligata*). *Behav. Ecol. Sociobiol.* 9:187–93.

Barry, J. C. 1983. *Herpestes* from the Miocene of Pakistan. *J. Paleontol.* 57:150–56.

Barth, F. 1956. Ecological relationships of ethnic groups in Swat, North Pakistan. *American Anthropologist* 58:1079–89.

Barth, F. 1966. *Models of Social Organization*. Royal Anthropological Institute, occasional paper no. 23. RAI, London.

Bateman, A. J. 1948. Intra-sexual selection in *Drosophila*. *Heredity* 2:349–68.

Bateson, P.P.G. 1982. Preferences for cousins in Japanese quail. *Nature* 295:236–37.

Bateson, P.P.G., ed. 1983. *Mate Choice*. Cambridge University Press.

Bauer, H. R. 1979. Agonistic and grooming behavior in the reunion context of Gombe Stream chimpanzees. In D. A. Hamburg and E. R. McCown, eds., *The Great Apes*, pp. 395–403. Benjamin/Cummings, Menlo Park, Calif.

Beall, C. M., and M. C. Goldstein. 1981. Tibetan fraternal polyandry: A test of sociobiological theory. *American Anthropologist* 83:5–12.

Beattie, J. 1960. *Bunyoro: An African Kingdom*. Holt, Rinehart & Winston, N.Y.

Beck, T.D.I., and C. E. Braun. 1978. Weights of Colorado sage grouse. *Condor* 80:241–43.

Beecher, M. D., and I. M. Beecher. 1979. Sociobiology of bank swallows: Reproductive strategy of the male. *Science* 205:1282–85.

Beecher, W. J. 1951. Adaptations for food-getting in the American blackbirds. *Auk* 68:411–40.

Begon, M. 1984. Density and individual fitness: Asymmetric competition. In B. Shorrocks, ed., *Evolutionary Ecology*, pp. 175–94. Blackwell, Oxford.

Bekoff, M., T. J. Daniels, and J. L. Gittleman. 1984. Life history patterns and the comparative social ecology of carnivores. *Ann. Rev. Ecol. Syst.* 15:191–232.

Bekoff, M., J. Diamond, and J. B. Mitton. 1981. Life-history patterns and sociality in canids: Body size, reproduction, and behavior. *Oecologia* 50:386–90.

Bekoff, M., and R. Jamieson. 1975. Physical development in coyotes (*Canis latrans*) with a comparison to other canids. *J. Mammal.* 56:685–92.

Bekoff, M., and M. Wells. 1980. The social ecology of coyotes. *Sci. Amer.* 242:130–51.

Bekoff, M., and M. Wells. 1982. Behavioral ecology of coyotes: Social organization, rearing patterns, space use, and resource defense. *Z. Tierpsychol.* 60:281–305.

Bell, R.H.V. 1970. The use of the herb layer by grazing ungulates in the Serengeti. In A. Wat-

Literature Cited

son, ed., *Animal Populations in Relation to Their Food Resources*, pp. 111–24. Blackwell, Oxford.

Bellrose, F. C. 1976. *Ducks, Geese, and Swans of North America*. Stackpole, Harrisburg, Pa.

Bellrose, F. C., T. G. Scott, A. S. Hawkins, and J. B. Low. 1961. Sex ratios and age ratios in North American ducks. *Ill. Nat. Hist. Surv. Bull.* 27:391–474.

Ben-Yaacov, R. and Y. Yom-Tov. 1982. On the biology of the Egyptian mongoose, *Herpestes ichneumon*, in Israel. *Z. Saugetierk.* 48:34–45.

Bent, A. C. 1958. *Life Histories of North American Blackbirds, Orioles, Tanagers, and Allies*. Bull. 211, U.S. National Museum, Smithsonian Inst., Washington, D.C.

Berg, W. E., and R. A. Chesness. 1978. Ecology of coyotes in northern Minnesota. In M. Bekoff, ed., *Coyotes: Biology, Behavior, and Management*, pp. 229–47. Academic Press, N.Y.

Bernstein, I. S. 1967. A field study of the pigtail monkey (*Macaca nemistina*). *Primates* 8:217–28.

Berremann, G. 1962. Pahari polyandry: A comparison. *American Anthropologist* 64:60–75.

Berrie, P. M. 1973. Ecology and status of the lynx in interior Alaska. In R. L. Eaton, ed., *The World's Cats*, Vol. 1, pp. 4–41. World Wildlife Safari, Winston, Or.

Berte, N. 1983. Relatedness and exchange among the K'ekchi'. Ph.D. diss., Northwestern University.

Bertram, B.C.R. 1975a. Social factors influencing reproduction in wild lions. *J. Zool.* 177:463–82.

Bertram, B.C.R. 1975b. Weights and measures of lions. *E. Afr. Wildl. J.* 13:141–43.

Bertram, B.C.R. 1976. Kin selection in lions and evolution. In P.P.G. Bateson and R. A. Hinde, eds., *Growing Points in Ethology*, pp. 281–301. Cambridge University Press.

Bertram, B.C.R. 1978. Living in groups: Predators and prey. In J. R. Krebs and N. B. Davies, eds., *Behavioural Ecology: An Evolutionary Approach*, 1st ed., pp. 64–96. Blackwell, Oxford.

Bertram, B.C.R. 1979. Serengeti predators and their social systems. In A.R.E. Sinclair and M. Norton-Griffiths, eds., *Serengeti: Dynamics of an Ecosystem*, pp. 159–79. University of Chicago Press.

Bertram, B.C.R. 1982. Leopard ecology as studied by radio tracking. *Symp. Zool. Soc. London* 49:341–52.

Betzig, L. L. 1983. Despotism and differential reproduction. Ph.D. diss., Northwestern University.

Betzig, L. L. 1986. *Despotism and Differential Reproductive Success: A Darwinian View of Human History*. Aldine Press, Hawthorne, N.Y.

Biben, M. 1981. Ontogeny of social behaviour related to feeding in the crab-eating fox (*Cerdocyon thous*) and the bush dog (*Speothos venaticus*). *J. Zool. London* 196:207–216.

Bindernagel, J. A. 1968. Game cropping in Uganda. Canadian International Agency. Mimeo.

Blake, E. R. 1968. Family Icteridae. In R. A. Paynter, ed., *Checklist of Birds of the World*, Vol. 14, pp. 138–202. Museum of Comparative Zoology, Cambridge, Mass.

Blohm, R. J. 1978. Migrational homing of male gadwalls to breeding grounds. *Auk* 95:763–66.

Blohm, R. J. 1979. The breeding ecology of the gadwall in southern Manitoba. Ph.D. diss., University of Wisconsin, Madison.

Blohm, R. J. 1982. Differential occurrence of yearling and adult male gadwalls in pair bonds. *Auk* 99:378–79.

Boag, D. A., and J. O. Murie. 1981. Population ecology of Columbia ground squirrels in southwestern Alberta. *Can. J. Zool.* 59:2230–40.

Literature Cited

Borgerhoff-Mulder, M. 1984. A Darwinian explanation for brideprice. Paper presented at the annual meeting of the American Anthropological Association, November 1984, Denver.

Borgerhoff-Mulder, M. 1987. Kipsigis bridewealth payments. In L. L. Betzig, M. Borgerhoff Mulder, and P. W. Turke, eds., *Human Reproductive Behavior: A Darwinian Perspective.* Cambridge University Press.

Borgia, G. 1979. Sexual selection and the evolution of mating systems. In M. S. Blum and N. A. Blum, eds., *Sexual Selection and Reproductive Competition in Insects*, pp. 19–80. Academic Press, N.Y.

Borner, M. 1978. Status and conservation of the Sumatran tiger. *Carnivore* 1:97–101.

Bossema, I., and R. F. Benus. 1985. Territorial defence and intra-pair cooperation in the carrion crow (*Corvus corone*). *Behav. Ecol. Sociobiol.* 16:99–104.

Bossema, I., and J. P. Kruijt. 1982. Male activity and female mate acceptance in the mallard (*Anas platyrhynchos*). *Behaviour* 79:313–24.

Bourliere, F. 1963. Specific feeding habits of African carnivores. *African Wild Life* 17:21–27.

Bowen, W. D. 1978. Social organization of the coyote in relation to prey size. Ph.D. diss., University of British Colombia, Vancouver.

Bowen, W. D. 1981. Coyote social organization and prey size. *Can. J. Zool.* 59:639–52.

Bowman, L. A., S. R. Dilley, and E. B. Keverne. 1978. Suppression of oestrogen-induced LH surges by social subordination in talapoin monkeys. *Nature* 275:56–58.

Boyd, H. 1962. Population dynamics and the exploitation of ducks and geese. In E. D. Le Cren and M. W. Holdgate, eds., *The Exploitation of Animal Populations*, pp. 85–95. Blackwell, Oxford.

Bradbury, J. W. 1981. The evolution of leks. In R. D. Alexander and D. Tinkle, eds., *Natural Selection and Social Behavior: Recent Research and New Theory*, pp. 138–69. Chiron Press, N.Y.

Bradbury, J. W., and R. M. Gibson. 1983. Leks and mate choice. In P.P.G. Bateson, ed., *Mate Choice*, pp. 109–138. Cambridge University Press.

Bradbury, J. W., R. M. Gibson, and I-M Tsai. 1986. Hotspots and the evolution of leks. *Anim. Behav.*, in press.

Bradbury, J. W., and S. L. Vehrencamp. 1977a. Social organization and foraging in emballonurid bats, II: A model for the determination of group size. *Behav. Ecol. Sociobiol.* 1:383–404.

Bradbury, J. W., and S. L. Vehrencamp. 1977b. Social organization and foraging in emballonurid bats, III: Mating systems. *Behav. Ecol. Sociobiol.* 2:1–17.

Bradbury, J. W., S. L. Vehrencamp, and R. M. Gibson. 1985. Leks and the unanimity of female choice. In M. Slatkin and P. J. Greenwood, eds., *Evolution: Essays in Honour of John Maynard Smith*, pp. 301–314. Cambridge University Press.

Brady, C. A. 1978. Reproduction, growth, and parental care in crab-eating foxes (*Cerdocyon thous*) at the National Zoological Park, Washington. *Int. Zoo. Yearb.* 18:130–34.

Brady, C. A. 1979. Observations on the behavior and ecology of the crab-eating fox (*Cerdocyon thous*). In J. F. Eisenberg, ed., *Vertebrate Ecology in the Northern Neotropics*, pp. 161–72. Smithsonian Institution Press.

Brady, C., and M. K. Ditton. 1979. Management breeding of maned wolves (*Chrysocyon brachyurus*) at the National Zoological Park, Washington. *Int. Zoo. Yearb.* 19:171–76.

Bray, O. E., J. Kennelly, and J. L. Guarino. 1975. Fertility of eggs produced on territories of vasectomized red-winged blackbirds. *Wilson Bull.* 87:187–95.

Breder, C. N., and D. E. Rosen. 1966. *Modes of Reproduction in Fishes.* Natural History Press, N.Y.

Bronson, M. T. 1979. Altitudinal variation in the life history of the golden-mantled ground squirrel (*Spermophilus lateralis*). *Ecology* 60:272–79.

480

Literature Cited

Brown, D., and D. Hotra. 1984. Measuring polygyny in a monogamous society. Paper presented at the annual meeting of the American Anthropological Association, November 1984, Denver.

Brown, J. L. 1964. The evolution of diversity in avian territorial systems. *Wilson Bull.* 76:160–69.

Brown, J. L. 1974. Alternate routes to sociality in jays with a theory for the evolution of altruism and communal breeding. *Amer. Zool.* 14:63–80.

Brown, J. L. 1978. Avian communal breeding systems. *Ann. Rev. Ecol. Syst.* 9:123–55.

Brown, J. L. 1980. Fitness in complex avian social systems. In H. Markl, ed., *Evolution of Social Behavior: Hypotheses and Empirical Tests*, pp. 115–28. Verlag Chemie, Basel.

Brown, J. L. 1982. Optimal size in territorial animals. *J. Theor. Biol.* 95:793–810.

Brown, J. L., and E. R. Brown. 1981a. Extended family system in a communal bird. *Science* 211:959–60.

Brown, J. L., and E. R. Brown. 1981b. Kin selection and individual selection in babblers. In R. D. Alexander and D. Tinkle, eds., *Natural Selection and Social Behavior: Recent Research and New Theory*, pp. 244–56. Chiron, N.Y.

Brown, J. L., and G. H. Orians. 1970. Spacing patterns in mobile animals. *Ann. Rev. Ecol. Syst.* 1:239–62.

Buechner, H. K. 1961. Territorial behavior in Uganda kob. *Science* 133:698–99.

Buechner, H. K. 1963. Territoriality as a behavioral adapation to environment in Uganda kob. *Proc. XVI Int. Congr. Zool.* 3:59–63.

Buechner, H. K. 1974. Implications of social behavior in the management of Uganda kob. In V. Geist and F. R. Walther, eds., *The Behaviour of Ungulates and Its Relation to Management*, pp. 853–70. IUCN, Morges, Switzerland.

Buechner, H. K., J. A. Morrison, and W. Leuthold. 1965. Reproduction in Uganda kob with special reference to behavior. *Symp. Zool. Soc. London* 15:69–88.

Buechner, H. K., and D. R. Roth. 1974. The lek system in Uganda kob antelope. *Amer. Zool.* 14:145–62.

Buechner, H. K., and R. Schloeth. 1965. Ceremonial mating behavior in Uganda kob (*Adenota kob thomasi* Neumann). *Z. Tierpsychol.* 22:209–225.

Burns, J. T., K. M. Cheng, and F. McKinney. 1980. Forced copulation in captive mallards, I: Fertilization of eggs. *Auk* 97:875–79.

Busse, C. D. 1978. Do chimpanzees hunt cooperatively? *Amer. Natur.* 112:767–70.

Butler, P. M. 1946. The evolution of carnassial dentitions in the mammalia. *Proc. Zool. Soc. London* 116:198–220.

Bygott, J. D. 1974. Agonistic behaviour in wild chimpanzees. Ph.D. diss., Cambridge University.

Bygott, J. D. 1979. Agonistic behavior, dominance, and social structure in wild chimpanzees of the Gombe National Park. In D. A. Hamburg and E. R. McCown, eds., *The Great Apes*, pp. 405–427. Benjamin/Cummings, Menlo Park, Calif.

Bygott, J. D., B.C.R. Bertram, and J. P. Hanby. 1979. Male lions in large coalitions gain reproductive advantages. *Nature* 282:838–40.

Cade, W. 1979. The evolution of alternative male reproductive strategies in field crickets. In M. S. Blum and N. A. Blum, eds., *Sexual Selection and Reproductive Competition in Insects*, pp. 343–79. Academic Press, N.Y.

Caithness, T. A., and W. J. Pengelly. 1973. Use of Pukepuke Lagoon by waterfowl. *Proc. N.Z. Ecol. Soc.* 20:1–6.

Calverley, B. K., and D. A. Boag. 1977. Reproductive potential in parkland- and arctic-nesting populations of mallards and pintails (Anatidae). *Can. J. Zool.* 55:1242–51.

Literature Cited

Camenzind, F. J. 1978. Behavioral ecology of coyotes (*Canis latrans*) on the National Elk Refuge, Jackson, Wyoming. Ph.D. diss. University of Wyoming, Laramie.

Caraco, T. 1979. Time budgeting and group size: A test of theory. *Ecology* 60:618–27.

Caraco, T., S. Martindale, and T. S. Whittam. 1980. An empirical demonstration of risk-sensitive foraging preferences. *Anim. Behav.* 28:820–30.

Caraco, T., and L. L. Wolf. 1975. Ecological determinants of group sizes of foraging lions. *Amer. Natur.* 109:343–52.

Cargill, S. M., and R. L. Jefferies. 1984. The effects of grazing by lesser snow geese on the vegetation of a sub-arctic salt marsh. *J. Appl. Ecol.* 21:669–86.

Carl, E. A. 1971. Population control in arctic ground squirrels. *Ecology* 52:395–413.

Carpenter, C. R. 1942. Sexual behavior of free-ranging rhesus monkeys (*Macaca mulatta*), I: Specimens, procedures, and behavioral characteristics of estrus. *J. Comp. Psychol.* 33:113–42.

Caughley, G. 1964. Social organization and daily activity of the red kangaroo and grey kangaroo. *J. Mammal.* 45:429–36.

Caughley, G. 1977. *Analysis of Vertebrate Populations*. Wiley, Chichester.

Chagnon, N. A. 1974. *Studying the Yąnomamö*. Holt, Rinehart & Winston, N.Y.

Chagnon, N. A. 1979a. Is reproductive success equal in egalitarian societies? In N. B. Chagnon and W. Irons, eds., *Evolutionary Biology and Human Social Behavior: An Anthropological Perspective*, pp. 374–401. Duxbury Press, N. Scituate, Mass.

Chagnon, N. A. 1979b. Mate competition, favoring close kin, and village fissioning among the Yąnomamö Indians. In N. A. Chagnon and W. Irons, eds., *Evolutionary Biology and Human Social Behavior: An Anthropological Perspective*, pp. 375–402. Duxbury Press, N. Scituate, Mass.

Chagnon, N. A. 1982. Sociodemographic attributes of nepotism in tribal populations: Man the rule-breaker. In King's College Sociology Group, eds., *Current Problems in Sociobiology*, pp. 291–318. Cambridge University Press.

Chagnon, N. A. 1983. *Yąnomamö: The Fierce People*. 3rd ed. Holt, Rinehart & Winston, N.Y.

Chagnon, N. A., M. V. Flinn, and T. F. Melancon. 1979. Sex-ratio variation among the Yąnomamö Indians. In N. A. Chagnon and W. Irons, eds., *Evolutionary Biology and Human Social Behavior*, pp. 290–320. Duxbury Press, N. Scituate, Mass.

Chagnon, N. A., and W. Irons, eds. 1979. *Evolutionary Biology and Human Social Behavior*. Duxbury Press, N. Scituate, Mass.

Chapman, F. M. 1928. The nesting habits of Wagler's oropendola (*Zarhynchus wagleri*) on Barro Colorado Island. *Bull. Amer. Mus. Nat. Hist.* 58:123–66.

Charles-Dominique, P. 1983. Ecology and social adaptations in didelphid marsupials: Comparisons with eutherians of similar ecology. In J. F. Eisenberg and D. G. Kleiman, eds., *Advances in the Study of Mammalian Behavior*, pp. 395–422. Special Publication No. 7. American Society of Mammalogists.

Charlesworth, B. 1980. *Evolution in Age-Structured Populations*. Cambridge University Press.

Cheney, D. L. 1983. Proximate and ultimate factors related to the distribution of male migration. In R. A. Hinde, ed., *Primate Social Relationships*, pp. 241–49. Blackwell, Oxford.

Cheney D. L. 1986. Interactions and relationships between groups. In B. B. Smuts, D. L. Cheney, R. M. Seyfarth, R. W. Wrangham, and T. T. Struhsaker, eds., *Primate Societies*. University of Chicago Press, in press.

Cheney, D. L., P. C. Lee, and R. M. Seyfarth. 1981. Behavioural correlates of non-random mortality among free-ranging female vervet monkeys. *Behav. Ecol. Sociobiol.* 9:153–61.

Literature Cited

Cheney, D. L., and R. M. Seyfarth. 1981. Selective forces affecting the predator alarm calls of vervet monkeys. *Behaviour* 76:25–61.

Cheney, D. L., R. M. Seyfarth, S. J. Andelman, and P. C. Lee. 1986. Reproductive success in vervet monkeys. In T. H. Clutton-Brock, ed., *Reproductive Success*. University of Chicago Press.

Cheney, D. L., and R. W. Wrangham. 1986. Predation. In B. B. Smuts, D. L. Cheney, R. M. Seyfarth, R. W. Wrangham, and T. T. Struhsaker, eds., *Primate Societies*. University of Chicago Press.

Cheng, K. M., R. N. Schoffner, R. E. Phillips, and F. B. Lee. 1978. Mate preference in wild and domesticated (game-farm) mallards *(Anas platyrhynchos)*, I: Initial preference. *Anim. Behav.* 26:996–1003.

Cheng, K. M., R. N. Schoffner, R. E. Phillips, and F. B. Lee. 1979. Mate preference in wild and domesticated (game-farm) mallards, II: Pairing success. *Anim. Behav.* 27:417–25.

Child, G., and W. Richter. 1969. Observations on ecology and behaviour of lechwe, puku, and waterbuck along the Chobe River, Botswana. *Z. Saugetierkde* 34:275–95.

Chism, J., and D. Olson. 1982. Reproductive strategies of male patas monkeys. Paper presented to the 9th Congr. Int. Primatol. Soc., Atlanta, Ga.

Clapp, R. B., M. K. Klimkiewicz, and J. H. Kennard. 1982. Longevity records of North American birds: Gaviidae through Alcidae. *J. Field Ornith.* 53:81–124.

Clark, T. W. 1977. Ecology and ethology of the white-tailed prairie dog *(Cynomys leucurus)*. *Publ. Biol. Geol. Milwaukee Pub. Mus.* 3:1–97.

Clarke, A. B. 1978. Sex ratio and local resource competition in a prosimian primate. *Science* 201:163–65.

Clarke, E. 1957. *My Mother Who Fathered Me*. Allen & Unwin, London.

Clutton-Brock, T. H. 1983. Selection in relation to sex. In D. S. Bendall, ed., *Evolution from Molecules to Man*, pp. 457–81. Cambridge University Press.

Clutton-Brock, T. H., and S. D. Albon. 1979. The roaring of red deer and the evolution of honest advertisement. *Behaviour* 69:145–70.

Clutton-Brock, T. H., S. D. Albon, and F. E. Guinness. 1982. Competition between female relatives in a matrilocal mammal. *Nature* 300:178–80.

Clutton-Brock, J., G. B. Corbett, and M. Hills. 1976. A review of the family Canidae, with a classification by numerical methods. *Bull. Mus. Nat. Hist.* 29:117–99.

Clutton-Brock, T. H., F. E. Guinness, and S. D. Albon. 1982. *Red Deer: The Ecology of Two Sexes*. University of Chicago Press.

Clutton-Brock, T. H., and P. Harvey. 1976. Evolutionary rules and primate societies. In *Growing Points in Ethology*, ed. P.P.G. Bateson and R. A. Hinde, pp. 195–237. Cambridge University Press.

Clutton-Brock, T. H., and P. H. Harvey. 1977. Primate ecology and social organization. *J. Zool. London* 183:1–39.

Clutton-Brock, T. H., and P. H. Harvey. 1984. Comparative approaches to investigating adaptation. In J. R. Krebs and N. B. Davies, eds., *Behavioural Ecology: An Evolutionary Approach*, 2nd ed., pp. 7–29. Blackwell, Oxford.

Cody, M. L. 1966. A general theory of clutch size. *Evolution* 20:174–84.

Coe, M. J., D. H. Cumming, and J. Phillipson. 1976. Biomass and production of large African herbivores in relation to rainfall and primary production. *Oecologia* 22:341–54.

Cohen, J. 1971. *Casual Groups of Monkeys and Men: Stochastic Models of Elemental Social Systems*. Harvard University Press.

Cohen, R. 1967. *The Kanuri*. Holt, Rinehart & Winston, N.Y.

Cohen, S. A. 1977. A review of the biology of the dhole or Asiatic wild dog *(Cuon alpinus Pallas)*. *Anim. Regul. Stud.* 1:141–58.

Literature Cited

Cooke, F., C. S. Findley, and R. F. Rockwell. 1984. Recruitment and the timing of reproduction in lesser snow geese (*Chen caerulescens caerulescens*). *Auk* 101:451–58.

Corbet, G. B., and H. N. Southern. 1977. *The Handbook of British Mammals*. 2nd ed. Blackwell, Oxford.

Corbett, L. K. 1979. Feeding ecology and social organization of wildcats (*Felis sylvestris*) and domestic cats (*Felis catus*) in Scotland. Ph.D. diss., Aberdeen University.

Cords, M. 1984. Mating patterns and social structure in redtail monkeys (*Cercopithecus ascanius*). *Z. Tierpsychol.* 64:313–29.

Coulter, M. W., and W. R. Miller. 1968. *Nesting Biology of Black Ducks and Mallards in Northern New England*. Vermont Fish Game Dept. Bull. No. 68-2.

Cox, C. R., and B. J. LeBoeuf. 1977. Female incitation of male competition: A mechanism of mate selection. *Amer. Natur.* 111:317–35.

Cox, J. A. 1983. Bartram's bird, the Florida scrub jay. *Florida Nat.* (June):7–10.

Cramp, S., and K.E.L. Simmons, eds. 1977. *Handbook of the Birds of Europe, the Middle East, and North Africa*, Vol. 1: *The Birds of the Western Palearctic*. Oxford University Press.

Crawford, R. D. 1977. The breeding biology of year-old and older female red-winged and yellow-headed blackbirds. *Wilson Bull.* 89:73–80.

Cronin, E. W., and P. W. Sherman. 1976. A resource-based mating system: The orange-rumped honeyguide. *Living Bird* 15:5–32.

Crook, J. H. 1964. The evolution of social organization and visual communication in weaverbirds (Ploceinae). *Behaviour (Suppl.)* 10:1–178.

Crook, J. H. 1965. The adaptive significance of avian social organisations. *Symp. Zool. Soc. London* 14:181–218.

Crook, J. H. 1970. The socio-ecology of primates. In J. H. Crook, ed., *Social Behaviour of Birds and Mammals*, pp. 103–166. Academic Press, London.

Crook, J. H. 1972. Sexual selection, dimorphism, and social organization in the primates. In B. G. Campbell, ed., *Sexual Selection and the Descent of Man*, pp. 231–81. Heinemann, London.

Crook, J. H., J. E. Ellis, and J. D. Goss-Custard. 1976. Mammalian social systems: Structure and function. *Anim. Behav.* 24:261–74.

Crook, J. H., and J. S. Gartlan. 1966. Evolution of primate societies. *Nature* 210:1200–1203.

Curio, E. 1983. Why do young birds reproduce less well? *Ibis* 125:400–404.

Dalke, P. D., D. B. Pyrah, D. C. Stanton, J. E. Crawford, and E. F. Schlatterer. 1963. Ecology, productivity, and management of sage grouse in Idaho. *J. Wildl. Mgmt.* 27:811–41.

Daly, M., and M. Wilson. 1983. *Sex, Evolution, and Behavior*. 2nd ed. Duxbury Press, Boston.

Dards, J. L. 1978. Home ranges of feral cats in Portsmouth dockyard. *Carnivore Genet. Newsl.* 3:242–55.

Darley, J. A. 1978. Pairing in captive brown-headed cowbirds (*Molothrus ater*). *Can. J. Zool.* 56:2249–52.

Darley, J. A. 1982. Territoriality and mating behavior of the male brown-headed cowbird. *Condor* 84:15–21.

Darwin, C. 1839. *Journal of Researches into the Geology and Natural History of the Various Countries Visited during the Voyage of H.M.S. Beagle Round the World*. London: John Murray.

Darwin, C. 1871. *The Decent of Man, and Selection in Relation to Sex*. D. Appleton and Co. Reprinted 1981, Princeton University Press.

Davenport, W. 1976. Sex in cross-cultural perspective. In F. A. Beach, ed., *Human Sexuality in Four Perspectives*, pp. 115–63. Johns Hopkins University Press.

Literature Cited

David, J.H.M. 1973. The behaviour of the Bontebok, *Damaliscus dorcas dorcas* (Pallas 1766), with special reference to territorial behaviour. *Z. Tierpsychol.* 33:38–107.

Davidar, E.R.E. 1975. Ecology and behaviour of the dhole or Indian wild dog (*Cuon alpinus* Pallan). In M. W. Fox, ed., *The Wild Canids*, pp. 109–119. Van Nostrand Reinhold, N.Y.

Davies, N. B. 1978. Ecological questions about territorial behaviour. In J. R. Krebs and N. B. Davies, eds., *Behavioural Ecology: An Evolutionary Approach*, 1st ed., pp. 317–50. Blackwell, Oxford.

Davies, N. B., and A. I. Houston. 1981. Owners and satellites: The economics of territory defence in the pied wagtail, *Motacilla alba. J. Anim. Ecol.* 50:157–80.

Davis, D. E. 1976. Hibernation and circannual rhythms of food consumption in marmots and ground squirrels. *Quart. Rev. Biol.* 51:477–514.

Davis, L. S. 1983. Behavioral interactions of Richardson's ground squirrels: Asymmetries based on kinship. In J. O. Murie and G. R. Michener, eds., *Biology of Ground-Dwelling Squirrels*, pp. 424–44. University of Nebraska Press.

Davis, L. S. 1984. Alarm calling in Richardson's ground squirrels (*Spermophilus richardsonii*). *Z. Tierpsychol.* 66:152–64.

Dawkins, R. 1976. *The Selfish Gene.* Oxford University Press.

Dawkins, R. 1980. Good strategy or evolutionarily stable strategy? In G. W. Barlow and J. Silverberg, eds., *Sociobiology: Beyond Nature and Nurture*, pp. 331–67. Westview Press, Boulder, Colo.

Dean, W.J.R., and D. M. Skead. 1977. The sex ratio in yellowbilled duck, redbilled teal, and southern pochard. *Ostrich* Suppl. 12:82–85.

Dean, W.R.J., and D. M. Skead. 1979. The weights of some southern African Anatidae. *Wildfowl* 30:114–17.

Dekker, D. 1968. Breeding the Cape hunting dog at Amsterdam Zoo. *Int. Zoo Yearb.* 8:27–30.

Derksen, D. V., and W. D. Eldridge. 1980. Drought-displacement of pintails to the arctic coastal plain, Alaska. *J. Wildl. Mgmt.* 44:224–29.

Derrickson, S. R. 1978. The mobility of breeding pintails. *Auk* 95:104–114.

Dervieux, A., and A. Tamisier. 1979. Quelques aspects éthologiques et physiologiques de l'activitié sexuelle de Sarcelles d'hiver en captivité. *L'Oiseau R.F.O.* 49:299–322.

DeVore, I. 1965. Male dominance and mating behavior in baboons. In F. Beach, ed., *Sex and Behavior*, pp. 266–89. Wiley, New York and London.

DeVore, I., and K.R.L. Hall. 1965. Baboon ecology. In I. DeVore, ed., *Primate Behavior: Field Studies of Monkeys and Apes*, pp. 20–52. Holt, Rinehart & Winston, N.Y.

deWaal, F. 1982. *Chimpanzee Politics.* Harper Colophon, N.Y.

Dickemann, M. 1979. Female infanticide, reproductive strategies, and social stratification: A preliminary model. In N. A. Chagnon and W. Irons, eds., *Evolutionary Biology and Human Social Behavior: An Anthropological Perspective*, pp. 321–67. Duxbury Press. N. Scituate, Mass.

Dickemann, M. 1981. Paternal confidence and dowry competition: A bio-cultural analysis of purdah. In R. D. Alexander and D. W. Tinkle, eds., *Natural Selection and Social Behavior: Recent Research and New Theory*, pp. 417–38. Chiron Press, N.Y.

Dickemann, M. 1982. Comments on "Polygyny and the inheritance of wealth." *Current Anthropology* 23:8-9.

Dietz, J. M. 1984. Ecology and social organization of the maned wolf (*Chrysocyon brachyurus*). *Smithsonian Contrib. Zool.* 392:1–51.

Dittus, W.P.J. 1977. The social regulation of population density and age-sex distribution of the Toque monkey. *Behaviour* 63:281–322.

Literature Cited

Dittus, W.P.J. 1979. The evolution of behaviour regulation density and age-specific sex ratios in a primate population. *Behaviour* 69:265–301.

Dittus, W.P.G. 1975. Population dynamics of the toque monkey, *Macaca* sinica. In R. H. Tuttle, ed., *Socioecology and Psychology of Primates*, pp. 125–51. Mouton, The Hague.

Dittus, W.P.J. 1980. The social regulation of primate populations: A synthesis. In D. G. Lindburg, ed., *The Macaques: Studies in Ecology, Behavior, and Evolution*, pp. 125–52. Van Nostrand Reinhold, N.Y.

Dobson, F. S. 1979. An experimental study of dispersal in the California ground squirrel. *Ecology* 60:1103–1109.

Dobson, F. S. 1983. Agonism and territoriality in the California ground squirrel. *J. Mammal.* 64:218–25.

Dobson, F. S. 1984. Environmental influences on sciurid mating systems. In J. O. Murie and G. R. Michener, eds., *Biology of Ground-Dwelling Squirrels*, pp. 229–49. University of Nebraska Press.

Dorst, J. and P. Dandelot. 1970. *A Field Guide to the Larger Mammals of Africa*. Houghton Mifflin, Boston.

Doty, H. A., and F. B. Lee. 1974. Homing to nest baskets by wild female mallards. *J. Wildl. Mgmt.* 38:714–19.

Downhower, J. F., and K. B. Armitage. 1971. The yellow-bellied marmot and the evolution of polygamy. *Amer. Natur.* 105:355–70.

Downhower, J. F., and K. B. Armitage. 1981. Dispersal of yearling yellow-bellied marmots (*Marmota flaviventris*). *Anim. Behav.* 29:1064–69.

Dowsett, R. J. 1966. Behaviour and population structure of hartebeest in the Kafue National Park. *The Puku* 4:147–54.

Drickamer, L. 1974. A ten-year summary of reproductive data for free-ranging *Macaca mulatta*. *Folia Primatol.* 21:61–80.

Drury, W. H., Jr. 1962. Breeding activities, especially nest building of the Yellowtail (*Ostinops decumanus*) in Trinidad, West Indies. *Zoologica* 47:39–58.

Dubost, G. 1970. L'organisation spatiale et sociale de *Muntiacus reevesi* Ogilby 1839 en semi-liberté. *Mammalia* 34:331–35.

Dubost, G. 1971. Notes on the ethology of the Muntjak (*Muntiacus muntjak*). *Z. Tierpsychol.* 28:401–408.

Dubost, G. 1980. L'ecologie et la vie sociale du Cephalophe bleu (*Cephalophus monticola* Thunberg), petit ruminant forestier africain. *Z. Tierpsychol.* 54:205–266.

Dubost, G., and F. Feer. 1981. The behaviour of the male *Antelope cervicapra* L., its development according to age and social rank. *Behaviour* 76:62–127.

DuBowy, P. J. 1985. Feeding ecology and behavior of postbreeding male blue-winged teal and northern shovelers. *Can. J. Zool.* 63:1292–97.

Duebbert, H. F., and J. T. Lokemoen. 1980. High duck nesting success in a predator-reduced environment. *J. Wildl. Mgmt.* 44:428–37.

Duebbert, H. F., J. T. Lokemoen, and D. E. Sharp. 1983. Concentrated nesting of mallards and gadwalls on Miller Lake Island, North Dakota. *J. Wildl. Mgmt.* 47:729–40.

Duffy, A. M., Jr. 1982. Movements and activities of radio-tracked brown-headed cowbirds. *Auk* 99:316–27.

Dunbar, R.I.M. 1973. The social dynamics of the gelada baboon. *Theropithecus gelada*. Ph.D. diss., University of Bristol.

Dunbar, R.I.M. 1977. Feeding ecology of gelada baboons: A preliminary report. In T. H. Clutton-Brock, ed., *Primate Ecology: Studies of Feeding and Ranging Behaviour in Lemurs, Monkeys, and Apes*, pp. 251–73. Academic Press, London and N.Y.

Literature Cited

Dunbar, R.I.M. 1978a. Sexual behaviour and social relationships among gelada baboons. *Anim. Behav.* 26:167–78.

Dunbar, R.I.M. 1978b. Competition and niche separation in a high altitude herbivore community. *E. Afr. Wildl. J.* 16:183–99.

Dunbar, R.I.M. 1979a. Structure of gelada baboon reproductive units, I: Stability of social relationships. *Behaviour* 69:72–87.

Dunbar, R.I.M. 1979b. Population demography, social organization, and mating strategies. In I. S. Bernstein and E. O. Smith, eds., *Primate Ecology and Human Origins: Ecological Influences on Social Organization*, pp. 65–88. Garland Press, N.Y.

Dunbar, R.I.M. 1980a. Determinants and evolutionary consequences of dominance among female gelada baboons. *Behav. Ecol. Sociobiol.* 7:253–65.

Dunbar, R.I.M. 1980b. Demographic and life history variables of a population of gelada baboons (*Theropithecus gelada*). *J. Anim. Ecol.* 49:485–506.

Dunbar, R.I.M. 1982a. Structure of social relationships in a captive group of gelada baboons: A test of some hypotheses derived from a wild population. *Primates* 23:89–94.

Dunbar, R.I.M. 1982b. Intraspecific variations in mating strategy. In P.P.G. Bateson and P. H. Klopfer, eds., *Perspectives in Ethology*, pp. 385–431. Plenum Press, N.Y.

Dunbar, R.I.M. 1983a. Life history tactics and strategies of reproduction. In P.P.G. Bateson, ed., *Mate Choice*, pp. 423–33. Cambridge University Press.

Dunbar, R.I.M. 1983b. Theropithecines and hominids: Contrasting solutions to the same ecological problem. *J. Human Evol.* 12:647–58.

Dunbar, R.I.M. 1983c. Structure of gelada baboon reproductive units, II: Social relationships between reproductive females. *Anim. Behav.* 31:556–64.

Dunbar, R.I.M. 1983d. Structure of gelada baboon reproductive units, III: The male's relationship with his females. *Anim. Behav.* 31:565–75.

Dunbar, R.I.M. 1983e. Structure of gelada baboon reproductive units, IV: Integration at group level. *Z. Tierpsychol.* 63:265–82.

Dunbar, R.I.M. 1984a. *Reproductive Decisions: An Economic Analysis of Gelada Baboon Social Strategies*. Princeton University Press.

Dunbar, R.I.M. 1984b. Use of infants by male gelada in agonistic contexts: Agonistic buffering, progeny protection or soliciting support? *Primates* 25:28–35.

Dunbar, R.I.M., and E. P. Dunbar. 1974a. The reproductive cycle of the gelada baboon. *Anim. Behav.* 22:203–210.

Dunbar, R.I.M., and E. P. Dunbar. 1974b. Social organization and ecology of the Klipspringer (*Oreotragus oreotragus*) in Ethiopia. *Z. Tierpsychol.* 35:481–93.

Dunbar, R.I.M., and E. P. Dunbar. 1975. *Social Dynamics of Gelada Baboons*. Contrib. Primatol. 6. Karger, Basel.

Dunbar, R.I.M., and M. Sharman. 1983. Female competition for access to males affects birth rate in baboons. *Behav. Ecol. Sociobiol.* 13:157–59.

Duncan, P. 1975. Topi and their food supply. Ph.D. diss., University of Nairobi.

Dunford, C. 1977a. Kin selection for ground squirrel alarm calls. *Amer. Natur.* 111:782–85.

Dunford, C. 1977b. Social system of round-tailed ground squirrels. *Anim. Behav.* 25:885–906.

Dunford, C. 1977c. Behavioral limitation of round-tailed ground squirrel density. *Ecology* 58:1254–68.

Dunn, P. O. 1983. Natal dispersal and lek fidelity in sage grouse. *Proc. 13th Western States Sage Grouse Workshop.*

Durham, W. 1986. *Coevolution: Genes, Culture, and Human Diversity*. Stanford University Press, in press.

Dwyer, T. J. 1974. Social behavior of breeding gadwalls in North Dakota. *Auk* 91:375–86.

Literature Cited

Dwyer, T. J. 1975. Time budget of breeding gadwalls. *Wilson Bull.* 87:335–43.

Dwyer, T. J., S. R. Derrickson, and D. S. Gilmer. 1973. Migrational homing by a pair of mallards. *Auk* 90:687.

Dwyer, T. J., G. L. Krapu, and D. M. Janke. 1979. Use of prairie pothole habitat by breeding mallards. *J. Wildl. Mgmt.* 43:526–31.

Dyson-Hudson, R., and E. A. Smith. 1978. Human territoriality: An ecological reassessment. *American Anthropologist* 80:21–41.

Dzubin, A. 1955. Some evidences of home range in waterfowl. *Trans. N. Amer. Wildl. Conf.* 20:278–98.

Dzubin, A. 1969. Comments on carrying capacity of small ponds for ducks and possible effects of density on mallard production. *Saskatoon Wetlands Seminar, Can. Wildl. Serv. Rept.*, Series No. 6, pp. 138–60.

Eadie, J. M., T. D. Nudds, and C. D. Ankney. 1979. Quantifying interspecific variation in foraging behavior of syntopic *Anas* (Anatidae). *Can. J. Zool.* 57:412–15.

Earle, R. A. 1981. Aspects of the social and feeding behaviour of the yellow mongoose *Cynictis penicillata*. *Mammalia* 45:143–52.

Eaton, R. L. 1974. *The Cheetah*. Van Nostrand Reinhold, N.Y.

Eaton, R. L. 1979. Interference competition among carnivores: A model for the evolution of social behavior. *Carnivore* 2:9–16.

Egoscue, H. J. 1962. Ecology and life history of the kit fox in Tooele County, Utah. *Ecology* 43:481–97.

Eisenberg, J. F. 1981. *The Mammalian Radiations*. University of Chicago Press.

Eisenberg, J. F., N. A. Muckenhirn, and R. Rudran. 1972. The relation between ecology and social structure in primates. *Science* 176:863–74.

Elder, W. H., and M. W. Weller. 1954. Duration of fertility in the domestic mallard hen after isolation from the drake. *J. Wildl. Mgmt.* 18:495–502.

Elliott, J. P., I. McT. Cowan, and C. S. Holling. 1977. Prey capture by the African lion. *Can. J. Zool.* 55:1811–28.

Elliott, P. F. 1975. Longevity and the evolution of polygamy. *Amer. Natur.* 109:281–87.

Elliott, P. F. 1980. Evolution of promiscuity in the brown-headed cowbird. *Condor* 82:138–41.

Eloff, F. C. 1973. Ecology and behavior of the Kalahari lion. In R. L. Eaton, ed., *The World's Cats*, Vol. 1., pp. 90–126. World Wildlife Safari, Winston, Or.

Ember, M., and C. R. Ember. 1971. The conditions favoring matrilocal versus patrilocal residence. *American Anthropologist* 73:571–94.

Ember, M., and C. R. Ember. 1983. *Marriage, Family, and Kinship*. HRAF Press, New Haven.

Emlen, S. T. 1982a. The evolution of helping, I: An ecological restraints model. *Amer. Natur.* 119:29–39.

Emlen, S. T. 1982b. The evolution of helping, II: The role of behavioral conflict. *Amer. Natur.* 119:40–53.

Emlem, S. T. 1984. Cooperative breeding in birds and mammals. In J. R. Krebs and N. B. Davies, eds., *Behavioural Ecology: An Evolutionary Approach*, 2nd ed., 305–339. Blackwell, Oxford.

Emlen, S. T., and L. W. Oring. 1977. Ecology, sexual selection, and the evolution of mating systems. *Science* 197:215–23.

Emlen, S. T., and S. L. Vehrencamp. 1983. Cooperative breeding strategies among birds. In A. H. Brush and G. A. Clark, Jr., eds., *Perspectives in Ornithology*, pp. 93–120. Cambridge University Press.

Literature Cited

Emmons, S. R. 1980. Lek attendance of male sage grouse in North Park, Colorado. M.S. thesis, Colorado State University, Fort Collins.

Eng, R. L., and P. Schladweiler. 1972. Sage grouse winter movements and habitat use in central Montana. *J. Wildl. Mgmt.* 36:141–46.

Erckmann, W. J. 1983. The evolution of polyandry in shorebirds: An evaluation of hypotheses. In S. K. Wasser, ed., *Social Behavior of Female Vertebrates*, pp. 112–68. Academic Press, N.Y.

Essock-Vitale, S. 1984. The reproductive success of wealthy Americans. *J. Ethol. and Sociobiol.* 5:45–49.

Estes, R. D. 1966. Behaviour and life history of the wildebeest (*Connochaetes taurinus* Burchell). *Nature* 212:999–1000.

Estes, R. D. 1967. The comparative behavior of Grant's and Thomson's gazelles. *J. Mammal.* 48:189–209.

Estes, R. D. 1969. Territorial behavior of the wildebeest (*Connochaetes taurinus* Burchell, 1823). *Z. Tierpsychol.* 26:284–370.

Estes, R. D. 1972. The role of the vomeronasal organ in mammalian reproduction. *Mammalia* 36:315–41.

Estes, R. D. 1974. Social organization of the African Bovidae. In V. Geist and F. R. Walther, eds., *The Behaviour of Ungulates and Its Relation to Management*, pp. 166–205. IUCN, Morges, Switzerland.

Estes, R. D., and R. K. Estes. 1979. The birth and survival of wildebeest calves. *Z. Tierpsychol.* 50:454–95.

Evans, C. D., A. S. Hawkins, and W. H. Marshall. 1952. Movements of waterfowl broods in Manitoba. *Special Scientific Report: Wildlife* 16:1–47.

Evans, F. C., and R. Holdenreid. 1943. A population study of the Beechey ground squirrel in Central California. *J. Mammal.* 24:231–60.

Ewald, P. W., and S. Rohwer. 1982. Effects of supplemental feeding on timing of breeding, clutch-size, and polygyny in red-winged blackbirds (*Agelaius phoeniceus*). *J. Anim. Ecol.* 51:429–50.

Ewer, R. F. 1973. *The Carnivores.* Cornell University Press.

Fathauer, G. 1961. Trobrianders. In D. Schneider and K. Gough, eds., *Matrilineal Kinship*, pp. 234–69. University of California Press.

Feduccia, A. 1980. *The Age of Birds.* Harvard University Press.

Feekes, F. 1981. Biology and colonial organization of two sympatric caciques, *Cacicus c. cela* and *Cacicus h. haemorrhous* (Icteridae: Aves) in Suriname. *Ardea* 69:83–107.

Fendley, T. T., and D. E. Buie. 1982. Seasonal home range sizes and movement patterns of adult bobcats on the Savannah River plant. Abstract from the International Cat Symposium, Kingsville, Texas.

Fentress, J. C., and J. Ryon. 1982. A long-term study of distributed pup feeding in captive wolves. In F. H. Harrington, ed., *Wolves of the World*, pp. 238–60. Noyes Publications, Park Ridge, N.J.

Ferguson, J.W.H., J.A.J. Nel, and M. J. DeWet. 1983. Social organization and movement patterns of black-backed jackals *Canis mesomelas* in South Africa. *J. Zool. London* 199:487–502.

Festa-Bianchet, M. 1981. Reproduction in yearling female Columbian ground squirrels (*Spermophilus columbianus*). *Can. J. Zool.* 59:1032–35.

Festa-Bianchet, M., and D. A. Boag. 1982. Territoriality in adult female Columbian ground squirrels. *Can. J. Zool.* 60:1060–66.

Festa-Bianchet, M., and W. J. King. 1984. Behavior and dispersal of yearling Columbian ground squirrels. *Can. J. Zool.* 62:161–67.

Literature Cited

ffrench, R. P. 1980. *A Guide to the Birds of Trinidad and Tabago*. Harrowood, Newton Square, Pa.

Field, C. R. 1968. A comparative study of the food habits of some wild ungulates in the Queen Elizabeth Park, Uganda. Preliminary report. *Symp. Zool Soc. London* 21:135–51.

Field, C. R. 1972. The food habits of wild ungulates in Uganda by analysis of stomach contents. *E. Afr. Wildl. J.* 10:17–42.

Fisher, R. A. 1930. *The Genetical Theory of Natural Selection*. 2nd ed., 1958. Dover, N.Y.

Fitch, H. S. 1948. Ecology of the California ground squirrel on grazing lands. *Amer. Midl. Nat.* 39:513–96.

Fitzgerald, J. P., and R. R. Lechleitner. 1974. Observations on the biology of Gunnison's prairie dog in central Colorado. *Amer. Midl. Nat.* 92:146–63.

Fitzpatrick, J. W., and G. E. Woolfenden. 1986. Demographic routes to cooperative breeding in some New World jays. In M. Nitecki and J. Kitchell, eds., *Evolution of Animal Behavior: Paleontological and Field Approaches*. Oxford University Press, New York.

Flannery, K. 1972. The cultural evolution of civilizations. *Ann. Rev. Ecol. Syst.* 3:399–426.

Flinn, M. V. 1981. Uterine vs. agnatic kinship variability and associated cousin marriage preferences. In R. D. Alexander and D. W. Tinkle, eds., *Natural Selection and Social Behavior*, pp. 439–75. Chiron Press, N.Y.

Flinn, M. V. 1983. Resources, mating, and kinship: The behavioral ecology of a Trinidadian village. Ph.D. diss., Northwestern University.

Flinn, M. V. 1986a. Mate guarding in a Trinidadian village. *Ethol. and Sociobiol.*, in press.

Flinn, M. V. 1986b. Parent-offspring interactions in a Trinidadian village. *Human Ecology*, in press.

Flinn, M. V. 1987. Correlates of reproductive success in a rural Trinidadian population. In L. L. Betzig, M. Borgerhoff-Mulder, and P. W. Turke, eds., *Human Reproductive Behavior: A Darwinian Perspective*. Cambridge University Press, Cambridge. Forthcoming.

Floody, O. R., and A. P. Arnold. 1975. Uganda kob (*Adenota kob thomasi*): Territoriality and the spatial distributions of sexual and agonistic behaviors at a territorial ground. *Z. Tierpsychol.* 37:192–212.

Folk, C., K. Hudec, and J. Toufar. 1966. The weight of the mallard, *Anas platyrhynchos*, and its changes in the course of the year. *Zool. Listy.* 15:249–60.

Foltz, D. W., and J. L. Hoogland. 1981. Analysis of the mating system in the black-tailed prairie dog (*Cynomys ludovicianus*) by likelihood of paternity. *J. Mammal.* 62:706–712.

Ford, N. I. 1983. Variations in mate fidelity in monogamous birds. In R. F. Johnston, ed., *Current Ornithology*, pp. 329–56. Plenum Press, N.Y.

Fortes, M. 1950. Kinship and marriage among the Ashanti. In A. R. Radcliffe-Browne and D. Forde, eds., *African Systems of Kinship and Marriage*, pp. 252–84. Oxford University Press.

Fortes, M. 1969. *Kinship and the Social Order*. Aldine, Chicago.

Fortune, R. 1963. *Sorcerers of Dobu*. (Reprint of 1932 ed.) Dutton, N.Y.

Fossey, D. 1983. *Gorillas in the Mist*. Houghton Mifflin, Boston.

Fossey, D. 1984. Infanticide in mountain gorillas (*Gorilla gorilla beringei*) with comparative notes on chimpanzees. In G. Hausfater and S. B. Hrdy, eds., *Infanticide: Comparative and Evolutionary Perspectives*, pp. 217–35. Aldine, N.Y.

Fossey, D., and A. H. Harcourt. 1977. Feeding ecology of free-ranging mountain gorilla (*Gorilla gorilla beringei*). In T. H. Clutton-Brock, ed., *Primate Ecology*, pp. 415–47. Academic Press, London.

Fraga, R. M. 1972. Cooperative breeding and a case of successive polyandry in the bay-winged cowbird. *Auk* 89:447–49.

490

Literature Cited

Frame, G. W. 1980. Cheetah social organization in the Serengeti ecosystem, Tanzania. Paper given at the Animal Behavior Society Meetings, Fort Collins, Colo.

Frame, G. W., and F. H. Wagner. 1981. Hares on t he Serengeti Plains, Tanzania. In K. Myers and C. D. MacInnes, eds., *Proceedings of the World Lagomorph Conference*, pp. 790–802. IUCN, Morges, Switzerland.

Frame, L. H., and G. W. Frame. 1977. Female African wild dogs emigrate. *Nature* 263:227–29.

Frame, L. H., J. R. Malcolm, G. W. Frame, and H. van Lawick. 1980. Social organization of African wild dogs (*Lycaon acrus*) on the Serengeti Plains, Tanzania 1967–1978. *Z. Tierpsychol.* 50:225–49.

Frase, B. A. 1983. Spatial and behavioral foraging patterns and diet selectivity in the social yellow bellied marmot. Ph.D. diss., University of Kansas, Lawrence.

Frase, B. A., and K. B. Armitage. 1984. Foraging patterns of yellow-bellied marmots: Role of kinship and individual variability. *Behav. Ecol. Sociobiol.* 16:1–10.

Frase, B. A., and R. S. Hoffmann. 1980. *Marmota flaviventris. Mammal. Sp.* 135:1–8.

Fredrickson, L. H., and R. D. Drobney. 1979. Habitat utilization by postbreeding waterfowl. In T. A. Bookhout, ed., *Waterfowl and Wetlands—An Integrated Review*, pp. 119–31. La Crosse Printing Co., La Crosse, Wisc.

Fretwell, S. D., and H. L. Lucas. 1970. On territorial behavior and other factors influencing habitat distribution in birds. *Acta Biotheoretica* 19:16–36.

Frith, H. J. 1956. Breeding habits of the family Megapodiidae. *Ibis* 98:620–40.

Fritzell, E. K., and K. J. Haroldson. 1982. *Urocyon cinereoargenteus. The American Society of Mammalogists* 189:1–8.

Fujioka, M., and S. Yamagishi. 1981. Extramarital and pair copulations in the cattle egret. *Auk* 98:134–44.

Furrer, R. K. 1975. Breeding success and nest site stereotype in a population of Brewer's blackbirds (*Euphagus cyanocephalus*). *Oecologia* 20:339–50.

Gadgil, M. 1972. Male dimorphism as a consequence of sexual selection. *Amer. Natur.* 106:547–80.

Gakathu, C. 1980. *Feeding and Predation in the Zebra.* Afr. Wildl. Lead. Found.

Galat-Luong, A. 1975. Notes preliminaires sur l'ecologie de *Cercopithecus ascanius schmidti* dans les environs de Bangui (R.C.A.) *Terre et Vie* 29:288–97.

Galdikas, B.M.F. 1979. Orangutan adaptation at Tanjung Puting Reserve: Mating and ecology. In D. A. Hamburg and E. R. McCown, eds., *The Great Apes*, pp. 195–233. Benjamin/Cummings, Menlo Park, Calif.

Gangloff, L. 1972. Breeding fennec foxes (*Fennecus zerda*) at Strasborg Zoo. *Int. Zoo Yearb.* 12:115–16.

Gashwiler, J. S., W. L. Robinette, and D. W. Morris. 1960. Foods of bobcats in Utah and eastern Nevada. *J. Wildl. Mgmt.* 24:226–29.

Gaston, A. J. 1978. The evolution of group territorial behavior and cooperative breeding. *Amer. Natur.* 112:1091–1100.

Gates, J. M. 1962. Breeding biology of the gadwall in northern Utah. *Wilson Bull.* 74:43–67.

Gates, S. 1979. A study of the home ranges of free-ranging Exmoor ponies. *Mammal Rev.* 9:3–18.

Gatti, R. C. 1983. Incubation weight loss in the mallard. *Can. J. Zool.* 61:565–69.

Gauthreaux, S. A., Jr. 1978. The ecological significance of behavioral dominance. In P.P.G. Bateson and P. H. Klopfer, eds., *Perspectives in Ethology*, Vol. 3, pp. 17–54. Plenum Press, N.Y.

Gautier, J. P. 1971. Étude morphologique et fonctionelle des annexes extralaryngees des Cercopithecinae: Liaison avec les cris d'espacement. *Biologia Gabonica* 7:229–67.

Literature Cited

Gautier-Hion, A. 1970. L'organisation sociale d'une bande de talapoions (*Miopithecus tala-poin*) dans le nord-est du Gabon. *Terre et Vie* 25:427–90.

Geist, V. 1974. On the relationship of social evolution and ecology in ungulates. *Amer. Zool.* 14:9–34.

Ghiglieri, M. P. 1979. The socioecology of chimpanzees in Kibale Forest, Uganda. Ph.D. diss., University of California, Davis.

Ghiglieri, M. P. 1984. *The Chimpanzees of Kibale Forest: A Field Study of Ecology and Social Structure*. Columbia University Press.

Gibson, R. M., and J. W. Bradbury. 1985. Sexual selection in lekking sage grouse: Phenotypic correlates of male mating success. *Behav. Ecol. Sociobio.* 18:117–23.

Gibson, R. M., and F. E. Guinness. 1980. Behavioural factors affecting male reproductive success in red deer (*Cervus elaphus*). *Anim. Behav.* 28:1163–74.

Gier, H. T. 1975. Ecology and social behavior of the coyote. In M. W. Fox, ed., *The Wild Canids*, pp. 247–62. Van Nostrand Reinhold, N.Y.

Gill, F. B., and L. L. Wolf. 1975. Economics of feeding territoriality in the golden-winged sunbird. *Ecology* 56:333–45.

Gilmer, D. S., R. E. Kirby, I. J. Ball, and J. H. Riechmann. 1977. Post-breeding activities of mallards and wood ducks in north-central Minnesota. *J. Wildl. Mgmt.* 41:345–59.

Gittleman, J. L. 1984. The behavioral ecology of carnivores. Ph.D. diss., University of Sussex.

Gittleman, J. L. 1985. Functions of communal care in mammals. In P. J. Greenwood and M. Slatkin, eds., *Evolution: Essays in Honour of John Maynard Smith*, pp. 187–205. Cambridge University Press.

Gladstone, D. E. 1979. Promiscuity in colonial monogamous birds. *Amer. Natur.* 114:545–57.

Glick, B. B. 1980. Ontogenetic and psychobiological aspects of the mating activities of male *Macaca radiata*. In D. G. Linburg, ed., *The Macaques: Studies in Ecology, Behavior and Evolution*, pp. 345–69. Van Nostrand Reinhold, N.Y.

Gochfeld, M. 1975. Comparative ecology and behavior of red-breasted meadowlarks (Aves, Icteridae) and their interactions in sympatry. Ph.D. diss., City University of New York.

Goldstein, M. 1978. Pahari and Tibetan polyandry revisited. *Ethnology* 17:325–37.

Goodall, J. 1968. The behaviour of free-living chimpanzees in the Gombe stream area. *Anim. Behav. Monogr.* 1:161–311.

Goodall, J. 1971. *In the Shadow of Man*. Collins, London.

Goodall, J. 1975. Chimpanzees of the Gombe National Park: Thirteen years of research. In G. Kurth and I. Eibl-Eibesfeldt, eds., *Hominisation und Verhalten*, pp. 74–136. Fischer, Stuttgart.

Goodall, J. 1977. Infant killing and cannibalism in free-living chimpanzees. *Folia primatol.* 28:259–82.

Goodall, J. 1983. Population dynamics during a fifteen-year period in one community of free-living chimpanzees in the Gombe National Park, Tanzania. *Z. Tierpsychol.* 61:1–60.

Goodall, J., A. Bandora, E. Bergmann, C. Busse, H. Matama, E. Mpongo, A. Pierce, and D. Riss. 1979. Intercommunity interactions in the chimpanzee population of the Gombe National Park. In D. A. Hamburg and E. R. McCown, eds., *The Great Apes*, pp. 12–53. Benjamin/Cummings, Menlo Park, Calif.

Goodburn, S. F. 1984. Mate guarding in the mallard *Anas platyrhynchos*. *Ornis Scand.* 15:261–65.

Goode, W. J. 1970. *World Revolution and Family Patterns*. Collier Macmillan Free Press, N.Y.

Goodwin, D. 1976. *Crows of the World*. Comstock Publ. Associates, Ithaca, N.Y.

Literature Cited

Goody, J., and S. J. Tambiah. 1973. *Bridewealth and Dowry*. Cambridge Papers in Social Anthropology No. 7. Cambridge University Press.

Gorman, M. L. 1975. The diet of feral *Herpestes auropunctatus* (Carnivora: Viverridae) in the Fijiian Islands. *J. Zool. London* 175:273–78.

Gorman, M. L. 1979. Dispersion and foraging of the small Indian mongoose, *Herpestes auropunctatus*, relative to the evolution of social viverrids. *J. Zool. London* 187:67–73.

Gosling, L. M. 1969. Parturition and related behavior in Coke's hartebeest, *Alcelaphus buselaphus cokei* Gunther. *J. Reprod. Fert. Suppl.* 6:265–86.

Gosling, L. M. 1974. The social behaviour of Coke's hartebeest (*Alcelaphus buselaphus cokei*). In V. Geist and F. R. Walther, eds., *The Behaviour of Ungulates and Its Relation to Management*, pp. 488–511. IUCN, Morges, Switzerland.

Gosling, L. M. 1975. The ecological significance of male behaviour in Coke's hartebeest, *Alcelaphus buselaphus cokei*. Ph.D. diss., University of Nairobi.

Gosling, L. M. 1980. Defense guilds of savannah ungulates as a context for scent communication. *Symp. Zool. Soc. London* 45:195–212.

Gosling, L. M. 1981. Demarkation in a gerenuk territory: An economic approach. *Z. Tierpsychol.* 56:305-322.

Gosling, L. M. 1982. A reassessment of the function of scent marking in territories. *Z. Tierpsychol.* 60:89–118.

Gosling, L. M., and M. Petrie. 1981. The economics of social organisation. In C. R. Townsend and P. Calow, eds., *Physiological Ecology: An Evolutionary Approach to Resource Use*, pp. 315–45. Blackwell, Oxford.

Goss-Custard, J. D., R.I.M. Dunbar, and P. Aldrich-Blake. 1972. Survival, mating, and rearing strategies in the evolution of primate social structure. *Folia Primatol.* 17:1–19.

Gough, K. 1961a. Central Kerala Nayars. In D. Schneider and K. Gough, eds., *Matrilineal Kinship*. University of California Press.

Gough, K. 1961b. Variation in residence. In D. Schneider and K. Gough, eds., *Matrilineal Kinship*, University of California Press.

Gough, K. 1961c. Variation in preferential marriage forms. In D. Schneider and K. Gough, eds., *Matrilineal Kinship*. University of California Press.

Gouzoules S., 1984. Primate mating systems, kin associations, and cooperative behavior: Evidence for kin recognition? *Yearb. Phys. Anthrop.* 27:99–134.

Gowaty, P. A., and A. A. Karlin. 1984. Multiple maternity and paternity in single broods of apparently monogamous eastern bluebirds (*Sialia sialis*). *Behav. Ecol. Sociobiol.* 15:91–95.

Graham, C. E. 1981. Menstrual cycle physiology of the great apes. In C. E. Graham, ed., *Reproductive Biology of the Great Apes*, pp. 1–43. Academic Press, New York and London.

Greenwood, P. J. 1980. Mating systems, philopatry, and dispersal in birds and mammals. *Anim. Behav.* 28:1140–62.

Greenwood, P. J. 1983. Mating systems and the evolutionary consequences of dispersal. In I. R. Swingland and P. J. Greenwood, eds., *The Ecology of Animal Movement*, pp. 116–31. Clarendon Press, Oxford.

Greenwood, P. J., and P. H. Harvey. 1976. The adaptive significance of variation in breeding area fidelity of the blackbird (*Turdus merula* L.). *J. Anim. Ecol.* 45:887–98.

Gregor, T. 1974. Privacy and extramarital affairs in a tropical forest community. In D. Gross, ed., *Peoples and Cultures of Native South America*. Doubleday Natural History Press, N.Y.

Grimsdell, J.J.R. 1969. The ecology of the buffalo, *Syncerus caffer*, in western Uganda. Ph.D. diss., University of Cambridge.

Grizzell, R. A. 1955. A study of the southern woodchuck, *Marmota monax monax*. *Amer. Midl. Nat.* 53:257–93.

Literature Cited

Groves, C. P. 1981. "Comment" on S. C. Johnson, Bonobos: generalized hominid prototypes or specialized insular dwarfs? *Current Anthropology* 22(4):366.

Guatier-Hion, A., and J. P. Gautier. 1976. Croissance, maturité sexuelle et sociale, reproduction chez les cercopithecines forestiers africains. *Folia Primatol.* 22:134–77.

Gwynne, M. D., and R.H.V. Bell. 1968. Selection of vegetation components by grazing ungulates in the Serengeti National Park. *Nature* 220:390–93.

Haartman, L. von. 1969. Nest-site and evolution of polygamy in European passerine birds. *Ornis Fenn.* 46:1–12.

Haglung, B. 1966. Winter habits of the lynx (*Lynx lynx*) and wolverine (*Gulo gulo*) as revealed by tracking in the snow. *Viltrevy* 4:245–83.

Hajnal, J. 1963. Random mating and the frequency of consanguineous marriages. *Proc. Roy. Soc. London B* 159:125–77.

Hall, K.R.L. 1962. The sexual, agonistic, and derived social behaviour patterns of the wild chacma baboon, *Papio ursinus. Proc. Zool. Soc. London* 139:283–327.

Hall, K.R.L. 1965. Behaviour and ecology of the wild patas monkey, *Erythrocebus patas*, in Uganda. *J. Zool. London* 148:15–87.

Hall, K.R.L., and I. DeVore. 1965. Baboon social behavior. In I. DeVore, ed., *Primate Behavior*, pp. 53–110. Holt, Rinehart & Winston, N.Y.

Halliday, T. R. 1983. The study of mate choice. In P.P.G. Bateson, ed., *Mate Choice*, pp. 3–29. Cambridge University Press.

Halperin, S. D. 1979. Temporary association patterns in free-ranging chimpanzees: An assessment of individual grouping preferences. In D. A. Hamburg and E. R. McCown, eds., *The Great Apes*, pp. 491-99. Benjamin/Cummings, Menlo Park, Calif.

Hames, R. 1979. Relatedness and interaction among the Ye'kwana: A preliminary analysis. In N. A. Chagnon and W. Irons, eds., *Evolutionary Biology and Human Socal Behavior*, pp. 238–50. Duxbury Press, N. Scituate, Mass.

Hamilton, A. C. 1981. *Environmental History of East Africa*. Academic Press, London.

Hamilton, W. D. 1964. The genetical evolution of social behavior, I, II. *J. Theor. Biol.* 7:1–52.

Hamilton, W. D. 1971. Geometry for the selfish herd. *J. Theor. Biol.* 31:295–311.

Hamilton, W. J., R. E. Buskirk, and W. H. Buskirk. 1976. Defence of space and resources by chacma (*Papio ursinus*) baboons in an African desert and swamp. *Ecology* 57:1264–72.

Hammond, M. C., and G. E. Mann. 1956. Waterfowl nesting islands. *J. Wildl. Mgmt.* 20:345–52.

Hanby, J. P., and J. D. Bygott. 1979. Population changes in lions and other predators. In A.R.E. Sinclair and M. Norton-Griffiths, eds., *Serengeti: Dynamics of an Ecosystem*, pp. 249–62. University of Chicago Press.

Hanby, J. P., L. T. Robertson, and C. H. Phoenix. 1971. The sexual behavior of a confined troop of Japanese macaques. *Folia Primatol.* 16:123–43.

Hanken, J., and P. W. Sherman. 1981. Multiple paternity in Belding's ground squirrel litters. *Science* 212:351–53.

Hanks, J., M. S. Price, and R. W. Wrangham. 1969. Some aspects of the ecology and behaviour of the Defassa waterbuck (*Kobus defassa*) in Zambia. *Mammalia* 33:471–94.

Harcourt, A. H. 1978. Strategies of emigration and transfer by primates, with particular reference to gorillas. *Z. Tierpsychol.* 48:401–420.

Harcourt, A. H. 1979a. Social relationships among adult female mountain gorillas. *Anim. Behav.* 27:251–64.

Harcourt, A. H. 1979b. Social relationships between adult male and female mountain gorillas in the wild. *Anim. Behav.* 27:325–42.

Literature Cited

Harcourt, A. H., D. Fossey, and J. Sabater-Pi. 1981. Demography of *Gorilla gorilla*. *J. Zool. London* 195:215–33.

Harrington, F. H., and L. D. Mech. 1982. Patterns of homesite attendance in two Minnesota wolf packs. In F. H. Harrington, ed., *Wolves of the World*, pp. 81–104. Noyes Publications, Park Ridge, N.J.

Harrington, F. H., L. D. Mech, and S. H. Fritts. 1983. Pack size and wolf pup survival: Their relationship under varying ecological conditions. *Behav. Ecol. Sociobiol.* 13:19–26.

Harrington, F. H., P. C. Paquet, J. Ryon, and J. C. Fentress. 1982. Monogamy in wolves: A review of the evidence. In F. H. Harrington, ed., *Wolves of the World*, pp. 209–222. Noyes Publications, Park Ridge, N.J.

Harris, H. J. 1970. Evidence of stress response in breeding blue-winged teal. *J. Wildl. Mgmt.* 34:747–55.

Harris, M. A., and J. O. Murie. 1984. Inheritance of nest sites in female Columbian ground squirrels. *Behav. Ecol. Sociobiol.* 15:97–102.

Hart, B. J. 1983. Flehmen behaviour and vomeronasal function. In D. Muller-Schwarze and R. M. Silverstein, eds., *Chemical Signals in Vertebrates*, Vol. 3, pp. 87–103. Plenum Press, N.Y.

Hart, C.W.M., and A. R. Pilling. 1960. *The Tiwi of North Australia*. Holt, Rinehart & Winston, N.Y.

Hartung, J. 1982. Polygyny and the inheritance of wealth. *Current Anthropology* 23:1–12.

Hartzler, J. E., 1972. An analysis of sage grouse lek behavior. Ph.D. diss., University of Montana, Missoula.

Hasegawa, T., and M. Hiraiwa-Hasegawa. 1983. Opportunistic and restrictive matings among wild chimpanzees in the Mahale Mountains, Tanzania. *J. Ethol.* 1:75–85.

Hausfater, G. 1975. *Dominance and Reproduction in Baboons (Papio cynocephalus)*. Contrib. to Primatol. 7. Karger, Basel.

Hausfater, G., and S. B. Hrdy. 1984. *Infanticide: Comparative and Evolutionary Perspectives*. Aldine, Hawthorne, N.Y.

Haverschmidt, F. 1968. *Birds of Surinam*. Oliver and Boyd Ltd., Edinburgh and London.

Hawkes, K., K. Hill, and J. F. O'Connell. 1982. Why hunters gather: Optimal foraging and the Ache of Eastern Paraguay. *American Ethnologist* 9(2):379–98.

Hays, H. 1972. Polyandry in the spotted sandpiper. *Living Bird* 11:43–57.

Heidemann, M. L., and L. W. Oring. 1976. Functional analysis of spotted sandpiper (*Actitis macularia*) song. *Behaviour* 56:181–93.

Helfman, G. S. 1984. School fidelity in fishes: The yellow perch pattern. *Anim. Behav.* 32:663–73.

Hemker, T. P. 1982. Population characteristics and movement patterns of cougars in southern Utah. M.Sc. thesis, Utah State University.

Hendrichs, H. 1972. Beobachtungen und Untersuchungen zur Okologie und Ethologie, insbesondere zur sozialen Organisation, ostafricanischer Saugetier. *Z. Tierpsychol.* 30:146–89.

Hendrichs, H. 1975. Changes in a population of dikdik, *Madoqua (Rhynchotragus) kirki* (Gunther 1880). *Z. Tierpsychol.* 38:55–69.

Hendrichs, H., and U. Hendrichs. 1971. *Dikdik und Elefanten*. Studies in Ethology. Piper and Co., Munich.

Henry, J. 1941. *Jungle People*. J. J. Augustin, New York.

Hepp, G. R., and J. D. Hair. 1983. Reproductive behaviour and pairing chronology in wintering dabbling ducks. *Wilson Bull.* 95:675–82.

Hepp, G. R., and J. D. Hair. 1984. Dominance in wintering waterfowl (Anatini): Effects on distribution of sexes. *Condor* 86:251–57.

Literature Cited

Hersteinsson, P. 1984. The behavioral ecology of the arctic fox in Iceland. Ph. D. diss., Oxford University.

Hersteinsson, P., and D. W. Macdonald. 1982. Some comparisons between red and arctic foxes, *Vulpes vulpes* and *Alopex lagopus*, as revealed by radio tracking. *Symp. Zool. Soc. London* 49:259–89.

Hiatt, L. R. 1981. Polyandry in Sri Landa: A test case for parental investment theory. *Man* 15:583–602.

Hildén, O. 1975. Breeding system of Temminck's stint *Calidris temminckii. Ornia Fenn.* 52:117–46.

Hildén, O., and S. Vuolanto. 1972. Breeding biology of the red-necked phalarope *Phalaropus lobatus* in Finland. *Ornis Fenn.* 49:57–85.

Hill, K. 1984. Prestige and reproductive success in man. *Ethology and Sociobiology* 5:77–95.

Hillman, C. 1976. The ecology and behavior of free-ranging Eland (*Taurotragus oryx* Pallas) in Kenya. Ph.D. diss., University of Nairoi.

Hillman, C. 1982. Eland. *Swara* 5:24–27.

Hinde, R. A., ed. 1983. *Primate Social Relationships*. Blackwell, Oxford.

Hiraiwa-Hasegawa, M., T. Hasegawa, and T. Nishida. 1984. Demographic study of a large-sized unit-group of chimpanzees in the Mahale Mountains, Tanzania. *Primates* 25(4):401–413.

Hladik, C. M. 1973. Alimentation et activité d'un groupe de chimpanzes reintroduits en foret Gabonaise. *Terre et vie* 27:343–413.

Hladik, C. M. 1975. Ecology, diet, and patterning in Old and New World primates. In R. H. Tuttle, ed., *Socioecology and Psychology of Primates*, pp. 3–36. Mouton, The Hague.

Hladik, C. M. 1977. Chimpanzees of Gabon and chimpanzees of Gombe: Some comparative data on the diet. In T. H. Clutton-Brock, ed., *Primate Ecology*, pp. 481–501. Academic Press, London.

Hochbaum, G. S., and E. F. Bossenmaier. 1972. Response of pintails to improved breeding habitat in southern Manitoba. *Can. Field Nat.* 86:79–81.

Hochbaum, H. A. 1944. *The Canvasback on a Prairie Marsh*. American Wildlife Institute, Washington, D.C.

Hochbaum, H. A. 1944. *The Canvasback on a Prairie Marsh*. American Wildlife Institute, Washington, D.C.

Hofmann, R. R. 1973. *The Ruminant Stomach: Stomach Structure and Feeding Habits of East African Game Ruminants*. East African Monographs in Biology, 2. East African Literature Bureau, Nairobi.

Höhn, E. O. 1947. Sexual behaviour and seasonal changes in the gonads and adrenals of the mallard. *Proc. Zool. Soc. Lond.* 117:281–304.

Holekamp, K. E. 1984. Dispersal in ground-dwelling sciurids. In J. O. Murie and G. R. Michener, eds., *Biology of Ground-Dwelling Squirrels*, pp. 297–320. University of Nebraska Press.

Holm, C. H. 1973. Breeding sex ratios, territoriality, and reproductive success in the red-winged blackbird (*Agelaius phoeniceus*). *Ecology* 54:356–65.

Holmes, W. G. 1984. The ecological basis of monogamy in Alaskan hoary marmots. In J. O. Murie and G. R. Michener, eds., *Biology of Ground-Dwelling Squirrels*, pp. 250–74. University of Nebraska Press.

Hoogland, J. L. 1979. Aggression, ectoparasitism, and other possible costs of prairie dog (Sciuridae, *Cynomys* spp.) coloniality. *Behaviour* 69:1–35.

Hoogland, J. L. 1981. Nepotism and cooperative breeding in the black-tailed prairie dog (Sciuridae: *Cynomys ludovicianus*). In R. D. Alexander and D. W. Tinkle, eds., *Natural Selection and Social Behavior*, pp. 283–310. Chiron Press, N.Y.

Literature Cited

Hoogland, J. L. 1983. Black-tailed prairie dog cotêries are cooperatively breeding units. *Amer. Natur.* 121:275–80.

Hopkins, R. A., M. J. Kutilek, and J. Shreve. 1982. The density and home range characteristics of mountain lions in the Diablo Range of California. Abstract from the International Cat Symposium, Kingsville, Texas.

Horn, A. D. 1979. The taxonomic status of the bonobo chimpanzee. *Amer. J. Phys. Anthro.* 51:273–82.

Horn, A. D. 1980. Some observations on the ecology of the bonobo chimpanzee (*Pan paniscus* Scwarz 1929) near Lake Tumba, Zaire. *Folia Primat.* 34:145–69.

Horn, H. S. 1968. The adaptive significance of colonial nesting in Brewer's blackbird (*Euphagus cyanocephalus*). *Ecology* 49:682–94.

Horn, H. S. 1970. Social behavior of nesting Brewer's blackbirds. *Condor* 72:15–23.

Hornocker, M. D. 1969. Winter territoriality in mountain lions. *J. Wildl. Mgmt.* 33:457–64.

Hornocker, M. D. 1970. *An Analysis of Mountain Lion Predation upon Mule Deer and Elk in the Idaho Primitive Area.* Wildl. Monogr. 21.

Howard R. D. 1983. Sexual selection and variation in reproductive success in a long-lived organism. *Amer Natur.* 122:301–25.

Hrdy, S. B. 1977. *The Langurs of Abu: Female and Male Strategies of Reproduction.* Harvard University Press.

Hrdy, S. B. 1981. *The Woman That Never Evolved.* Harvard University Press.

Hrdy, S. B., and P. Whitten. 1986. The patterning of sexual activity among primates. In B. B. Smuts, D. L. Cheney, R. M. Seyfarth, R. W. Wrangham, and T. T. Struhsaker, eds., *Primate Societies.* University of Chicago Press.

Humburg, D. H., H. H. Prince, and R. A. Bishop. 1978. The social organization of a mallard population in northern Iowa. *J. Wildl. Mgmt.* 42:72–80.

Hupp, J. 1983. Spring changes in lipid reserves of adult male sage grouse. *Proc. of the 13th Western States Sage Grouse Workshop.*

Hurd, J. 1981. Mate choice among the Nebraska Amish. Ph.D. diss., Pennsylvania State University.

Hurtado, M., K. Hawkes, K. Hill, and H. Kaplan. 1984. Female subsistence strategies among the Ache hunter gatherers of Eastern Paraguay. *Human Ecology,* in press.

Huxley, C. R., and N. A. Wood. 1976. Aspects of the breeding of the moorhen in Britain. *Bird Study* 23:1–10.

Ikeda, H. 1983. Development of young and parental care of the raccoon dog *Nyctereutes procyonoides viverrinus* Temmick, in captivity. *J. Mammal. Japan* 9:229–36.

Inglis, J. M. 1976. Wet season movements of individual wildebeests of the Serengeti migratory herd. *E. Afr. Wildl. J.* 14:17–33.

Irons, W. 1975. *The Yomut Turkmen: A Study of Social Organization among a Central Asian Turkic-Speaking Population.* Anthropological Paper No. 58, Museum of Anthropology, University of Michigan.

Irons, W. 1979a. Investment and primary social dyads. In N. A. Chagnon and W. Irons, eds., *Evolutionary Biology and Human Social Behavior: An Anthropological Perspective.* Duxbury Press, N. Scituate, Mass.

Irons, W. 1979b. Cultural and biological success. In N. A. Chagnon and W. Irons, eds., *Evolutionary Biology and Human Social Behavior: An Anthropological Perspective.* Duxbury Press, N. Scituate, Mass.

Irons, W. 1981. Why lineage exogamy? In R. D. Alexander and D. W. Tinkle, eds., *Natural Selection and Social Behavior: Recent Research and New Theory.* Chiron Press, N.Y.

Irons, W. 1983. Human female reproductive strategies. In S. K. Wasser, ed., *Social Behavior of Female Vertebrates.* Academic Press, N.Y.

Literature Cited

Iwamoto, T., and R.I.M. Dunbar. 1983. Thermoregulation, habitat quality, and the behavioural ecology of gelada baboons. *J. Anim. Ecol.* 53:357–66.

Izawa, K. 1970. Unit-groups of chimpanzees and their nomadism in the savannah woodland. *Primates* 11:1–46.

Jackson, J. F. 1973. Distribution and population phenetics of the Florida scrub lizard, *Sceloporus woodi. Copeia* 1973:746–61.

Janetos, A. C. 1980. Strategies of female mate choice: a theoretical analysis. *Behav. Ecol. Sociobiol.* 7:107–112.

Janzen, D. H. 1979. How to be a fig. *Ann. Rev. Ecol. Syst.* 10:13–51.

Jarman, M.V. 1979. Impala social behaviour. Territory, hierarchy, mating, and the use of space. *Adv. Ethol.*, Paul Parey, Berlin.

Jarman, P. J. 1974. The social organisation of antelope in relation to their ecology. *Behaviour* 48:215–67.

Jarman, P. J. 1982. Prospects for interspecific comparison in sociobiology. In King's College Sociobiology Group, eds. *Current Problems in Sociobiology*, pp. 323–42. Cambridge University Press.

Jarman, P. J. 1983. Mating system and sexual dimorphism in large, terrestrial, mammalian herbivores. *Biol. Rev.* 58:485–520.

Jarman, P. J., and M. V. Jarman. 1979. The dynamics of ungulate social organization. In A.R.E. Sinclair and M. Norton-Griffiths, eds., *Serengeti: Dynamics of an Ecosystem*, pp. 185–220. University of Chicago Press.

Jarman, P. J., and A.R.E. Sinclair. 1979. Feeding strategies and the pattern of resource partitioning in ungulates. In A.R.E. Sinclair and M. Norton-Griffiths, eds., *Serengeti: Dynamics of an Ecosystem*, pp. 130–63. University of Chicago Press.

Jarvis, J.V.M. 1981. Eusociality in a mammal: Cooperative breeding in naked mole-rat colonies. *Science* 212:571–73.

Jehl, J. R., Jr. 1968. *Relationships in the Charadrii (Shorebirds): A Taxonomic Study Based on Color Patterns of Downy Young.* Memoir 3, San Diego Nat. Hist. Soc.

Jenness, D. 1959. *The People of the Twilight.* University of Chicago Press.

Jenni, D. A. 1974. Evolution of polyandry in birds. *Amer. Zool.* 14:129–44.

Jewell, P. A. 1972. Social organisation and movements of topi (*Damiliscus korrigum*) during the rut at Inshasha, Queen Elizabeth Park, Uganda. *Zool. Afr.* 7:233–55.

Johns, D. W., and K. B. Armitage. 1979. Behavioral ecology of alpine yellow-bellied marmots. *Behav. Ecol. Sociobiol.* 5:133–57.

Johnsgard, P. A. 1960a. A quantitative study of sexual behavior of mallards and black ducks. *Wilson Bull.* 72:135–55.

Johnsgard, P. A. 1960b. Pair-formation mechanisms in *Anas* (Anatidae) and related genera. *Ibis* 102:616–18.

Johnsgard, P. A. 1965. *Handbook of Waterfowl Behavior.* Cornell University Press.

Johnsgard, P. A. 1978. *Ducks, Geese, and Swans of the World.* University of Nebraska Press.

Johnsingh, A.J.T. 1982. Reproductive and social behaviour of the dhole, *Cuon alpinus* (Canidae). *J. Zool. London* 198:443–63.

Johnson, C. N., and P. G. Bayliss. 1981. Habitat selection by sex, age, and reproductive class in the red kangaroo, *Macropus rufus*, in western New South Wales. *Austr. Wildl. Res.* 8:465–74.

Johnson, D. H., and A. B. Sargeant. 1977. *Impact of Red Fox Predation on the Sex Ratio of Prairie Mallards.* U.S. Fish and Wildlife Service, Wildlife Research Report 6.

Johnson, D. M., G. L. Stewart, M. Corley, R. Ghrist, J. Hagner, A. Ketterer, B. McDonnell, W. Newsom, E. Owen, and P. Samuels. 1980. Brown-headed cowbird (*Molothrus ater*) mortality in an urban winter roost. *Auk* 97:299–320.

Literature Cited

Johnson, K. A. 1977. Ecology and management of the red-necked pademelon *Thylogale thetis* on the Dorrigo Plateau of northern New South Wales. Ph.D. diss., University of New England, Armidale, Australia.

Johnson, L. K., and S. P. Hubbell. 1975. Contrasting foraging strategies and coexistence of two bee species on a single resource. *Ecology* 56:1398–1406.

Johnson, S. C. 1981. Bonobos: Generalized hominid prototypes or specialized insular dwarfs? *Current Anthropology* 22:363–75.

Jolly, C. J. 1970. The large African monkeys as an adaptive array. In J. R. Napier and P. H. Napier, eds., *Old World Monkeys*, pp. 139–74. Academic Press, London.

Jolly, C. J. 1972. The classification and natural history of *Theropithecus (Simopithecus)* (Andrews, 1916), baboons of the African Plio-Pleistocene. *Bull. Brit. Mus. Nat. Hist., Geol.* 22:1–123.

Jordan, P. A., P. C. Shelton, and D. L. Allen. 1967. Numbers, turnover, and social structure of the Isle Royale wolf population. *Amer. Zool.* 7:233–52.

Jorgensen, W. 1980. *Western Indians*. Freeman, San Francisco.

Joste, N., J. D. Ligon, and P. B. Stacey. 1985. Shared paternity in the acorn woodpecker (*Melanerpes formicivorus*). *Behav. Ecol. Sociobiol.* 17:39–41.

Joubert, S.C.T. 1974. The social organization of the Roan antelope *Hippotragus equinus* and its influence on the special distribution of herds in the Kruger National Park. In V. Geist and F. R. Walther, eds., *The Behaviour of Ungulates and Its Relation to Management*, pp. 661–75. IUCN, Morges, Switzerland.

Joubert, S.C.T. 1975. The mating behaviour of the Tsessebe *(Damaliscus lunatus lunatus)* in the Kruger National Park. *Z. Tierpsychol.* 37:182–91.

Jungers, W. L., and R. L. Susman. 1984. Body size and skeletal allometry in African apes. In R. L. Susman, ed., *The Pygmy Chimpanzee: Evolutionary Biology and Behavior*, pp. 131–78. Plenum Press, New York.

Jungius, H. 1971. *The Biology and Behaviour of the Reedbuck (Redunca arundinum Boddaert 1785) in the Kruger National Park. Mamm. depicta*. Paul Parey, Berlin.

Kano, T. 1971. The chimpanzee of Filabanga, Western Tanzania. *Primates* 12:229–46.

Kano, T. 1979. A field study on the ecology of pygmy chimpanzees *Pan paniscus*. In D. A. Hamburg and E. R. McCown, eds., *The Great Apes*, pp. 123–35. Benjamin/Cummings, Menlo Park, Calif.

Kano, T. 1980. Social behavior of wild pygmy chimpanzees (*Pan paniscus*) of Wamba: A preliminary report. *J. Human Evol.* 9:243–60.

Kano, T. 1982. The social group of pygmy chimpanzees (*Pan paniscus*) of Wamba. *Primates* 23:171–88.

Kano, T. 1983. An ecological study of the pygmy chimpanzees (*Pan paniscus*) of Yalosidi, Republic of Zaire. *Int. J. Primatol.* 4(1):1–31.

Kano, T., and M. Mulavwa. 1984. Feeding ecology of the pygmy chimpanzees (*Pan paniscus*) of Wamba. In R. L. Sussman, ed., *Evolutionary Morphology and Behavior of the Pygmy Chimpanzee*, pp. 233–74. Plenum Press, N.Y.

Kaufmann, J. H. 1962. Ecology and social behavior of the coati, *Nasua narica*, on Barro Colorado Island, Panama. *Univ. Calif. Publ. Zool.* 60:95–222.

Kaufmann, J. H. 1965. A three-year study of mating behavior in a free-ranging band of rhesus monkeys. *Ecology* 46:500–512.

Kaufmann, J. H. 1974. The ecology and evolution of social organization in the kangaroo family. (Macropodidae). *Amer. Zool.* 14:51-62.

Kaufmann, J. H. 1975. Field observations of the social behaviour of the eastern grey kangaroo, *Macropus giganteus. Anim. Behav.* 23:214–21.

Literature Cited

Kawai, M., ed. 1979. *Ecological and Sociobiological Studies of Gelada Baboons.* Karger, Basel.

Kawai, M., R.I.M. Dunbar, U. Mori, and H. Ohsawa. 1983. Social organisation of gelada baboons: Social units and definitions. *Primates* 24:13–24.

Kawanaka, K. 1981. Infanticide and cannibalism in chimpanzees—with special reference to the newly observed case in the Mahale Mountains. *Afr. Stud. Monogr.* 1:69–100.

Kear, J. 1970. The adaptive radiation of parental care in waterfowl. In J. H. Crook, ed., *Social Behaviour in Birds and Mammals*, pp. 357–92. Academic Press, London.

Keith, L. B. 1981. Population dynamics of wolves. Department of Wildlife Ecology, University of Wisconsin. Typescript.

Kemp, A. C. 1971. Some observations on the sealed-in nesting method of hornbills (Family: Bucerotidae). *Ostrich* 8 (Suppl.):149–55.

Kemp, A. C. 1976a. A study of the ecology, behavior, and systematics of *Tockus* hornbills (Aves: Bucerotidae). *Transv. Mus. Mem.* 20.

Kemp, A. C. 1976b. Factors affecting the onset of breeding in African hornbills. *Proc. 16th Int. Ornithol. Congr.* pp. 248–57.

Kemp, A. C. 1979. A review of the hornbills: Biology and radiation. *Living Bird* 17:105–136.

Kemp, A. C., and M. I. Kemp. 1975. *Report on a Study of Hornbills in Sarawak, with Comments on Their Conservation.* World Wildlife Fund Project 2/74 Report.

Kemp, A. C., and M. I. Kemp. 1980. The biology of the southern ground hornbill *Bucorvus leadbeateri* (Vigors) (Aves: Bucerotidae). *Ann. Trans. Mus.* 32:65–100.

Kennedy, R. 1975. *The Irish.* University of California Press.

Kennelly, J. J. 1978. Coyote reproduction. In M. Bekoff, ed., *Coyotes*, pp. 73–93. Academic Press, N.Y.

Kenward, R. E. 1978. Hawks and doves: Factors affecting success and selection in goshawk attacks on wood pigeons. *J. Anim. Ecol.* 47:449–60.

Keverne, E. B., J. A. Eberhart, and R. E. Meller. 1982. Dominance and subordination: Concepts or physiological states? In B. Chiarelli, ed., *Advanced Views in Primate Biology*, pp. 81–94. Springer-Verlag, Berlin.

Kiell, D. J., and J. S. Millar. 1980. Reproduction and nutrient reserves of arctic ground squirrels. *Can. J. Zool.* 58:416–21.

Kiley-Worthington, M. 1965. The waterbuck (*Kobus defassa* Ruppel 1835 and *K. ellipsiprimnus* Ogilby 1833) in East Africa: Spatial distribution. A study of the sexual behaviour. *Mammalia* 29:177–204.

Kilgore, D. L., Jr., and K. B. Armitage. 1978. Energetics of yellow-bellied marmot populations. *Ecology* 59:78–88.

King, J. A. 1955. Social behavior, social organization, and population dynamics in a black-tailed prairiedog town in the Black Hills of South Dakota. *Contr. Lab. Vert. Biol. Univ. Mich.* 67:1–123.

King, J. R. 1973. Energetics of reproduction in birds. In D. S. Farner, ed., *Breeding Biology of Birds*, pp. 78–107. Natl. Acad. Sci., Washington, D.C.

King, J. R. 1974. Seasonal allocation of time and energy resources in birds. In R. A. Paynter, Jr., ed., *Avian Energetics*, pp. 4–70. Nuttall Ornithol. Club, Cambridge, Mass.

Kingdon, J. 1977. *East African Mammals, Vol. 3A: Carnivores.* Academic Press, N.Y.

Kingdon, J. 1982. *East African Mammals, Vol. 3C: Bovids.* Academic Press, N.Y.

King's College Sociobiology Group. 1982. *Current Problems in Sociobiology.* Cambridge University Press.

Kinzey, W. G. 1984. The dentition of the pygmy chimpanzee, *Pan paniscus.* In R. L. Susman, ed., *Evolutionary Morphology and Behavior of the Pygmy Chimpanzee*, pp. 65–88. Plenum Press, N.Y.

Literature Cited

Kirkpatrick, T. H. 1966. Studies of the Macropodidae in Queensland, 4: Social organization of the grey kangaroo (*Macropus giganteus*). *Q. J. Agri. Anim. Sci.* 23:317–22.

Kitamura, K. 1983. Pygmy chimpanzee association patterns in ranging. *Primates* 24:1–12.

Kitchen, D. W. 1974. Social behavior and ecology of the pronghorn. *Wildl. Monogr.* 38:5–96.

Kitchener, S. L. 1971. Observations on the breeding of the bush dog, *Speothus venaticus*, at Lincoln Park Zoo, Chicago. *Int. Zoo. Yearb.* 11:99–101.

Kleiman, D. G. 1977. Monogamy in mammals. *Quart. Rev. Biol.* 52:39–69.

Kleiman, D. G., and J. Eisenberg. 1973. Comparisons of canid and felid social systems from an evolutionary perspective. *Anim. Behav.* 21:637–59.

Klein, L. L., and D. Klein. 1975. Social and ecological contrasts between four taxa of Neotropical primates. In R. Tuttle, ed., *Sociobiology and Psychology of Primates*. pp. 59–85. Mouton, The Hague.

Klingel, H. 1974. Social organization and reproduction in equids. *J. Reprod. Fert. Suppl.* 23:7–11.

Knowlton, F. F. 1972. Preliminary interpretations of coyote population mechanics with some management implications. *J. Wildl. Mgmt.* 36:369–82.

Koenig, W. D. 1981. Reproductive success, group size, and the evolution of cooperative breeding in the acorn woodpecker. *Amer. Natur.* 117:421–43.

Koenig, W. D., R. L. Mumme, and F. A. Pitelka. 1983. Female roles in cooperatively breeding acorn woodpeckers. In S. K. Waser, ed., *Social Behavior of Female Vertebrates*. Academic Press, N.Y.

Koenig, W. D., and F. A. Pitelka. 1981. Ecological factors and kin selection in the evolution of cooperative breeding in birds. In R. D. Alexander and D. W. Tinkle, eds., *Natural Selection and Social Behavior: Recent Research and New Theory*. Chiron Press, N.Y.

Koepcke, M. 1972. Uber die Resistenzformen der Vogeinester in einem begrenzten Gebeit des tropischen Regenwaldes in Peru. *J. Ornith.* 113:138–60.

Kok, O. B. 1972. Breeding success and territorial behavior of male boat-tailed grackles. *Auk* 89:528–40.

Kortlandt, A. 1962. Chimpanzees in the wild. *Sci. Amer.* 206(5):128–38.

Krapu, G. L. 1974. Foods of breeding pintails in North Dakota. *J. Wildl. Mgmt.* 38:408–417.

Krapu, G. L. 1979. Nutrition of female dabbling ducks during reproduction. In T. A. Bookhout, ed., *Waterfowl and Wetlands—An Integrated Review*, pp. 59–70. La Crosse Printing Co., La Crosse, Wisc.

Krapu, G. L. 1981. The role of nutrient reserves in mallard reproduction. *Auk* 98:29–38

Krapu, G. L., and H. A. Doty. 1979. Age-related aspects of mallard reproduction. *Wildfowl* 30:35–39.

Krebs, J. R. 1974. Colonial nesting and social feeding as strategies for exploiting food resources in the great blue heron (*Ardea herodias*). *Behaviour* 51:99–134.

Krebs, J. R., and N. B. Davies. 1978. *Behavioural Ecology: An Evolutionary Approach*. Blackwell, Oxford.

Krebs, J. R., and N. B. Davies. 1984. *Behavioural Ecology: An Evolutionary Approach*, 2nd edition. Blackwell, Oxford.

Kruijt, J. P., G. de Vos, A. Bossema, and O. Bruinsma. Sexual selection in black grouse. Typescript.

Kruijt, J. P., I. Bossema, and G. J. Lammers. 1982. Effects of early experience and male activity on mate choice in mallard females (*Anas platyrhynchos*). *Behaviour* 80:32–43.

Kruuk, H. 1972. *The Spotted Hyaena: A Study of Predation and Social Behaviour*. University of Chicago Press.

Kruuk, H. 1975. Functional aspects of social hunting in carnivores. In G. Baerends, C. Beer, and A. Manning, eds., *Function and Evolution in Behaviour*, pp. 119–41. Oxford Univ. Press, Oxford.

Kruuk, H. 1976. Feeding and social behaviour of the striped hyaena (*Hyaena vulgaris* Desmarest). *E. Afr. Wildl. J.* 14:91–111.

Kruuk, H. 1978a. Spatial organization and territorial behaviour of the European badger *Meles meles. J. Zool. London* 184:1–19.

Kruuk, H. 1978b. Foraging and spatial organization of the European badger, *Meles meles* L. *Behav. Ecol. Sociobiol.* 4:75–89.

Kruuk, H., and M. Turner. 1967. Comparative notes on predation by lion, leopard, cheetah, and wild dog in the Serengeti area, East Africa. *Mammalia* 31:1–27.

Kummer, H. 1968. *Social Organization of Hamadryas Baboons: A Field Study.* University of Chicago Press.

Kummer, H., W. Angst, and W. Goetz. 1974. Triadic differentiation: An inhibitory process protecting pair bonds in baboons. *Behaviour* 49:62–87.

Kummer, H., W. Goetz, and W. Angst. 1970. Cross-species modification of social behaviour in baboons. In J. R. Napier and P. H. Napier, eds., *Old World Monkeys*, pp. 351–64. Academic Press, London.

Kurland, J. A. 1979. Paternity, mother's brother, and human sociality. In N. A. Chagnon and W. Irons, eds., *Evolutionary Biology and Human Social Behavior: An Anthropological Perspective.* Duxbury Press, N. Scituate, Mass.

Kuroda, S. 1979. Grouping of the pygmy chimpanzees. *Primates* 20:161–83.

Kuroda, S. 1980. Social behavior of the pygmy chimpanzees. *Primates* 21:181–97.

Lack, D. 1968. *Ecological Adaptations for Breeding in Birds.* Methuen, London.

Lack, D., and J. T. Emlen. 1939. Observations of breeding behavior in tricolored redwings. *Condor* 41:225–30.

Lamprecht, J. 1978a. On diet, foraging behaviour, and interspecific food competition of jackals in the Serengeti National Park, East Africa. *Z. Saugetier.* 43:210–23.

Lamprecht, J. 1978b. The relationship between food competition and foraging group size in some larger carnivores. *Z. Tierpsychol.* 46:337–43.

Lamprecht, J. 1979. Field observations on the behaviour and social system of the bat-eared fox (*Otocyon megalotis* Desmarest). *Z. Tierpsychol.* 49:260–84.

Lamprecht, J. 1981. The function of social hunting in larger terrestrial carnivores. *Mammal. Rev.* 11:169–79.

Lancaster, J. 1971. Play-mothering: The relations between juvenile females and young infants among free-ranging vervet monkeys (*Cercopithecus aethiops*). *Folia Primatol.* 15:161–82.

Lancia, R. A., D. W. Woodward, and S. D. Miller. 1982. Summer movement patterns and habitat use of bobcats (*Lynx rufus*) on Croatan National Forest, North Carolina. Abstract from the International Cat Symposium, Kingsville, Texas.

Lank, D. B., L. W. Oring, and S. J. Maxson. 1985. Mate and nutrient limitation of egg-laying in the spotted sandpiper (*Actitis macularia*), a polyandrous shorebird. *Ecology*, in press.

Lanyon, W. E. 1957. The comparative biology of the meadowlarks (*Sturnella*) in Wisconsin. *Pub. Nuttall Ornith. Club* 1:1–67.

LaPrade, H. R., and H. B. Graves. 1982. Polygyny and female-female aggression in redwinged blackbirds (*Agelaius phoeniceus*). *Amer. Natur.* 120:135–38.

La Rivers, L. 1948. Some Hawaiian ecological notes. *Wasmann Coll.* 7:85–110.

Larkin, P., and M. Roberts. 1979. Reproduction in the ring-tailed mongoose. *Int. Zoo Yearb.* 19:188–93.

Laurie-Ahlberg, C. C., and F. McKinney. 1979. The nod-swim display of male green-winged teal (*Anas crecca*). *Anim. Behav.* 27:165–72.

Lawick, H. van. 1973. *Solo.* Collins, London.

Literature Cited

Lawick, H. van, and J. van Lawick-Goodall. 1970. *The Innocent Killers*. Collins, London.

Leach, E. 1961. *Pul Eliya, A Village in Ceylon: A Study of Land Tenure and Kinship*. Cambridge University Press.

LeBoeuf, B. J. 1974. Male-male competition and reproductive success in elephant seals. *Amer. Zool.* 14:163–76.

Lebret, T. 1961. The pair formation in the annual cycle of the mallard, *Anas platyrhynchos* L. *Ardea* 49:97–158.

LeCroy, M. 1980. The genus *Paradisea*: Display and evolution. *Amer. Mus. Novitates* 2714:1–52.

LeCroy, M. A. Kulupi, and W. S. Peckover. 1980. Goldie's bird of paradise: Display, natural history, and traditional relationships of people to the bird. *Wilson Bull.* 92:289–301.

Lee, R. 1979. *The !Kung San: Men, Women and Work in a Foraging Society*. Cambridge University Press.

Lee, R. 1982. Politics, sexual and non-sexual, in an egalitarian society. In E. Leacock and R. Lee, eds., *Politics and History in Band Societies*. Cambridge University Press.

Lee, R., and I. DeVore. 1968. *Man the Hunter*. Aldine, Chicago.

Leighton, M. 1982. Fruit resources and patterns of feeding, spacing, and grouping among sympatric Bornean hornbills (Bucerotidae). Ph.D. diss., University of California, Davis.

Leighton, M., and D. R. Leighton. 1982. The relationship of size of feeding aggregate to size of food patch: Howler monkeys (*Alouatta palliata*) feeding in *Trichilia cipo* fruit trees on Barro Colorado Island. *Biotropica* 14:81–90.

Leighton, M., and D. R. Leighton. 1983. Vertebrate responses to fruiting seasonality within a Bornean rain forest. In S. L. Sutton, T. C. Whitmore, and A. C. Chadwick, eds., *Tropical Rain Forest: Ecology and Management*, pp. 181-96. Blackwell, Oxford.

Lembeck, M. 1982. Long-term behavior and population dynamics of an unharvested bobcat population in San Diego County. Abstract from the International Cat Symposium, Kingsville, Texas.

Lenarz, M. S. 1982. Habitat partitioning in feral horses: The value of being dominant. Ph.D. diss., University of Chicago.

Lenington, S. 1980. Female choice and polygyny in red-winged blackbirds. *Anim. Behav.* 28:347–61.

Lent, P. C. 1974. Mother-infant relationships in ungulates. In V. Geist and F. R. Walther, eds., *The Behaviour of Ungulates and Its Relation to Management*, pp. 14–55. IUCN, Morges, Switzerland.

Leutenegger, W., and J. T. Kelly. 1977. Relationship of sexual dimorphism in canine size and body size to social, behavioral, and ecological correlates in Anthropoid primates. *Primates* 18:117–36.

Leuthold, W. 1966. Variations in territorial behavior of Uganda kob *Adenota kob thomasi* (Neumann 1896). *Behaviour* 27:214–57.

Leuthold, W. 1970. Preliminary observations on food habits of gerenuk in Tsavo National Park, Kenya. *E. Afr. Wildl J.* 8:73–84.

Leuthold, W. 1974. Observations on home range and social organization of lesser kudu, *Tragelaphus imberis* (Blyth, 1869). In V. Geist and F. R. Walther, eds., *The Behaviour of Ungulates and Its Relation to Management*, pp. 206–234. IUCN, Morges, Switzerland.

Leuthold, W. 1977. *African Ungulates: A Comparative Review of Their Ethology and Behavioural Ecology*. Springer-Verlag, Berlin.

Leuthold, W. 1978a. On the ecology of the gerenuk, *Litocranius walleri* (Brooke, 1878). *J. Anim. Ecol.* 47:471–90.

Leuthold, W. 1978b. On social organization and behaviour of the gerenuk, *Litocranius walleri* (Brooke 1878). *Z. Tierpsychol.* 47:194–216.

Literature Cited

Lévi-Strauss, C. 1949. *Les Structures Elementaires de la Parente.* Plon, Paris.

Liberg, O. 1981. Predation and social behaviour in a population of domestic cats: An evolutionary perspective. Ph.D. diss., University of Lund.

Lieberman, D., J. B. Hall, M. D. Swaine, and M. Lieberman. 1979. Seed dispersal by baboons in the Shai Hills, Ghana. *Ecology* 60:65–75.

Ligon, J. D. 1981. Demographic patterns and communal breeding in the green woodhoopoes, *Phoeniculus purpureus.* In R. D. Alexander and D. W. Tinkle, eds., *Natural Selection and Social Behavior,* pp. 231–43. Chiron Press, N.Y.

Ligon, J. D., and S. H. Ligon. 1978. Communal breeding in green woodhoopoes as a case for reciprocity. *Nature* 276:496–98.

Lindburg, D. G. 1971. The rhesus monkey in north India: An ecological and behavioral study. In L. A. Rosenblum, ed., *Primate Behavior: Developments in Field and Laboratory Research,* pp. 1–106. Academic Press, N.Y.

Lindburg, D. G. 1977. Feeding behaviour and diet of rhesus monkeys (*Macaca mulatta*) in Siwalik forest in north India. In T. H. Clutton-Brock, ed., *Primate Ecology,* pp. 223–49. Academic Press, London.

Lindburg, D. G. 1983. Mating behavior and estrus in the Indian rhesus monkey. In S. K. Seth, ed., *Perspectives in Primate Biology,* pp. 45–61. Today and Tomorrow, New Delhi.

Litvaitis, J. A., J. T. Major, and J. A. Sherburne. 1982. Bobcat movements in relation to snowshoe hare density. Abstract from the International Cat Symposium, Kingsville, Texas.

Loeb, E. M. 1934. Patrilineal and matrilineal organization in Sumatra, Pt. 2: The Menagkebau. *American Anthropologist* 36:25–56.

Lokemoen, J. T., H. F. Duebbert, and D. E. Sharp. 1984. Nest spacing, habitat selection, and behavior of waterfowl on Miller Lake Island, North Dakota. *J. Wildl. Mgmt.* 48:309–321.

Lombaard, L. J. 1971. Age determination and growth curves in the black-backed jackal, *Canis mesomelas* Schreber, 1775 (Carnivore: *canidae*). *Ann. Transvaal Mus.* 27(7):135–85.

Longhurst, W. 1944. Observations on the ecology of the Gunnison prairie dog in Colorado. *J. Mammal.* 25:24–36.

Low, B. S. 1976. Evolution of amphibian life histories in the desert. In D. W. Goodall, ed., *Evolution of Desert Biotas.* University of Texas Press, Austin.

Low, B. S. 1978. Environmental uncertainty and the parental strategies of marsupials and placentals. *Amer. Natur.* 112:197–213.

Low, B. S. 1979. Sexual selection and human ornamentation. In N. A. Chagnon and W. Irons, eds., *Evolutionary Biology and Human Social Behavior: An Anthropological Perspective.* Duxbury Press, N. Scituate, Mass.

Low, B. S. 1987. In L. Betzig, M. Bergerhoff Mulder, and P. Turke, eds., *Human Reproductive Behavior: A Darwinian Perspective.* Cambridge University Press. Forthcoming.

Lowther, P. E. 1975. Geographic and ecological variation in the family Icteridae. *Wilson Bull.* 87:481–95.

Lumpkin, S. 1981. Avoidance of cuckoldry in birds: The role of the female. *Anim. Behav.* 29:303–304.

Lumpkin, S. 1983. Female manipulation of male avoidance of cuckoldry behavior in the ring dove. In S. K. Wasser, ed., *Social Behavior of Female Vertebrates,* pp. 91–112. Academic Press, N.Y.

Lumsden, H. G. 1968. The displays of the sage grouse. *Ontario Dept. of Lands and Forests, Research Report (Wildlife)* 83:1–94.

Lynch, C. 1980. Ecology of the suricate, *Suricata suricatta* and yellow mongoose, *Cynictis penicillata* with special reference to their reproduction. *Memoirs van die Nasionale Museum* [South Africa] 14:1–145.

Literature Cited

Macdonald, D. W. 1979a. Flexibility of the social organization of the golden jackal, *Canis aureus*. *Behav. Ecol. Sociobiol.* 5:17–38.

Macdonald, D. W. 1979b. 'Helpers' in fox society. *Nature* 282:69–71.

Macdonald, D. W. 1980. Social factors affecting reproduction among red foxes. In E. Zimen, ed., *Biogeographica, Vol. 18: The Red Fox*, p. 123–75. Dr. W. Junk Publ., The Hague.

Macdonald, D. W. 1981. Resource dispersion and the social organization of the red fox (*Vulpes vulpes*). In J. A. Chapman and D. Pursley, eds., *Worldwide Furbearer Conference Proceedings*, pp. 918–49. Frostburg, Md.

Macdonald, D. W. 1983. The ecology of carnivore social behavior. *Nature* 301:379–84.

Macdonald, D. W., and P. D. Moehlman. 1983. Cooperation, altruism, and restraint in the reproduction of carnivores. In P.P.G. Bateson and P. Klopfer, eds., *Perspectives in Ethology*, pp. 433–67. Plenum Press, N.Y.

Mace, G. M., and P. H. Harvey. 1983. Energetic constraints on home-range size. *Amer. Natur.* 121:120–132.

MacKinnon, J. R. 1974. The behaviour and ecology of wild orangutans (*Pongo pygmaeus*). *Anim. Behav.* 22:3–74.

MacPherson, A. H. 1969. The dynamics of Canadian arctic fox populations. *Can. Wildl. Serv. Rep. Ser.* 8:1–49.

Madge, S. C. 1969. Notes on the breeding of the bushy-crested hornbill *Anorrhinus galeritus*. *Malay Nat. J.* 23:1–6.

Malcolm, J. R. 1979. Social organization and communal rearing of pups in African wild dogs (*Lycaon pictus*). Ph.D. diss., Harvard University.

Malcolm, J. R., and K. Marten. 1982. Natural selection and the communal rearing of pups in African wild dogs (*Lycaon pictus*). *Behav. Ecol. Sociobiol.* 10:1–13.

Manson, J. 1983. Parallel cousin marriage: A review and some predictions from evolutionary biology. Paper presented at the annual meeting of the American Anthropological Association, December, 1983, Washington, D.C.

Maple, T., and M. Hoff. 1982. *Gorilla Behavior*. Van Nostrand Reinhold, N.Y.

Maroney, R. J., J. M. Warrne, and M. M. Sinha. 1959. Stability of social dominance hierarchies in monkeys (*Macaca mulatta*). *J. Soc. Psychol.* 50:285–93.

Martin, P. S. 1966. Africa and the Pleistocene overkill. *Nature* 212:339–42.

Martin, S. G. 1974. Adaptations for polygynous breeding in the bobolink, *Dolichonyx oryzivorous*. *Amer. Zool.* 14:109–119.

Maxson, S. J. 1977. Common grackle preys on spotted sandpiper chick. *Prairie Nat.* 9:53–54.

Maxson, S. J., and L. W. Oring. 1978. Mice as a source of egg loss among ground-nesting birds. *Auk* 95:582–84.

Maxson, S. J., and L. W. Oring. 1980. Breeding season time and energy budgets of the polyandrous spotted sandpiper. *Behaviour* 74:200–263.

Maynard Smith, J. 1974. The theory of games and the evolution of animal conflict. *J. Theor. Biol.* 47:209–222.

Maynard Smith, J. 1977. Parental investment—a prospective analysis. *Anim. Behav.* 25:1–9.

Maynard Smith, J. 1979. Game theory and the evolution of behaviour. *Proc. Roy. Soc. London B.* 205:475–88.

Maynard Smith, J., and G. A. Parker. 1976. The logic of asymmetric contests. *Anim. Behav.* 24:159–75.

Maynard Smith, J., and G. R. Price. 1973. The logic of animal conflict. *Nature* 246:15–18.

McCarley, H. 1966. Annual cycle, population dynamics, and adaptive behavior of *Citellus tridecemlineatus*. *J. Mammal.* 47:294–316.

McCarley, H. 1970. Differential reproduction in *Spermophilus tridecemlineatus*. *Southwest Nat.* 14:293–96.

Literature Cited

McCommon, C. 1983. Female reproductive strategies among the Garifuna. Ph.D. diss., Pennsylvania State University.

McCort, W. D. 1979. The feral asses (*Equus usinus*) of Ossabaw Island, Georgia. In R. H. Denniston, ed., *Symposium on the Ecology and Behavior of Wild and Feral Equids*, Wyoming University Press, pp. 71–83.

McGrew, J. C. 1979. *Vulpes macrotis*. *Mammal. Sp.* 123:1–6.

McGrew, W. C. 1983. Animal foods in the diets of wild chimpanzees (*Pan troglodytes*): Why cross-cultural variation? *J. Ethol.* 1:46–61.

McHenry, M. G. 1971. Breeding and post-breeding movements of blue-winged teal (*Anas discors*) in southwestern Manitoba. Ph.D. diss., University of Oklahoma, Norman.

McKinney, F. 1965. Spacing and chasing in breeding ducks. *Wildfowl Trust Ann. Rep.* 16:92–106.

McKinney, F. 1975. The evolution of duck displays. In G. Baerends, C. Beer, and A. Manning, eds., *Function and Evolution in Behaviour*, pp. 331–57. Clarendon Press, Oxford.

McKinney, F. 1985. Primary and secondary male reproductive strategies of dabbling ducks. In *Avian Monogamy*, ed. P. A. Gowaty and D. W. Mock, Ornithological Monographs, 37, pp. 68–82.

McKinney, F., and D. J. Bruggers. 1983. Status and breeding behavior of the Bahama pintail and the New Zealand blue duck. *Proc. 1983 Jean Delacour/IFCB Symposium on Breeding Birds in Captivity*, pp. 211–21. Hollywood, Calif.

McKinney, F., K. M. Cheng, and D. J. Bruggers. 1984. Sperm competition in apparently monogamous birds. In R. L. Smith, ed., *Sperm Competition and the Evolution of Animal Mating Systems*, pp. 523–45. Academic Press, N.Y.

McKinney, F., S. R. Derrickson, and P. Mineau. 1983. Forced copulation in waterfowl. *Behaviour* 86:250–94

McKinney, F., W. R. Siegfried, I. J. Ball, and P.G.H. Frost. 1978. Behavioral specializations for river life in the African black duck (*Anas sparsa* Eyton). *Z. Tierpsychol.* 48:349–400.

McKinney, F., and P. Stolen. 1982. Extra-pair-bound courtship and forced copulation among captive green-winged teal (*Anas crecca carolinensis*). *Anim. Behav.* 30:461–74.

McLandress, M. R., and D. G. Raveling. 1981. Changes in diet and body composition of Canada geese before spring migration. *Auk* 98:65–79.

McLean, I. G. 1982. The association of female kin in the arctic ground squirrel *Spermophilus parryii*. *Behav. Ecol. Sociobiol.* 10:91–99.

McLean, I. G. 1983. Paternal behaviour and killing of young in arctic ground squirrels. *Anim. Behav.* 31:32–44.

McVittie, R. 1979. Changes in the social behaviour of South West African cheetah. *Madoqua* 11:171–84.

Mech, L. D. 1970. *The Wolf: Ecology and Social Behavior of an Endangered Species*. Natural History Press, N.Y.

Meyer, P. L. 1970. *Introductory Probability and Statistical Applications*. Addison-Wesley, Reading, Mass.

Michener, G. R. 1973. Field observations on the social relationships between adult female and juvenile Richardson's ground squirrels. *Can. J. Zool.* 51:33–38.

Michener, G. R. 1979a. Yearling variations in the population dynamics of Richardson's ground squirrels. *Can. Field Nat.* 93:363–70.

Michener, G. R. 1979b. Spatial relationships and social organization of adult Richardson's ground squirrels. *Can. J. Zool.* 57:125–39.

Michener, G. R. 1980. Differential reproduction among female Richardson's ground squirrels and its relation to sex ratio. *Behav. Ecol. Sociobiol.* 7:173–78.

Michener, G. R. 1983. Kin identification, matriarchies, and the evolution of sociality in

Literature Cited

ground-dwelling sciurids. In J. F. Eisenberg and D. G. Kleiman, eds., *Advances in the Study of Mammalian Behavior*, pp. 528–72. Amer. Soc. Mammalogists Sp. Pub. No. 7.

Michener, G. R., and D. R. Michener. 1977. Population structure and dispersal in Richardson's ground squirrels. *Ecology* 58:359–68.

Michener, G. R., and J. O. Murie. 1983. Black-tailed prairie dog coteries: Are they cooperatively breeding units? *Amer. Natur.* 121:266–74.

Millar, J. S. 1977. Adaptive features of mammalian reproduction. *Evolution*: 31:370–86.

Millar, J. S. 1981. Prepartum reproductive characteristics of eutherian mammals. *Evolution* 35:1149–63.

Miller, R. and R. H. Denniston. 1979. Interband dominance in feral horses. *Z. Tierpsychol.* 51:41–47.

Mills, M.G.L. 1982. The mating system of the brown hyaena, *Hyaena brunnea*, in the southern Kalahari. *Behav. Ecol. Sociobiol.* 10:131–36.

Mineau, P., and F. Cooke. 1979. Rape in the lesser snow goose. *Behaviour* 70:280–91.

Mitani, J. C. 1985. Mating behavior of male orangutans in the Kutai Game Reserve, East Kalimantan, Indonesia. *Anim. Behav.* 33:392–402.

Mitchell, B., J. Shenton, and J. Uys. 1965. Predation on large mammals in the Kafue National Park, Zambia. *Zool. Afr.* 1:297–318.

Mloszewski, M. J. 1983. *The Behaviour and Ecology of the African Buffalo*. Cambridge University Press.

Moehlman, P. D. 1979a. Behavior and ecology of feral asses (*Equus asinus*). *Nat. Geo. Soc. Res. Rep.*, pp. 405–411.

Moehlman, P. D. 1979b. Jackal helpers and pup survival. *Nature* 277:382–83.

Moehlman, P. D. 1981. Why do jackals help their parents? Reply to Montgomerie. *Nature* 289:824–25.

Moehlman, P. D. 1983. Socioecology of silver-backed and golden jackals (*Canis mesomelas, C. aureus*). In J. F. Eisenberg and D. G. Kleiman, eds., *Recent Advances in the Study of Mammalian Behavior*. Special Publication No. 7, Amer. Soc. of Mammalogists.

Montfort-Braham, N. 1975. Variations dans la structure sociale du topi, *Damaliscus korrigum* Ogilby, au Parc National de l'Akagera, Rwanda. *Z. Tierpsychol.* 39:332–64.

Moore, J. C. 1959. Relationships among living squirrels of the Sciurinae. *Bull. Amer. Mus. Nat. Hist.* 118:159–206.

Moore, J. and R. Ali. 1984. Are dispersal and inbreeding avoidance related? *Anim. Behav.* 32:94–112.

Mori, A. 1979. *Reproductive Behaviour: Ecological and Sociological Studies of Gelada Baboons*. Contrib. to Primatol. 16. Karger, Basel.

Mori, U., and R.I.M. Dunbar. 1985. Changes in the reproductive condition of female gelada baboons following the takeover of one-male units. *Z. Tierpsychol.* 67:215–24.

Morrison, P., and W. Galster. 1975. Patterns of hibernation in the arctic ground squirrel. *Can. J. Zool.* 53:1345–55.

Morton, M. L., and R. L. Parmer. 1975. Body size, organ size, and sex ratios in adult and yearling Belding ground squirrels. *Gt. Basin Nat.* 35:305–309.

Morton, M. L., and P. W. Sherman. Effects of a spring snowstorm on behavior, reproduction, and survival of Belding's ground squirrels. *Can. J. Zool.* 56:2578–90.

Muckenhirn, N. A., and J. F. Eisenberg. 1973. Home ranges and predation of the Ceylon leopard. In R. L. Eaton, ed., *The World's Cats*, Vol. 1, pp. 142–75. World Wildlife Safari, Winston, Or.

Mulvany, P. 1977. *Dairy Cow Condition Scoring*. National Institute for Research in Dairying, No. 4468.

Murdock, G. P. 1949. *Social Structure*. Macmillan, N.Y.

Literature Cited

Murdock, G. P. 1967. *Ethnographic Atlas*. Pittsburgh University Press.

Murie, J. O., and M. A. Harris. 1978. Territoriality and dominance in male Columbian ground squirrels (*Spermophilus columbianus*). *Can. J. Zool.* 56:2402–12.

Murie, J. O., and M. A. Harris. 1982. Annual variation of spring emergence and breeding in Columbian ground squirrels (*Spermophilus columbianus*). *J. Mammal.* 63:431–39.

Murray, M. G. 1981. Structure of association in Impala, *Aepyceros melampus*. *Behav. Ecol. Sociobiol.* 9:23–33.

Murray, M. G. 1982a. Home range, dispersal, and the clan system of impala. *Afr. J. Ecol.* 20:253–69.

Murray, M. G. 1982b. The rut of impala: Aspects of seasonal mating under tropical conditions. *Z. Tierpsychol.* 59:319–37.

Myers, J. P. 1981. A test of three hypotheses for latitudinal segregation of the sexes in wintering birds. *Can. J. Zool.* 59:1527–34.

National Research Council. [NRC] 1982. *Wild and Free Roaming Horses and Burros*. National Academy Press.

Nagel, U. 1971. Social organisation in a baboon hybrid zone. *Proc. 3rd Int. Congr. Primatol.* 3:48–57.

Neal, E. 1970. The banded mongoose, *Mungos mungo* Gmelin. *E. Afr. Wildl. J.* 8:63-72.

Nee, J. 1969. Reproduction in a population of yellow-bellied marmots (*Marmota flaviventris*). *J. Mammal.* 50:756–65.

Needham, R. 1962. *Structure and Sentiment*. University of Chicago Press.

Nel, J.A.J. 1978. Notes on the food and foraging behavior of the bat-eared fox, *Otocyon megalotis*. *Bull. Carnegie Mus. Nat. Hist.* 6:132–37.

Nel, J.A.J., and M. H. Bester. 1983. Communication in the southern bat-eared fox *Otocyon m. megalotis* (Demarest, 1822). *Z. Saugetierk*.

Nel, J.A.J., M.G.L. Mills, and R. J. van Aarde. 1984. Fluctuating group size in bat-eared foxes (*Otocyon m. megalotis*) in the south-western Kalahari. *Notes from the Mammal Society* 48:294–98.

Nellis, C. H., and L. B. Keith. 1968. Hunting activities and success of lynxes in Alberta. *J. Wildl. Mgmt.* 32:718–22.

Nellis, C. H., and L. B. Keith. 1976. Population dynamics of coyotes in Central Alberta. *J. Wildl. Mgmt.* 40:389–99.

Nellis, D. and C.O.R. Everard. 1983. *The Biology of the Mongoose in the Caribbean*. Studies on the Fauna of Curacao and Other Caribbean Islands, 64. Utrecht.

Nishida, T. 1974. Ecology of wild chimpanzees. In *Human Ecology*, ed. T. Ohtsuka and T. Nishida, pp. 15–60. Kyoritsu-Shuppan, Tokyo. (In Japanese.)

Nishida, T. 1979. The social structure of the chimpanzees of the Mahale Mountains. In D. A. Hamburg and E. R. McCown, eds., *The Great Apes*, pp. 72–121. Benjamin/Cummings, Menlo Park, Calif.

Nishida, T. 1983a. Alpha status and agonistic alliance in wild chimpanzees (*Pan troglodytes schweinfurthii*). *Primates* 24(3):318–36.

Nishida, T. 1983b. Alloparental behavior in wild chimpanzees of the Mahale Mountains, Tanzania. *Folia Primat.* 41:1–33.

Nishida, T., and M. Hiraiwa-Hasegawa. 1986. Chimpanzees and bonobos. In B. B. Smuts, D. L. Cheney, R. M. Seyfarth, R. W. Wrangham, and T. T. Struhsaker, eds., *Primate Societies*. University of Chicago Press.

Nishida, T., M. Hiraiwa-Hasegawa, T. Hasegawa, and Y. Takahata. 1985. Group extinction and female transfer in wild chimpanzees in the Mahale Mountains. *Z. Tierpsychol.* 67:284–301.

Literature Cited

Nishida, T., and K. Kawanaka. 1972. Inter-unit-group relationships among wild chimpanzees of the Mahali Mountains. *Kyoto Univ. Afr. Stud.* 7:131–69.

Nishida, T., and S. Uehara. 1983. Natural diet of chimpanzees (*Pan troglodytes schweinfurthii*): Long-term record from the Mahale Mountains, Tanzania. *Afr. Stud. Monogr.* 3:109–30.

Noë, R. 1984. Coalition formation among males in three free-ranging baboon groups. *Int. J. Primatol.* 5(4):369.

Nordenskiold, E. 1949. The Cuna. In J. Steward, ed., *Handbook of South American Indians*, Vol. 4. U.S. Government Printing Office, Washington, D.C.

Norikoshi, K., and N. Koyama. 1974. Group shifting and social organization among Japanese monkeys. In S. Kondo, M. Kawai, A. Ehara, and S. Kawamura, eds., *Symp. 5th Cong. Int. Primate Soc.*, pp. 43–61. Japan Science Press, Tokyo.

Novellie, P. A. 1979. Courtship behaviour of the blesbok (*Damaliscus dorcas phillipsi*). *Mammalia* 43:263–74.

Nudds, T. D., and C. D. Ankney. 1982. Ecological correlates of territory and home range size in North American dabbling ducks. *Wildfowl* 33:58–62.

Nudds, T. D., and J. N. Bowlby. 1984. Predator-prey size relationships in North American dabbling ducks. *Can. J. Zool.* 62:2002–2008.

Ohde, B. R., R. A. Bishop, and J. J. Dinsmore. 1983. Mallard reproduction in relation to sex ratios. *J. Wildl. Mgmt.* 47:118–26.

Ohsawa, H. 1979. Herd dynamics. In M. Kawai, ed., *Ecological and Sociological Studies of Gelada Baboons*, pp. 47–80. Kodansha, Tokyo, and Karger, Basel.

Ohsawa, H., and R.I.M. Dunbar. 1984. Variations in the demographic structure and dynamics of gelada baboon populations. *Behav. Ecol. Sociobiol.* 15:231–40.

Orians, G. H. 1961. The ecology of blackbird (*Agelaius*) social systems. *Ecol. Monogr.* 31:285–312.

Orians, G. H. 1969. On the evolution of mating systems in birds and mammals. *Amer. Natur.* 103:589–603.

Orians, G. H. 1972. The adaptive significance of mating systems in the Icteridae. In K. H. Voous, ed., *Proc. 15th Int. Ornithol. Congr.*, pp. 389–98. E. J. Brill, Leiden.

Orians, G. H. 1980. *Some Adaptations of Marsh-Nesting Blackbirds*. Princeton University Press.

Orians, G. H. 1983. Notes on the behavior of the melodious blackbird (*Dives dives*). *Condor* 85:453–60.

Orians, G. H., and H. S. Horn. 1969. Overlap in foods and foraging of four species of blackbirds in the potholes of central Washington. *Ecology* 50:930–938.

Orians, G. H., C. E. Orians, and K. J. Orians. 1977. Helpers at the nest in some Argentine blackbirds. In B. Stonehouse and C. Perrins, eds., *Evolutionary Ecology*, pp. 137–51. Macmillan, London.

Oring, L. W., 1986. Avian polyandry: A review. In R. F. Johnson, ed., *Current Ornithology*, Vol. 3, pp. 309–351. Plenum Press, N.Y.

Oring, L. W. 1964a. Behavior and ecology of certain ducks during the postbreeding period. *J. Wildl. Mgmt.* 28:223–33.

Oring, L. W. 1964b. Predation upon flightless ducks. *Wilson Bull.* 76:190.

Oring, L. W. 1982. Avian mating systems. In D. Farner, J. King, and K. Parkes, eds., *Avian Biology*, Vol. 6, pp. 1–92. Academic Press, N.Y.

Oring, L. W., and M. L. Knudson. 1972. Monogamy and polyandry in the spotted sandpiper. *Living Bird* 11:59–73.

Oring, L. W., and D. B. Lank. 1982. Sexual selection, arrival times, philopatry, and site fidelity in the polyandrous spotted sandpiper. *Behav. Ecol. Sociobiol.* 10:185–91.

Literature Cited

Oring, L. W., and D. B. Lank. 1984. Breeding area fidelity, natal philopatry, and the social systems of sandpipers. In J. Burger and B. L. Olla, eds., *Behavior of Marine Animals: Shorebirds' Breeding Behavior and Populations*, Vol. 5, pp. 125–47. Plenum Press, N.Y.

Oring, L. W., D. B. Lank, and S. J. Maxson. 1983. Population studies of the polyandrous spotted sandpiper. *Auk* 100:272–85.

Oring, L. W., and S. J. Maxson. 1978. Instances of simultaneous polyandry by a spotted sandpiper, *Actitis macularis. Ibis* 120:349–53.

Osmond, M. W. 1966. Towrd monogamy: A cross-cultural study of correlates of types of marriage. *Social Forces* 45:8–16.

Otterbein, K. F., and C. S. Otterbein. 1965. An eye for an eye and a tooth for a tooth: A cross-cultural study of feuding. *American Anthropologist* 67:1470–82.

Owen, M. 1980. *Wild Geese of the World: Their Life History and Ecology*. Batsford, London.

Owen-Smith, N. 1977. On territoriality in ungulates and an evolutionary model. *Quart. Rev. Biol.* 52:1–38.

Owen-Smith, N. 1979. Assessing the foraging efficiency of a large herbivore, the kudu. *S. Afr. J. Wildl. Res.* 9:102–110.

Owen-Smith, N. 1984. Spatial and temporal components of the mating systems of kudu bulls and red deer stags. *Anim. Behav.* 32:321–32.

Owen-Smith, N., and P. Novellie, 1982. What should a clever ungulate eat? *Amer. Natur.* 119:151–78.

Owens, M. J., and D. D. Owens. 1978. Feeding ecology and its influence on social organization in brown hyaenas (*Hyaena brunnea* Thunberg). *E. Afr. Wildl. J.* 16:112–36.

Owens, M. J., and D. D. Owens. 1984. Kalahari lions break the rules. *Int. Wildl.* 14:4–13.

Owings, D. H., M. Borchert, and R. Virginia. 1977. The behavior of California ground squirrels. *Anim. Behav.* 25:221–30.

Packer, C. 1977. Inter-troop transfer and inbreeding and avoidance in *Papio anubis* in Tanzania. Ph.D. diss., University of Sussex.

Packer, C. 1979a. Male dominance and reproductive activity in *Papio anubis. Anim. Behav.* 27:37–45.

Packer, C. 1979b. Inter-troop transfer and inbreeding avoidance in *Papio anubis. Anim. Behav.* 27:1–36.

Packer, C., L. Herbst, A. E. Pusey, J. D. Bygott, J. P. Hanby, S. J. Cairns, and M. Bergerhoff-Mulder. 1987. Reproductive success of lions. In T. H. Clutton-Brock, ed., *Reproductive Success*. University of Chicago Press.

Packer, C., and A. E. Pusey. 1982. Cooperation and competition within coalitions of male lions: Kin selection or game theory? *Nature* 296:740–42.

Packer, C., and A. E. Pusey. 1983a. Adaptations of female lions to infanticide by incoming males. *Amer. Natur.* 121:716–28.

Packer, C., and A. E. Pusey. 1983b. Cooperation and competition in lions. *Nature* 302:356.

Packer, C., and A. E. Pusey. 1983c. Male takeovers and female reproductive parameters: A simulation of oestrous synchrony in lions (*Panthera leo*). *Anim. Behav.* 31:334–40.

Packer, C., and A. E. Pusey. 1984. Infanticide in carnivores. In G. Hausfater and S. B. Hrdy, eds., *Infanticide: Comparative and Evolutionary Perspectives*, pp. 31–42. Aldine Press, N.Y.

Packer, C., and A. E. Pusey. 1985. Asymmetric contests in social mammals: Respect, manipulation, and age-specific aspects. In *Evolution: Essays in Honor of John Maynard Smith*, ed. P. Greenwood, M. Slatkin and P. H. Harvey, pp. 173–86. Cambridge University Press.

Paige, K. E., and J. M. Paige. 1981. *The Politics of Reproductive Ritual*. University of California Press.

Literature Cited

Parker, G. A. 1978. Searching for mates. In J. R. Krebs and N. B. Davies, eds., *Behavioural Ecology, an Evolutionary Approach*, pp. 214–44. Blackwell, Oxford.

Parker, G. A. 1983. Mate quality and mating decisions. In P.P.G. Bateson, ed., *Mate Choice*, pp. 141–66. Cambridge University Press.

Parker, G. A., and D. I. Rubenstein. 1981. Role assessment, reserve strategy, and acquisition of information in asymmetric animal conflicts. *Anim. Behav.* 29:221–40.

Patterson, I. J. 1977. Aggression and dominance in winter flocks of shelduck *Tadorna tadorna* (L.) *Anim. Behav.* 25:447–59.

Patterson, R. L. 1952. *The Sage Grouse in Wyoming.* Sage Books, Denver.

Paulus, S. L. 1982. Feeding ecology of Gadwalls in Louisiana in winter. *J. Wildl. Mgmt.* 46:71–79.

Paulus, S. L. 1983. Dominance relations, resource use, and pairing chronology of gadwalls in winter. *Auk* 100:947–52.

Paulus, S. L. 1984. Activity budgets of nonbreeding gadwalls in Louisiana. *J. Wildl. Mgmt.* 48:371–80.

Payne, R. B. 1984. Sexual selection, lek and arena behavior, and sexual size dimorphism in birds. *Ornithol. Monogr.* 33:1–52.

Peacock, N. 1985. Time allocation, work, and fertility among Efe pygmy women of northeast Zaire. Ph.D. diss., Harvard University.

Pennycuick, L. 1975. Movements of the migratory wildebeest population in the Serengeti area between 1960 and 1973. *E. Afr. Wildl. J.* 13:65–87.

Perrins, C. M. 1970. The timing of birds' breeding seasons. *Ibis* 112:242–55.

Peter, Prince of Greece and Denmark. 1963. *A Study of Polyandry.* Mouton, The Hague.

Peterson, J. G. 1970. The food habits and summer distribution of juvenile sage grouse in central Montana. *J. Wildl. Mgmt.* 34:147–55.

Peterson, R. O., J. D. Woolington, and T. N. Bailey. 1984. Wolves of the Kenai Peninsula, Alaska. *Wildl. Monogr.* 88:1–52.

Petrie, M. 1982. Winter flocking in moorhens. Ph.D. diss., University of East Anglia, Norwich.

Petrie, M. 1983a. Female moorhens compete for small fat males. *Science* 220:413–15.

Petrie, M. 1983b. Mate choice in role-reversed species. In P.P.G. Bateson, ed. *Mate Choice*, pp. 167–79. Cambridge University Press.

Petrie, M. 1984. Territory size in the moorhen (*Gallinula chloropus*): An outcome of RHP asymmetry between neighbours. *Anim. Behav.* 32:861–70.

Petter, G. 1969. Interpretation evolutive des caracteres de la denture des viverrides africains. *Mammalia* 33:607–625.

Pfeiffer, S. R. 1982. Variability in reproductive output and success of *Spermophilus elegans* ground squirrels. *J. Mammal.* 63:284–89.

Picman, J. 1981. The adaptive value of polygyny in marsh-nesting red-winged blackbirds: Renesting, territory tenacity, and mate fidelity of females. *Can. J. Zool.* 59:2284–96.

Pitelka, F. A., R. T. Holmes, and S. F. MacLean, Jr. 1974. Ecology and evolution of social organization in arctic sandpipers. *Amer. Zool.* 14:185–204.

Pleasants, B. Y. 1979. Adaptive significance of variable dispersion pattern of breeding northern orioles. *Condor* 81:28–34.

Pollack, J. 1980. *Behavioural ecology and body condition changes in new forest ponies.* Farm Livestock Advisory Committee, RSPCA.

Popp, J. L. 1978. Male baboons and evolutionary principles. Ph.D. diss., Harvard University.

Popp, J. L. 1983. Ecological determinism in the life histories of baboons. *Primates* 24:198–210.

Literature Cited

Post, W. 1981. Biology of the yellow-shouldered blackbird (*Agelaius xanthomus*) on a tropical island. *Bull. Florida State Mus. Biol. Sciences* 26:125–202.

Post, W., and J. W. Wiley. 1976. The yellow-shouldered blackbird—present and future. *Amer. Birds* 30:13–20.

Poston, H. J. 1974. Home range and breeding biology of the shoveler. *Can. Wildl. Ser. Rep. Ser.* 25:1–48.

Power, H. W. 1980. Male escorting and protecting females at the nest cavity in mountain bluebirds. *Wilson Bull.* 92:509–511.

Power, H. W., E. Litkovich, and M. P. Lombardo. 1981. Male starlings delay incubation to avoid being cuckolded. *Auk* 98:386–87.

Prater, S. H. 1948. *The Book of Indian Animals.* Bombay Natural History Society.

Prince, H. H., and D. H. Gordon. Postbreeding strategies of dabbling ducks. In M. W. Weller, ed., *Waterfowl in Winter,* in press.

Provost, E. E., C. A. Nelson, and A. D. Marshall. 1973. Population dynamics and behavior in the bobcat. In R. L. Eaton, ed., *The World's Cats,* Vol. 1, pp. 42–67. World Wildlife Safari, Winston, Or.

Pulliam, H. R. 1973. On the advantages of flocking. *J. Theor. Biol.* 38:419–22.

Pulliam, H. R., K. A. Anderson, A. Misztal, and N. Moore. 1974. Temperature-dependent social behavior in juncos. *Ibis* 116:360–64.

Pulliam, H. R., and T. Caraco. 1984. Living in groups: Is there an optimal group size? In J. R. Krebs and N. B. Davies, eds., *Behavioural Ecology: An Evolutionary Approach,* 2nd ed., pp. 122–47. Blackwell, Oxford.

Pusey, A. E. 1980. Inbreeding avoidance in chimpanzees. *Anim. Behav.* 28:543–52.

Pusey, A. E., and C. Packer. 1983. Once and future kings. *Nat. Hist.* 92:54–63.

Pusey, A. E., and C. Packer. 1986. Dispersal and philopatry. In B. B. Smuts, D. L. Cheney, R. M. Seyfarth, R. W. Wrangham, and T. T. Struhsaker, eds. *Primate Societies.* University of Chicago Press.

Pusey, A. E., and C. Packer. Dispersal and group fissions in lions. *Behaviour,* in press.

Pyrah, D. B., H. E. Jorgensen, and R. O. Wallestad. 1972. *Effects of Chemical and Mechanical Sagebrush Control on Sage Grouse.* Montana Fish and Game Department, Job Progress Report, Project W-105-R-6.

Rabenold K. N. 1985. Cooperation in breeding by non-reproductive wrens: Kinship, reciprocity, and demography. *Behav. Ecol. Sociobiol.* 17:1–17.

Radcliffe-Brown, A. R. 1952. *Structure and Function in Human Society.* Cohen and West, London.

Raikow, R. J. 1973. Locomotor mechanisms in North American ducks. *Wilson Bull.* 85:295–307.

Ransom, T. W., and B. S. Ransom. 1971. Adult-male-infant relations among baboons (*Papio anubis*). *Folia Primatol.* 16:179–95.

Ransom, T. W., and T. E. Rowell. 1972. Early social development of feral baboons. In F. E. Poirier, ed., *Primate Socialization,* pp. 105–144. Random House, N.Y.

Rappaport, R. A. 1968. *Pigs for the Ancestors: Ritual in the Ecology of a New Guinea People.* Yale University Press.

Rasa, O.A.E. 1983a. Dwarf mongoose and hornbill mutualism in the Taru Desert, Kenya. *Behav. Ecol. Sociobiol.* 12:181–90.

Rasa, O.A.E. 1983b. A case of invalid care in wild dwarf mongooses. *Z. Tierpsychol.* 62:235–40.

Rasmussen, D. R. 1979. Correlates of patterns of range use of a troop of yellow baboons (*Papio cynocephalus*), I: Sleeping sites, impregnable females, births, and male emigrations and immigrations. *Anim. Behav.* 27:1098–1112.

Literature Cited

Rasmussen, K.L.R. 1980. Consort behaviour and mate selection in yellow baboons (*Papio cynocephalus*). Ph.D. diss., Cambridge University.

Rattray, R. S. 1927. *Religion and Art in Ashanti*. Oxford University Press.

Rausch, R. A. 1967. Some aspects of the population ecology of wolves, Alaska. *Amer. Zool.* 7:253–65.

Reich, A. 1981. The behavior and ecology of the African wild dog in Kruger National Park. Ph.D. diss., Yale University.

Reighord, J. 1920. The breeding behaviour of the suckers and minnows. *Biol. Bull.* 38:1–32.

Reynolds, V. 1975. How wild are the Gombe chimpanzees? *Man* 10:123–25.

Reynolds, V., and F. Reynolds. 1965. Chimpanzees of the Budongo Forest. In I. DeVore, ed., *Primate Behavior*, pp. 368–424. Holt, Rinehart & Winston, N.Y.

Richards, A. I. 1950. Some types of family structure amongst the Central Bantu. In A. R. Radcliffe-Browne and D. Forde, eds., *African Systems of Kinship and Marriage*. Cambridge University Press.

Ricklefs, R. E. 1974. Energetics of reproduction in birds. In R. A. Paynter, Jr., ed., *Avian Energetics*, pp. 152–297. Pub. Nuttall Ornith. Club No. 15.

Ricklefs, R. E. 1975. The evolution of co-operative breeding in birds. *Ibis* 117:531–34.

Riechert, S. 1978. Games spiders play: Behavioural variability in territorial dispute. *Behav. Ecol. Sociobiol.* 3:135–62.

Rijksen, H. D. 1978. A field study on Sumatran orangutans (*Pongo pygmaeus abelii* Lesson 1827). H. Veenman and B. V. Zonen, Wageningen, The Netherlands.

Riss, D. C., and C. D. Busse. 1977. Fifty-day observation of a free-ranging adult male chimpanzee. *Folia Primatol.* 28:283–97.

Riss, D. C., and J. Goodall. 1977. The recent rise to the alpha-rank in a population of free-living chimpanzees. *Folia Primatol.* 27:134–51.

Ritter, L. V. 1983. Nesting ecology of scrub jays in Chico, California. *Western Birds* 14:147–58.

Roberts, A. 1951. *The Mammals of South Africa*. Central News Agency, Cape Town.

Roberts, T. A., and J. J. Kennelly. 1977. Assessing promiscuity among female red-winged blackbirds in Massachusetts. *Trans. Northeast Fish Wildl. Conf.* 34:99–105.

Robertson, R. J. 1972. Optimal niche space of the red-winged blackbird: Spatial and temporal patterns of nesting activity and success. *Ecology* 54:1085–1093.

Robinson. J. G. 1982. Intersexual competition and mate choice in primates. *Amer. J. Primatol. Supp.* 1:131–44.

Robinson, S. K. 1984. Social behavior and sexual selection in a neotropical oriole. Ph.D. diss., Princeton University.

Robinson, S. K. 1985a. Coloniality in the yellow-rumped cacique as a defense against nest predators. *Auk* 102:506–19.

Robinson, S. K. 1985b. The yellow-rumped cacique and its associated nest pirates. In P. A. Buckley, M. S. Foster, E. S. Morton, R. S. Ridgely, and F. G. Buckley, eds., *Neotropical Ornithology*, pp. 898–907. Ornithological Monographs No. 36. American Ornithologists' Union, Washington, D. C.

Robinson, S. K. 1985c. Benefits, costs, and determinants of dominance in a neotropical oriole. *Anim. Behav.*, in press.

Robinson, S. K. 1985d. Fighting and assessment in the yellow-rumped cacique (Icterinae: *Cacicus cela*). *Behav. Ecol. Sociobiol.*, in press.

Robinson, S. K. 1985e. Competitive and mutualistic interactions among females of a neotropical oriole. *Anim. Behav.*, in press.

Robinson, S. K., and R. T. Holmes. 1982. Foraging behavior of forest birds: The relationships among search tactics, diet, and habitat structure. *Ecology* 63:1918–31.

Literature Cited

Rodman, P. S. 1973. Population composition and adaptive organization among orangutans of the Kutai Nature Reserve. In R. P. Michael and J. H. Crook, eds., *Comparative Ecology and Behaviour of Primates*, pp. 171–209. Academic Press, London.

Rodman, P. S. 1977. Feeding behavior of orangutans of the Kutai Nature Reserve, East Kalimantan. In T. H. Clutton-Brock, ed., *Primate Ecology*, pp. 383–413. Academic Press, London.

Rodman, P. S. 1981. Inclusive fitness and group size with a reconsideration of group size in lions and wolves. *Amer. Natur.* 118:275–83.

Rodman, P. S. 1984. Foraging and social systems of orang-utans and chimpanzees. In P. S. Rodman and J.G.H. Cant, eds., *Adaptations for Foraging in Non-Human Primates*, pp. 134–60. Columbia University Press.

Rongstad, O. J. 1965. A life history study of thirteen-lined ground squirrels in southern Wisconsin. *J. Mammal.* 46:76–87.

Rood, J. P. 1974. Banded mongoose males guard young. *Nature*. 248:176.

Rood, J. P. 1975. Population dynamics and food habits of the banded mongoose. *E. Afr. Wildl. J.* 13:89–111.

Rood, J. P. 1978. Dwarf mongoose helpers at the den. *Z. Tierpsychol.* 48:277–87.

Rood, J. P. 1980. Mating relationships and breeding suppression in the dwarf mongoose. *Anim. Behav.* 28:143–50.

Rood, J. P. 1983a. The social system of the dwarf mongoose. In J. F. Eisenberg and D. G. Kleiman, eds., *Recent Advances in the Study of Mammalian Behavior*. Special Publication No. 7, Amer. Soc. of Mammalogists.

Rood, J. P. 1983b. Banded mongoose rescues pack member from eagle. *Anim. Behav.* 31:1261–62.

Rood, J. P. 1986. Reproductive success in the dwarf mongoose. In *Reproductive Success*, ed. T. H. Clutton-Brock, University of Chicago Press.

Rood, J. P. 1987. Patterns of dispersal in the dwarf mongoose. In B. D. Chepko-Sade and Z. Halpin, eds., *Patterns of Dispersal among Social Mammals and Their Effect on the Genetic Structure of Populations*. University of Chicago Press.

Rood, J. P., and D. W. Nellis. 1980. Freeze marking mongooses. *J. Wildl.* 44:500–502.

Rood, J. P., and P. Waser. 1978. The slender mongoose, *Herpestes sanquineus*, in the Serengeti. *Carnivore* 1:54–58.

Rosaldo, M., and L. Lamphere, eds., 1974. *Woman, Culture, and Society*. Stanford University Press.

Rose, F.G.G. 1960. *Classification of Kin, Age Structure, and Marriage amongst the Groote Eylandt Aborigines: A Study in Method and a Theory of Australian Kinship*. Akademic Verlag, Berlin.

Rosenzweig, M. L. 1977. Net primary productivity of terrestrial communities: Prediction from climatological data. *Amer. Natur.* 102:67–74.

Rosevear, D. R. 1974. *The Carnivores of West Africa*. The British Museum (Natural History), London.

Rothstein, S. I., J. Verner, and E. Stevens. 1980. Range expansion and diurnal changes in dispersion of the brown-headed cowbird in the Sierra Nevada. *Auk* 97:253–67.

Rothstein, S. I., J. Verner, and E. Stevens. 1984. Radio-tracking confirms a unique diurnal pattern of spatial occurrence in the parasitic brown-headed cowbird. *Ecology* 65:77–88.

Rowe-Rowe, D. T. 1976. Food of the black-backed jackal in nature conservation and farming areas in Natal. *E. Afr. Wildl. J.* 14:345–48.

Rowe-Rowe, D. T. 1977. Food ecology of otters in Natal, South Africa. *Oikos* 28:210–19.

Rowe-Rowe, D. T. 1978. The small carnivores of Natal. *The Lammergeyer* 25:5–41.

Rowell, T. E. 1966. Forest living baboons in Uganda. *J. Zool. London* 149:344–64.

Literature Cited

Rowell, T. E. 1968. Long-term changes in a population of Ugandan baboons. *Folia Primatol.* 11:241–54.

Rowell, T. E. 1970. The reproductive cycles of two *Cercopithecus* monkeys. *J. Reprod. Fert.* 21:133–41.

Rowell, T. E. 1972a. Female reproductive cycles and social behavior in primates. *Adv. Stud. Behav.* 4:69–105.

Rowell, T. E. 1972b. *Social Behaviour of Monkeys*. Penguin, Harmondsworth.

Rowell, T. E. 1973. Social organization of wild talapoin monkeys. *Amer. J. Phys. Anthro.* 38:593–97.

Rowley, I. 1983. Re-mating in birds. In P.P.G. Bateson, ed., *Mate Choice*, pp. 331–60. Cambridge University Press.

Rubbelke, D. L. 1976. Distribution and relative abundance of potential prey of spotted sandpipers (*Actitia macularia*) on Little Pelican Island, Leech Lake, Cass Co., Minnesota. M.Sc. thesis, University of North Dakota, Grand Forks.

Rubenstein, D. I. 1978. On predation, competition, and the advantages of group living. *Persp. Ethol.* 3:205–231.

Rubenstein, D. I. 1980. On the evolution of alternative mating strategies. In J.E.R. Staddon, ed., *Limits to Action: The Allocation of Individual Behavior*, pp. 1–44. Academic Press, N.Y.

Rubenstein, D. I. 1981a. Individual variation and competition in the Everglades pygmy sunfish. *J. Anim. Ecol.* 50:337–50.

Rubenstein, D. I. 1981b. Population density, resource patterning, and territoriality in the Everglades pygmy sunfish. *Anim. Behav.* 29:155–72.

Rubenstein, D. I. 1981c. Behavioral ecology of island feral horses. *Equine Vet. J.* 13:27–34.

Rubenstein, D. I. 1982. Risk, uncertainty, and evolutionary strategies. In King's College Sociobiology Group, eds., *Current Problems in Sociobiology*, pp. 91–111. Cambridge University Press.

Rubenstein, D. I., and R. W. Wrangham. 1980. Why is altruism towards kin so rare? *Z. Tierpsychol.* 54:381–87.

Rudnai, J. 1974. *The Social Life of the Lion*. Medical and Technical Publishing, St. Leonardsgate, U.K.

Rudran, R. 1976. Socio-ecology of the blue monkeys (*Cercopithecus mitis stuhlmanni*) of the Kibale Forest, Uganda. Ph.D. diss., University of Maryland.

Rudran, R. 1978. Socio-ecology of the blue monkeys (*Cercopithecus mitis stuhlmanni*) of the Kibale Forest, Uganda. *Smithson. Contrib. Zool.* 249:1–88.

Ryan, M. J., M. D. Tuttle, and L. K. Taft. 1981. The costs and benefits of frog chorusing behaviour. *Behav. Ecol. Sociobiol.* 8:273–78.

Ryder, J. P. 1970. A possible factor in the evolution of clutch size in Ross' goose. *Wilson Bull.* 82:5–13.

Saayman, G. S. 1970. The menstrual cycle and sexual behaviour in a troop of free-ranging chacma baboons (*Papio ursinus*). *Folia Primatol.* 12:81–110.

Sabater-Pi, J. 1977. Contribution to the study of alimentation of lowland gorillas in the natural state, in Rio Muni, Republic of Equatorial Guinea (West Africa). *Primates* 18:183–204.

Sabater-Pi, J. 1979. Feeding behaviour and diet of chimpanzees (*Pan troglodytes troglodytes*) in the Okorobiko Mountains of Rio Muni (West Africa). *Z. Tierpsychol.* 50:265–81.

Sade, D. S., K. Cushing, P. Cushing, J. Dunale, A. Figueroa, J. R. Kaplan, C. Lauer, D. Rhodes, and J. Schneider. 1976. Population dynamics in relation to social structure on Cayo Santiago. *Yearb. Phys. Anthro.* 20:253–62.

Sadlier, R.M.F.S. 1969. *The Ecology of Reproduction in Wild and Domestic Mammals*. Methuen, London.

Sahlins, M. D. 1972. *Stone Age Economics*. Aldine, Chicago.

Sahlins, M. D. 1976. *The Use and Abuse of Biology*. University of Michigan Press.

Salomonsen, F. 1968. The moult migration. *Wildfowl* 19:5–24.

Sampson, R. J. 1975. *Surface II Graphics System*. Kansas Geological Survey, Lawrence.

Sanft, K. 1960. Bucerotidae. *Das Tierreich* 76.

Sargeant, A. B. 1972. Red fox spatial characteristics in relation to waterfowl predation. *J. Wildl. Mgmt.* 36:225–36.

Sargeant, A. B., S. H. Allen, and R. T. Eberhardt. 1984. Red fox predation on breeding ducks in midcontinent North America. *Wildl. Monogr.* 89:1–41.

Saunders, J. K. 1963. Food habits of the lynx in Newfoundland. *J. Wildl. Mgmt.* 27:384–90.

Saunders, J. K. 1964. Physical characteristics of the Newfoundland lynx. *J. Mammal.* 45:36–47.

Savage, R.J.G. 1978. Carnivora. In V. J. Maglio and H.B.S. Cooke, eds., *Evolution of African Mammals*. Harvard University Press.

Savage-Rumbaugh, E. S., and B. J. Wilkerson. 1978. Socio-sexual behavior in *Pan paniscus* and *Pan troglodytes*: A comparative study. *J. Human Evol.* 7:327–44.

Savard, J.-P. L. 1985. Evidence of long-term pair-bonds in Barrow's goldeneye (*Bucephala islandica*). *Auk* 102:389–91.

Schaik, C. P. van. 1983. Why are diurnal primates living in groups? *Behaviour* 87:120–44.

Schaik, C. P. van, and J.A.R.A.M. van Hooff. 1983. On the ultimate causes of primate social systems. *Behaviour* 85:91–117.

Schaik, C. P. van, and M. A. van Noordwijk. Reproductive seasonality in Sumatran long-tailed macaques (in prep.).

Schaller, G. B. 1963. *The Mountain Gorilla: Ecology and Behavior*. University of Chicago Press.

Schaller, G. B. 1967. *The Deer and the Tiger*. University of Chicago Press.

Schaller, G. B. 1972. *The Serengeti Lion*. University of Chicago Press.

Schaller, G. B. 1977. *Mountain Monarchs*. University of Chicago Press.

Schaller, G. B., and P. G. Crawshaw. 1980. Movement patterns of jaguar. *Biotropica* 12:161–68.

Schaller, G. B., and J.M.C. Vasconcelos. 1978. Jaguar predation on capybara. *Z. Saugetierk.* 43:296–301.

Schantz, T. von. 1981. Female cooperation, male competition, and dispersal in the red fox *Vulpes vulpes*. *Oikos* 37:63–68.

Schapera, I. 1950. Kinship and marriage among the Tswana. In A. R. Radcliffe-Browne and D. Forde, eds., *African Systems of Kinship and Marriage*. Cambridge University Press.

Schenkel, R. 1966. On sociology and behaviour in impala (*Aepyceros melampus suara* Matschie). *Z. Saugetierk.* 31:177–205.

Schlegal, A. 1972. *Male Dominance and Female Autonomy*. HRAF Press, New Haven.

Schlitter, D. A. 1974. Notes on the Liberian mongoose, *Liberiictis kuhni* Hayman 1958. *J. Mammal.* 55:438–42.

Schoener, T. W. 1968. Sizes of feeding territories among birds. *Ecology* 49:123–41.

Schoener, T. W. 1971. The theory of feeding strategies. *Ann. Rev. Ecol. Syst.* 2:369–404.

Schuster, R. H. 1976. Lekking behaviour of Kafue lechwe. *Science* 192:1240–42.

Schwagmeyer, P. R. 1980. Alarm calling behavior of the thirteen-lined ground squirrel, *Spermophilus tridecemlineatus*. *Behav. Ecol. Sociobiol.* 7:195–200.

Schwagmeyer, P. R., and C. H. Brown. 1983. Factors affecting male-male competition in thirteen-lined ground squirrels. *Behav. Ecol. Sociobiol.* 13:1–6.

Schwartz, E. 1934. On the local races of the chimpanzee. *Annals and Magazine of Natural History* 10:576–83.

Literature Cited

Schwartz, O. A., and K. B. Armitage. 1980. Genetic variation in social mammals: The marmot model. *Science* 207:665–67.

Scott, D. K. 1980. Functional aspects of the pair bond in winter in Bewick's swans (*Cygnus columbianus bewickii*). *Behav. Ecol. Sociobiol.* 7:323–27.

Scott, D. M., and C. D. Ankney. 1980. Fecundity of the brown-headed cowbird in southern Ontario. *Auk* 97:677–83.

Scott, J. W. 1942. Mating behavior of the sage grouse. *Auk* 59:477–98.

Searcy, W. A. 1979. Male characteristics and pairing success in red-winged blackbirds. *Auk* 96:353–63.

Searcy, W. A., and K. Yasukawa. 1983. Sexual selection and red-winged blackbirds. *Amer. Scientist* 71:166–74.

Seidensticker, J. C., IV, M. G. Hornocker, W. V. Wiles, and J. P. Messick. 1973. *Mountain Lion Social Organization in the Idaho Primitive Area.* Wildl. Monogr. 35.

Seitz, A. V. 1959. Beobachtungen an handauf gezogenen Goldschakalen (*Canis aureus algirensis* Wagner 1843). *Z. Tierpsychol.* 16:747–71.

Selander, R. K. 1958. Age determination and molt in the boat-tailed grackle. *Condor* 60:353–76.

Selander, R. K. 1964. Speciation in wrens of the genus *Campylorhynchus. Univ. Calif. Publ. Zool.* 74:1–224.

Selander, R. K. 1966. Sexual dimorphism and differential niche utilization in birds. *Condor* 68:113–51.

Selander, R. K. 1972. Sexual selection and dimorphism in birds. In B. Campbell, ed., *Sexual Selection and Descent of Man*, pp. 180–230. Aldine Press, Chicago.

Selander, R. K., and D. R. Giller. 1961. Analysis of sympatry of great-tailed and boat-tailed grackles. *Condor* 63:29–86.

Senzota, R.B.M. 1978. Some aspects of the ecology of two dominant rodents in the Serengeti Ecosystem. M.Sc. thesis, University of Dar Es Salaam.

Senzota, R.B.M. 1982. The habitat and food habits of the grass rats (*Arvicanthis niloticus*) in the Serengeti National Park, Tanzania. *Afr. J. Ecol.* 20:241–52.

Seyfarth, R. M. 1978a. Social relationships among adult male and female baboons, I: Behaviour during sexual consortship. *Behaviour* 64:204–226.

Seyfarth, R. M. 1978b. Social relationships among adult male and female baboons, II: Behaviour throughout the female reproductive cycle. *Behaviour* 64:227–47.

Seymour, N. R. 1974. Territorial behaviour of wild shovelers at Delta, Manitoba. *Wildfowl* 25:49–55.

Seymour, N. R., and R. D. Titman. 1978. Changes in activity patterns, agonistic behavior, and territoriality of black ducks (*Anas rubripes*) during the breeding season in a Nova Scotia tidal marsh. *Can. J. Zool.* 56:1773–85.

Shelford, R. 1899. On some hornbill embryos and nestlings. *Ibis* 5:538–49.

Sherman, P. W. 1977. Nepotism and the evolution of alarm calls. *Science* 197:1246–53.

Sherman, P. W. 1980. The limits of ground squirrel nepotism. In G. W. Barlow and J. Silverberg, eds., *Sociobiology: Beyond Nature/Nurture?*, pp. 505–544. Westview Press, Boulder.

Sherman, P. W. 1981. Reproductive competition and infanticide in Belding's ground squirrels and other animals. In R. D. Alexander and D. W. Tinkle, eds., *Natural Selection and Social Behavior*, pp. 311–31. Chiron Press, N.Y.

Sherman, P. W. 1982. Infanticide in ground squirrels. *Anim. Behav.* 30:938–39.

Sherman, P. W., and W. G. Holmes. 1985. Kin recognition: Issues and evidence. In B. Holldobler and M. Lindauer, eds., *Experimental Behavioral Ecology*, pp. 437–60. G. Fisher Verlag, Stuttgart.

Literature Cited

Sherman, P. W., and M. L. Morton. 1984. Demography of Belding's ground squirrels. *Ecology* 65:1617–28.

Shields, W. M. 1982. *Philopatry, Inbreeding, and the Evolution of Sex*. State University of New York Press.

Shields, W. M. 1983. Optimal inbreeding and the evolution of philopatry. In I. R. Swingland and P. J. Greenwood, eds., *The Ecology of Animal Movement*, pp. 132–59. Oxford University Press.

Shotake, T. 1980. Genetic variability within and between herds of gelada baboons in central Ethiopian highlands. *Anthro. Contemporanea* 3:270.

Sibley, C. G. 1957. The evolutionary and taxonomic significance of sexual dimorphism and hybridization in birds. *Condor* 59:166–91.

Sibley, R. M. 1983. Optimal group size is unstable. *Anim. Behav.* 31:947–48.

Siegfried, W. R., and P.G.H. Frost. 1975. Continuous breeding and associated behaviour in the moorhen (*Gallinula chloropus*). *Ibis* 117:102–109.

Sigg, H. 1980. Differentiation of female positions in hamadryas one-male units. *Z. Tierpsychol.* 53:265–302.

Sigg, H., and A. Stolba. 1981. Home range and daily march in a hamadryas baboon troop. *Folia Primatol.* 36:40–75.

Sigg, H., A. Stolba, J.-J. Abbeglen, and V. Dasser. 1982. Life history of hamadryas baboons: Physical development, infant mortality, reproductive parameters, and family relationships. *Primates* 23:473–87.

Silk, J. 1986. Social behavior in evolutionary perspective. In B. B. Smuts, D. L. Cheney, R. M. Seyfarth, R. W. Wrangham, and T. T. Struhsaker, eds., *Primate Societies*. University of Chicago Press.

Simonds, P. 1965. The bonnet macaque in south India. In I. DeVore, ed., *Primate Behavior: Field Studies of Monkeys and Apes*, pp. 175–95. Holt, Rinehart & Winston, N.Y.

Simpson, G. G. 1945. The principles of classification and a classification of the mammals. *Bull. Amer. Mus. Nat. Hist.* 85:1–350.

Sinclair, A.R.E. 1974a. The natural regulation of buffalo populations in East Africa, I: Introduction and resource requirements. *E. Afr. Wildl. J.* 12:135–54.

Sinclair, A.R.E. 1974b. The natural regulation of buffalo populations in East Africa, IV: The food supply as a regulating factor, and competition. *E. Afr. Wildl. J.* 12:291–311.

Sinclair, A.R.E. 1977. *The African Buffalo: A Study of Resource Limitation of Populations*. University of Chicago Press.

Skutch, A. F. 1954. *Life Histories of Central American Birds, Vol. 1*. Pacific Coast Avifauna 31.

Skutch, A. F. 1972. *Studies of Tropical American Birds*. Publ. Nuttall Ornith. Club 10.

Slade, N. A., and D. F. Balph. 1974. Population ecology of Uinta ground squirrels. *Ecology* 55:989–1003.

Smith, M. G. 1962. *West Indian Family Structure*. University of Washington Press.

Smith, N. G. 1968. The advantage of being parasitized. *Nature* 219:690–94.

Smith, N. G. 1983. *Zarhynchus wagleri*. In D. H. Janzen, ed., *Costa Rican Natural History*, pp. 614–16. University of Chicago Press.

Smith, R. I. 1968. The social aspects of reproductive behavior in the pintail. *Auk* 85:381–96.

Smith, R. I. 1970. Response of pintail breeding populations to drought. *J. Wildl. Mgmt.* 34:943–46.

Smith, R. T. 1956. *The Negro Family in British Guiana*. Humanities Press, N.Y.

Smithers, R.H.N. 1971. *The Mammals of Botswana*. Museum Memoirs No. 4, National Museum of Rhodesia, Salisbury.

Literature Cited

Smithers, R.H.N. 1983. *The Mammals of the Southern African Subregion*. University of Pretoria Press.

Smuts, B. B. 1982. Special relationships between adult male and female olive baboons (*Papio anubis*). Ph.D. diss., Stanford University.

Smuts, B. B. 1985. *Sex and Friendship in Baboons*. Aldine, N.Y.

Snyder, R. L. 1962. Reproductive performance of a population of woodchucks after a change in sex ratio. *Ecology* 43:506–515.

Sokal, R. R., and F. J. Rohlf. 1969. *Biometry*. W. H. Freeman & Co., San Francisco.

Sokal, R. R., and F. J. Rohlf. 1981. *Biometry*, 2nd ed. W. H. Freeman & Co., San Francisco.

Sosnovskii, I. P. 1967. Breeding of the red dog or dhole, *Cuon alpinus*, at Moscow Zoo. *Int. Zoo. Yearb.* 7:120–22.

Southwell, C. J. 1981. Sociobiology of the eastern grey kangaroo, *Macropus giganteus*. Ph.D. diss., University of New England, Armidale, Australia.

Southwell, C. J. 1984a. Variability in grouping in the eastern grey kangaroo *Macropus giganteus*, I: Group density and group size. *Austral. Wildl. Res.* 11:423–36.

Southwell, C. J. 1984b. Variability in grouping in the eastern grey kangaroo *Macropus giganteus*, II: Dynamics of group formation. *Austral. Wildl. Res.* 11:437–50.

Southwick, C. H., M. A. Beg, and M. R. Siddiqi. 1965. Rhesus monkeys in north India. In I. DeVore, ed., *Primate Behavior: Field Studies of Monkeys and Apes*, pp. 111–59. Holt, Rinehart & Winston, N.Y.

Soutiere, E. C., H. S. Myrick, and E. C. Bolen. 1972. Chronology and behavior of American widgeon wintering in Texas. *J. Wildl. Mgmt.* 36:752–58.

Sowls, L. K. 1955. *Prairie Ducks*. Wildlife Management Institute, Washington, D.C.

Spinage, C. A. 1969. Territoriality and social organization of the Uganda defassa waterbuck *Kobus defassa ugandae*. *J. Zool.* 159:329–61.

Spurr, E. B., and H. Milne. 1976. Adaptive significance of autumn pair formation in the common eider *Somateria mollissima* (L.). *Ornis Scand.* 7:85–89.

Stähli, P., and M. Zurbuchen. 1979. Two topographic maps 1:25000 of Simen, Ethiopia. In M. Messerli and K. Aerni, eds., *Simen Mountains—Ethiopia, Vol. I: Cartography and Its Application for Geographical and Ecological Problems*, pp. 11–31. Geographical Institute, University of Bern, Switzerland.

Stallcup, J. A., and G. E. Woolfenden. 1978. Family status and contribution to breeding by Florida scrub jays. *Anim. Behav.* 26:1144–56.

Stammbach, E. 1978. On social differentiation in groups of captive female hamadryas baboons. *Behaviour* 67:322–38.

Standen, P. J. 1980. The social display of the Chilean teal *Anas flavirostris flavirostris*. *J. Zool. London* 191:293–313.

Stanley Price, M. R. 1974. The feeding ecology of Coke's hartebeest, *Alcelaphus buselaphus cokii*, Günther, in Kenya. Ph.D. diss., Oxford University.

Stanley Price, M. R. 1977. The estimation of food intake, and its seasonal variation, in the hartebeest. *E. Afr. Wildl. J.* 15:107–124.

Stanley Price, M. R. 1978. The nutritional ecology of Coke's hartebeest (*Alcelaphus buselaphus cokei*) in Kenya. *J. Appl. Ecol.* 15:33–49.

Stephens, W. N. 1963. *The Family in Cross-Cultural Perspective*. Holt, Rinehart & Winston, N.Y.

Stewart, D.R.M. 1967. Analysis of plant epidermis in faeces: A technique for studying the food preferences of grazing herbivores. *J. Appl. Ecol.* 4:83–111.

Stewart, G. R., and R. D. Titman. 1980. Territorial behaviour by prairie pothole blue-winged teal. *Can. J. Zool.* 58:639–49.

Stewart, K., and A. H. Harcourt. 1986. Gorillas: Variations in female relationships. In B. B.

Literature Cited

Smuts, D. L. Cheney, R. W. Seyfarth, R. W. Wrangham, and T. T. Struhsaker, eds., *Primate Societies*. Chicago University Press.

Stockard, A. H. 1929. Observations on reproduction in the white-tailed prairie dog (*Cynomys leucurus*). *J. Mammal.* 10:209–212.

Stolba, A. 1979. Entscheidungsfindung in Verbänden von Papio hamadryas. Ph.D. diss., University of Zurich.

Stolen, P., and F. McKinney. 1983. Bigamous behaviour of captive Cape teal. *Wildfowl* 34:10–13.

Stoltz, L. P., and G. S. Saayman. 1970. Ecology and behaviour of baboons in the Northern Transvaal. *Ann. Transv. Mus.* 26:99–143.

Storm, G. I., and E. D. Ables. 1966. Notes on newborn and full-term wild red foxes. *J. Mammal.* 47:116–18.

Storm, G. L., R. D. Andrews, R. L. Phillips, R. A. Bishop, D. B. Siniff, and J. R. Tester. 1976. *Morphology, Reproduction, Dispersal, and Mortality of Midwestern Red Fox Populations*. Wildl. Monogr. 49.

Stotts, V. D., and D. E. Davis. 1960. The Black duck in the Chesapeake Bay of Maryland: Breeding behavior and biology. *Chesapeake Science* 1:127–54.

Strauch, J. G. 1978. The phylogeny of the Charadriiformes (Aves): A new estimate using the method of character compatability analysis. *Trans. Zool. Soc. London* 34:263–345.

Struhsaker, T. T. 1967. Social structure among vervet monkeys (*Cercopithecus aethiops*). *Behaviour* 29:6–121.

Struhsaker, T. T. 1969. Correlates of ecology and social organization among African cercopithecines. *Folia Primatol.* 11:80–118.

Struhsaker, T. T. 1975. *The Red Colobus Monkey*. University of Chicago Press.

Struhsaker, T. T. 1977. Infanticide and social organization in the redtail monkey (*Cercopithecus ascanius schmidti*) in the Kibale Forest, Uganda. *Z. Tierpsychol.* 45:75–84.

Struhsaker, T. T. 1978. Food habits of five monkey species in the Kibale Forest, Uganda. In D. J. Chivers and J. Herbert, eds., *Recent Advances in Primatology, Vol. 1: Behaviour*, pp. 225–48. Academic Press, N.Y.

Struhsaker, T. T. 1980. Comparison of the behavior and ecology of red colobus and redtail monkeys in the Kibale Forest, Uganda. *Afr. J. Ecol.* 18:33–51.

Struhsaker T. T., and L. Leland. 1979. Socioecology of five sympatric monkey species in the Kibale Forest, Uganda. *Adv. Stud. Behav.* 9:159–225.

Strum, S. C. 1982. Agonistic dominance in male baboons: An alternative view. *Int. J. Primatol.*, 3:175–202.

Strum, S. C., and J. O. Western. 1982. Variations in fecundity with age and environment in olive baboons (*Papio anubis*). *Amer. J. Primatol.* 3:61–76.

Stuart, C. T. 1977. Analysis of *Felis libyca* and *Genetta genetta* scats from the central Namib Desert, South West Africa. *Zool. Afr.* 12:239–41.

Suarez, B., and D. R. Ackerman. 1971. Social dominance and reproductive behavior in male rhesus monkeys. *Amer. J. Phys. Anthro.* 35:219–22.

Sugden, L. G. 1973. Feeding ecology of pintail, gadwall, American wigeon, and lesser scaup ducklings in southern Alberta. *Canad. Wildl. Ser. Rep. Ser.* No. 24, pp. 1–45.

Sugiyama, Y. 1968. Social organization of chimpanzees in the Budongo Forest, Uganda. *Primates* 9:225–58.

Sugiyama, Y. 1973. The social structure of wild chimpanzees: A review of field studies. In R. P. Michael and J. H. Crook, eds., *Comparative Ecology and Behaviour of Primates*, pp. 375–410. Academic Press, London.

Sugiyama, Y. 1981. Observations on the population dynamics and behavior of wild chimpanzees at Bossou, Guinea, in 1979–1980. *Primates* 22:435–44.

Literature Cited

Sugiyama, Y., and J. Koman 1979. Social structure and dynamics of wild chimpanzees at Bossou, Guinea. *Primates* 20:323–39.

Sugiyama, Y., and H. Osawa. 1982. Population dynamics of Japanese monkeys with special reference to the effect of artificial feeding. *Folia Primatol.* 3:61–76.

Sunquist, M. E. 1981. *The Social Organization of Tigers Panthera tigris in Royal Chitawan National Park, Nepal.* Smithsonian Institution Press, Washington, D.C.

Susman, R. L., N. Badrian, and A. Badrian. 1980. Locomotor behavior of *Pan paniscus* in Zaire. *Amer. J. Phys. Anthro.* 53:69–80.

Suzuki, A. 1969. An ecological study of chimpanzees in a savanna woodland. *Primates* 10:103–148.

Suzuki, A. 1975. The origin of hominid hunting: A primatological perspective. In R. H. Tuttle, ed., *Socioecology and Psychology of Primates*, pp. 259–78. Mouton, The Hague.

Svendsen, G. E. 1974. Behavioral and environmental factors in the spatial distribution and population dynamics of a yellow-bellied marmot population. *Ecology* 55:760–71.

Svendsen, G. E., and K. B. Armitage. 1973. Mirror-image stimulation applied to field behavioral studies. *Ecology* 54:623–27.

Swanson, G. A., G. L. Krapu, and J. R. Serie. 1979. Foods of laying female dabbling ducks on the breeding grounds. In T. A. Bookhout, ed., *Waterfowl and Wetlands—An Integrated Review*, pp. 47–57. La Crosse Printing Co., La Crosse, Wisc.

Symons, D. 1979. *The Evolution of Human Sexuality.* Oxford University Press, London.

Tambiah, S. J. 1966. Polyandry in Ceylon. In C. Von Fuhrer-Haimendorf, ed., *Caste and Kin in Nepal, India, and Ceylon.* Asia Publishing House, London.

Tamisier, A. 1972. Rythmes nycthermeraux des sarcelles d'hiver pendent leur hivernage en Camargue. *Alauda* 40:109–159.

Taryannikov, V. I. 1976. Reproduction of the jackal (*Canis aureus aureus* L.) in Central Asia. *Ekologiya* 2:107.

Tashian, R. E. 1957. Nesting behavior of the crested oropendola (*Psarocolius decumanus*) in northern Trinidad. *B.W.I. Zoologica* 42:87–97.

Taylor, P. D., and G. C. Williams. 1982. The lek paradox is not resolved. *Theor. Pop. Biol.* 22:392–409.

Taylor, R. J. 1981. The comparative ecology of the eastern grey kangaroo and wallaroo in the New England tablelands of New South Wales. Ph.D. diss., University of New England, Armidale, Australia.

Taylor, R. J. 1982. Group size in the eastern grey kangaroo, *Macropus giganteus*, and the wallaroo, *Macropus robustu. Austral. Wildl. Res.* 9:229–37.

Terborgh, J. 1983. *Five New World Primates.* Princeton University Press.

Terborgh, J., and A. Wilson Goldizen. 1985. On the mating system of the cooperatively breeding saddle-backed tamarin. *Behav. Ecol. Sociobiol.* 16:293–99.

Thomas, G. J. 1980. The ecology of breeding waterfowl at the Ouse Washes, England. *Wildfowl* 31:73–88.

Thomas, G. J. 1982. Autumn and winter feeding ecology of waterfowl at the Ouse Washes, England. *J. Zool. London* 197:131–72.

Thompson, L. 1940. Southern Lau, Fiji: An ethnography. Bernice Bishop Museum Bulletin 162, Honolulu (also in the HRAF, 1959).

Thompson, S. E. 1979. Socioecology of the yellow-bellied marmot (*Marmota flaviventris*) in central Oregon. Ph.d. diss., University of California, Berkeley.

Thornhill, R., and J. Alcock. 1983. *The Evolution of Insect Mating Systems.* Harvard University Press.

Tilson, R. L. 1977. Social organization of Simakobu monkeys (*Nasalis concolor*) in Seiberut Island, Indonesia. *J. Mammal.* 58:202–212.

Literature Cited

Tinbergen, N. 1957. The functions of territory. *Bird Study* 4:14–27.

Titman, R. D. 1981. A time-activity budget for breeding mallards (*Anas platyrhynchos*) in Manitoba. *Can. Field Nat.* 95:266–71.

Titman, R. D. 1983. Spacing and three-bird flights of mallards breeding in pothole habitat. *Can J. Zool.* 61:839–47.

Titman, R. D., and N. R. Seymour. 1981. A comparison of pursuit flights by six North American ducks of the genus *Anas. Wildfowl* 32:11–18.

Tokuda, K., and G. D. Jensen. 1969. Determinants of dominance hierarchy in a captive group of pigtail monkeys (*Macaca nemestrina*). *Primates* 10:227–35.

Trapp, G. R., and D. L. Hallberg. 1975. Ecology of the gray fox (*Urocyon cinereoargenteus*): A review. In M. W. Fox, ed., *The Wild Canids*, pp. 164–78. Van Nostrand Reinhold, N.Y.

Triesman, M. 1975. Predation and the evolution of gregariousness, II: An economic model for predator-prey interaction. *Anim. Behav.* 23:801–825.

Trivers, R. L. 1972. Parental investment and sexual selection. In B. Campbell, ed., *Sexual Selection and the Descent of Man, 1871–1971*, pp. 136–79. Aldine Press, Chicago.

Turke, P. W. 1985. Tests of Darwinian and economic models of reproductive behavior on Ifaluk and Yap. Ph.D. diss., Northwestern University.

Turke, P. W. 1984. Effects of ovulatory concealment and synchrony of protohominid mating systems and parental roles. *Ethol. and Sociobiol.* 5:33–44.

Turke, P. W., and L. L. Betzig. 1985. Those who can do: Wealth, status, and reproductive success on Ifaluk. *Ethol. and Sociobiol.* 6:79–88.

Tutin, C.E.G. 1975. Sexual behaviour and mating patterns in a community of wild chimpanzees (*Pan troglodytes schweinfurthii*). Ph.D. diss. University of Edinburgh.

Tutin, C.E.G. 1979. Mating patterns and reproductive strategies in a community of wild chimpanzees (*Pan troglodytes schweinfurthii*). *Behav. Ecol. Sociobiol.* 6:29–38.

Tutin, C.E.G., and M. Fernandez. 1985. Foods consumed by sympatric populations of *Gorilla g. goriilla* and *Pan t. troglodytes* in Gabon: Some preliminary data. *Int. J. Primatol.* 6:27–44.

Tutin, C.E.G., and P. R. McGinnis. 1981. Chimpanzee reproduction in the wild. In C. E. Graham, ed., *Reproductive Biology of the Great Apes*, pp. 239–64. Academic Press, N.Y.

Tutin, C.E.G., W. C. McGrew, and P. J. Baldwin. 1981. Responses of wild chimpanzees to potential predators. In A. B. Chiarelli and R. S. Corruccini, eds., *Primate Behavior and Sociobiology*, pp. 136–41. Springer Verlag, Berlin.

Tutin, C.E.G., W. C. McGrew, and P. J. Baldwin 1983. Social organization of savanna-dwelling chimpanzees, *Pan troglodytes verus*, at Mt. Assirik, Senegal. *Primates* 24(2):154–73.

Tylor, E. B. 1889. On a method of investigating the development of institutions: Applied to laws of marriage and descent. *J. Roy. Anthro. Inst.* 18:245–69.

Uehara, S. 1981. The social unit of wild chimpanzees: A reconsideration based on the diachronic data accumulated at Kasoje in the Mahale Mountains, Tanzania. *Africa Kenkyu (Journal of African Studies)* 20:15–32. (In Japanese, with English summary.)

van Bellenberghe, V., and L. D. Mech. 1975. Weights, growth, and survival of timber wolf pups in Minnesota. *J. Mammal.* 56:44–63.

Van den Berghe, P. 1979. *Human Family Systems*. Elsevier, N.Y.

van der Merwe, N. S. 1953. The jackal. *Fauna and Flora* 4:3–77.

Van Orsdol, K. G. 1981. Lion predation in Rwenzori National Park, Uganda. Ph.D. diss., Cambridge University.

Vehrencamp, S. L. 1978. The adaptive significance of communal nesting in grooved-billed Anis (*Crotophaga sulcirostris*). *Behav. Ecol. Sociobiol.* 4:1–33.

Vehrencamp, S. L. 1983. A model for the evolution of despotic versus egalitarian societies. *Anim. Behav.* 31:667–82.

Literature Cited

Vehrencamp, S. L., and J. W. Bradbury. 1984. Mating systems and ecology. In J. R. Krebs and N. B. Davies, eds., *Behavioural Ecology: An Evolutionary Approach*, 2nd edition. Blackwell, Oxford.

Verbeek, N.A.M. 1973. The exploitation system of the yellow-billed magpie. *Univ. Calif. Publ. Zool.* 99:1–58.

Verner, J. 1964. Evolution of polygamy in the long-billed marsh wren. *Evolution* 18:252–61.

Verner, J. 1975. Interspecific aggression between yellow-headed blackbirds and long-billed marsh wrens. *Condor* 77:328–30.

Verner, J., and M. F. Willson. 1966. The influence of habitats on mating systems of North American passerine birds. *Ecology* 47:143–47.

Vesey-Fitzgerald, D. F. 1960. Grazing succession among East African game animals. *J. Mammal.* 41:161–72.

von de Wall, W. 1965. ''Gesellschaftsspiel'' und Balz der Anatini. *J. Ornith.* 106:65–80.

Von Richter, W. 1972. Territorial behaviour of the black wildebeest *Connochaetes gnou. Zool. Afr.* 7:207–231.

Vos, de A. 1965. Territorial behavior among Puku in Zambia. *Science* 148:1752–53.

Vos, de A., and R. J. Dowsett. 1966. The behaviour and population structure of three species of the genus Kobus. *Mammalia* 30:30–55.

Wade, M. D., and S. J. Arnold. 1980. The intensity of sexual selection in relation to male sexual behavior, female choice, and sperm precedence. *Anim. Behav.* 28:446–61.

Walker, E. P. 1983. *Mammals of the World*, Vol. 2. 4th ed. Johns Hopkins University Press.

Wallestad, R. O., J. G. Peterson, and R. L. Eng. 1975. Food of adult sage grouse in central Montana. *J. Wildl. Mgmt.* 39:628–30.

Wallestad, R. and D. Pyrah. 1974. Movement and nesting of sage grouse hens in central Montana. *J. Wildl. Mgmt.* 38:630–33.

Walters, J. R. 1984. The evolution of parental behavior and clutch size in shorebirds. In J. Burger and B. L. Olla, eds., *Behavior of Marine Animals: Shorebirds Breeding Behavior and Populations*, Vol. 5, pp. 243–87. Plenum Press, N.Y.

Walther, F. R. 1964. Einige verhaltensbeobachtungen an Thomsongazellen (*Gazella thomsoni* Gunther, 1884) im Ngorongoro-Krater. *Z. Tierpsychol.* 21:871–90.

Walther, F. R. 1965. Verhaltensstudien an der Grantgazelle (*Gazella granti* Brooke, 1872) im Ngorongoro-Krater. *Z. Tierpsychol.* 22:167–208.

Walther, F. R. 1969. Flight behaviour and avoidance of predators in Thomson's gazelle (*Gazella thomsoni* Guenther 1884). *Behaviour* 34:184–221.

Walther, F. R. 1972a. Territorial behaviour in certain horned ungulates, with special reference to the examples of Thomson's and Grant's gazelles. *Zool. Afr.* 7:303–307.

Walther, F. R. 1972b. Social grouping in Grant's gazelle (*Gazella granti* Brooke 1827) in the Serengeti National Park. *Z. Tierpsychol.* 31:348–403.

Walther, F. R. 1977. Sex and activity dependency of distances between Thomson's gazelles (*Gazella thomsoni* Gunther 1884). *Anim. Behav.* 25:713–19.

Walther, F. R. 1978a. Behavioural observations on oryx antelope (*Oryx beisa*) invading Serengeti National Park, Tanzania. *J. Mammal.* 59:243–60.

Walther, F. R. 1978b. Quantitative and functional variations of certain behaviour patterns in male Thomson's gazelle of different social status. *Behaviour* 65:212–40.

Walther, F. R. 1978c. Forms of aggression in Thomson's gazelle; their situational motivation and their relative frequency in different sex, age, and social classes. *Z. Tierpsychol.* 47:113–72.

Walther, F. R. 1979. Das verhalten der Horntrager (Bovidae). *Handb. Zool.* 10:1–184.

Walther, F. R. 1981. Remarks on behaviour of springbok, *Antidorcas marsupialis* Zimmermann, 1780. *Zool. Gart.* 51:81–103.

Literature Cited

Walther, F. R., E. C. Mungall, and G. A. Grau. 1983. *Gazelles and Their Relatives*. Noyes Publications, Park Ridge, N.J.

Wandrey, R. 1975. Contribution to the study of the social behaviour of captive golden jackals (*Canis aureus* L.). *Z. Tierpsychol.* 39:365–402.

Ward, J. M., Jr., and K. B. Armitage. 1981. Circannual rhythms of food consumption, body mass, and metabolism in yellow-bellied marmots. *Comp. Biochem. Physiol.* 69A:621–26.

Ward, P., and A. Zahavi 1973. The importance of certain assemblages of birds as 'information-centres' for food finding. *Ibis* 115:517–34.

Waser, P. M. 1974. Spatial associations and social interactions in a "solitary" ungulate: The bushbuck *Tragelaphus scriptus* (Pallas). *Z. Tierpsychol.* 37:24–36.

Waser, P. M. 1977. Feeding, ranging, and group size in the mangabey *Cercocebus albigena*. In T. H. Clutton-Brock, ed., *Primate Ecology: Studies of Feeding and Ranging Behaviour in Lemurs, Monkeys, and Apes*, pp. 183–222. Academic Press, London and New York.

Waser, P. M. 1980. Small nocturnal carnivores: Ecological studies in the Serengeti. *Afr. J. Ecol.* 18:167–85.

Waser, P. M. 1981. Sociality or territorial defense? The influence of resource renewal. *Behav. Ecol. Sociobiol.* 8:231–37.

Waser, P. M., and O. Floody. 1974. Ranging patterns of the mangabey *Cercocebus albigena*, in the Kibale Forest, Uganda. *Z. Tierpsychol.* 35:85–101.

Waser, P. M., and W. T. Jones. 1983. Natal philopatry among solitary mammals. *Quart. Rev. Biol.* 58:355–90.

Waser, P. M., and M. S. Waser. 1985. *Ichneumia albicauda* and the evolution of viverrid gregariousness. *Z. Tierpsychol.* 68:137–51.

Wasser, S. K., and D. P. Barash. 1983. Reproductive suppression among female mammals: Implications for biomedicine and sexual selection theory. *Quart. Rev. Biol.* 58:513–35.

Watanabe, K. 1981. Variations in group composition and population density of the two sympatric mentawaian leaf-monkeys. *Primates* 22:145–60.

Waterman, P. G., A. Vedder, and D. P. Watts. 1983. Digestibility, digestion-inhibitors, and nutrients in herbaceous foliage from an African montane flora and its comparison with other tropical flora. *Oecologica* 60:244–49.

Watson, R. M. 1969. Reproduction of wildebeest, *Connochaetes taurinus albojubatus* Thomas, in the Serengeti region, and its significance to conservation. *J. Reprod. Fert. Suppl.* 6:287–310.

Watt, J. M., and M. G. Breyer-Brankwijk. 1962. *Medicinal and Poisonous Plants of Southern and Eastern Africa*. Edinburgh and London.

Watts, D. P. 1983. Feeding strategy and socioecology of mountain gorillas (*Pan gorilla beringei*). Ph.D. diss., University of Chicago.

Weatherhead, P. J., and R. J. Robertson. 1979. Offspring quality and the polygyny threshold: The "sexy son hypothesis." *Amer. Natur.* 113:201–208.

Weidmann, U. 1956. Verhaltensstudien an der Stockente (*Anas platyrhynchos* L.), I: Das Aktionssystem. *Z. Tierpsychol.* 13:208–271.

Weidmann, U., and J. Darley. 1971. The role of the female in the social display of mallards. *Anim. Behav.* 19:287–98.

Weller, M. W. 1964. The reproductive cycle. In J. Delacour, ed., *The Waterfowl of the World*. Vol. 4, pp. 35–79. Country Life, London.

Weller, M. W. 1967. Notes on some marsh birds of Cape San Antonio, Argentina. *Ibis* 109:391–411.

Weller, M. W. 1972. Ecological studies of Falkland Islands' waterfowl. *Wildfowl* 23:25–44.

Weller, M. W. 1975. Habitat selection by waterfowl of Argentine Isla Grande. *Wilson Bull.* 87:83–90.

Literature Cited

West, M. J., A. P. King, and D. H. Eastger. 1981. Validating the female bioassay of cowbird song: Relating differences in song potency to mating success. *Anim. Behav.* 29:490–501.

Westermark, E. 1891. *The History of Human Marriage*. Macmillan, N.Y.

Western, D. 1973. The structure, dynamics, and changes of the Amboseli ecosystem. Ph.D. diss., University of Nairobi.

Western, D. 1979. Size, life history, and ecology in mammals. *Afr. J. Ecol.* 17:185–204.

White, D. H., and D. James. 1978. Differential use of fresh water environments by wintering waterfowl of coastal Texas. *Wilson Bull.* 90:99–111.

Whitney, G. 1976. Genetic substrates for the initial evolution of human sociality. I: Sex chromosome mechanisms. *Amer. Natur.* 110:867–75.

Whitten, P. L. 1982. Female reproductive strategies among vervet monkeys. Ph.D. diss., Harvard University.

Whitten, P. L. 1983. Females, flowers, and fertility. *Amer. J. Phys. Anthro.* 60(2):269–70.

Whyte, M. K. 1978. Cross-culture codes dealing with the relative status of women. *Ethnology* 17:211–37.

Wickler, W., and U. Seibt. 1983. Monogamy: An ambiguous concept. In P.P.G. Bateson, ed., *Mate Choice*, pp. 33–52. Cambridge University Press.

Wiessner, P. 1982. Risk, reciprocity, and social influences on !Kung San economics. In E. Leacock and R. Lee, eds., *Politics and History in Band Societies*. Cambridge University Press.

Wilbur, H. M. 1977. Density-dependent aspects of growth and metamorphosis in *Bufo americanus*. *Ecology* 58:196–200.

Wilbur, H. M., D. W. Tinkle, and J. P. Collins. 1974. Environmental certainty, trophic level, and resource availability in life history evolution. *Amer. Natur.* 108:805–817.

Wildt, D. E., U. Doyle, S. C. Stone, and R. M. Harrison. 1977. Correlation of perineal swelling with serum ovarian hormone levels, vaginal cytology, and ovarian follicular development during the baboon reproductive cycle. *Primates* 18:261–70.

Wiley, R. H. 1973a. Territoriality and non-random mating in sage grouse. *Anim. Behav. Monogr.* 6:85–169.

Wiley, R. H. 1973b. The strut display of the male sage grouse: A "fixed" action pattern. *Behaviour* 47:129–52.

Wiley, R. H. 1974. Evolution of social organization and life history patterns among grouse. *Quart. Rev. Biol.* 49:201–227.

Wiley, R. H. 1976. Communication and spatial relationships in a colony of common grackles. *Anim. Behav.* 24:570–84.

Wiley, R. H., and M. S. Wiley. 1980. Spacing and timing in the nesting ecology of a tropical blackbird: Comparison of populations in different environments. *Ecol. Monogr.* 50:153–78.

Williams, D. M. 1982. Agonistic behaviour and mate selection in the mallard (*Anas platyrhynchos*). Ph.D. diss., University of Leicester.

Williams, D. M. 1983. Mate choice in the mallard. In P.P.G. Bateson, ed., *Mate Choice*, pp. 297–309. Cambridge University Press.

Williams, G. C. 1966. *Adaptation and Natural Selection*. Princeton University Press.

Williams, L. 1952. Breeding behavior of the Brewer blackbird. *Condor* 54:3–47.

Willson, M. F. 1966. The breeding ecology of the yellow-headed blackbird. *Ecol. Monogr.* 36:51–77.

Willson, M. F., and G. H. Orians. 1963. Comparative ecology of red-winged and yellow-headed Blackbirds during the breeding season. *Proc. XVI Int. Congr. Zool.* 3:342–46.

Wilson, E. O. 1975. *Sociobiology: The New Synthesis*. Harvard University Press.

Wirtz, P. 1981. Territorial defence and territory take-over by satellite males in the waterbuck *Kobus ellipsiprymnus* (Bovidae). *Behav. Ecol. Sociobiol.* 8:161–62.

Literature Cited

Wirtz, P. 1982. Territory holders, satellite males, and bachelor males in a high-density population of waterbuck (*Kobus ellipsiprymnus*) and their associations with conspecifics. *Z. Tierpsychol.* 58:277–300.

Wishart, R. A. 1983a. Pairing chronology and mate selection in the American wigeon. *Can. J. Zool* 61:1733–43.

Wishart, R. A. 1983b. The behavioral ecology of the American wigeon (*Anas americana*) over its annual cycle. Ph.D. diss., University of Manitoba.

Wistrand, H. 1974. Individual, social, and seasonal behavior of the thirteen-lined ground squirrel (*Spermophilus tridecemlineatus*). *J. Mammal.* 55:329–47.

Wittenberger, J. F. 1978. The evolution of mating systems in grouse. *Condor* 80:126–37.

Wittenberger, J. F. 1979. The evolution of mating systems in birds and mammals. In P. Marler and J. G. Vandenburg, eds., *Handbook of Behavioral Neurobiology, Vol. 3: Social Behavior and Communication*, pp. 271–349. Plenum Press, N.Y.

Wittenberger, J. F. 1980a. Vegetation structure, food supply, and polygyny in bobolinks (*Dolichonyx oryzivorous*). *Ecology* 61:140–50.

Wittenberger, J. F. 1980b. Group size and polygamy in social mammals. *Amer. Natur.* 115:197–222.

Wittenberger, J. F. 1983. Tactics of mate choice. In P.P.G. Bateson, ed., *Mate Choice*, pp. 435–47. Cambridge University Press.

Wittenberger, J. F., and R. L. Tilson. 1980. The evolution of monogamy: Hypotheses and evidence. *Ann. Rev. Ecol. Syst.* 11:197–232.

Wolf, L. L. 1975. "Prostitution" behavior in a tropical hummingbird. *Condor* 77:140–44.

Wood, N. A. 1974. The breeding behaviour and biology of the moorhen. *British Birds* 67:104–115 and 137–58.

Woolfenden, G. E. 1967. Selection for a delayed simultaneous wing molt in loons (Gaviidae). *Wilson Bull.* 79:416–20.

Woolfenden, G. E. 1981. Selfish behavior by Florida scrub jay helpers. In R. D. Alexander and D. W. Tinkle, eds., *Natural Selection and Social Behavior: Recent Research and New Theory*, pp. 237–60. Chiron Press, N.Y.

Woolfenden, G. E., and J. W. Fitzpatrick. 1977. Dominance in the Florida scrub jay. *Condor* 79:1–12.

Woolfenden, G. E., and J. W. Fitzpatrick. 1978. The inheritance of territory in group-breeding birds. *BioScience* 28:104–108.

Woolfenden, G. E., and J. W. Fitzpatrick 1984. *The Florida Scrub Jay: Demography of a Cooperative-Breeding Bird*. Princeton University Press.

Wozencraft, W. C. 1982. Reviewer for the family Herpestidae. In J. H. Honacki, K. E. Kinman, and J. W. Koeppl, eds., *Mammal Species of the World*, pp. 271–76. Allen Press, Lawrence, Kansas.

Wrangham, R. W. 1974. Artificial feeding of chimpanzees and baboons in their natural habitat. *Anim. Behav.* 22:83–93.

Wrangham, R. W. 1975. Behavioural ecology of chimpanzees in Gombe National Park, Tanzania. Ph.D. diss., Cambridge University.

Wrangham, R. W. 1977. Feeding behaviour of chimpanzees in Gombe National Park, Tanzania. In T. H. Clutton-Brock, ed., *Primate Ecology*, pp. 504–538. Academic Press, New York.

Wrangham, R. W. 1979a. On the evolution of ape social systems. *Soc. Sci. Inform.* 18:335–68.

Wrangham, R. W. 1979b. Sex differences in chimpanzee dispersion. In D. A. Hamburg and E. R. McCown, eds., *The Great Apes*, pp. 481–90. Benjamin/Cummings, Menlo Park, Calif.

Literature Cited

Wrangham, R. W. 1980. An ecological model of female-bonded primate groups. *Behaviour* 75:262–300.

Wrangham, R. W. 1982. Mutualism, kinship, and social evolution. In King's Sociobiology Group, eds., *Current Problems in Sociobiology*, pp. 269–90. Cambridge University Press.

Wrangham, R. W. 1983. Social relationships in comparative perspective. In R. A. Hinde, ed., *Primate Social Relationships*, pp. 255–62. Blackwell, Oxford.

Wrangham, R. W. 1986. The evolution of social structure. In B. B. Smuts, D. L. Cheney, R. M. Seyfarth, R. W. Wrangham, and T. T. Struhsaker, eds. *Primate Societies*. University of Chicago Press.

Wrangham, R. W. and E. Ross. Individual differences in activities, family size, and food production among Lese horitculturalists of Northeast Zaire. Paper presented at the annual meeting of the American Anthropological Association, December 1983, Washington, D.C.

Wrangham R. W., and B. B. Smuts. 1980. Sex differences in the behavioral ecology of chimpanzees in the Gombe National Park, Tanzania. *J. Reprod. Fert. Suppl.* 28:13–31.

Wyman, J. 1967. The jackals of the Serengeti. *Animals* 10:79–83.

Yalman, N. 1967. *Under the Bo Tree*. University of California Press.

Yasukawa, K. 1979. Territory establishment in red-winged blackbirds: Importance of aggressive behavior and experience. *Condor* 81:358–64.

Yasukawa, K. 1981. Male quality and female choice of mate in the red-winged blackbird (*Agelaius phoeniceus*). *Ecology* 62:922–29.

Yasukawa, K., and W. A. Searcy. 1981. Nesting synchrony and dispersion in red-winged blackbirds. Is the harem competitive or cooperative? *Auk* 98:659–68.

Yasukawa, K., and W. A. Searcy. 1982. Aggression in female red-winged blackbirds: A strategy to ensure male parental investment. *Behav. Ecol. Sociobiol.* 11:13–17.

Ydenberg, R. C., and H. H. Th. Prins. 1981. Spring grazing and the manipulation of food quality by barnacle geese. *J. Appl. Ecol.* 18:443–54.

Yeaton, R. I. 1972. Social behavior and social organization in Richardson's ground squirrel (*Spermophilus richardsonii*) in Saskatchewan. *J. Mammal.* 53:139–47.

Zahavi, A. 1974. Communal nesting by the Arabian babbler: A case of individual selection. *Ibis* 116:84–87.

Zimen, E. 1976. On the regulation of pack size in wolves. *Z. Tierpsychol.* 40:300–341.

Author Index

529

Taxonomic Index

●

Subject Index

•

Library of Congress Cataloging-in-Publication Data

Ecological aspects of social evolution.

Bibliography: p.

Includes index.

1. Mammals—Behavior. 2. Mammals—Ecology. 3. Mammals—Evolution.
4. Birds—Behavior. 5. Birds—Ecology. 6. Birds—Evolution.
7. Social behavior in animals. 8. Social evolution.
I. Rubenstein, Daniel I., 1950–   . II. Wrangham, Richard W., 1948–

QL739.3.E24 1986   559'.051   86-9371
ISBN 0–691–08439–4   ISBN 0–691–08440–8 (pbk.)

Daniel I. Rubenstein is Associate Professor of Biology
at Princeton University.
Richard W. Wrangham is Associate Professor of Anthropology
and Biology at the University of Michigan.

639417

Made in the USA